OXFORD **READERS**

Evolution

Second Edition

Edited by Mark Ridley

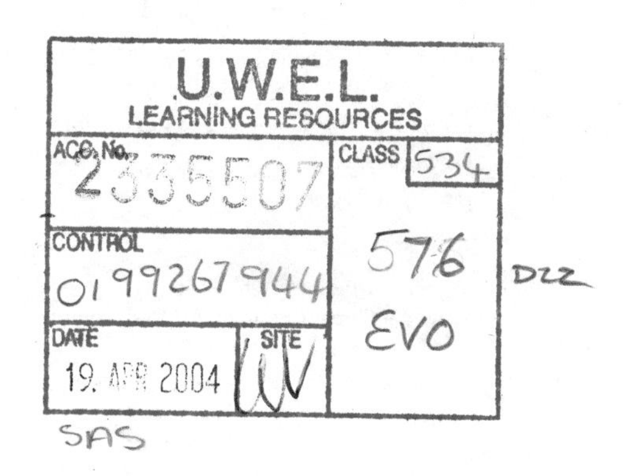

OXFORD
UNIVERSITY PRESS

OXFORD
UNIVERSITY PRESS

Great Clarendon Street, Oxford OX2 6DP

Oxford University Press is a department of the University of Oxford.
It furthers the University's objective of excellence in research, scholarship,
and education by publishing worldwide in

Oxford New York

Auckland Bangkok Buenos Aires Cape Town Chennai
Dar es Salaam Delhi Hong Kong Istanbul Karachi Kolkata
Kuala Lumpur Madrid Melbourne Mexico City Mumbai Nairobi
São Paulo Shanghai Taipei Tokyo Toronto

Oxford is a registered trade mark of Oxford University Press
in the UK and in certain other countries

Published in the United States
by Oxford University Press Inc., New York

First edition 1997

British Library Cataloguing in Publication Data

Data available

ISBN 0–19–926794–4

10 9 8 7 6 5 4 3 2 1

Typeset by RefineCatch Limited, Bungay, Suffolk
Printed in Great Britain by Ashford Colour Press Ltd, Gosport, Hampshire

Contents

Introduction 1

A. From Darwin to the modern synthesis

Section introduction 7

1. CHARLES DARWIN (1858), Extract from an unpublished work on species 9
2. CHARLES DARWIN (1858), Abstract of a letter from C. Darwin, Esq., to Prof. Asa Gray, Boston, USA 13
3. JOHN MAYNARD SMITH (1987), Weismann and modern biology 15
4. R. A. FISHER (1930), The nature of inheritance 20
5. SEWALL WRIGHT (1932), The roles of mutation, inbreeding, crossbreeding, and selection in evolution 29
6. J. B. S. HALDANE (1949), Disease and evolution 37

B. Natural selection and random drift in populations

Section introduction 44

7. H. B. D. KETTLEWELL (1958), A résumé of investigations on the evolution of melanism in the Lepidoptera 49
8. L. M. COOK, R. L. H. DENNIS, and G. S. MANI (1999), Melanic morph frequency in the peppered moth in the Manchester area 53
9. MARY N. KARN and L. S. PENROSE (1951), Birth weight and gestation time in relation to maternal age, parity, and infant survival 57
10. L. ULIZZI and L. TERRENATO (1992), Natural selection associated with birth weight: towards the end of the stabilizing component 59
11. H. LISLE GIBBS and PETER R. GRANT (1987), Oscillating selection on Darwin's finches 63
12. R. C. LEWONTIN (1974), The paradox of variation 67
13. MOTOO KIMURA (1990), Recent development of the neutral theory viewed from the Wrightian tradition of theoretical population genetics 75

C. Adaptation

Section introduction 82

14. R. A. FISHER (1930), The nature of adaptation 85
15. G. C. WILLIAMS (1966), Adaptation and natural selection 89

16. A. GRAFEN (1986), Adaptation versus selection in progress 91
17. H. K. REEVE and P. W. SHERMAN (1991), An operational, nonhistorical definition of adaptation 94
18. H. ALLEN ORR and JERRY A. COYNE (1992), The genetics of adaptation: a reassessment 96
19. A. J. CAIN (1964), The perfection of animals 100
20. S. J. GOULD and R. C. LEWONTIN (1979), The spandrels of San Marco and the Panglossian paradigm: a critique of the adaptationist programme 114
21. RICHARD DAWKINS (1976), The selfish gene 123

D. Speciation and biodiversity

Section introduction 131
22. E. MAYR (1958), Typological versus population thinking 134
23. E. MAYR (1963), Species concepts and their application 137
24. CHARLES DARWIN (1859), The sterility of hybrids 147
25. THEODOSIUS DOBZHANSKY (1970), Reproductive isolation as a product of genetic divergence and natural selection 151
26. WILLIAM R. RICE and ELLEN E. HOSTERT (1993), Laboratory experiments on speciation: what have we learned in 40 years? 155
27. JERRY A. COYNE and H. ALLEN ORR (2000), The evolutionary genetics of speciation 161
28. DOLPH SCHLUTER (2000), Ecological basis of postmating isolation 175
29. V. GRANT (1981), Hybrid speciation 178

E. Macroevolution

Section introduction 182
30. DOUGLAS H. ERWIN and ROBERT L. ANSTEY (1995), Speciation in the fossil record 185
31. GAVIN DE BEER (1971), Homology: an unsolved problem 197
32. RICHARD DAWKINS (1996), The *ey* gene 205
33. W. J. DICKINSON (1995), Molecules and morphology: where's the homology? 207
34. E. HAECKEL (1905), The fundamental law of organic evolution 211
35. W. GARSTANG (1951), Three poems 216

F. Evolutionary genomics

Section introduction 220
36. HOWARD OCHMAN, JEFFREY G. LAWRENCE, and EDUARDO A. GROISMAN (2000), Lateral gene transfer and the nature of bacterial innovation 221

37. TODD J. VISION, DANIEL G. BROWN, and STEVEN D. TANKSLEY (2000), The origins of genomic duplications in *Arabidopsis* 231
38. INTERNATIONAL HUMAN GENOME SEQUENCING CONSORTIUM (2001), Initial sequencing and analysis of the human genome 237
39. SEAN B. CARROLL (2003), Genetics and the making of *Homo sapiens* 244
40. R. A. RAFF (1996), Co-option of eye structures and genes 249
41. STEVEN A. BENNER, M. DANIEL CARACO, J. MICHAEL THOMSON, and ERIC A. GAUCHER (2002), Planetary biology—paleontological, geological, and molecular histories of life 250

G. The history of life

Section introduction 258

42. JOHN MAYNARD SMITH and EÖRS SZATHMÁRY (1999), From chemistry to heredity 259
43. J. WILLIAM SCHOPF (1994), Disparate rates, differing fates: tempo and mode of evolution changed from the Precambrian to the Phanerozoic 265
44. ALAN COOPER and RICHARD FORTEY (1998), Evolutionary explosions and the phylogenetic fuse 275
45. DAVID DILCHER (2000), Towards a new synthesis: major evolutionary trends in the angiosperm fossil record 284

H. Case studies

Section introduction 292

46. P. B. MEDAWAR (1951), An unsolved problem of biology 293
47. F. H. C. CRICK (1968), The origin of the genetic code 299
48. J. MAYNARD SMITH (1971), The maintenance of sex 307
49. D. H. JANZEN (1983), A caricature of seed dispersal by animal guts 310
50. DAN-E. NILSSON and SUSANNE PELGER (1994), A pessimistic estimate of the time required for an eye to evolve 317
51. JOHN GERHART and MARK KIRSCHNER (1997), Evolutionary novelty: the example of lactose synthetase 326
52. PAUL D. SNIEGOWSKI, PHILIP J. GERRISH, TOBY JOHNSON, and AARON SHAVER (2000), The evolution of mutation rates: separating causes from consequences 328

I. Human evolution

Section introduction 337

53. VINCENT M. SARICH and ALLAN C. WILSON (1967), Immunological time scale for hominid evolution 340

54. MARY-CLAIRE KING and A. C. WILSON (1975), Evolution at two levels in humans and chimpanzees 345
55. ROY J. BRITTEN (2002), Divergence between samples of chimpanzee and human DNA sequences is 5%, counting indels 350
56. H. J. MULLER (1950), Our load of mutations 354
57. FRANK B. LIVINGSTONE (1962), On the non-existence of human races 361
58. WILTON M. KROGMAN (1951), The scars of human evolution 363
59. STEVEN PINKER (1994), The big bang 368

J. Evolution and human affairs

Section introduction 383

60. MICHAEL F. ANTOLIN and JOAN M. HERBERS (2001), Evolution's struggle for existence in America's public schools 385
61. THEODOSIUS DOBZHANSKY (1973), Nothing in biology makes sense except in the light of evolution 400
62. DAVID HUME (1779), The argument from design 410
63. J. L. MONOD (1974), On the molecular theory of evolution 412
64. THOMAS HENRY HUXLEY (1893), Evolution and ethics 418
65. STEPHEN R. PALUMBI (2001), Humans as the world's greatest evolutionary force 421

Select bibliography 434
Biographical notes 437
Acknowledgements 441
Index 447

Introduction

Most scientific theories either deal with questions that only interest specialists or (like fundamental physics) require a forbidding set of technical skills before you can understand them. The theory of evolution is scientifically unique in its combination of universal interest and accessibility. It is also fortunate in the number of superb scientific writers that it has attracted. In editing this anthology, I was spoiled for choice, not scraping the barrel.

The contents of the anthology mainly illustrate the range of modern scientific work in the subject, but we start with a more historical section (Section A). Darwin's own writing is historical almost only in a chronological sense—Darwin was mainly interested in questions that people are still interested in and he thought about them in a modern way. This is not because biologists have been thinking the same things for 150 years: it is more because in the twentieth century they were rediscovering Darwin's ideas beneath the layers of accumulated rubbish. After Darwin published *The Origin of Species*, the idea of evolution was (in a sense) accepted, but his explanation for it—natural selection—was rejected. The kind of evolution people came to believe in was a linear, ladder-climbing ascent from simple life to humans: this fitted in well with the standard idea of a Great Chain of Being, connecting all life from top to bottom. It was a developmental view of evolution, in which evolution resembled the orderly and predestined way in which an egg grows up into an adult. This was not Darwin's own view of evolution—he dismissed the distinction between lower and higher animals as meaningless—but in some ironical historical sense he contributed to its appeal. If evolution is like development, it is not easily explained by natural selection, which is a contingent, short-term process that works with accidental, rather than progressive, variation.

Darwin lacked a theory of inheritance. The main ideas he used—they included the inheritance of acquired characteristics, blending inheritance, and particulate inheritance—had serious flaws that were well known by the end of the nineteenth century. Weismann (see Chapter 3) had destroyed in the 1880s the theory that individually acquired characteristics are inherited, though reputable biologists (if not experts on inheritance) continued to take it seriously well into the 1920s. Blending inheritance means that, if an organism inherits one hereditary molecule or factor (call it A) from its father and another (call it A') from its mother, then the two blend ($A+A' \rightarrow A''$) and when the organism reproduces it passes the new kind of molecule (A'') to its offspring. Fisher (Chapter 4) beautifully explains the difficulty created by this kind of theory: the variation needed for evolution disappears rapidly as it is blended out of existence. The problem was solved when Mendelism was rediscovered and confirmed after 1900. In the Mendelian process, in terms of the notation just given, if an organism inherits A from its father and A' from its mother, that organism is a heterozygote that preserves the two molecules in combined AA' form; and, when it reproduces, it sends A to half its offspring and A' to the other half. Monod (Chapter 63) argues that Mendelism, or something like it, was a scientific prediction, now vindicated, of Darwin's theory, even though Darwin himself did not realize it.

Given Mendelism, the scientific stage can retrospectively be seen to be set for the synthesis between Mendel's theory of inheritance and Darwin's theory of natural selection in the 1910s and 1920s. Evolution could now be reimagined in almost its Darwinian form, as selection among accidental variants, though there was the important difference that the inherited variants were scientifically understood and Mendelian rather than the hypothetical and erroneous particles that Darwin thought about.

In the following decades the new understanding of evolution spawned a series of research programmes. One was on the process of selection itself, which had been left almost uninvestigated since Darwin's time. Section B is about this work, and shows how it has medical and economic, in addition to its scientific, importance. In the 1960s, the research programme became more complicated when it encountered the first ripples of molecular evidence. Evolution at the molecular level appeared to have properties that would not be predicted if it were driven by natural selection, and much of molecular evolution is now widely (if not universally) thought to be non-adaptive. Adaptation, however, remains as important as ever in supramolecular biology. Evolutionary theory aims, as it has since Darwin, to give an account of why living creatures are well organized and adapted for life on Earth. The Darwinian account is theologically subversive, as is touched on in the extract from Hume (Chapter 62), but that is no longer its main interest, as the modern work in Section C reveals.

The modern work arguably represents another stage in the rediscovery of Darwin's ideas. There is a tension between two concepts of natural selection. What are the entities that natural selection selects between? One conception is more (in nineteenth century language) 'racial'; it supposes that selection is between competing groups of organisms ('group selection' in modern terms). In the age of nationalist power philosophies, there was more stress on selection in the form of an offensive struggle between groups. More recently, 'group selection' has been used to explain how selection favours cooperation by its benefit at the group level, despite the individual advantage of selfishness. Natural selection in Darwin's own theory occurs between individuals within a population; adaptations then evolve for the benefit of organisms rather than larger entities. In the 1960s and 1970s, individual selection was rediscovered, and group selection by and large rejected. That revolution is under-represented in this book, because most of the pieces in Section C are 'post-revolutionary': the authors assume that selection produces individual adaptations and they are concerned with other issues. However, I include Dawkins' (Chapter 21) argument, from *The selfish gene*, that genes are units of selection. Also, the rejection of group selection in turn created a series of problems, such as senescence, sex, and mutation rates, which are discussed as case studies in section H (Chapters 46, 48, and 52 respectively). I have not included papers by one of the main figures, W. D. Hamilton, because they are appearing in a superb anthology of their own, complete with delightful autobiographical introductions; I give the reference in the Select Bibliography, for Section C.

Students of biodiversity might see a relation between the individual/group selection debate and Mayr's distinction between what he calls 'population thinking' and typology, or essentialism (Section D, Chapter 22). Mayr's thinking has been mainly concerned with the nature of species and higher classificatory groups, and of speciation, rather than with adaptation. He argued that Darwin's theory implied a revolutionary new view of how individuals are defined into classificatory groups. Taxonomists before Mayr and like thinkers were mainly typologists; they assigned individuals to species (or higher classificatory groups) on the basis of observable physical attributes. An individual is a mammal, for instance, if it is an animal that lactates, is viviparous, and warm-blooded (together with some clauses such that males are not ruled out). For species membership, Mayr argues that what matters is not what attributes an individual possesses, but whom it can interbreed with. A species is a reproductive community, not a set of individuals with certain observable attributes. Mayr calls an interbreeding community a population, and population thinking therefore means defining species membership by interbreeding rather than resemblance to some 'type' set of attributes. This viewpoint became widely accepted, and was called 'the new systematics' when it was introduced in the 1940s.

At about the same time, our understanding of how new species originate underwent important advances. If an ancestral species is divided into two separate populations, each of which evolves adaptations to its local conditions, the two will accumulate different genes—and those genes will probably prove incompatible if members of the two populations later hybridize. Darwin (Chapter 24) realized that speciation occurs as an incidental consequence of evolutionary change; but his own understanding of the conditions in which speciation occurs, and the genetic changes that drive it, look limited or even erroneous compared with the modern theory that began to emerge in the 1940s. The extracts in Section D show how our modern understanding built up.

From the origin of species, the next step up in the scale of evolutionary change is the origin of higher taxa such as mammals. At this stage a new set of questions is raised. One concerns the reconstruction of evolutionary history. We look at this problem in Sections E and G. Another set of questions concerns the relation between short-term evolutionary processes of the sort that can be observed directly and evolutionary events, such as the transition of reptiles into mammals, that occurred over much longer time periods; this is included in Section E, and touched on in some case studies (Section H). In the past decade or so, the study of evolution has increasingly benefited from the 'molecular revolution'. The study of macroevolution, and of the reconstruction of evolutionary history, is particularly benefitting from the huge quantities of molecular sequences that are coming out of the human genome project and related research. The molecular evidence can be used to study long-standing questions in the narrative of life's history, as we see in Section G. Likewise, as biologists learn more about the molecular genetic control of individual development, they can look again at the old question of the relation between development and evolution and the deep evolutionary concept of homology (Section E).

For the second edition, I have added a section on evolutionary genomics (Section F). The subject hardly existed before the year 2000, but it is already a safe prediction that genomics will rapidly grow into a huge and innovative area of evolutionary biology. You can hear people say that evolutionary theory is finally coming into its own now that we are obtaining genomic sequences for several forms of life. The new evidence probably forces us to revise our view of evolution as tree-like: genes seem to move quite freely between the branches, particularly the microbial branches (Chapter 36). But many of the best stories in genomics concern the evolution of particular genes, or families of related genes. The inferences of this kind in Section F are preliminary and could prove false in their details. But there is no doubt that inferences about genomic history can be made; in time the details should be settled. The Darwinian genomic treasure-hunt has begun.

Evolution is important beyond pure biology. At a fairly straightforward (if not uncontroversial) level, we can apply the methods or reconstructing evolutionary history to reconstruct human evolutionary history. Evolution also influences the way we think about the human future, medicine, human races, and about such apparently distinctive human features as language. All these topics are discussed in Section I. And yet further from the lab bench lies the relations between evolution and human politics, philosophy, and values: we look at some of these in the final Section (J).

I should like to finish with a few editorial remarks. In choosing the papers I had the following aims, or constraints, in mind (ignoring constraints outside my control, such as the permission of copyright owners). One was that I wanted to include empirical science wherever possible. I suspected the selection would tend towards abstract and theoretical argument, because a paper full of facts could easily look more specialized than a competing paper of broad-ranging argument (and also perhaps because I may myself be biased towards theory, though I should of course deny it). The anthology could then end up portraying evolution as a mass of controversial theoretical debates. Those debates exist all right, and they are fascinating; but most evolutionary research is nevertheless concerned with making empirical measurements, or doing experiments, to test theories, and I wanted a good representation of research in the hard-facts tradition. Secondly, I wanted as many authors as possible to be great minds. It is always more stimulating to read, and think with, one of the greatest human intellects than with—how can I put it diplomatically?—well, more purely derivative authors. I therefore have included classics rather than secondary sources, even where a secondary source might cover a topic in a more balanced, or up-to-date, way. Thirdly, I wanted some mix of 'ancient' and 'modern': of the work that formed modern evolutionary biology and of the science that is now being done. These aims had to be traded off at various points with what was there in the literature and with how much space I had, but may help to explain in part what is on offer.

I have grouped the extracts into ten sections. Some are more naturally unified than others, and I had to make plenty of arbitrary decisions as to which section to put some of the material in. There is a section of case studies, for example; but I have included various case studies in other sections. Some of the extracts form obvious combinations and could beneficially be read together, but for many of them it does not matter in which order they are read and the section groupings can be ignored as appropriate. In the section introductions I have pointed out related extracts in other sections.

The extracts vary in the proportion of the original that I have included. One consideration was that I did not want one line of thought or research to take over too much of the book. At one extreme, some of the extracts are from books, and we clearly could not reproduce the whole thing. I aimed to make each extract a readable unit, and have therefore minimized my editorial violence within each extract (the traces of that violence are to be seen in the symbols of elision [. . .]); but some of the subtlety, at least, of the original will inevitably have sometimes been lost. My first aim here is to provide viewpoints for people to think about, and not to do justice to the experimental set-ups, the pre-emptive replies to critics, the argumentative range of the original authors. When readers are provoked to disagree with any of the authors (as opposed to purely with the ideas) I have included, I recommend they check the original, particularly if there are lots of [. . .] signs in the version here. The extracts also vary in how easy they are to read and in the technical knowledge they presuppose. My editorial introductions to each section are intentionally introductory: I have used them to sketch in what context I could. About the only thing that is everywhere assumed is some knowledge of elementary genetics; but that is not difficult to pick up elsewhere. Some of the more demanding papers assume a knowledge of how to measure selection, or of the names of animal and plant taxa. I have used the footnotes to explain some points, but it would have been impossible to explain every point to an introductory level and I have been forced here and there into editorial guesswork about what would be most useful to say or not to say. I believe that almost any reader will be able to follow the outline of what is going on within every extract—I have excluded papers in which some readers would be sunk at the outset—but some readers may wish to glaze over (or consult a reference work) among the details of the more advanced work. I hope, though, that problems of that kind will be rare: I have certainly tried to make them so. The kind of reader I had in mind is not a professional biologist who routinely scans the journal contents pages (though I hope they will find something interesting in the book), but anyone else who wishes to sample the delights of 150 years of thought about evolution.

Section A

From Darwin to the modern synthesis

The mail delivery of 17 June 1858 at Down House, Kent, England, is one reasonable starting-date for the literature of evolution. Charles Darwin (who lived at Down House) then received a letter and manuscript from Alfred Russel Wallace, a British naturalist travelling in the Malay archipelago. (The original letter and manuscript are both lost and the 17 June date of receipt is conjectural.) Darwin had invented the theory of evolution by natural selection about twenty years earlier, and since then had studiously avoided publishing a single word of it. He was saving himself to write a Big Work on the subject. Now Wallace, as the mailed manuscript revealed, had invented much the same theory and seemed poised to scoop him. However, an arrangement was made and papers by Darwin and Wallace were presented simultaneously later in 1858 (and printed in 1859). Darwin need not have worried about being scooped. The world ignored the 1858 papers and it was the subsequent 1859 publication of *On the Origin of Species* that caused the intellectual earthquake.

The 1858 papers (Chapters 1, 2) are little read even now but they contain, in Darwin's own hand, as good a short summary as any of his ideas. He derives the theory of natural selection from Malthus's ideas about population, and he applies it to explain evolution and adaptation; he also describes his principle of divergence, which he used to explain the branching, tree-like, diverging pattern of living diversity. Wallace's own paper I have not included here—it is rather longer and not always as clear a read; it is also questionable whether he had Darwin's (and our modern) concept of natural selection between individuals within a population.

After Darwin's death his theory went into something of an eclipse. One of the main reasons was that the mechanism of heredity was not yet satisfactorily understood. Darwin tried various ideas, but none of them worked. Blending theories, according to which the hereditary materials of the two parents physically blended to produce an offspring of intermediate hereditary composition (see the Main Introduction to this book), seemed attractive; but they created difficulties that Darwin knew about and were classically described by Fisher (Chapter 4). It was also thought that attributes acquired by an individual during its lifetime—increased or decreased muscular strength due to more or less physical work, for example—were passed on to the offspring. This theory of the inheritance of acquired characteristics, or Lamarckian inheritance, is now known to be wrong, but few doubted it in the late nineteenth and early twentieth centuries. One biologist who did doubt it was August Weismann (Chapter 3). Weismann is increasingly seen as the main torch-bearer of Darwinism in this period, and Maynard Smith's essay discusses a number of his contributions. One is that Weismann had an 'information science' concept of biology: see his analogy between a telegraphic translator and a Lamarckian mechanism.

Looking back, the next step towards the modern theory of evolution was the Mendelian theory of heredity. It became established after the turn of the century; but its initial effect was anti-Darwinian, because the early Mendelians all disliked Darwin's theory. In the 1910s and 1920s a new generation grew up of biologists who accepted Mendelism and were able to show that the Mendelian theory, far from counting against Darwinism, was exactly what Darwinism needed. The combination of the Mendelian theory of heredity and the Darwinian theory of evolution has come to be called variously the Modern Synthesis, or the synthetic theory of evolution, or neo-Darwinism. The most famous names in this synthesis of Darwin and Mendel were Fisher, Haldane, and Wright. I have included one paper by each of them (Chapters 4–6). Fisher and Wright are not the easiest of authors, and they tended to write technically, with technically skilled readers in mind. The short paper included by Wright (Chapter 5) is generally regarded as the most readable of his papers from the heroic era. Haldane, by contrast, was a great science popularizer, and I simply picked out one possibility among the many in his brilliant and readable output.

The main point to make about Fisher, Haldane, and Wright in their historical context is the similarity of their understanding of evolution. Lamarckism and a myriad of strange ideas (often suggesting some inherent tendency of species to change progressively) that had grown up in the previous half-century were cut away and forgotten. Evolution was the change in frequency of Mendelian genes. However, as the synthetic theory gained ground, what would previously have been minor differences of viewpoint started to become more conspicuous. Fisher and Wright, in particular, have come to represent different lines of evolutionary thinking. Fisher and his followers have tended to see natural selection as much the most important evolutionary mechanism. Wright and his followers see a larger number of forces at work, particularly random genetic drift. This difference is clear enough in the two extracts here, as Fisher sets up an argument for the importance of selection and includes a short dismissive paragraph about genetic drift; Wright sets up a more general model of the evolutionary process, with selection as only one of a number of factors. The contrast will play out in a number of debates later in this reader: in Section B, where we look at 'selectionist' and 'neutralist' ideas about molecular evolution, and in Section C, where we find some authors who think that the importance of adaptation has been exaggerated and others who do not. There are related debates about the importance of natural selection in the origin of new species and new major groups. Haldane is less clearly positioned in this controversy (indeed his followers like to claim that they alone are able to see both sides of the argument). Haldane wrote a classic series of mathematical papers in the 1920s, in which he worked out how natural selection operated given the Mendelian system. He also wrote a famous summary book, *The Causes of Evolution* (1932). But his intellect characteristically turned to isolated, innovative, and usually highly influential contributions in a wide variety of subjects. His paper on disease (Chapter 6) is a much-cited instance, but little read because of the obscurity of the source. It discusses several topics that we shall follow up in Section B; it also contains the word 'Panglossist', which features in more than one sense in the debates of evolutionary biology. The allusion is to Dr Pangloss, Voltaire's caricature of Leibniz in *Candide*. Dr Pangloss's philosophy was that 'in this best of possible worlds . . . all is for the best'. Haldane used it to describe 'group selectionism', according to which natural selection favours adaptations that benefit the group, or the species, over the individual; but it has been borrowed in another sense by Gould and Lewontin (Chapter 20). When biologists debate adaptation, they often find that the view they disagree with (whatever it might be) is Panglossian. It is another of Haldane's many influences in the subject.

CHARLES DARWIN

1 Extract from an unpublished work on species, by C. Darwin, Esq., consisting of a portion of a chapter entitled, 'On the Variation of Organic Beings in a State of Nature; on the Natural Means of Selection; on the Comparison of Domestic Races and True Species'

Darwin argues: (1) that living creatures produce more offspring than can survive; and (2) that, when conditions change, those individuals that are better adapted to the new conditions will survive better. Darwin's account of natural selection is ecological; he concentrates on the relations of organisms and their environment. Modern accounts tend to be more genetic, describing changes in the frequencies of genes over time. The two sorts of account are complementary. Darwin ends by noticing that natural selection can work by 'the struggle of the males for females', or sexual selection as it is now called. [Editor's summary.]

De Candolle, in an eloquent passage, has declared that all nature is at war, one organism with another, or with external nature. Seeing the contented face of nature, this may at first well be doubted; but reflection will inevitably prove it to be true. The war, however, is not constant, but recurrent in a slight degree at short periods, and more severely at occasional more distant periods; and hence its effects are easily overlooked. It is the doctrine of Malthus applied in most cases with tenfold force. As in every climate there are seasons, for each of its inhabitants, of greater and less abundance, so all annually breed; and the moral restraint which in some small degree checks the increase of mankind is entirely lost. Even slow-breeding mankind has doubled in twenty-five years; and if he could increase his food with greater ease, he would double in less time. But for animals without artificial means, the amount of food for each species must, *on an average*, be constant, whereas the increase of all organisms tends to be geometrical, and in a vast majority of cases at an enormous ratio. Suppose in a certain spot there are eight pairs of birds, and that *only* four pairs of them annually (including double hatches) rear only four young, and that these go on rearing their young at the same rate, then at the end of seven years (a short life, excluding violent deaths, for any bird) there will be 2048 birds, instead of the original sixteen. As this increase is quite impossible, we must conclude either that birds do not rear nearly half their young, or that the average life of a bird is, from accident, not nearly seven years. Both checks probably concur. The same kind of calculation applied to all plants and animals affords results more or less striking, but in very few instances more striking than in man.

Many practical illustrations of this rapid tendency to increase are on record, among which, during peculiar seasons, are the extraordinary numbers of certain animals; for instance, during the years 1826 to 1828, in La Plata, when from drought some millions of cattle perished, the whole country actually *swarmed* with mice. Now I think it cannot be doubted that during the breeding-season all the mice (with the exception of a few males or females in excess) ordinarily pair, and therefore that this astounding increase during three years must be attributed to a greater number than usual surviving the first year, and then breeding, and so on till the third year, when their numbers were brought down to their usual limits on the return of wet weather. Where man has introduced plants and animals into a new and favourable country, there are many accounts in how surprisingly few years the whole country has become stocked with them. This increase would necessarily stop as soon as the country was fully stocked; and yet we have every reason to believe, from what is known of wild animals, that *all* would pair in the spring. In the majority of cases it is most difficult to imagine where the checks fall—though generally, no doubt, on the seeds, eggs, and young; but when we remember how impossible, even in mankind (so much better known than any other animal), it is to infer from repeated casual observations what the average duration of life is, or to discover the different percentage of deaths to births in different countries, we ought to feel no surprise at our being unable to discover where the check falls in any animal or plant. It should always be remembered, that in most cases the checks are recurrent yearly in a small, regular degree, and in an extreme degree during unusually cold, hot, dry, or wet years, according to the constitution of the being in question. Lighten any check in the least degree, and the geometrical powers of increase in every organism will almost instantly increase the average number of the favoured species. Nature may be compared to a surface on which rest ten thousand sharp wedges touching each other and driven inwards by incessant blows. Fully to realize these views much reflection is requisite. Malthus on man should be studied; and all such cases as those of the mice in La Plata, of the cattle and horses when first turned out in South America, of the birds by our calculation, &c., should be well considered. Reflect on the enormous multiplying power *inherent and annually in action* in all animals; reflect on the countless seeds scattered by a hundred ingenious contrivances, year after year, over the whole face of the land; and yet we have every reason to suppose that the average percentage of each of the inhabitants of a country usually remains constant. Finally, let it be borne in mind that this average number of individuals (the external conditions remaining the same) in each country is kept up by recurrent struggles against other species or against external nature (as on the borders of the Arctic regions, where the cold checks life), and that ordinarily each individual of every species holds its place, either by its own struggle and capacity of acquiring nourishment in some period of its life, from the egg upwards; or by the struggle of its parents (in short-lived organisms, when the main check occurs at longer intervals) with other individuals of the *same* or *different* species.

But let the external conditions of a country alter. If in a small degree, the relative proportions of the inhabitants will in most cases simply be slightly changed; but let the number of inhabitants be small, as on an island, and free access to it from other countries be circumscribed, and let the change of conditions continue progressing (forming new stations), in such a case the original inhabitants must cease to be as perfectly adapted to the changed conditions as they were originally. It has been shown in a former part of this work, that such changes of external conditions would, from their acting on the reproductive system, probably cause the organization of those beings which were most affected to become, as under domestication, plastic.[1] Now, can it be doubted, from the struggle each individual has to obtain subsistence, that any minute variation in structure, habits, or instincts, adapting that individual better to the new conditions, would tell upon its vigour and health? In the struggle it would have a better *chance* of surviving; and those of its offspring which inherited the variation, be it ever so slight, would also have a

[1] Darwin here argues that a change in the environmental conditions will cause the range of variation within a population to increase. Crudely speaking, this is no longer thought to be so or, at least, it is no longer part of a standard account of natural selection. There are special cases in which environmental change can increase the amount of variation, but modern thinking does not rely on them. The amount of variation that normally exists in a population is more than enough to fuel evolution by natural selection. Fisher discusses why Darwin made this argument, and why it is unnecessary with Mendelian inheritance (see Chapter 4, opening section).

better *chance*. Yearly more are bred than can survive; the smallest grain in the balance, in the long run, must tell on which death shall fall, and which shall survive. Let this work of selection on the one hand, and death on the other, go on for a thousand generations, who will pretend to affirm that it would produce no effect, when we remember what, in a few years, Bakewell effected in cattle, and Western in sheep, by this identical principle of selection?

To give an imaginary example from changes in progress on an island:—let the organization of a canine animal which preyed chiefly on rabbits, but sometimes on hares, become slightly plastic; let these same changes cause the number of rabbits very slowly to decrease, and the number of hares to increase; the effect of this would be that the fox or dog would be driven to try to catch more hares: his organization, however, being slightly plastic, those individuals with the lightest forms, longest limbs, and best eyesight, let the difference be ever so small, would be slightly favoured, and would tend to live longer, and to survive during that time of the year when food was scarcest; they would also rear more young, which would tend to inherit these slight peculiarities. The less fleet ones would be rigidly destroyed. I can see no more reason to doubt that these causes in a thousand generations would produce a marked effect, and adapt the form of the fox or dog to the catching of hares instead of rabbits, than that greyhounds can be improved by selection and careful breeding. So would it be with plants under similar circumstances. If the number of individuals of a species with plumed seeds could be increased by greater powers of dissemination within its own area (that is, if the check to increase fell chiefly on the seeds), those seeds which were provided with ever so little more down, would in the long run be most disseminated; hence a greater number of seeds thus formed would germinate, and would tend to produce plants inheriting the slightly better-adapted down.

Besides this natural means of selection, by which those individuals are preserved, whether in their egg, or larval, or mature state, which are best adapted to the place they fill in nature, there is a second agency at work in most unisexual animals, tending to produce the same effect, namely, the struggle of the males for the females. These struggles are generally decided by the law of battle, but in the case of birds, apparently, by the charms of their song, by their beauty or their power of courtship, as in the dancing rock-thrush of Guiana. The most vigorous and healthy males, implying perfect adaptation, must generally gain the victory in their contests. This kind of selection, however, is less rigorous than the other; it does not require the death of the less successful, but gives to them fewer descendants. The struggle falls, moreover, at a time of year when food is generally abundant, and perhaps the effect chiefly produced would be the modification of the secondary sexual characters, which are not related to the power of obtaining food, or to defence from enemies, but to fighting with or rivalling other males. The result of this struggle amongst the males may be compared in some respects to that produced by those agriculturists who pay less attention to the careful selection of all their young animals, and more to the occasional use of a choice mate.

CHARLES DARWIN

2 Abstract of a letter from C. Darwin, Esq., to Prof. Asa Gray, Boston, USA, dated Down, September 5th, 1857

Darwin explains natural selection by analogy with artificial selection—the way humans breed for desirable qualities in domestic and agricultural plants and animals. He also re-describes the struggle for life, due to the excess production of offspring. He finishes with the 'principle of divergence', as it came to be called: it is Darwin's explanation for the diversity of life. Species tend to evolve differences between each other, because competition is disadvantageous. [Editor's summary.]

1. It is wonderful what the principle of selection by man, that is the picking out of individuals with any desired quality, and breeding from them, and again picking out, can do. Even breeders have been astounded at their own results. They can act on differences inappreciable to an uneducated eye. Selection has been *methodically* followed in *Europe* for only the last half century; but it was occasionally, and even in some degree methodically, followed in most ancient times. There must have been also a kind of unconscious selection from a remote period, namely in the preservation of the individual animals (without any thought of their offspring) most useful to each race of man in his particular circumstances. The 'roguing', as nurserymen call the destroying of varieties which depart from their type, is a kind of selection. I am convinced that intentional and occasional selection has been the main agent in the production of our domestic races; but however this may be, its great power of modification has been indisputably shown in later times. Selection acts only by the accumulation of slight or greater variations, caused by external conditions, or by the mere fact that in generation the child is not absolutely similar to its parent. Man, by this power of accumulating variations, adapts living beings to his wants—may be said to make the wool of one sheep good for carpets, of another for cloth, etc.

2. Now suppose there were a being who did not judge by mere external appearances, but who could study the whole internal organization, who was never capricious, and should go on selecting for one object during millions of generations; who will say what he might not effect? In nature we have some *slight* variation occasionally in all parts; and I think it can be shown that changed conditions of existence is the main cause of the child not exactly resembling its parents; and in nature geology shows us what changes have taken place, and are taking place. We have almost unlimited time; no one but

a practical geologist can fully appreciate this. Think of the Glacial period, during the whole of which the same species at least of shells have existed; there must have been during this period millions on millions of generations.

3. I think it can be shown that there is such an unerring power at work in *Natural Selection* (the title of my book), which selects exclusively for the good of each organic being. The elder De Candolle, W. Herbert, and Lyell have written excellently on the struggle for life; but even they have not written strongly enough. Reflect that every being (even the elephant) breeds at such a rate, that in a few years, or at most a few centuries, the surface of the earth would not hold the progeny of one pair. I have found it hard constantly to bear in mind that the increase of every single species is checked during some part of its life, or during some shortly recurrent generation. Only a few of those annually born can live to propagate their kind. What a trifling difference must often determine which shall survive, and which perish!

4. Now take the case of a country undergoing some change. This will tend to cause some of its inhabitants to vary slightly—not but that I believe most beings vary at all times enough for selection to act on them. Some of its inhabitants will be exterminated; and the remainder will be exposed to the mutual action of a different set of inhabitants, which I believe to be far more important to the life of each being than mere climate. Considering the infinitely various methods which living beings follow to obtain food by struggling with other organisms, to escape danger at various times of life, to have their eggs or seeds disseminated, &c. &c., I cannot doubt that during millions of generations individuals of a species will be occasionally born with some slight variation, profitable to some part of their economy. Such individuals will have a better chance of surviving, and of propagating their new and slightly different structure; and the modification may be slowly increased by the accumulative action of natural selection to any profitable extent. The variety thus formed will either coexist with, or, more commonly, will exterminate its parent form. An organic being, like the woodpecker or misseltoe, may thus come to be adapted to a score of contingences—natural selection accumulating those slight variations in all parts of its structure, which are in any way useful to it during any part of its life.

5. Multiform difficulties will occur to every one, with respect to this theory. Many can, I think, be satisfactorily answered. *Natura non facit saltum* answers some of the most obvious. The slowness of the change, and only a very few individuals undergoing change at any one time, answers others. The extreme imperfection of our geological records answers others.

6. Another principle, which may be called the principle of divergence, plays, I believe, an important part in the origin of species. The same spot will support more life if occupied by very diverse forms. We see this in the many generic forms in a square yard of turf, and in the plants or insects on any little uniform islet, belonging almost invariably to as many genera and families as

species. We can understand the meaning of this fact amongst the higher animals, whose habits we understand. We know that it has been experimentally shown that a plot of land will yield a greater weight if sown with several species and genera of grasses, than if sown with only two or three species. Now, every organic being, by propagating so rapidly, may be said to be striving its utmost to increase in numbers. So it will be with the offspring of any species after it has become diversified into varieties, or sub-species, or true species. And it follows, I think, from the foregoing facts, that the varying offspring of each species will try (only few will succeed) to seize on as many and as diverse places in the economy of nature as possible. Each new variety or species, when formed, will generally take the place of, and thus exterminate its less well-fitted parent. This I believe to be the origin of the classification and affinities of organic beings at all times; for organic beings always *seem* to branch and sub-branch like the limbs of a tree from a common trunk, the flourishing and diverging twigs destroying the less vigorous—the dead and lost branches rudely representing extinct genera and families.

This sketch is *most* imperfect; but in so short a space I cannot make it better. Your imagination must fill up very wide blanks.

[*Journal of the Proceedings of the Linnean Society (Zoology)*, 3 (1858, pub. 1859), 45–62.]

JOHN MAYNARD SMITH

3 Weismann and modern biology

The extract looks at Weismann's: (1) rejection of the theory of the inheritance of acquired characteristics; (2) ideas about sexual reproduction; and (3) status in the history of biology. Weismann gave both theoretical and empirical reasons for ruling out the theory that acquired characteristics are inherited. One of the theoretical reasons made use of an 'informational' concept of heredity. [Editor's summary.]

In the preface of *Back to Methuselah*, Shaw describes how Weismann cut the tails off mice, in order to demonstrate the non-inheritance of acquired characters. Not only cruel, Shaw exclaims, but stupid. Weismann should have known that dog fanciers have been docking the tails of bitches for generations without the smallest effect on their offspring. In any case, Lamarckists would expect only actively acquired changes to be inherited: the surgical loss of a tail is not such a change. As it happens, Shaw is being unfair to Weismann's brains, if not to his heart. Weismann describes how, when he first put forward the view that acquired characteristics are not inherited, he was met by a chorus of objections from critics who pointed to the known inheritance of mutilations, including the birth of tailless puppies to mothers whose tails had been docked. He adds 'even students' fencing scars were said to have been occasionally transmitted to their sons (happily not to their daughters)' (2,65).[1] In refuting this criticism, Weismann referred to the absence of documented evidence for the effects claimed, and to the ineffectiveness of such human practices as circumcision; he also performed his famous mouse experiment. [. . .]

I have gradually become aware, that, after Darwin, Weismann was the greatest evolutionary biologist of the nineteenth century. Further, the problems he was concerned with are often the same problems that concern us today. I have neither the linguistic nor the historical skills to write a proper history of his ideas. Instead, I have attempted to summarize what he thought towards the end of his life, when he wrote *The Evolution Theory*, and to explain why he thought as he did. In the light of present knowledge, it is sometimes easy to see where he went wrong; it is harder to see what he got right, because when he did so, his ideas are now so widely accepted that we do not appreciate that people once thought otherwise. For this reason, the following account may appear unduly critical. It is not intended to be. Indeed, the main impression I am left with is of a man who spent his life thinking about precisely the problems that have concerned me during my own work, who often got it exactly right, and who was admirably willing to admit that he had been wrong and to try again. [. . .]

Are functional modifications transmitted?

Weismann was well aware that functional modifications—for example in the growth of bone, muscle, and tendon—can occur during ontogeny. He explained these modifications in terms of Roux's idea that 'an organ increases through its own specific activity'. He thought that this came about by a kind of selection operating between the cells of an individual, each cell type being stimulated to multiply by appropriate conditions. But he is clear that the properties of the cells that cause them to react appropriately 'are themselves adaptations of the organism, and can therefore be referred to personal selection' (1,248), and 'not from a struggle between the cells themselves' (1,250). [. . .]

But although Weismann accepted the reality of such functional modifications, he was adamant that they were not transmitted via the germ cells.[2] He defends this position by four arguments:

[1] This and other references in this piece are to the volume and page numbers of A. Weismann, *The Evolution Theory*, 2 vols. (London: Edward Arnold, 1904).

[2] Germ cells are the reproductive cell line—the cell lines that produce eggs in females and sperm in males. Weismann divided the cells of a body into the 'somatic line' (or soma) and 'germ line'. The soma is all the parts of the body that are destined to die when the organism dies; the germ line reproduces down the generations and is potentially immortal.

1. There is no empirical evidence for the inheritance of acquired characters.
2. Many adaptations are of a kind that could not arise as functional modifications during ontogeny.
3. Adaptations of the sterile castes of insects could not evolve by a Lamarckian mechanism.
4. There is a theoretical difficulty about Lamarckian inheritance, which we would today describe as the problem of reverse translation.

The first point was discussed briefly above, and the third is persuasive and needs no elaboration. Point 2 is also persuasive. If an organ is fully formed before it is used, and cannot then be modified, there will be no functional modifications to be inherited. Examples of such structures are the colour of many animals (in particular, of his beloved Lepidoptera), the cuticle of arthropods, and the protective thorns and hairs of plants. Of particular interest are instincts, as the idea that instincts might be inherited habits has often been seen as favouring Lamarckism. Weismann points out that there are many instincts that could not have arisen as a learnt habit in the first place, for several reasons. First, the animal in question may be of too low intelligence to learn a complex habit: for example, it is inconceivable that a solitary wasp should learn by trial and error its complex sequence of nest-building and provisioning behaviour. Secondly, many instinctive acts are performed only once in a lifetime, and so afford no opportunity for learning: for example, the acts leading to the suspension of some lepidopteran pupae. Finally, the success or failure of some instinctive acts is decided only after the performer is dead, and so could not guide learning; for example, oviposition in many insects.

If, then, many complex structures and behaviours have evolved without the transmission of functional modifications, and if there is no evidence for such transmission in any case, there is no need to assume that the Lamarckian mechanism has ever been important. But Weismann has a final theoretical argument: in fact, although I have listed it above as a final argument, I have little doubt that, in the genesis of Weismann's own ideas, it came first, and the empirical evidence supporting it followed. The argument can best be put in Weismann's own words: oddly, both quotations use the analogy of translation into Chinese.

'But, as these primary constituents [i.e. the genetic determinants, J.M.S.] are quite different from the parts themselves [i.e. the adult organs, J.M.S.], they would require to vary in quite a different way from that in which the finished parts had varied: which is very like supposing that an English telegram to China is there received in the Chinese language' (2,63).

And, discussing whether instincts are inherited habits: 'How could it happen that the constant exercise of memory throughout a lifetime . . . could influence the germ cells in such a way that in the offspring the same brain cells which preside over memory will likewise be more highly developed? . . . if we take our stand upon the theory of determinants, it would be necessary to a transmission of acquired strength of memory that the states of these brain cells should be communicated by the telegraphic path of the nerve cells to the germ cells, and should there modify only the determinants of the brain cells, and should do so in such a way that, in the subsequent development of an embryo from the germ cell, the corresponding brain cells should turn out to be capable of increased functional activity . . . I can only compare the assumption of the transmission of the results of memory-exercise to the telegraphing of a poem, which is handed in in German, but at the place of arrival appears on the paper translated into Chinese' (2,107).

It is hard to imagine a clearer expression of the theoretical difficulty of Lamarckian inheritance. In particular, the use of the information analogy is clear and modern. Not surprisingly, since information-transducing machines were rare in 1900, and as there was no scientific definition of information, Weismann did not maintain a consistently informational concept of the gene, a failure that cost him dear when he came to formulate his ideas of germinal selection and the origin of new variation. Despite the persuasiveness of the theoretical argument, Weismann did not regard it as decisive. There is so much in biology that we do not understand, he says, that we cannot rule out a process merely because we cannot imagine how it could happen: we need empirical arguments as well. He would have agreed with Jacob's[3] remark about reverse translation: 'Not that such a mechanism is theoretically impossible—simply it does not exist'.

To me, the most surprising thing about the chapters dealing with the transmission of functional modification is the argument that is not there—the dog that did not bark in the night. There is no mention, in this context, of the segregation of the germ track. I do not know how far Weismann's research on the origin of germ cells influenced him in reaching his conclusion (first published in 1883) that acquired characters are not inherited, but it is clear that the argument played little part in his final conclusions. Nor do I see how it could have done. Darwin had proposed his pangenesis theory to account for the supposed 'effect of use and disuse', according to which 'gemmules' carry information from the soma to the germ cells. Weismann starts his discussion of Lamarckism by outlining this theory. He rejects it on theoretical grounds (the first of the two Chinese language quotations immediately follows his discussion of Darwin's theory): the segregation of the germ line would be wholly irrelevant. [. . .]

[3] F. Jacob, *The Possible and the Actual* (New York: Pantheon, 1982). 'Reverse translation' refers to the production of DNA (or mRNA) from protein. In a genetic sense, translation is the final stage in which protein is read off from DNA. Reverse translation would proceed in the opposite direction.

Sex and breeding systems

It is not to be expected that Weismann should have had a full understanding of the evolution of sex, as we still are feeling our way to such an understanding, but there is much that he does understand. He is clear that sex is not needed for reproduction, and that, at the cellular level, it is the opposite of reproduction. He understands that meiotic segregation generates variability. He suggests that sex is needed for 'coadaptation'—that is, for the bringing together in a single individual of 'harmonious' variations that originated in different individuals. In effect, this is the idea proposed by Fisher and Muller[4] that sex accelerates evolution by bringing together, by genetic recombination, favourable mutations from different ancestors. He recognizes that this is a long-term rather than a short-term explanation (2,198), and (2,199) suggests a more immediate advantage in terms of normalizing selection, which, he suggests, can only act effectively to maintain adaptation in a sexual population. I am unable to follow his argument on this point—it is not one that can adequately be presented in verbal terms—and I suspect that it is fallacious. It is true that, for different reasons, Muller and Kondrashov[5] have argued that recombination can reduce the genetic load needed to eliminate harmful mutation, but I do not think that Weismann can be credited with reaching this conclusion. However, it is interesting that he does discuss the advantages of sex in the two contexts of directional selection; and the elimination of harmful variation.

He argues that the origin of sex requires a 'direct' (i.e. immediate) selective advantage. He rejects the idea that sex has a rejuvenating function, because of the survival of parthenogens (he cultured a parthenogenetic Ostracod for 80 generations). He suggests that the selective advantage responsible for the origin of cellular fusion may have been that each partner contributed genetic material deficient in the other: an explanation in terms of complementation that most of us today would accept. He is aware of inbreeding depression, and explains it by saying that, in an inbred population, 'the germ plasm may then consist entirely of identical ids' (2,273). He contrasts this with the absence of deterioration in parthenogens which lack a reduction division. He argues that hermaphroditism is favoured in sedentary organisms because of the difficulty gonochorists[6] would experience in finding mates. Finally, and for me annoyingly, he writes 'By the occurrence of parthenogenesis, the number

[4] R. Fisher, *The Genetical Theory of Natural Selection* (Oxford: Oxford University Press, 1930), and H. J. Muller, *American Naturalist*, 66 (1932), 118–38.

[5] H. J. Muller, *Mutation Research*, I (1964), 2–9, and A. Kondrashov, *Nature*, 336 (1988), 435–40.

[6] Gonochorists are organisms with separate sexes, in which a body is either male or female. Humans are gonochorists. For further discussion of the point in the next sentence, and to see why Maynard Smith might be annoyed to find himself anticipated, see the other Maynard Smith extract (Chapter 48).

of ova produced by a particular colony of animals may be doubled, because each individual is a female' (2,243). This is a fairly impressive list of insights for a book published in 1904.

Weismann and theoretical biology

Darwin's theory of evolution by natural selection is the central theory of biology. Yet Darwin was a shame-faced theorist. He found it necessary to pretend to others, and even to himself, that he reached his conclusions by induction from a vast array of facts. Weismann, like Darwin, was first and foremost a naturalist, but he is a less ashamed theorist. In a sense, he is the first conscious theoretical biologist. His failing eyesight may have contributed to this tendency. However, he is a theoretician without the tools of a theoretician's trade. There is, I think, not one line of algebra, although he does use symbols—A's and B's and C's—to refer to differing determinants. Perhaps more surprising, he does not use diagrams to represent his ideas, and to deduce their consequences. The diagram mentioned above, that represents the consequences of segregation in meiosis, is a unique exception. The appearance of a complete theoretical biology, resting on mathematics, had to wait for Fisher, Haldane, and Wright.

Occasionally, Weismann's lack of mathematics lets him down, as in his treatment of normalizing selection and sex that was mentioned above, or in his attempted explanation of Gaussian distribution of phenotypes (2,206). Faced with the achievements of Darwin and Weismann, one cannot claim that mathematics, or even diagrams, are needed for successful theoretical work in biology. But they certainly make it easier.

['Weismann and Modern Biology', *Oxford Surveys in Evolutionary Biology*, 6 (1989), 1–12.]

R. A. FISHER

4 The nature of inheritance

Mendel's theory of inheritance arguably 'saved' Darwin's theory of evolution, producing the synthetic theory of evolution, or neo-Darwinism. Fisher here explores the consequences of blending and particulate theories of inheritance.[1] *He first shows that variation is rapidly*

[1] In a blending theory, as explained in my editorial introduction to this section, the parental hereditary factors (or gene-equivalents) blend in the offspring to form some new intermediate form. The offspring sends this new form of hereditary factor to its offspring. In a particulate theory, the parental genes are preserved within the offspring and passed on to the next generation in essentially the same form. Blending theories are now rejected for factual reasons. Mendelism is a particulate theory.

destroyed with blending inheritance. Variation can only be maintained if mutation introduces huge amounts of new variation each generation. He then shows that with Mendelian, particulate inheritance variation is preserved. Fisher goes on to consider whether mutation alone could drive evolution, and argues it cannot. He also argues that population sizes are too large for random genetic drift to have much effect. This leaves natural selection as the main factor driving evolution. [Editor's summary.]

The consequences of the blending theory

That Charles Darwin accepted the fusion or blending theory of inheritance, just as all men accept many of the undisputed beliefs of their time, is universally admitted. That his acceptance of this theory had an important influence on his views respecting variation, and consequently on the views developed by himself and others on the possible causes of organic evolution, was not, I think, apparent to himself, nor is it sufficiently appreciated in our own times. In the course of the present chapter I hope to make clear the logical consequences of the blending theory, and to show their influence, not only on the development of Darwin's views, but on the change of attitude towards these, and other suppositions, necessitated by the acceptance of the opposite theory of particulate inheritance.

It is of interest that the need for an alternative to blending inheritance was certainly felt by Darwin, though probably he never worked out a distinct idea of a particulate theory. In a letter to Huxley probably dated in 1857 occur the sentences (*More Letters*, vol. i, Letter 57).

> Approaching the subject from the side which attracts me most, viz., inheritance, I have lately been inclined to speculate, very crudely and indistinctly, that propagation by true fertilization will turn out to be a sort of mixture, and not true fusion, of two distinct individuals, or rather of innumerable individuals, as each parent has its parents and ancestors. I can understand on no other view the way in which crossed forms go back to so large an extent to ancestral forms. But all this, of course, is infinitely crude.

The idea apparently was never developed, perhaps owing to the rush of work which preceded and followed the publication of the *Origin*. Certainly he did not perceive that the arguments on variation in his rough essays of 1842 and 1844, which a year later (1858) he would be rewriting in the form of the first chapter of the *Origin*, required, on a particulate theory, a complete reformulation. The same views indeed are but little changed when 'The Causes of Variability' came to be discussed in Chapter XXII of *Variation of Animals and Plants* published in 1868.

The argument which can be reconstructed from these four sources may be summarized as follows:

(*a*) with blending inheritance bisexual reproduction will tend rapidly to produce uniformity; therefore

(*b*) if variability persists, causes of new variation must be continually at work; hence
(*c*) the causes of the great variability of domesticated species, of all kinds and in all countries, must be sought for in the conditions of domestication; but
(*d*) the only characteristics of domestication sufficiently general to cover all cases are changed conditions and increase of food;
(*e*) some changes of conditions seem to produce definite and regular effects, e.g. increased food causes (hereditary) increase in size, yet the important effect is an indefinite variability in all directions, ascribable to a disturbance, by change of conditions, of the regularity of action of the reproductive system;
(*f*) wild species also will occasionally, by geological changes, suffer changed conditions, and occasionally also a temporary increase in the supply of food; they will therefore, though perhaps rarely, be caused to vary. If on these occasions no selection is exerted the variations will neutralize one another by bisexual reproduction and die away, but if selection is acting, the variations in the right direction will be accumulated and a permanent evolutionary change effected.

To modern readers this will seem a very strange argument with which to introduce the case for Natural Selection; all that is gained by it is the inference that wild as well as domesticated species will at least occasionally present heritable variability. Yet it is used to introduce the subject in the two essays and in the *Origin*. It should be remembered that, at the time of the essays, Darwin had little direct evidence on this point, which, since the power of human selection to modify domesticated races was widely admitted, was a cardinal point in the original argument. Even in the *Origin* the second chapter on 'Variation under Nature' deals chiefly with natural varieties sufficiently distinct to be listed by botanists, and these were certainly regarded by Darwin not as the materials but as the products of evolution. During the twenty-six years between 1842 and 1868 evidence must have flowed in sufficiently at least to convince him that heritable variability was as widespread, though not nearly so great, in wild as in domesticated species. The line of reasoning in question seems to have lost its importance sufficiently for him to introduce the subject in 1868 (*Variation*, Chapter XXII) with the words 'The subject is an obscure one; but it may be useful to probe our ignorance.'

It is the great charm of the essays that they show the *reasons* which led Darwin to his conclusions, whereas the later works often only give the *evidence* upon which the reader is to judge of their truth. The antithesis is not an unnatural one, for every active mind must form opinions without direct evidence, else the evidence too often would never be collected. Impartiality and scientific discipline come into action effectively in submitting the opinions formed to as much relevant evidence as can be made available. The earlier steps in the argument set out above appear only in the two essays, while the conclusions continue almost unchanged up to the *Variation of Animals and Plants*. Indeed the first step (a), logically the most important of all, appears explicitly only in 1842. In 1844 it is clearly implied by its necessary consequences. I believe its significance for the argument of the *Origin* would scarcely ever be detected from a study only of that book. The passage in the 1842 MS. is (*Foundations*, p. 2):

Each parent transmits its peculiarities, therefore if varieties allowed freely to cross, except by the *chance* of two characterized by same peculiarity happening to marry, such varieties will be constantly demolished. All bisexual animals must cross, hermaphrodite plants do cross, it seems very possible that hermaphrodite animals do cross—conclusion strengthened:

together with a partly illegible passage of uncertain position,

If individuals of two widely different varieties be allowed to cross, a third race will be formed—a most fertile source of the variation in domesticated animals. If freely allowed, the characters of pure parents will be lost, number of races thus [illegible] but differences [?] besides the [illegible]. But if varieties differing in very slight respects be allowed to cross, such small variation will be destroyed, at least to our senses—a variation just to be distinguished by long legs will have offspring not to be so distinguished. Free crossing is a great agent in producing uniformity in any breed.

The proposition is an important one, marking as it does the great contrast between the blending and the particulate theories of inheritance. [. . .]

The important consequence of the blending is that, if not safeguarded by intense marital correlation, the heritable variance is approximately halved in every generation. To maintain a stationary variance fresh mutations must be available in each generation to supply the half of the variance so lost. If variability persists, as Darwin rightly inferred, causes of new variability must continually be at work. Almost every individual of each generation must be a mutant, i.e. must be influenced by such causes, and moreover must be a mutant in many different characters.

An inevitable inference of the blending theory is that the bulk of the heritable variance present at any moment is of extremely recent origin. One half is new in each generation, and of the remainder one half is only one generation older, and so on. Less than one-thousandth of the variance can be ten generations old; even if by reason of selective mating we ought to say twenty generations, the general conclusion is the same; the variability of domesticated species must be ascribed by any adherent of the blending theory to the conditions of domestication much as they now exist. If variation is to be used by the human breeder, or by natural selection, it must be snapped up at once, soon after the mutation has appeared, and before it has had time to die away. [. . .]

Conservation of the variance

Particulate inheritance differs from the blending theory in an even more important fact. There is no inherent tendency for the variability to diminish. In a population breeding at random in which two alternative genes of any factor, exist in the ratio p to q, the three genotypes will occur in the ratio $p^2:2pq:q^2$, and thus ensure that their characteristics will be represented in fixed proportions of the population, however they may be combined with characteristics determined by other factors, provided that the ratio $p:q$ remains unchanged. This ratio will indeed be liable to slight changes; first by the chance survival and reproduction of individuals of the different kinds; and secondly by selective survival, by reason of the fact that the genotypes are probably unequally fitted, at least to a slight extent, to their task of survival and reproduction. The effect of chance survival is easily susceptible of calculation, and it appears [. . .] that in a population of n individuals breeding at random the variance will be halved by this cause acting alone in 1.4 n generations. Since the number of individuals surviving to reproduce in each generation must in most species exceed a million, and in many is at least a million-fold greater, it will be seen that this cause of the diminution of hereditary variance is exceedingly minute, when compared to the rate of halving in one or two generations by blending inheritance.

The circumstance that smaller numbers, even less than 100, are sometimes found to reproduce themselves locally, does not, as has been supposed, add to the frequency of random extinction, or to the importance of the so-called 'genetic drift'. For this, perfect isolation is required over a number of generations equally numerous with the population isolated. Even if perfect isolation could be postulated, which is always questionable, it is still improbable that the small isolated population would not ordinarily die out altogether before a period of evolutionary significance could elapse, or that it would not be later absorbed in other populations with a different genetic constitution.

[. . .] Selection is a much more important agency in keeping the variability of species within limits. But even relatively intense selection will change the ratio $p:q$ of the gene frequencies relatively slowly, and no reasonable assumptions could be made by which the diminution of variance due to selection, in the total absence of mutations, would be much more than a ten-thousandth of that ascribable to blending inheritance. The immediate consequence of this enormous contrast is that the mutation rate needed to maintain a given amount of variability is, on the particulate theory, many thousand times smaller than that which is required on the blending theory.

Theories, therefore, which ascribe to agencies believed to be capable of producing mutations, as was 'use and disuse' by Darwin, a power of governing the direction in which evolution is taking place, appear in very different lights, according as one theory of inheritance, or the other, is accepted. For any evolutionary tendency which is supposed to act by favouring mutations in one direction rather than another, and a number of such mechanisms have from time to time been imagined, will lose its force many thousand-fold, when the particulate theory of inheritance, in any form, is accepted; whereas the directing power of Natural Selection, depending as it does on the amount of heritable variance maintained, is totally uninfluenced by any such change. This consideration, which applies to all such theories alike, is independent of the fact that a great part of the reason, at least to Darwin, for ascribing to the environment any considerable influence in the production of mutations, is swept away when we are no longer forced to consider the great variability of domestic species as due to the comparatively recent influence of their artificial environment.

The striking fact, of which Darwin was well aware, that whole brothers and sisters, whose parentage, and consequently whose entire ancestry is identical, may differ greatly in their hereditary composition, bears under the two theories two very different interpretations. Under the blending theory it is clear evidence of new and frequent mutations, governed, as the greater resemblance of twins suggests, by temporary conditions acting during conception and gestation. On the particulate theory it is a necessary consequence of the fact that for every factor a considerable fraction, often not much less than one half, of the population will be heterozygotes, any two offspring of which will be equally likely to receive unlike as like genes from their parents. In view of the close analogy between the statistical concept of variance and the physical concept of energy, we may usefully think of the heterozygote as possessing variance in a potential or latent form, so that instead of being lost it is merely stored in a form from which it will reappear when the heterozygous genotypes are mated. A population mating at random immediately establishes the condition of statistical equilibrium between the latent and the apparent form of variance. The particulate theory of inheritance resembles the kinetic theory of gases with its perfectly elastic collisions, whereas the blending theory resembles a theory of gases with inelastic collisions, and in which some outside agency would be required to be continually at work to keep the particles astir.

The property of the particulate theory of conserving the variance for an indefinite period explains at once the delayed or cumulative effect of domestication in increasing the variance of domesticated species, to which Darwin calls attention. Many of our domesticated varieties are evidently ill-fitted to survive in the wild condition. The mutations by which they arose may have been occurring for an indefinite period prior to domestication without establishing themselves, or appreciably affecting the variance of the wild species. In domestication, however, not only is the rigour of Natural Selection relaxed so that mutant types can survive, and each such survival add something to the store of heritable variance, but novelties of form or colour, even if semi-monstrous, do undoubtedly attract human attention and interest, and are valued by man for their peculiarity. The rapidity with which new variance is accumulated will thus be enhanced. Without postulating any change in the mutation rates due to domestication, we should necessarily infer from what is known of the conditions of domestication that the variation of domesticated species should be greater than that of similar wild species, and that this contrast should be greatest with those species most anciently domesticated. Thus one of the main difficulties felt by Darwin is resolved by the particulate theory.

Theories of evolution worked by mutations

The theories of evolution which rely upon hypothetical agencies, capable of modifying the frequency or direction in which mutations are taking place, fall into four classes. In stating these it will be convenient to use the term 'mutation', to which many meanings have at different times been assigned, to denote simply the initiation of any heritable novelty.

(A) It may be supposed, as by Lamarck in the case of animals, that the mental state, and especially the desires of the organism, possess the power of producing mutations of such a kind, that these desires may be more readily gratified in the descendants. This view postulates (i) that there exists a mechanism by which mutations are caused, and even designed, in accordance with the condition of the nervous system, and (ii) that the desires of animals in general are such that their realization will improve the aptitude of the species for life in its natural surroundings, and also will maintain or improve the aptitude of its parts to co-operate with one another, both in maintaining the vital activity of the adult animal, and in ensuring its normal embryological development. The desires of animals must, in fact, be very wisely directed, as well as being effective in provoking suitable mutations.

(B) A power of adaptation may be widely observed, both among plants and animals, by which particular organs, such as muscles or glands, respond by increased activity and increased size, when additional physiological calls are made upon them. It may be suggested, as it was by Darwin, that such responses of increased functional activity induce, or are accompanied by, mutations of a kind tending to increase the size or activity of the organ in question in future generations, even if no additional calls were made upon this organ's activity. This view implies (i) that the power which parts of organisms possess, of responding adaptively to increased demands upon them, is not itself a product of evolution, but is postulated as a primordial property of living matter: and requires (ii) that a mechanism exists by which the adaptive response shall itself tend to cause, or to be accompanied by, an appropriate mutation.

Both these two suggested means of evolution expressly aim at explaining, not merely the progressive change of organic beings, but the aptitude of the organism to its place in nature, and of its parts to their function in the organism.

(C) It may be supposed that the environment in which the organism is placed controls the nature of the mutations which occur in it, and so directs its evolutionary course; much as the course of a projectile is controlled by the field of force in which it flies.

(D) It may be supposed that the mutations which an organism undergoes are due to an 'inner urge' (not necessarily connected with its mental state) implanted in its primordial ancestors, which thereby directs its predestined evolution.

The two last suggestions give no particular assistance towards the understanding of adaptation, but each contains at least this element of truth; that however profound our ignorance of the causes of mutation may be, we cannot but ascribe them, within the order of Nature as we know it, either to the nature of the organism, or to that of its surrounding environment, or, more generally, to the interaction of the two. What is common, however, to all four of these suppositions, is that each one postulates that the direction of evolutionary change is governed by the predominant direction in which mutations are taking place. However reasonable such an assumption might have seemed when, under the blending theory of inheritance, every individual was regarded as a mutant, and probably a multiple mutant, it is impossible to let it pass unquestioned, in face of the much lower mutation rates appropriate to the particulate theory. [. . .]

Nature and frequency of observed mutations

The assumption that the direction of evolutionary change is actually governed by the direction in which mutations are occurring is not easily compatible with the nature of the numerous mutations which have now been observed to occur. For the majority of these produce strikingly disadvantageous deformities, and indeed much the largest class are actually lethal. If we had to admit, as has been so often assumed in theory, that these mutations point the direction of evolution, the evolutionary prospects of the little fruit-fly *Drosophila* would be deplorable indeed. Nor is the position apparently different with man and his domesticated animals and plants; as may be judged from the frequency with which striking recessive defects, producing, for example, albinism, deaf-mutism, and feebleness of mind in man, must have occurred in the comparatively recent past, as mutations. Mutant defects seem to attack the human eye as much as that of *Drosophila*, and in general the mutants which occur in domesticated races are often monstrous and predominantly defective, whereas we know in many cases that the evolutionary changes which these creatures have undergone under human selection have been in the direction of a manifest improvement.

In addition to the defective mutations, which by their conspicuousness attract attention, we may reasonably suppose that other less obvious mutations are occurring which, at least in certain surroundings, or in certain genetic combinations, might prove themselves to be beneficial. It would be unreasonable, however, to assume that such mutations appear individually with a frequency much greater than that which is observed in the manifest defects. The frequency of individual mutations in *Drosophila* is certainly seldom greater than one in 100,000 individuals, and we may take this figure to illustrate the inefficacy of any agency, which merely controls the predominant direction of mutation, to determine the predominant direction of evolutionary change. For even if selection were totally absent, a lapse of time of the order of 100,000 generations would be required to produce an important change with respect to the factor concerned, in the heritable nature of the species. Moreover, if the mutant gene were opposed, even by a very minute selective disadvantage, the change would be brought to a standstill at a very early stage.

[. . .] It will be readily understood that if we speak of a selective advantage of one per cent., with the meaning that animals bearing one gene have an expectation of offspring only one per cent. greater than those bearing its allelomorph,[2] the selective advantage in question will be a very minute one; at least in the sense that it would require an enormous number of experimental animals, and extremely precise methods of experimentation, to demonstrate so small an effect experimentally. Such a selective advantage would, however, greatly modify the genetic constitution of the species, not in 100,000 but in 100 generations.[3] If, moreover, we imagine these two agencies opposed in their tendencies, so that a mutation which persistently occurs in one in 100,000 individuals, is persistently opposed by a selective advantage of only one per cent., it will easily be seen that an equilibrium will be arrived at when only about one individual in 1,000 of the population will be affected by the mutation. This equilibrium, moreover, will be stable; for if we imagine that by some chance the number of mutants is raised to a higher proportion than this, the proportion will immediately commence to diminish under the action of selection, and evolution will proceed in the direction contrary to

[2] We should now say 'allele'.

[3] This is a slight exaggeration; still in 200 generations the percentage would change from 26.9% to 73.1% [Fisher's note].

the mutation which is occurring, until the proportion of mutant individuals again reaches its equilibrium value. For mutations to dominate the trend of evolution it is thus necessary to postulate mutation rates immensely greater than those which are known to occur, and of an order of magnitude which, in general, would be incompatible with particulate inheritance.

[*The Genetical Theory of Natural Selection* (Oxford: OUP, 1930).]

SEWALL WRIGHT

5 The roles of mutation, inbreeding, crossbreeding, and selection in evolution

Wright begins by establishing that populations contain huge amounts of variation. He then introduces his concept of an adaptive landscape, which shows how the adaptive quality of living creatures depends on their genetic make-up. The adaptive landscape will contain many peaks and valleys. He then uses the landscape image to think about the evolutionary effects of several factors: mutation, selection, genetic drift, migration. He argues that selection drives populations to local peaks, and genetic drift allows populations to evolve from one peak to another. [Editor's summary.]

The enormous importance of biparental reproduction as a factor in evolution was brought out a good many years ago by East. The observed properties of gene mutation—fortuitous in origin, infrequent in occurrence and deleterious when not negligible in effect—seem about as unfavorable as possible for an evolutionary process. Under biparental reproduction, however, a limited number of mutations which are not too injurious to be carried by the species furnish an almost infinite field of possible variations through which the species may work its way under natural selection.

Estimates of the total number of genes in the cells of higher organisms range from 1000 up.[1] Some 400 loci have been reported as having mutated in *Drosophila* during a laboratory experience which is certainly very limited compared with the history of the species in nature. Presumably, allelomorphs[2] of all type genes are present at all times in any reasonably numerous species. Judging from the frequency of multiple allelomorphs in those organisms which have been studied most, it is reasonably certain that many different allelomorphs of each gene are in existence at all times. With 10 allelomorphs in each of 1000 loci, the number of possible combinations is 10^{1000} which is a very large number. It has been estimated that the total

[1] Estimates now come from genome sequences and are higher. The fruitfly *Drosophila* has about 12,000 genes. Humans have about 25,000–30,000. The higher figures magnify the point that Wright is making.

[2] Now usually called alleles.

number of electrons and protons in the whole visible universe is much less than 10^{100}.

However, not all of this field is easily available in an interbreeding population. Suppose that each type gene is manifested in 99 percent of the individuals, and that most of the remaining 1 percent have the most favorable of the other allelomorphs, which in general means one with only a slight differential effect. The average individual will show the effects of 1 per cent of the 1000, or 10 deviations from the type, and since this average has a standard deviation of $\sqrt{10}$ only a small proportion will exhibit more than 20 deviations from type where 1000 are possible. The population is thus confined to an infinitesimal portion of the field of possible gene combinations, yet this portion includes some 10^{40} homozygous combinations, on the above extremely conservative basis, enough so that there is no reasonable chance that any two individuals have exactly the same genetic constitution in a species of millions of millions of individuals persisting over millions of generations. There is no difficulty in accounting for the probable genetic uniqueness of each individual human being or other organism which is the product of biparental reproduction.

If the entire field of possible gene combinations be graded with respect to adaptive value under a particular set of conditions, what would be its nature? [. . .] The two dimensions of Figure 5.1 are a very inadequate representation of such a field.[3] The contour lines are intended to represent the scale of adaptive value.

One possibility is that a particular combination gives maximum adaptation and that the adaptiveness of the other combinations falls off more or less regularly according to the number of removes. A species whose individuals are clustered about some combination other than the highest would move up the steepest gradient toward the peak, having reached which it would remain unchanged except for the rare occurrence of new favorable mutations.

But even in the two factor case it is possible that there may be two peaks, and the chance that this may be the case greatly increases with each additional locus. With something like 10^{1000} possibilities it may be taken as certain that there will be an enormous number of widely separated

[3] Wright here introduces the concept, or picture, variously described as an adaptive landscape, a fitness surface, or an adaptive topography. The two axes drawn on the page represent the frequencies of genes, or of genotypes. The frequency of one gene (or genotype) is shown up the page; the frequency of another across the page. The contour lines within the picture show the adaptive quality of the organisms in a population with a given set of gene (or genotype) frequencies. For instance, Figure 5.1 has a peak at the top left corner—a population with this combination of genes would have well adapted members. At the lower right corner is a valley; a population with this combination of genes would have poorly adapted members. Natural selection will usually cause a population to climb the local hill until it reaches a peak. Wright's controversial claim is that the landscape has multiple peaks.

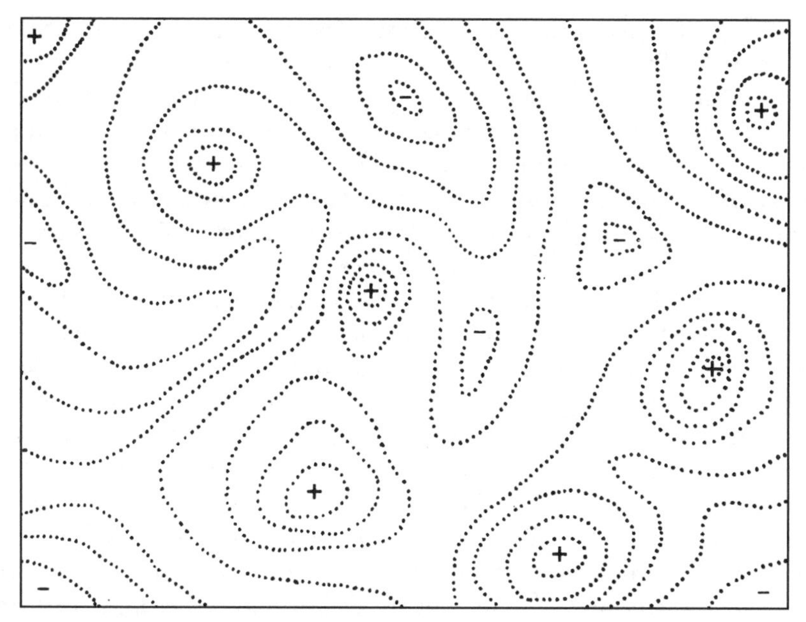

Figure 5.1: Diagrammatic representation of the field of gene combinations in two dimensions instead of many thousands. Dotted lines represent contours with respect to adaptiveness.

harmonious combinations. The chance that a random combination is as adaptive as those characteristic of the species may be as low as 10^{-100} and still leave room for 10^{800} separate peaks, each surrounded by 10^{100} more or less similar combinations. In a rugged field of this character, selection will easily carry the species to the nearest peak, but there may be innumerable other peaks which are higher but which are separated by 'valleys.' The problem of evolution as I see it is that of a mechanism by which the species may continually find its way from lower to higher peaks in such a field. In order that this may occur, there must be some trial and error mechanism on a grand scale by which the species may explore the region surrounding the small portion of the field which it occupies. To evolve, the species must not be under strict control of natural selection. Is there such a trial and error mechanism?

At this point let us consider briefly the situation with respect to a single locus. [. . .] The elementary evolutionary process is, of course, change of gene frequency, a practically continuous process. Owing to the symmetry of the Mendelian mechanism, any gene frequency tends to remain constant in the absence of disturbing factors. If the type gene mutates at a certain rate, its frequency tends to [decrease], but at a continually decreasing rate. The type gene would ultimately be lost from the population if there were no opposing factor. But the type gene is in general favored by selection. Under selection, its frequency tends to [increase]. The rate is greatest at some point near the middle of the range. At a certain gene frequency the opposing pressures are equal and opposite, and at this point there is consequently equilibrium. There are other mechanisms of equilibrium among evolutionary factors which need not be discussed here. Note that we have here a theory of the stability of species in spite of continuing mutation pressure, a continuing field of variability so extensive that no two individuals are ever genetically the same, and continuing selection.

If the population is not indefinitely large, another factor must be taken into account: the effects of accidents of sampling among those that survive and become parents in each generation and among the germ cells of these, in other words, the effects of inbreeding.[4] Gene frequency in a given generation is in general a little different one way or the other from that in the preceding, merely by chance. In time, gene frequency may wander a long way from the position of equilibrium, although the farther it wanders the greater the pressure toward return. The result is a frequency distribution within which gene frequency moves at random. There is considerable spread even with very slight inbreeding and the form of distribution becomes U-shaped with close inbreeding. The rate of movement of gene frequency is very slow in the former case but is rapid in the latter (among unfixed genes). In this case, however, the tendency toward complete fixation of genes, practically irrespective of selection, leads in the end to extinction.

In a local race, subject to a small amount of crossbreeding with the rest of the species, the tendency toward random fixation is balanced by immigration pressure instead of by mutation and selection. In a small sufficiently isolated group all gene frequencies can drift irregularly back and forth about their mean values at a rapid rate, in terms of geologic time, without reaching fixation and giving the effects of close inbreeding. The resultant differentiation of races is of course increased by any local differences in the conditions of selection.

Let us return to the field of gene combinations (Fig. 5.1). In an indefinitely large but freely interbreeding species living under constant conditions, each gene will reach ultimately a certain equilibrium. The species will occupy a certain field of variation about a peak in our diagram (heavy broken contour in upper left of each figure). The field occupied remains constant although no two individuals are ever identical. Under the above conditions further evolution can occur only by the appearance of wholly new (instead of recurrent) mutations, and ones which happen to be favorable from the first. Such mutations would change the character of the field itself, increasing the

[4] Wright says inbreeding, but he could just as well have said random genetic drift. Random genetic drift is mathematically equivalent to, indeed a form of, inbreeding.

elevation of the peak occupied by the species. Evolutionary progress through this mechanism is excessively slow since the chance of occurrence of such mutations is very small and, after occurrence, the time required for attainment of sufficient frequency to be subject to selection to an appreciable extent is enormous.

The general rate of mutation may conceivably increase for some reason. For example, certain authors have suggested an increased incidence of cosmic rays in this connection. The effect (Fig. 5.2A) will be as a rule a spreading of the field occupied by the species until a new equilibrium is reached. There will be an average lowering of the adaptive level of the species. On the other hand, there will be a speeding up of the process discussed above, elevation of the peak itself through appearance of novel favorable mutations. Another possibility of evolutionary advance is that the spreading of the field occupied may go so far as to include another and higher peak, in which case the species will move over and occupy the region about this. These mechanisms do not appear adequate to explain evolution to an important extent.

The effects of reduced mutation rate (Fig. 5.2B) are of course the opposite: a rise in average level, but reduced variability, less chance of novel favorable mutation, and less chance of capture of a neighboring peak.

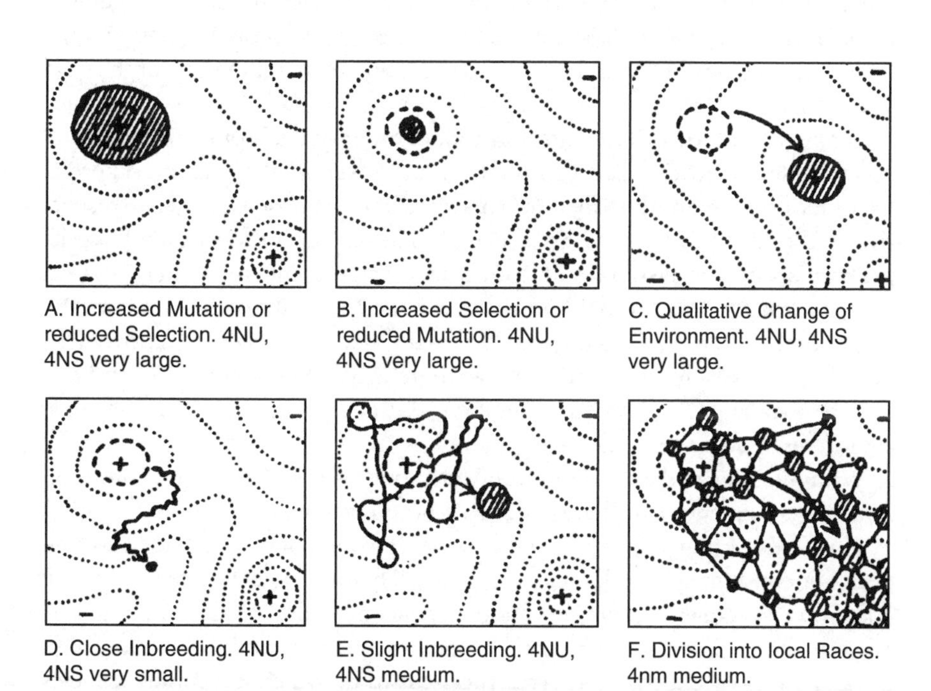

Figure 5.2: Field of gene combinations occupied by a population within the general field of possible combinations. Type of history under specified conditions indicated by relation to initial field (heavy broken contour) and arrow.

The effect of increased severity of selection (also 5.2B) is, of course, to increase the average level of adaptation until a new equilibrium is reached. But again this is at the expense of the field of variation of the species and reduces the chance of capture of another adaptive peak. The only basis for continuing advance is the appearance of novel favorable mutations which are relatively rapidly utilized in this case. But at best the rate is extremely slow even in terms of geologic time, judging from the observed rates of mutation.

Relaxation of selection has of course the opposite effects and thus effects somewhat like those of increased mutation rate (Fig. 5.2A).

The environment, living and non-living, of any species is actually in continual change. In terms of our diagram this means that certain of the high places are gradually being depressed and certain of the low places are becoming higher (Fig. 5.2C). A species occupying a small field under influence of severe selection is likely to be left in a pit and become extinct, the victim of extreme specialization to conditions which have ceased, but if under sufficiently moderate selection to occupy a wide field, it will merely be kept continually on the move. Here we undoubtedly have an important evolutionary process and one which has been generally recognized. It consists largely of change without advance in adaptation. The mechanism is, however, one which shuffles the species about in the general field. Since the species will be shuffled out of low peaks more easily than high ones, it should gradually find its way to the higher general regions of the field as a whole.

Figure 5.2D illustrates the effect of reduction in size of population below a certain relation to the rate of mutation and severity of selection. There is fixation of one or another allelomorph in nearly every locus, largely irrespective of the direction favored by selection. The species moves down from its peak in an erratic fashion and comes to occupy a much smaller field. In other words there is the deterioration and homogeneity of a closely inbred population. After equilibrium has been reached in variability, movement becomes excessively slow, and, such as there is, is nonadaptive. The end can only be extinction. Extreme inbreeding is not a factor which is likely to give evolutionary advance.

With an intermediate relation between size of population and mutation rate, gene frequencies drift at random without reaching the complete fixation of close inbreeding (Fig. 5.2E). The species moves down from the extreme peak but continually wanders in the vicinity. There is some chance that it may encounter a gradient leading to another peak and shift its allegiance to this. Since it will escape relatively easily from low peaks as compared with high ones, there is here a trial and error mechanism by which in time the species may work its way to the highest peaks in the general field. The rate of progress, however, is extremely slow since change of gene frequency is of the order of the reciprocal of the effective population size and this reciprocal must be of the order of the mutation rate in order to meet the conditions for this case.

Finally (Fig. 5.2F), let us consider the case of a large species which is subdivided into many small local races, each breeding largely within itself but occasionally crossbreeding. The field of gene combinations occupied by each of these local races shifts continually in a nonadaptive fashion (except in so far as there are local differences in the conditions of selection). The rate of movement may be enormously greater than in the preceding case since the condition for such movement is that the reciprocal of the population number be of the order of the proportion of crossbreeding instead of the mutation rate. With many local races, each spreading over a considerable field and moving relatively rapidly in the more general field about the controlling peak, the chances are good that one at least will come under the influence of another peak. If a higher peak, this race will expand in numbers and by crossbreeding with the others will pull the whole species toward the new position. The average adaptiveness of the species thus advances under intergroup selection, an enormously more effective process than intragroup selection. The conclusion is that subdivision of a species into local races provides the most effective mechanism for trial and error in the field of gene combinations.

It need scarcely be pointed out that with such a mechanism complete isolation of a portion of a species should result relatively rapidly in specific differentiation, and one that is not necessarily adaptive. [. . .]

How far do the observations of actual species and their subdivisions conform to this picture? This is naturally too large a subject for more than a few suggestions.

That evolution involves nonadaptive differentiation to a large extent at the subspecies and even the species level is indicated by the kinds of differences by which such groups are actually distinguished by systematists. It is only at the subfamily and family levels that clear-cut adaptive differences become the rule.[5] The principal evolutionary mechanism in the origin of species must thus be an essentially nonadaptive one. [. . .]

Subdivision into numerous local races whose differences are largely non-adaptive has been recorded [. . .] wherever a sufficiently detailed study has been made. Among the land snails of the Hawaiian Islands, Gulick (sixty years ago) found that each mountain valley, often each grove of trees, had its own characteristic type, differing from others in 'nonutilitarian' respects.

[5] Wright here cites G. C. Robson, *The Species Problem* (Edinburgh, Oliver & Boyd, 1928) and A. P. Jocot, *American Naturalist*, 66 (1932), 346–64. In the 1920s it seems to have been common to imagine species differences non-adaptive; it is less common now, though there are still plenty of people who think that way. See likewise Haldane, p. 42 below. The extracts on adaptation in Section C below give a feel for the nature of the controversy.

Gulick attributed this differentiation to inbreeding. More recently Crampton has found a similar situation in the land snails of Tahiti and has followed over a period of years evolutionary changes which seem to be of the type here discussed. I may also refer to the studies of fishes by David Starr Jordan, garter snakes by Ruthven, bird lice by Kellogg, deer mice by Osgood, and gall wasps by Kinsey as others which indicate the role of local isolation as a differentiating factor. Many other cases are discussed by Osborn and especially by Rensch in recent summaries. Many of these authors insist on the nonadaptive character of most of the differences among local races. Others attribute all differences to the environment, but this seems to be more an expression of faith than a view based on tangible evidence.

An even more minute local differentiation has been revealed when the methods of statistical analysis have been applied. Schmidt demonstrated the existence of persistent mean differences at each collecting station in certain species of marine fish of the fjords of Denmark, and these differences were not related in any close way to the environment. That the differences were in part genetic was demonstrated in the laboratory. David Thompson has found a correlation between water distance and degree of differentiation within certain fresh water species of fish of the streams of Illinois. Sumner's extensive studies of subspecies of Peromyscus (deer mice) reveal genetic differentiations, often apparently nonadaptive, among local populations and demonstrate the genetic heterogeneity of each such group. [. . .]

Summing up: I have attempted to form a judgment as to the conditions for evolution based on the statistical consequences of Mendelian heredity. The most general conclusion is that evolution depends on a certain balance among its factors. There must be gene mutation, but an excessive rate gives an array of freaks, not evolution; there must be selection, but too severe a process destroys the field of variability, and thus the basis for further advance; prevalence of local inbreeding within a species has extremely important evolutionary consequences, but too close inbreeding leads merely to extinction. A certain amount of crossbreeding is favorable but not too much. In this dependence on balance the species is like a living organism. At all levels of organization life depends on the maintenance of a certain balance among its factors.

More specifically, under biparental reproduction a very low rate of mutation balanced by moderate selection is enough to maintain a practically infinite field of possible gene combinations within the species. The field actually occupied is relatively small though sufficiently extensive that no two individuals have the same genetic constitution. The course of evolution through the general field is not controlled by direction of mutation and not directly by selection, except as conditions change, but by a trial and error mechanism consisting of a largely nonadaptive differentiation of local races (due to inbreeding balanced by occasional crossbreeding) and a determination of long time trend by intergroup selection. The splitting of species depends on the effects of more complete isolation, often made permanent by the accumulation of chromosome aberrations, usually of the balanced type. Studies of natural species indicate that the conditions for such an evolutionary process are often present.

['The Roles of Mutation, Inbreeding, Crossbreeding, and Selection in Evolution', *Proc. of the VI International Congress of Genetics*, 1 (1932), 356–66.]

J. B. S. HALDANE

6 Disease and evolution

Parasitic disease is a major evolutionary factor, with distinctive evolutionary properties. Disease tends to inflict more damage on hosts as host population density increases. Parasites can benefit their hosts by excluding competitors. Hosts have a wide diversity of resistance genotypes, and rapidly evolving parasites continually find ways round host resistance: in this way parasites promote genetic polymorphism in their hosts. But, in other circumstances, parasites can promote speciation in their hosts. [Editor's summary.]

It is generally believed by biologists that natural selection has played an important part in evolution. When however an attempt is made to show how natural selection acts, the structure or function considered is almost always one concerned either with protection against natural 'forces' such as cold or against predators, or one which helps the organism to obtain food or mates. I want to suggest that the struggle against disease, and particularly infectious disease, has been a very important evolutionary agent, and that some of its results have been rather unlike those of the struggle against natural forces, hunger, and predators, or with members of the same species.

Under the heading infectious disease I shall include, when considering animals, all attacks by smaller organisms, including bacteria, viruses, fungi, protozoa, and metazoan parasites. In the case of plants it is not so clear whether we should regard aphids or caterpillars as a disease. Similarly there is every gradation between disease due to a deficiency of some particular food constituent and general starvation.

The first question which we should ask is this. How important is disease as a killing agent in nature? On the one hand what fraction of members of a species die of disease before reaching maturity? On the other, how far does disease reduce the fertility of those members which reach maturity? Clearly the answer will be very different in different cases. A marine species producing millions of small eggs with planktonic larvae will mainly be eaten by predators. One which is protected against predators will lose a larger proportion from disease.

There is however a general fact which shows how important infectious disease must be. In every species at least one of the factors which kills it or lowers its fertility must increase in efficiency as the species becomes denser. Otherwise the species, if it increased at all, would increase without limit. A predator cannot in general be such a factor, since predators are usually larger than their prey, and breed more slowly. Thus if the numbers of mice increase, those of their large enemies, such as owls, will increase more slowly. Of course the density-dependent check may be lack of food or space. Lack of space is certainly effective on dominant species such as forest trees or animals like *Mytilus*. Competition for food by the same species is a limiting factor in a few phytophagous animals such as defoliating caterpillars, and in very stenophagous animals such as many parasitoids. I believe however that the density-dependent limiting factor is more often a parasite whose incidence is disproportionately raised by over-crowding.

As an example of the kind of analysis which we need, I take Varley's[1] remarkable study on *Urophora jaceana*, which forms galls on the composite *Centaurea nigra*. In the year considered 0.5% of the eggs survived to produce a mature female. How were the numbers reduced to 1/200 of the initial value?

If we put $200 = e^k$, we can compare the different killing powers of various environmental agents, writing $K = k_1 + k_2 + k_3 + \ldots$, where k_r is a measure of the killing power of each of them. Surprisingly, the main killers appear to be mice and voles (*Mus, Microtus*, etc.) which eat the fallen galls and account for at least 22%, and perhaps 43% of *k*. Parasitoids account for 31% of the total kill, and the effect of *Eurytoma curta* was shown to be strongly dependent on host density, and probably to be the main factor in controlling the numbers of the species, since the food plants were never fully occupied.

When we have similar tables for a dozen species we shall know something about the intensity of possible selective agencies. Of course in the case of *Urophora jaceana* analysis is greatly simplified by the fact that the imaginal period is about 2% of the whole life cycle, so that mortality during it is unimportant.

A disease may be an advantage or a disadvantage to a species in competition with others. It is obvious that it can be a disadvantage. Let us consider an ecological niche which has recently been opened, that of laboratories where the genetics of small insects are studied. A number of species of *Drosophila* are well adapted for this situation. Stalker attempted to breed the related genus *Scaptomyza* under similar conditions, and found that his

[1] G. C. Varley, *Journal of Animal Ecology*, 16 (1947), 139. *Urophora jaceana* is a trypetid dipteran fly with 'picture'-patterned wings; its vernacular name is the knapweed gallfly. *Centaurea nigra* is knapweed. The species mentioned a few lines on, *Eurytoma curta*, is a hymenopteran (wasp) that parasitizes *U. jaceana*.

cultures died of bacterial disease. Clearly the immunity of *Drosophila* to such diseases must be of value to it in nature also.

Now let us take an example where disease is an advantage. Most, if not all, of the South African artiodactyls are infested by trypanosomes such as *T. rhodesiense* which are transmitted by species of *Glossina* to other mammals and, sometimes at least, to men. It is impossible to introduce a species such as *Bos taurus* into an area where this infection is prevalent. Clearly these ungulates have a very powerful defence against invaders. The latter may ultimately acquire immunity by natural selection, but this is a very slow process, as is shown by the fact that the races of cattle belonging to the native African peoples have not yet acquired it after some centuries of sporadic exposure to the infection. Probably some of the wild ungulates die of, or have their health lowered by the trypanosomes, but this is a small price to pay for protection from other species.

A non-specific parasite to which partial immunity has been acquired, is a powerful competitive weapon. Europeans have used their genetic resistance to such viruses as that of measles (rubeola) as a weapon against primitive peoples as effective as fire-arms. The latter have responded with a variety of diseases to which they are resistant. It is entirely possible that great and, if I may say so, tragic episodes in evolutionary history such as the extinction of the Noto-ungulata and Litopterna may have been due to infectious diseases carried by invaders such as the ungulates, rather than to superior skeletal or visceral developments of the latter.

A suitable helminth parasite may also prove a more efficient protection against predators than horns or cryptic coloration, though until much more is known as to the power of helminths in killing vertebrates or reducing their fertility, this must remain speculative.

However it may be said that the capacity for harbouring a non-specific parasite without grave disadvantage will often aid a species in the struggle for existence. An ungulate species which is not completely immune to *Trypanosoma rhodesiense* has probably (or had until men discovered the life history of this parasite) a greater chance of survival than one which does not harbour it, even though it causes some mortality directly or indirectly.

I now pass to the probably much larger group of cases where the presence of a disease is disadvantageous to the host. And here a very elementary fact must be stressed. In all species investigated the genetical diversity as regards resistance to disease is vastly greater than that as regards resistance to predators.

Within a species of plant we can generally find individuals resistant to any particular race of rust (Uredineae) or any particular bacterial disease. Quite often this resistance is determined by a single pair, or a very few pairs, of genes. In the same way there are large differences between different breeds of mice and poultry in resistance to a variety of bacterial and virus diseases.

To put the matter rather figuratively, it is much easier for a mouse to get a set of genes which enable it to resist *Bacillus typhi murium* than a set which enable it to resist cats. The genes commonly segregating in plants have much more effect on their resistance to small animals which may be regarded as parasites, than to larger ones. Thus a semiglabrous mutant of *Primula sinensis* was constantly infested by aphids, which however are never found on the normal plant. I suppose thornless mutants of *Rubus* are less resistant to browsing mammals than the normal type, but such variations are rarer.

Anyone with any experience of plant diseases will of course point out that the resistance of which I have spoken is rarely very general. When a variety of wheat has been selected which is immune to all the strains of *Puccinia graminis* in its neighbourhood, a new strain to which it is susceptible usually appears within a few years, whether by mutation, gene recombination, or migration. Doubtless the same is true for bacterial and virus disease. The microscopic and sub-microscopic parasites can evolve so much more rapidly than their hosts that the latter have little chance of evolving complete immunity to them. It is very remarkable that *Drosophila* is as generally immune as it is. I venture to fear that some bacillus or virus may yet find a suitable niche in the highly overcrowded *Drosophila* populations of our laboratories, and that if so this genus will lose its proud position as a laboratory animal. The most that the average species can achieve is to dodge its minute enemies by constantly producing new genotypes, as the agronomists are constantly producing new rust-resistant wheat varieties.

Probably a very small biochemical change will give a host species a substantial degree of resistance to a highly adapted microorganism. This has an important evolutionary effect. It means that it is an advantage to the individual to posses a rare biochemical phenotype. For just because of its rarity it will be resistant to diseases which attack the majority of its fellows. And it means that it is an advantage to a species to be biochemically diverse, and even to be mutable as regards genes concerned in disease resistance. For the biochemically diverse species will contain at least some members capable of resisting any particular pestilence. And the biochemically mutable species will not remain in a condition where it is resistant to all the diseases so far encountered, but an easy prey to the next one. A beautiful example of the danger of homogeneity is the case of the cultivated banana clone 'Gros Michel' which is well adapted for export and has been widely planted in the West Indies. However it is susceptible to a root infection by the fungus *Fusarium cubense* to which many varieties are immune, and its exclusive cultivation in many areas has therefore had serious economic effects.

Now every species of mammal and bird so far investigated has shown a quite surprising biochemical diversity revealed by serological tests. The antigens concerned seem to be proteins to which polysaccharide groups are attached. We do not know their functions in the organism, though some of them seem to be part of the structure of cell membranes. I wish to suggest that they may play a part in disease resistance, a particular race of bacteria or virus being adapted to individuals of a certain range of biochemical constitution, while those of other constitutions are relatively resistant. I am quite aware that attempts to show that persons of a particular blood group are specially susceptible to a particular disease have so far failed. This is perhaps to be expected, as a disease such as diphtheria or tuberculosis is caused by a number of biochemically different races of pathogens. The kind of investigation needed is this. In a particular epidemic, say of diphtheria, are those who are infected (or perhaps those who are worst affected) predominantly drawn from one serological type (for example *AB*, *MM*, or *BMM*)? In a different epidemic a different type would be affected.

In addition, if my hypothesis is correct, it would be advantageous for a species if the genes for such biochemical diversity were particularly mutable, provided that this could be achieved without increasing the mutability of other genes whose mutation would give lethal or sublethal genotypes. Dr P. A. Gorer informs me that there is reason to think that genes of this type are particularly mutable in mice. Many pure lines of mice have split up into sublines which differ in their resistance to tumour implantation. This can only be due to mutation. The number of loci concerned is comparable, it would seem, with the number concerned with coat colour. But if so their mutation frequency must be markedly greater.

We have here, then, a mechanism which favours polymorphism, because it gives a selective value to a genotype so long as it is rare. Such mechanisms are not very common. Among others which do so are a system of self-sterility genes of the *Nicotiana* type. Here a new and rare gene will always be favoured because pollen tubes carrying it will be able to grow in the styles of all plants in which it is absent, while common genes will more frequently meet their like. However this selection will only act on genes at one locus, or more rarely at two or three. A more generally important mechanism is that where a heterozygote is fitter than either homozygote [. . .] This does not, however, give an advantage to rarity as such. It need hardly be pointed out that, in the majority of cases where it has been studied, natural selection reduces variance.

I wish to suggest that the selection of rare biochemical genotypes has been an important agent not only in keeping species variable, but also in speciation. We know, from the example of the *Rh* locus[2] in man, that

[2] These genes control the 'Rhesus' blood group. In this case, as with some other blood groups, crosses between individuals of different groups have lower fertility than crosses between individuals of the same group. Natural selection then favours which ever blood type is commoner in the population. Haldane here argues that the situation also tends to promote speciation. The form of selection is called frequency-dependent, and in this case it is positively frequency-dependent, unlike the negatively frequency-dependent selection of the previous paragraph.

biochemical differentiation of this type may lower the effective fertility of matings between different genotypes in mammals. Wherever a father can induce immunity reactions in a mother the same is likely to be the case. If I am right, under the pressure of disease, every species will pursue a more or less random path of biochemical evolution. Antigens originally universal will disappear because a pathogen had become adapted to hosts carrying them, and be replaced by a new set, not intrinsically more valuable, but favouring resistance to that particular pathogen. Once a pair of races is geographically separated they will be exposed to different pathogens. Such races will tend to diverge antigenically, and some of this divergence may lower the fertility of crosses. It is very striking that Irwin[3] finds that related, and still crossable, species of *Columba*, *Streptopelia*, and allied genera differ in respect of large numbers of antigens. I am quite aware that random mutation would ultimately have the same effect. But once we have a mechanism which gives a mutant gene as such an advantage, even if it be only an advantage of one per thousand, the process will be enormously accelerated, particularly in large populations. [. . .]

We see then that in certain circumstances, parasitism will be a factor promoting polymorphism and the formation of new species. And this evolution will in a sense be random. Thus any sufficiently large difference in the times of emergence or oviposition of two similar insect species will make it very difficult for the same parasitoid to attack both of them efficiently. So will any sufficiently large difference in their odours. We may have here a cause for some of the apparently unadaptive differences between related species.

Besides these random effects, disease will of course have others. It is clear that natural selection will favour the development of all kinds of mechanisms of resistance, including tough cuticles, phagocytes, the production of immune bodies, and so on. It will have other less obvious effects. It will be on the whole an antisocial agency. Disease will be less of a menace to animals living singly or in family groups than to those which live in large communities. Thus it is doubtful if all birds could survive amid the faecal contamination which characterises the colonies of many sea birds. A factor favouring dispersion will favour the development of methods of sexual recognition at great distances such as are found in some Lepidoptera.

Again, disease will set a premium on the finding of radically new habitats. When our ancestors left the water, they must have left many of their parasites behind them. A predator which ceases to feed on a particular prey, either through migration or changed habits, may shake off a cestode which depends on this feeding habit. When cerebral development has gone far enough to

[3] M. R. Irwin, *Advances in Genetics*, 1 (1949), 133.

make this possible, it will favour a negative reaction to faecal odours and an objection to cannibalism, and will so far be of social value. A vast variety of apparently irrelevant habits and instincts may prove to have selective value as a means of avoiding disease.

A few words may be said on non-infectious diseases. These include congenital diseases due to lethal and sub-lethal genes. Since mutation seems to be non-specific as between harmful and neutral or beneficial genes, and mutation rate is to some extent inherited, it follows that natural selection will tend to lower the mutation rate, and this tendency may perhaps go so far as to slow down evolution. It will also tend to select other genes which neutralise the effect of mutants, and thus to make them recessive or even ineffective, as Fisher has pointed out. Whether the advantage thus given to polyploids is ever important, we do not know. But the evolution of dominance must tend to make the normal genes act more intensely and thus probably earlier in ontogeny, so that a character originally appearing late in the life cycle will tend to develop earlier as time goes on.

Deaths from old age are due to the breakdown of one organ or another, in fact to disease, and the study of the mouse has shown that senile diseases such as cancer and nephrosis are often congenital. In animals with a limited reproductive period senile disease does not lower the fitness of the individual, and increases that of the species. A small human community where every woman died of cancer at 55, would be more prosperous and fertile than one where this did not occur. Senile disease may be an advantage wherever the reproductive period is limited; and even where it is not, a genotype which leads to disease in the 10% or so of individuals which live longest may be selected if it confers vigour on the majority. [. . .]

In this brief communication I have no more than attempted to suggest some lines of thought. Many or all of them may prove to be sterile. Few of them can be followed profitably except on the basis of much field work.

Comment by Haldane in the discussion recorded after his paper

Perhaps the theory that most diseases evolve into symbioses is somewhat Panglossist. I doubt if it occurs as a general rule, though it may do so. The position for the original host is however best.

[*La ricercha scientifica*, 19, suppl. (1949), 68–76.]

Section B

Natural selection and random drift in natural populations

Evolution is driven by two main mechanisms: natural selection and random drift. This section looks at both mechanisms, but the kind of material we look at differs for the two. For selection, we look at some examples of selection in action. Research on drift has been less concerned with particular examples. The main argument for the importance of drift comes from theoretical arguments, using molecular evidence that has been accumulating since the 1960s. Before the 1960s, it could reasonably be doubted whether drift was an important evolutionary mechanism. But today almost no one doubts that drift is the main evolutionary mechanism for certain kinds of molecular change. Evidence about molecular evolution is the main reason for this change in viewpoint.

Studies of selection in action

Darwin described in 1859 how selection might operate in nature, but he was forced to reason theoretically and by analogy with artificial selection; he had no evidence of natural selection. It was well into the next century before research on the topic gained momentum, following the theoretical population genetic work (Section A) that established the modern synthesis. Research on selection in action has two main aims: to understand the factors that influence the relative survival and reproductive success of different kinds of individuals; and to estimate the quantitative values for the 'fitness' of the different kinds of individual. (The word 'fitness' in evolutionary biology has nothing to do with athletic prowess; it refers to the relative contributions to the next generation made by the different kinds of individual. In the work included in this section, the measurements are mainly for relative survival: individuals with higher survival have higher fitness.)

The work can be divided into two types. In one, the genetic basis of the characters concerned is fairly simple and also understood. In the other, the genetic basis is more or less unknown. Kettlewell's (Chapter 7) famous study of melanism in peppered moths is an example of the first type. It is not a gross oversimplification to treat light and dark coloration in these moths as controlled by one genetic locus with two alleles (though it is really more complicated). Natural selection favours the gene for dark, melanic colour in polluted areas and the gene for light, peppered colour in unpolluted areas. Birds eat more of the less well camouflaged forms in any one area.

The peppered moth story is thought in some quarters to be 'controversial'. The controversy is, I believe, mainly bogus, but it has fed on some interesting research developments that have occurred during, and subsequent to, Kettlewell's work. In early experiments, Kettlewell pinned out moths on treetrunks. He measured the rate at which birds took the two forms of moth, in polluted and in unpolluted areas: he used these measurements as fitness estimates. The method was unsatisfactory for various reasons, and he went on to use mark-recapture experiments, as described in Chapter 7. In all, the fitnesses of the two forms have been measured in many experiments. The results of all the experiments qualitatively agree: the melanic form has higher fitness in polluted areas, and the light form in unpolluted areas. The original pin-out experiments, though superseded, were not misleading. Subsequently, it has been empasized that the moths do not settle on treetrunks but in twigs higher up (Kettlewell was aware of this fact). However, the place where the moths settle is irrelevant for the mark-recapture results. (Cook (2000), listed in the 'Select bibliography' at the end of this book, reviews all the experimental fitness measurements.)

Through the 1970s, an increasing number of biologists came to doubt whether bird predation was the main factor at work. They argued that the melanic form had increased in frequency since the industrial revolution because it had some 'inherent' advantage relative to the light form. In the 1980s it was quite often said other more or less mysterious factors were at work along with bird predation. That view, however, is now in retreat, mainly because of the kind of result described by Cook *et al.* (Chapter 8; similar results have been recorded elsewhere). As air pollution decreased, so too did the frequency of the melanic form. This would not be expected if the melanic form had an inherent advantage, independent of bird predation. In consequence, many biologists now hold views closer to those Kettlewell originally proposed. For instance, Majerus in his book on *Melanism* (1998) concludes in favour of Kettlewell's view, and against the hypothesis that other factors are at work. (This is noteworthy, because in the 'controversy', Majerus is often described as one of Kettlewell's critics.) Bird predation, dependent on industrial pollution, appears to be the main factor at work in the rise, and fall, of the melanic form of the peppered moth. Kettlewell's ideas are more, not less, strongly supported than 50 years ago, now that we have new observations on the decline of the melanic form.

The second type of work on selection in action uses attributes for which the genetic control is not understood. The attributes are things like body size, or body weight, or leg length—easily measurable attributes of an organism. It is known that genetic differences underlie the differences in the measured attribute, but the numbers of genetic loci and alleles is unknown. In the case where the genetics was known, it was possible to measure the survivals of individuals of different genotypes, and to investigate why they differed. In this case, where the genetics is unknown, we measure the attribute in many organisms; we also measure the survival rate of those organisms; and we look at the relation between the two.

Karn and Penrose (Chapter 9) looked at birth weight and infant survival in humans. Their most famous result is the picture showing stabilizing selection—stabilizing selection means that natural selection works against the extreme forms in a population, and favours some intermediate form. Thus, babies that are at the light or heavy extreme have lower survival, probably for different reasons (these reasons have not been scientifically investigated, but it is thought that lighter babies have lower survival because they are poorly nourished, whereas the heavier ones are liable to injure themselves, their mother, or both, during birth and have lower subsequent survival in consequence). Karn and Penrose's research was done in the 1940s, and probably illustrates a pattern of natural selection that has been going on throughout human evolution since our brains enlarged one million or so years ago. Ulizzi and Terrenato's (Chapter 10) follow-up paper shows how, in wealthy countries, medical care has almost eliminated this kind of selection: we are living through the end of a great era of natural selection. Karn and Penrose's U-shaped curve has now almost flattened and (except at the extremes) babies of all birth weights have the same chance of survival. It is a fascinating example of 'relaxed selection', which Muller writes about in Section I (Chapter 56).

Karn and Penrose also demonstrated a directional component to selection on birth weight, but you have to keep your wits about you to notice it (much more so in the original, where the two key numbers are several pages apart in a paper composed with Spartan scientific asceticism). One number is the average birth weight (7.2 lb); the other is the birth weight with maximum survival (8 lb). Natural selection would seem to favour a higher birth weight than is observed. Things may not be as simple as they seem, however, and much has been written on the discrepancy, without any firm conclusion being reached. Ulizzi and Terrenato find that the directional selection has remained, if flickeringly, in the past 50 years while the stabilizing selection has disappeared.

The short paper by Gibbs and Grant (Chapter 11) also concerns attributes, the exact genetic control of which is unknown. Their study concerns a species of Darwin's finches, a group of birds that live in the Galápagos Islands. It is but one paper in what is probably the most important long-term study of natural selection that is being done at present; the project is directed by Peter Grant and concerns many other subjects in addition to selection. A fine popular science book, Weiner's *Beak of the finch*, has been written about the work. Gibbs and Grant's paper contains some technicalities, but anyone can follow the main story. Selection favours larger body size in most years, but through the El Niño disturbance it favoured smaller size; the result is a kind of stabilizing selection over long time periods.

Random genetic drift and Kimura's neutral theory

Evolution could be studied at the molecular level by about the 1960s. The amino acid sequences of the same protein in a number of different species were then becoming available; and the amount of difference in the sequence suggested the existence of a 'molecular clock': that proteins evolve at a constant rate. The rates differed between proteins—some evolve faster than others—but the rate for any one protein appeared to be approximately constant. (The same conclusion emerged when some proxy for sequence difference, rather than the sequences themselves, was measured. For example, an immunological measure of sequence differences was used in a classic study of the time of human origins (Chapter 53).)

Another technical advance—gel electrophoresis—made it possible to measure the amount of protein variation in a natural population. It was immediately found that many more variants exist for each protein than had been suspected. Lewontin (Chapter 12) was one of the first to apply gel electrophoresis in this way, and I have included an extract from his superb book *The genetical basis of evolutionary change* (1974), in which he describes the consequences of what they found. Biologists were used to the idea that a protein could exist in more than one form. It was well known, for instance, that humans can have many different forms of haemoglobin. One rare form of haemoglobin (sickle cell haemoglobin) was known to be maintained because heterozygotes had an advantage in malarial zones. What was awkward was the sheer amount of variation that was revealed. If you tried to extrapolate up the sort of process seen in the sickle cell example, you soon arrived at a paradox that is the subject of a famous calculation at the end of the passage from Lewontin extracted here.

The important consequence of these early observations was that they led a Japanese theoretician, Motoo Kimura, to propose a radical hypothesis about the force driving molecular evolution. In 1968 Kimura suggested molecular evolution was mainly driven not by natural selection but by random drift among equally well-adapted sequence variants. He was not suggesting that proteins are non-adaptive: that would be absurd, for it is known that enzymes are almost thermodynamically optimal catalysts. However, it could be that for any one well-adapted protein there are a number of equally well-adapted forms, and evolution would then consist of random drift among them. Lewontin (Chapter 12) argues that Kimura's neutral theory was an extension of the earlier 'classical' (as opposed to 'balance') school of thought about biological variation. (Some hint of the classical viewpoint is contained in Muller's paper, Section I (Chapter 56).) Not everyone agrees with Lewontin's analysis, but it is an influential way of thinking about the neutral theory and places that theory in a deep conceptual context.

Sequences of the DNA itself began to appear in the 1970s and to accumulate at an accelerating rate through the 1980s and 1990s. The debate about neutralism moved more from proteins to DNA. Kimura enjoyed a triumph that is rare in science when certain patterns that he had previously predicted were found to exist in DNA. One concerned the rates of evolution at synonymous and non-synonymous sites in the DNA. The DNA sequences codes for the protein amino acid sequences, but the DNA code contains redundancies. Only 20 kinds of amino acid are found in our proteins, but there are 64 symbols for those 20 at the DNA level. Therefore, some evolutionary changes in the DNA do not change the protein it codes for: these changes are called synonymous (or silent). DNA changes that do change the protein are called non-synonymous or meaningful. Both these sets of terms appear in the papers in this section, and in Section F on evolutionary genomics. Kimura predicted rapid evolution in synonymous DNA, and he turned out to be right.

Synonymous and non-synonymous changes apply in the coding regions of the DNA—the regions that contain genes and code for proteins. Much of DNA is 'non-coding'—it does not code for anything. Kimura also predicted rapid evolution in non-coding DNA, and he again turned out to be right. Biologists generally accept that most evolution in non-coding DNA is by drift. We also now know that, in eukaryotic organisms (that is, all life except viruses, bacteria, and the recently discovered Archaea), most DNA is non-coding. For this reason, Kimura's original claim—that most evolutionary change at the molecular level is by drift—has become an orthodoxy. The evolutionary forces in the coding part of the DNA remain more controversial.

I have included one of Kimura's later papers (Chapter 13) in which he describes some of the development of his thinking, reflects on evolution in DNA sequences, and places the theory in the 'Wrightian' tradition—a tradition that stresses drift in small, divided populations in addition to selection (see the introduction to Section A and Chapter 5).

Some consequences, and cross-references

One observation explained by Kimura's neutral theory—the molecular clock—has proved most fruitful in newly emerging and in long-established areas of evolutionary biology: in evolutionary genomics (Section F) and in reconstructing the history of life (Section G). We meet the molecular clock in two papers about genomes (Chapters 37 and 41) and in two phylogenetic case studies (Chapters 44 and 53). All sorts of extraordinary findings are dropping out of new molecular evidence: the *ey* gene (Chapters 32 and 33) and the amazing story of lens proteins (Chapter 40) are two represented in this book.

Lewontin's (Chapter 12) extract is about the amount of variation in matural populations. Two factors that influence it are mutation rates and whether reproduction is sexual or asexual. Two extracts discuss these topics, as case studies (Chapters 48 and 52).

H. B. D. KETTLEWELL

7 A résumé of investigations on the evolution of melanism in the Lepidoptera

While industrial pollution darkened the branches of trees, a dark form of the peppered moth increased in frequency relative to the formerly common light form. The change in frequency is almost certainly an example of natural selection. Kettlewell estimated the relative advantage of the two forms using mark-recapture experiments, in which samples of the light and the dark forms of the moth were released both in polluted and in unpolluted areas. [Editor's summary.]

Apart from flight, the wings in the Lepidoptera are used for developing various pattern and colour devices enabling them to pass the daytime protected from predators.

In the cryptic species, complicated patterns have been built up suitable for concealment on lichened tree trunks, boughs, rocks, reeds or posts, and the phenomenon of 'industrial melanism' is found among this group only. Of the 780 species of Macro-lepidoptera which occur in the British Isles, about seventy are in the process[1] of replacing their populations with dark or black individuals. In the majority of cases the black mutant is inherited as a simple Mendelian dominant. The position of the melanics occurring in other groups, however, is completely different. In those insects which gain protection from mechanisms other than crypsis (warning, threat and flash coloration), or even in those cryptic moths which benefit from resembling dead leaves, the melanics are rare, recessive, sub-lethal and are probably maintained by recurrent mutation. A third group of melanics can be referred to as 'geographic melanics', and they are confined to the Highlands of Scotland, the west coast of Great Britain and Ireland. Their inheritance in the few

[1] Kettlewell was referring to the mid-twentieth century. Industrial pollution declined in the second half of the century and the process went into reverse. By the 1990s, the light form of the peppered moth had re-replaced the dark melanic form even in the formerly most polluted areas, such as near Manchester; Chapter 8 illustrates the reverse trend. This is strong evidence for Kettlewell's underlying idea, that industrial pollution and melanism were linked.

cases known is dominant (complete or incomplete) or in others multifactorial; so far never recessive. It will be noted that these areas are comprised of primeval forest, 'unspoilt' countryside, or environments little affected by civilization. In recent years we have been attempting to analyze the relative advantages and disadvantages of the melanics and their light forms and, to date, these studies have been confined to investigations on the so-called industrial melanics. These have, for the most part, been conducted on *Biston betularia* L., the Peppered Moth (Selidosemidae), and one of its melanics *carbonaria* Jordan; its other melanic, *insularia* Th.-Mieg, is complicated and probably determined by a series of alleles and is, except for population frequencies, omitted from this paper.

The life history of *B. betularia* is as follows: the ova are laid from May to August and the larvae feed from May to October. The species overwinters as a pupa, the imago hatching from May to August. The eggs are laid in masses or 'cakes' in crevices in bark. Dispersal takes place immediately after hatching, the young larvae being airborne on a long thread. They feed on most deciduous trees. The female moths rarely come to light, but the males do so freely and they assemble readily to newly hatched females from 9 p.m. (G.M.T.) to dawn. For these reasons the males only are used for mark-release experiments. Investigations have been, and are being, carried out on the following special problems involved:

Relative cryptic advantages of the black and the light forms

Aviary experiments were carried out using a pair of great tits (*Parus major* L.) at the Madingley Research Station, Cambridge. The following points were shown:

(1) The birds ate both forms of the Peppered Moth.
(2) They did this selectively according to the cryptic advantage of the insects and in the order of their conspicuousness as previously scored by us.
(3) The birds rapidly became conditioned to looking for this species and, to overcome this, other species of moths had to be released at the same time.

These aviary experiments were sufficiently encouraging to demand a large-scale mark-release in nature. Accordingly, in the summer of 1953 this was undertaken near the industrial centre of Birmingham where 85% of the *betularia* population is of the *carbonaria* form. A total of 630 *betularia* were marked and released on to a sample of the only available trunks and boughs. Of these, 447 were *carbonaria*, 137 *typical* and 46 *insularia*. Each was scored for its cryptic efficiency. In this area, it appeared to man that the light form

(= the *typical*) was five times more conspicuous than *carbonaria*. Table I(a)[2] shows detailed figures for captures, releases and subsequent recaptures, taken within 24h of being released. More than twice as many *carbonaria* were recovered as *typicals*. Furthermore, two species of birds (hedge sparrow, *Prunella modularis* L., and robin, *Erithacus rubecula* L.) were seen to take and eat our releases, and they did this on the majority of occasions in order of conspicuousness as previously scored by me. It was possible to exclude factors other than bird predation as being responsible for the differential figures of the recaptures.

A similar experiment was undertaken in the summer of 1955 in Dean End, a heavily lichened and pollution-free wood in Dorset, where the *carbonaria* form, if it exists, is under 1% of the population. Of 984 *betularia* released, 496 were *typical* and 473 *carbonaria* (*insularia* 15). Of the *typicals* 12.5% were retrapped, but only 6.34% of the *carbonaria*. Fewer individuals were captured here and, assuming equally efficient collecting techniques to those used in the Birmingham experiments, it would appear that the pitch of predation was higher than in the Birmingham experiments. In Table 7.1 are given the number of recaptures during the whole of both experiments (and not just those released within 24 h), together with their expected values calculated from the frequency of releases of the three phenotypes. It will again be seen that for each day taken separately there is always an excess of *carbonaria* and a deficiency of the light form in the Birmingham experiments (with f. *insularia* lying between both). The reverse is true for the unpolluted wood of Dean End for the majority of releases, and for their sum (Table 7.2). A comparison of the two tables gives valuable information. In the first place, for the Birmingham releases, it will be seen that extending the period over which recaptures were made, the number of *carbonaria* taken is materially increased, whereas this is not so for the other two phenotypes. We have given reasons already for believing that there is no differential migration from an area so that Table 7.1 clearly reflects a difference in the death-rates for the three phenotypes, consequently from this table it can be said that *carbonaria* is at an advantage to *typical* in the Birmingham area. Furthermore, the direct observations of Dr Tinbergen and myself show that the birds ate more

[2] The table is not included here. It gives raw numbers for recaptured moths of the three types in the two places. I have included only a later table (see below) for the recaptures for the whole experiment, not just the first 24 hours after release. The interest of the first 24 hours is that the local birds might learn to recognize the experimentally released moths and mortalities over time might be biased. However, the figures for 24 hours and for longer time periods were not interestingly different. The numbers also bear upon a more recent criticism. The peppered moth does not naturally settle on tree trunks, but on higher twigs of the trees. The initial mortalities might have been biased by the unnatural release site. However, any such bias was probably small, because the relative mortalities of the two forms were much the same over shorter and longer measuring periods.

Table 7.1: Birmingham recaptures

DATE	OBSERVED				EXPECTED		
	C	*T*	*I*	TOTAL	*C*	*T*	*I*
25 June	5	1	2	8	2.5000	3.0000	2.5000
26 June	0	0	0	0	0	0	0
27 June	0	1	2	3	1.6780	0.5593	0.7627
28 June	9	4	2	15	8.8095	5.0000	1.1905
29 June	0	0	0	0	0	0	0
30 June	17	2	1	20	13.888	5.0980	1.5686
1 July	41	6	0	47	37.1053	8.6579	1.2368
2 July	30	2	1	33	24.9184	7.0714	1.0102
3 July	26	2	0	28	22.9398	5.0602	0
4 July	12	0	0	12	10.1772	1.5190	0.3038
	140	18	8	166	121.4615	35.9658	8.5726

Notes: C = *Carbonaria* (Melanic); T = *typical* (light-coloured); I = *insularia*.

carbonaria than *typicals* at Dean End and the reverse at Birmingham. We succeeded in filming seven species of birds doing this.

It therefore seems almost certain that, not only is selection acting in the two woods, but that it is sufficiently intense for it to have a marked effect on the expectation of life of the three phenotypes. [. . .]

Summary

It appears that in the recent past, prior to one hundred years ago, the so-called industrial melanics were absent from collections and records. They are still absent from large areas of western and south-western England, where lichen-covered deciduous trees predominate. Evidence has been provided that, in these areas, melanic mutations would be rapidly eliminated because of their cryptic disadvantages. Nevertheless, in the changed countryside of central and eastern England, the melanic forms are rapidly replacing the light-coloured individuals and even managing to re-establish themselves in their melanic forms in centres from which they have long been absent (*Procus literosa* and *Apamea characterea* in Sheffield). Industrial melanism probably represents an example of transient polymorphism.

Table 7.2: Dean End recaptures

DATE	OBSERVED				EXPECTED		
	C	*T*	*I*	TOTAL	*C*	*T*	*I*
13. vi.	6	3	3	12	5.2857	5.4286	1.2857
14. vi.	1	7	2	10	2.9630	6.2963	0.7407
15. vi.	0	1	0	1	0.2353	0.7353	0.0294
16. vi.	8	8	0	16	5.6774	10.3226	0
17. vi.	0	0	0	0	0	0	0
18. vi.	2	8	0	10	3.8182	5.9091	0.2727
19. vi.	1	2	0	3	1.0541	1.9459	0
20. vi.	2	4	0	6	1.7778	4.2222	0
21. vi.	4	4	0	8	4.7324	3.2676	0
22. vi.	—	—	—	—	0	0	0
23. vi.	—	—	—	—	0	0	0
24. vi.	—	—	—	—	0	0	0
25. vi.	2	2	0	4	2.6240	1.3760	0
26. vi.	—	—	—	—	0	0	0
27. vi.	1	5	0	6	3.8734	2.1266	0
28. vi.	0	7	0	7	3.5000	3.5000	0
29. vi.	3	8	0	11	5.3429	5.6571	0
30. vi.	2	1	0	3	2.0571	0.9429	0
4. vii	0	7	0	7	4.4767	2.5233	0
	32	67	5	104	47.4180	54.2535	2.3285

[*Proc. of the Royal Society of London*, Ser. B, 145 (1956), 297–303.]

L. M. COOK, R. L. H. DENNIS, AND G. S. MANI

8 Melanic morph frequency in the peppered moth in the Manchester area

Data are presented for the Manchester area, showing the recent change in frequency of the melanic morph carbonaria *of the peppered moth* Biston betulari *(L.). The frequency has fallen from 90% in 1983 to below 10% at present; this decline shows that the phenomenon of industrial melanism, first noted in this species in Manchester, is now almost past. [. . .] Records from north-west Kent, published by B. K. West, also show a less intense decline from a lower peak several years in advance of the Manchester decline. [. . .] When all available data are compared, there is a negative relation between estimated fitness of* carbonaria *over the period of decline and initial level of atmospheric pollution.* [Authors' summary.]

Introduction

Increase in frequency of melanic forms of the peppered moth *Biston betularia* was first recorded from the Manchester area, and inheritance of the forms and causes of spread were discussed, in the 19th and early 20th centuries. Whatever the precise causes, high frequencies of melanic morphs were clearly associated with urbanization and industrialization. Three categories of moth were distinguished: speckled black and white typicals, the uniform dominant black morph *carbonaria*, and an intermediate category *insularia*. This variation is due to several alleles producing different degrees of darkening. The moth is annual, and the rate of increase in melanic frequency in Manchester in the 19th century showed that *carbonaria* was then strongly advantageous. [. . .]

By the mid-1960s it was clear that the industrialized environment would change. Oil and electricity replaced coal as the main fuel, large cities underwent massive redevelopment and the old heavy industries declined. Smoke-control legislation consolidated a change already in progress. There was rapid reduction in atmospheric smoke pollution in built-up areas and slower but nevertheless marked reduction in atmospheric sulphur dioxide. The geographic pattern of high and low SO_2 concentrations has always been more spread-out and less tied to urbanization than that of smoke, because the gas is more mobile and because some is produced by single large sources, such as power stations, which are good dispersers and not always in urban areas. Kettlewell's data show a plateau of high *carbonaria* frequency in urban and industrial regions with a low frequency to the west and south-west; this pattern fits better with the distribution of known SO_2 concentrations than with smoke levels.

A further survey was made in north-west England and north Wales to record the beginning of the anticipated change in morph frequency, and to investigate relative fitness. [. . .]

We have assembled data for the Manchester area to try to establish when and by how much the decline has occurred there. [. . .]

Estimates for 1952 to the present are shown in Figure 8.1. [. . .]

Decline has also occurred in other parts of the country. The pattern in northwest Kent is illustrated in Figure 8.2. At the turn of the century *carbonaria* had not been observed there, but by the 1952–70 surveys it had risen to 65–89% in the north of the county. Evidently, it too achieved a lower peak and has declined at a lower rate than at Manchester.

Discussion

Haldane pointed out that, to achieve the 19th century change, *carbonaria* must have had an advantage over typical of 50% per generation. This may be

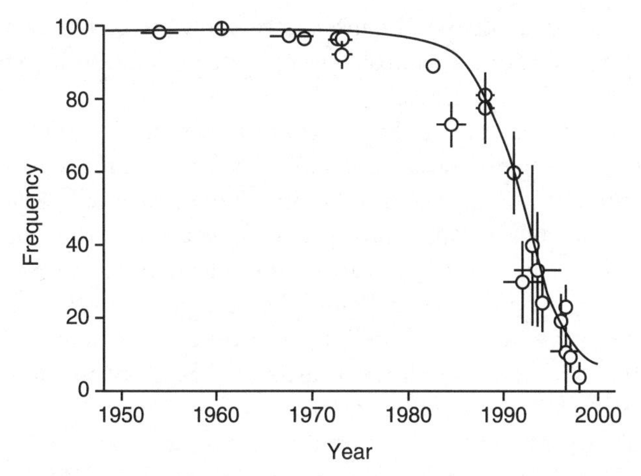

Figure 8.1: Change in frequency of the carbonaria *form of the peppered moth* Biston betularia *(L.) in the Manchester area since 1950. Vertical lines show standard error, horizontal lines range of years included. The curve is the theoretical prediction from the model of Mani.*

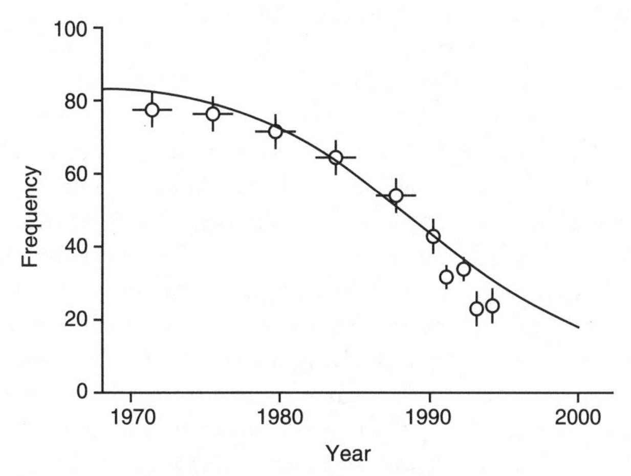

Figure 8.2: Change in frequency of carbonaria *form in north-west Kent in records published by B. K. West. The curve is the theoretical prediction from the model of Mani. Other details as for Figure 8.1.*

largely because of greater protection of melanic adults from predation as the environment became uniformly dark. If such selection was completely responsible for the change the population would have been entirely melanic by the 20th century. That did not occur, one possible reason, suggested by Haldane, being that there was heterozygote advantage. When data from breeding experiments were examined, however, there appeared instead to be a non-visual advantage to *carbonaria* homozygotes while heterozygotes and typicals showed equal fitness. The insects are known to be highly mobile, and the steady level of about 95% probably indicates migration of typical forms from less industrialized regions.

The recent declines also indicate strong selection, this time against the melanics. If the simplest model is assumed, with constant selection and no migration, the Caldy data estimate the fitness of *carbonaria* as 85% of that of typicals, the Manchester data from 1954 as 79%, and the Kent data as 89%. These figures give us a rough picture of the selection. However, in all cases the constant selection curves lag below the data points in the middle of the sequence. The more realistic modelling procedure removes this effect. When extracted, the SO_2-associated fitness is estimated as about 55% in Manchester compared with 85% for the other localities.

The results for Manchester and Kent and those of Mani and Majerus and Grant *et al.* allow 19 comparisons between early records with high frequencies of *carbonaria* and later records with low ones. Taken together, they suggest that selection against *carbonaria* may be greater in localities that initially were more heavily polluted. [. . .] SO_2 concentration was a good predictor of *carbonaria* frequency when the pattern was more or less stable, and it is possible that the association indicates a direct effect on the polymorphism. Evidence from some other industrial melanic species seems to favour this interpretation. Alternatively or in addition, SO_2 may act by altering the composition of the flora on surfaces used as resting sites by the moths, so as to modify relative crypsis. It is well known that lichens and bryophytes are sensitive to acid rain deposition, and that they disappeared from industrialized areas while unpolluted localities continued to have rich and diverse epiphyte floras. Although the process is not a simple reversion, lichens have increased in extent and diversity where SO_2 concentration has fallen. In the north-west, epiphyte patterns changed after the decline in atmospheric pollution. A similar association of changes has occurred in The Netherlands. Grant *et al.* question epiphyte change as a possible factor explaining the change in frequency, but if selective predation is involved, then anything likely to affect background colour or heterogeneity may be important. [. . .]

As Mani noted, 'The decline in the melanic frequencies in response to falling levels of air pollution presents an excellent opportunity for an intensive study both in the laboratory and in the field of gene action and non-visual selection effects. It would be a pity for the understanding of evolutionary ecology if we missed it.' The opportunity has now nearly gone.

[*Proc. of the Royal Society of London*, Ser. B, 266 (1999), 297–303.]

MARY N. KARN AND L. S. PENROSE

9 Birth weight and gestation time in relation to maternal age, parity, and infant survival

The paper is a classic study of natural selection on human birth weight. The paper looks at the survival, and frequency of stillbirths among newborn infants, in relation to birth weight. The evidence shows 'normalizing' or 'stabilizing' selection: intermediate-sized infants have the highest survival, with larger and smaller birth weights being selected against. Also, the average birth weight is found to differ from the birth weight at which survival is highest. [Editor's summary.]

The present communication continues a statistical analysis of the relation between birth weight and gestation time, taking into account the mother's age and parity.

The large body of data assembled from the records of U.C.H. Obstetric Hospital for the years 1935–46 contains information about 13,730 infants (7037 male, 6693 female) and their mothers.

A number of twins was recorded for these years, but, as the present investigation was concerned with the variations of normal birth weight and gestation time, these were not included. The data contained information on stillbirths and neonatal deaths (non-survivors at 28 days), 340 males, 274 females. Thus, the complete data were as in [Table 9.1].

Table 9.1

	MALES	FEMALES
Survivors (beyond 28 days)	6697	6419
Non-survivors (stillbirths and neonatal deaths)	340	274
Totals	7037	6693

The means and standard deviations for the four variables are set out in Table 1,[1] for males and females, for the groups of survivors, non-survivors, together and separately. The means for non-survivors can be compared with those for the survivors. The non-survivors are thus seen to be significantly lighter at birth with considerably shorter mean gestation time and somewhat older maternal age than survivors. Parity in the two groups is the same. [. . .]

The mean birth weight for the sexes together and for all parties (7.24 lb.) falls within the range of means for numerous data from diverse groups extending over the period from 1866 up to date, recorded by Murray (1924) and Henderson (1945).

[1] Table not included here.

In Table 7[2] the combined still birth and neonatal mortality rates per cent and the survival rates per mille, for males and females, are given according to weight.

The distributions of survival rates in relation to birth weights show highest values in the group above average weight, 7½–8½ lb. (979 for males and 985 for females); next in order comes the group with weights 6½–7½ lb. (with values of 973 for males and 981 for females). [. . .]

It is natural to assume that in consequence of the action of natural selection the mean value of any biological measurement would be the most normal value and associated with the most favourable survival rate. In the case of birth weight the optimal value from the point of view of avoiding stillbirth or neonatal mortality is clearly greater than the mean.

The optimal birth weight can be estimated with fair precision on the assumption that both survivors and non-survivors have weight distributions approximating to Gaussian curves. The logarithm of the ratio of probability of surviving to that of not surviving would then describe a parabola with its apex at the birth weight value where the odds on survival was maximal. From the data under consideration, parabolas were fitted to the observed probability ratios at different birth weights both for males and females. In doing this, the contribution of each observed point was made proportional to the product, divided by their sum, of the numbers of survivors and non-survivors in each weight group. Figure 9.1 shows the result of fitting these

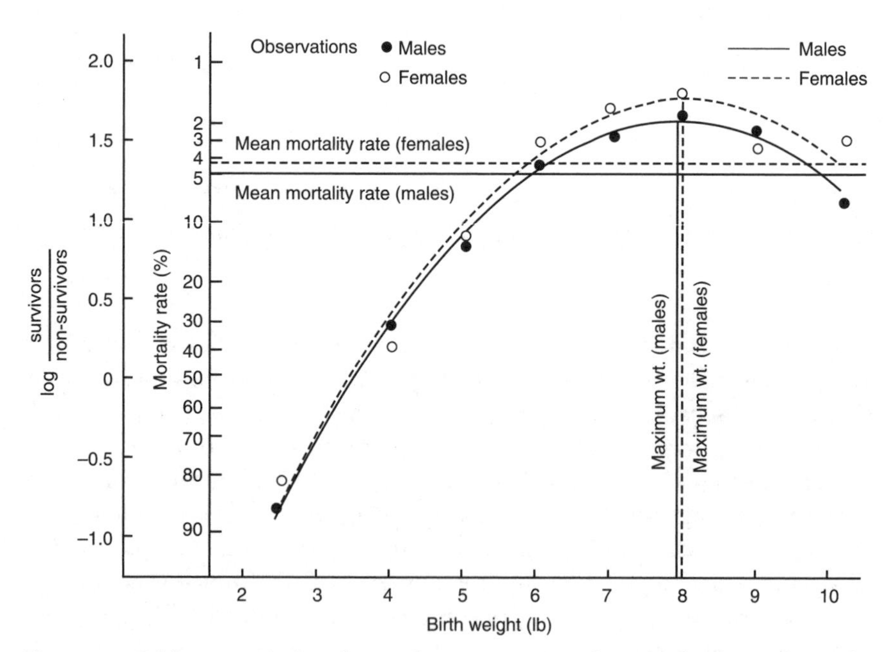

Figure 9.1: Odds on survival and mortality rate per cent for given birth weight. Males and females.

[2] Table not included here.

parabolas. [. . .] The odds on survival are maximal at about 8 lb. in each case (x = 7.93 for males and x = 8.01 for females), and the mortality rate there is less than 2%. The average mortality rate for males, 4.8%, occurs at 5.91 lb. and also at 9.95 lb. Between these critical points mortality rate is less than the average, and outside them it is in excess of the average. For females the critical points, where average mortality is 4.1%, are 5.85 and 10.05 lb.

[*Annals of Eugenics*, 16 (1951), 147–64.]

L. ULIZZI AND L. TERRENATO

10 Natural selection associated with birth weight: towards the end of the stabilizing component

A similar analysis to that of Karn and Penrose (Chapter 9) shows that stabilizing selection on human birth weight all but disappeared between 1950 and 1985, at least in Italy. This is an example of relaxed selection in wealthy human populations—improved medical care has relaxed a force of selection that used to operate. [Editor's summary.]

Birth weight is one of the best predictors of the probability of surviving the hazards of perinatal life. Keeping apart the grossly abnormal case of the preterm baby, full-term babies of very light and very heavy birth weights exhibit a much higher perinatal mortality when compared with those of intermediate birth weight. The selective relevance of this phenomenon is proportional to the genetic component of birth weight variability; in industrialized countries the progressively increasing percentage which the perinatal mortality rate contributes to the total pre-reproductive mortality rate makes this selective phenomenon even more relevant.

Selection against both tails of the birth weight distribution has a stabilizing effect. As a consequence of this mortality distribution, birth weight variance is reduced in surviving babies in comparison with the birth weight variance seen in the entire cohort of babies who complete intrauterine life. After the pioneering work of Karn and Penrose,[1] [. . .] several authors studied stabilizing selection on birth weight in different populations.

In industrialized countries programmes of regular prenatal care attempt to control the conditions in which pregnancies are brought to term, with the result of decreased birth weight variability on the one hand, and with an increased survival rate, primarily of babies with extreme birth weights, on the other. This complex of phenomena has been proposed to be a cause of relaxation of selection during pregnancy and perinatal life. Therefore, the trend over time of stabilizing selection on birth weight in industrialized

[1] See Chapter 9 above.

countries should be analysed carefully. The available scientific literature only reports studies performed in a given population during a given year, but does not follow the phenomenon over time. We have been monitoring stabilizing selection on birth weight for many years in different populations and have demonstrated a very rapid relaxation of this component of natural selection. In this paper we extend data on the Italian cohorts of newborns up to 1985, a critical period of dramatic changes in environmental conditions, i.e. the transition from a mainly rural to an industrialized country.

Data on birth weight for all single, full-term stillbirths and livebirths in Italy in the years 1954, 1961, 1967, 1974, 1980 and 1985 were drawn from the Demographic Yearbook (Istituto Centrale di Statistica, Roma). The first and the last years represent the least and the most recent data made available by the Italian Census Office.

The stillbirth rates from 1954 to 1985 are plotted against birth weight in Fig. 1.[2] [. . .]

The uppermost curve refers to the 1954 cohort and is clearly U-shaped, the 3.75 kg class being associated with minimal mortality. The lowest curve, however, which refers to the 1985 cohort, is almost completely flat. During this thirty-year interval the selective advantage of being in the central-birth weight class (which in the past had allowed the use of the denomination 'optimal birth weight') is progressively disappearing since nowadays the mortality in this class is practically identical to that of the neighbouring weight classes. Therefore, while the curve relative to the 1954 cohort has a shape similar to that originally described by Karn and Penrose in 1951, after only one generation, in the 1985 cohort almost complete relaxation of the phenomenon is observed. The intermediate steps (shown by the 1961, 1967, 1974 and 1980 cohorts) reveal that relaxation has been progressively achieved, at a decreasing rate. In fact during the last interval of time only minimal mortality variation for each birth weight class is observed.

The modification of selection parameters [. . .] are shown in Table 10.1. Selection intensity[3] decreases by about four times, while the percentage

[2] The Figure is not included here, but is similar in appearance to Karn and Penrose's Fig 9.1 above. Ulizzi and Terrenato plotted the results the other way up from Karn and Penrose, with mortality increasing up the y-axis: that is why the picture is, as the next sentence says, U-shaped.

[3] J. B. S. Haldane, *Caryologia*, Suppl. (1954), 480–7. To understand selection intensity, imagine dividing the infants up into a series of birth weight classes. One of these classes (the one with intermediate birth weights) will have the lowest mortality of all the classes—that is, the 'minimum mortality'. We can also calculate the 'average mortality' for all the infants. The relation between the minimum mortality and the average mortality in the population is a measure of how much selection is going on. If the minimum mortality = average mortality, there is no selection; if the minimum is much less than the average, selection is strong. In Fig. 10.1 below, the 45° line is for no selection and the further a point is to the right of that line, the stronger selection is.

Table 10.1 Selection characteristics from 1954 to 1985

	1954	1961	1967	1974	1980	1985
			Selection intensity			
Males	0.0073	0.0051	0.0039	0.0029	0.0023	0.0013
Females	0.0060	0.0040	0.0038	0.0023	0.0018	0.0015
			Selective deaths			
Males	34.29	30.54	31.40	36.25	50.00	43.33
Females	37.04	28.57	34.91	30.26	41.86	48.39

Note: Selection intensity and selective deaths are calculated according to Haldane (1954) as:

$$\text{selection intensity} = \log(s) - \log(S)$$

$$\text{percentage of selective deaths} = [(1 - S) - (1 - s)/(1 - S)] \times 100$$

where s is the maximal survival (the fraction which survives among optimal phenotypes) and S is the survival in the total population.

of selective mortalities remained practically unchanged. It has been shown in a previous paper that the two selection parameters behave in this manner when a parallel reduction of average and minimal mortality occurs. This is appreciable in Figure 10.1 where the relationship between minimal mortality (i.e. that associated with the central birth weight class) and average mortality is reported. For these parameters it is also worth noting that the last interval (1980–1985) does not show variation comparable to that observed during the preceding intervals of time. Moreover, while thirty years ago there was an appreciable difference between male and female mortality rates (minimal and average mortality being constantly higher in males), in the last few years any sex-specific differences tend to disappear.

The efficiency of stabilizing selection is measured by the reduction of birth weight variance between all births (still- plus livebirths) and livebirths. [. . .] During the last thirty years a marked reduction of birth weight variance has been observed and the slope does not show any flattening in recent years. The room available for selection (birth weight variance in the exposed cohort) drops by about 25%; however, in comparison with other metric traits (e.g. adult stature) birth weight variation is still relatively high. As for the stabilizing effect, it is progressively vanishing: in 1954 the reduction of variance due to stabilizing selection equals 2.24 and 1.79% in males and females respectively, while in 1985 the reduction is equal to 0.87 and to 0.94% respectively. Therefore the environmental changes in the course of thirty years have brought about a reduction of 25% of the birth weight variance of the cohort exposed to stabilizing selection (all births) and in the same period the

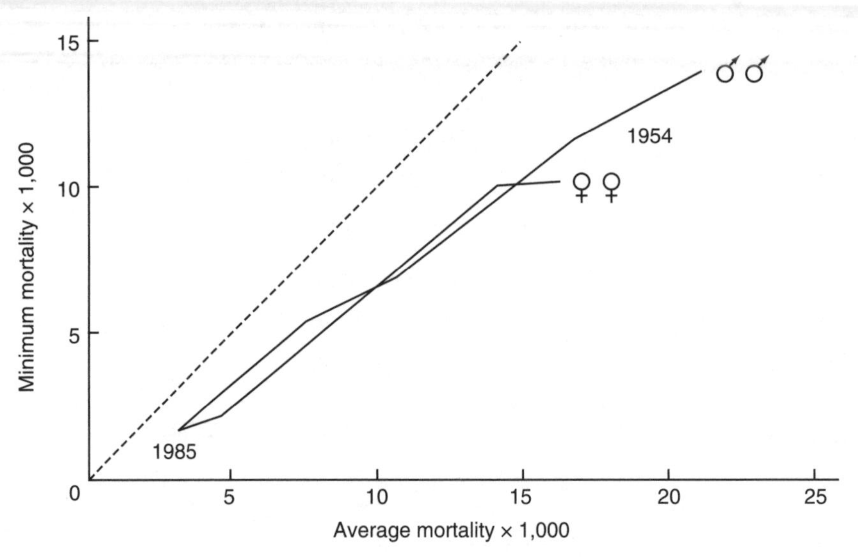

Figure 10.1: Relationship between average and minimal mortality, which is always found in the central birth weight class (3.5–4.0 kg) both in males and in females.

efficiency of stabilizing selection in reducing birth weight variance is more than halved.

As to a possible directional component, average values[4] demonstrate that it is constantly present, even if at only a very low extent, the increase after selection being in the range of 0.03–0.09%. [. . .]

In conclusion, to answer the question 'Is stabilizing selection on birth weight still operating?', one must consider separately 'average' and 'extreme' birth weights. It can be claimed that the selection against extreme weights remains unchanged, suggesting that genotypes grossly affecting birth weight are still selected against. At the same time the effect of increased prenatal care is likely to have reduced the selective relevance of the many genes which possibly cooperate in controlling minimal birth weight variations; therefore these genes become in this aspect progressively neutral. This appears to be a relatively final stage and no drastic variation of the phenomenon is expected in the near future.

[*Annals of Human Genetics*, 56 (1992), 113–18.]

[4] Directional selection refers to the way the optimal birth weight is slightly higher than the average birth weight (see Chapter 9). Natural selection seems to favour larger infants than are born on average. The detailed numbers of Ulizzi and Terrenato are not given here, but directional selection seems not to have declined along with stabilizing selection.

H. LISLE GIBBS AND PETER R. GRANT

11 Oscillating selection on Darwin's finches

In 1983, food supplies for Darwin's finches on the Galápagos Islands increased following wet weather. In the following years (1984–85), natural selection favoured smaller birds and average bird size evolutionarily decreased. Following droughts, in 1976–77 and 1981–82, natural selection favoured larger birds and size evolved up. The form of selection oscillates over time. [Editor's summary.]

An important goal in the study of evolution is to determine the occurrence, causes and possible micro-evolutionary consequences of selection in natural populations. Darwin's finches (Geospizinae) are suitable organisms for investigation because their morphological traits are highly heritable, and they live in a climatically variable environment (Galápagos Islands). It has been suggested that selection fluctuates in direction and intensity, favouring different morphological optima in different years, because strong annual variation in rainfall causes changes in food supply composition. This suggestion has been supported in part by studies of the medium ground finch, *Geospiza fortis*, on the island of Daphne Major, which have shown that large adult size is favoured under drought conditions, when the overall food supply is low and large hard seeds are disproportionately abundant. Here we document a reversal in the direction of selection following the opposite climatic extreme, and demonstrate the connection between oscillating selection and fluctuations in food supply.

From December 1982 until July 1983 exceptionally heavy and prolonged rain fell on Daphne Major, in association with the most severe El Niño event (an oceanic disturbance) of the century. A total of 1,359 mm of rain was recorded, ten times the previously recorded maximum of 137 mm in 1978. Birds bred continuously for eight months, instead of the usual one to three months, and by the time the rains ceased both food level (seeds) and bird density were exceptionally high. We examined the survival of individually ringed and measured birds over the next two years when rainfall was low (53 mm, 1984) to very low (4 mm, 1985). During this time seed biomass steadily declined but was nevertheless higher than in the dry years of 1977 and 1982.

We measured directional selection on adults by comparing the sizes of survivors and non-survivors. Results are given in Table 11.1 under two headings. Selection gradients (β) represent the direct effects of selection on each trait and are measured as the partial regression coefficients of relative fitness (survival) on each trait. Selection differentials (s) measure the combined direct effects and indirect effects of selection caused by phenotypic correlations between traits, and are calculated as the difference in the mean value of the trait before and after selection, here measured in standard deviation units.

Small adults survived better than large ones in the two years following the exceptionally wet El Niño year (Table 11.1, Fig. 11.1); selection differentials for all measured traits were negative and often significant. These are in marked contrast to selection on all traits in the opposite direction under the contrasting conditions of drought and low overall food supply (Fig. 11.1).

Selection was consistent over sexes and cohorts. For example, all selection differentials for males and females were negative, although not all were significantly different from zero. Males, the larger sex, were apparently subject to stronger selection than females. Selection differentials for overall body size (principal component I, PCI in Table 11.1) were almost twice as large for males (-0.41, $P < 0.001$) as for females (-0.22, $P < 0.05$).

Significant β values identify the targets of direct selection among the set of intercorrelated morphological traits subject to analysis. No single trait, with the possible exception of bill width, was an obvious target of selection (Table 11.1). In one year selection weakly favoured increased wing length, but no net selection occurred because of a positive correlation with overall size which was selected in the opposite direction. These results differ from those during droughts and conditions of low food supply, when individual targets and directions of selection could be clearly identified (weight and bill depth selected to increase, bill width selected to decrease). Thus we interpret the strong differentials and weak gradients to indicate selection on overall body size, as manifested by highly significant negative selection differentials for PCI scores (Table 11.1).

Large adult size is favoured when food is scarce because the supply of small and soft seeds is depleted first, and only those birds with large bills can crack open the remaining large and hard seeds. In contrast, small adult size is favoured in years following very wet conditions, possibly because the food supply is dominated by small soft seeds. In 1984 and 1985, the proportion of total seed biomass made up by small soft seeds was 21% to 80%, which is two to ten times greater than the previous maximum (8% in May 1978) following a normal wet year. Despite an initially large standing crop of seeds, finch mortality was high and correlated with size, possibly because large birds had difficulty finding enough large seeds. As has been found in other years, finches seen feeding on large seeds were significantly larger in all traits except bill length, and in PCI scores, than those feeding solely on small seeds (t-tests, $P \leqslant 0.05$) in 1984 and 1985 ($N = 108 - 124$ birds). Thus morphological contraints on foraging behaviour may underlie the fluctuating selection on body size under the contrasting feeding conditions induced by droughts and El Niño events.

In contrast to adults, juveniles were not subjected to directional selection on body size or bill size during the first six to nine months following their

Table 11.1: Standardized selection differentials (s) and gradients (β) for adult G. fortis *from 1984 to the beginning of 1986*

	PERIOD OF SELECTION					
	1984		1985		1984–85 COMBINED	
CHARACTER	β	s	β	s	β	s
Weight	−0.12 ± 0.07	−0.12**	−0.03 ± 0.08	−0.08	−0.09 ± 0.10	−0.18**
Wing	−0.07 ± 0.06	−0.09*	0.15 ± 0.07*	0.03	0.11 ± 0.08	−0.06
Tarsus	0.03 ± 0.06	−0.06	−0.06 ± 0.06	−0.09	−0.09 ± 0.07	−0.17**
Bill length	0.10 ± 0.06	−0.03	−0.10 ± 0.07	−0.05	0.08 ± 0.09	−0.09
Bill depth	0.13 ± 0.09	−0.08*	−0.02 ± 0.11	−0.10*	0.03 ± 0.13	−0.18**
Bill width	−0.21 ± 0.09*	−0.12**	−0.15 ± 0.10	−0.11**	−0.29 ± 0.12*	−0.21***
PC I		−0.12**		−0.10**		−0.21***
PC II		0.03		0.00		0.03
N	314 (0.61)	320–391	316 (0.56)	318–341	489 (0.36)	496–537
r^2	0.06*	—	0.05*	—	0.05**	—

Notes: Values of β are shown ±s.e.m. Before each analysis data were $\log_e$-transformed and standardized to have zero mean and unit variance. Principal component (PC) scores were extracted from a correlation matrix calculated from all individuals for which measurements of all six univariate traits were available. PC I is interpreted as a measure of overall body size because all characters load positively on it, whereas PC II is interpreted as a shape index in terms of bill size relative to tarsus length. N is the number of birds alive at the beginning of the period, and varies according to the number of measurements taken. The proportion surviving is given in parentheses. The 1984–85 combined sample includes birds born in 1984. Significance of selection differentials was assessed by t-tests comparing survivors with those who disappeared (*$P \leqslant 0.05$; **$P \leqslant 0.01$; ***$P \leqslant 0.001$).

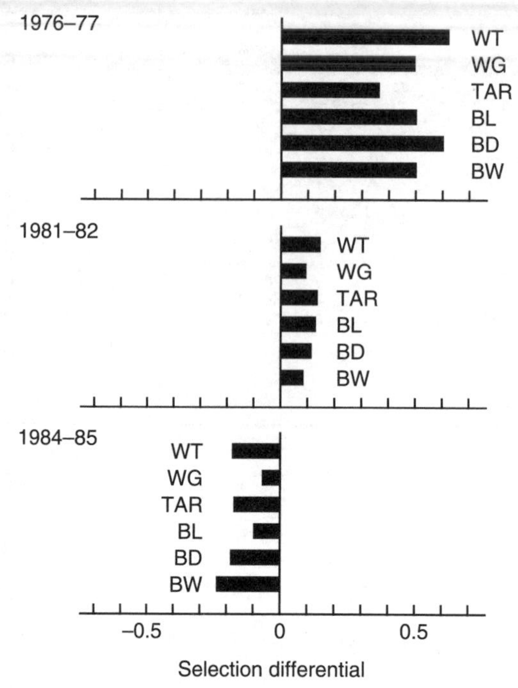

Figure 11.1: Selection differentials for traits of adult G. fortis *during two years of drought and low food supply, 1976–77 and 1981–82, and during the years that followed the El Niño event of 1982–83. Traits measured are shown as follows. WT, weight; WG, wing length; TAR, tarsus size; BL, bill length; BD, bill depth; BW, bill width.*

birth in 1983 and 1984. Juveniles were not present in 1985 because no successful breeding occurred in this year. Small size is favoured among juveniles when small seeds are scarce following a breeding season.

The chief implication of the results of this long-term study is that natural populations subject to the effects of rare climatic perturbations are not at a fixed equilibrium state demographically, phenotypically, or genetically. Selection in opposite directions [. . .] in different years and generations, may result in weak overall stabilizing selection on morphological traits over a period greater than a single generation.

[*Nature*, 327 (1987), 511–13.]

R. C. LEWONTIN

12 The paradox of variation

The extract comes from a book published in 1974, as new molecular techniques allowed genetic variation to be measured in natural populations—measurements that led to the 'neutral' theory of molecular evolution (though Lewontin argues that 'neutralism' is a misnomer). Lewontin begins with a summary of the factors that influence genetic variation. He then distinguishes two theories. According to the classical theory, populations have little genetic variation and natural selection mainly acts to remove disadvantageous genetic forms. According to the balance theory, populations have extensive genetic variation, actively maintained by selection. The findings of extensive genetic variation in nature might seem to refute the classical theory, but it developed itself into the neoclassical (that is, 'neutral') theory. Lewontin finishes by arguing that the balance theory runs into absurdities if it tries to account for the observed facts. [Editor's summary.]

For many years population genetics was an immensely rich and powerful theory with virtually no suitable facts on which to operate. It was like a complex and exquisite machine, designed to process a raw material that no one had succeeded in mining. Occasionally some unusually clever or lucky prospector would come upon a natural outcrop of high-grade ore, and part of the machinery would be started up to prove to its backers that it really would work. But for the most part the machine was left to the engineers, forever tinkering, forever making improvements, in anticipation of the day when it would be called upon to carry out full production.

Quite suddenly the situation has changed. The mother-lode has been tapped and facts in profusion have been poured into the hoppers of this theory machine. And from the other end has issued—nothing. It is not that the machinery does not work, for a great clashing of gears is clearly audible, if not deafening, but it somehow cannot transform into a finished product the great volume of raw material that has been provided. The entire relationship between the theory and the facts needs to be reconsidered.

The modulators of variation

The genotypic distribution in a population is subject to a complex of forces that act separately and in interaction to increase, decrease, or stabilize the amount of variation. [. . .]

Variation is introduced into a population by either mutations or the immigration of genes from other populations with different alleles. [. . .] Once allelic variation has been introduced by mutation or immigration, the total array of genotypes may be considerably increased by recombination, but of course this source of genetic variance is completely limited by the allelic variation that is being recombined. [. . .]

The third force that can increase or at least conserve variation is natural selection. If an initially rare allele is favored by selection, either because the environment has changed, or because the allele has only recently been introduced into the population, the increase in allelic frequency toward 0.5 means an increase in genetic variation. However, if the allele is unconditionally more fit and is eventually fixed in the population, or at least driven to a very high frequency, then variation again decreases as the allele passes from a frequency of 0.5 to near 1. Selection for the replacement of an old 'wild type' by a new one is then a temporary producer of heterozygosity.[1] As we shall see, this transient heterozygosity cannot account for much of the genetic variation in a population.

There is one form of selection, however, that is a powerful preserver of genetic variation: balancing selection. In its simplest form, balancing selection arises from an unconditional superiority in fitness of heterozygotes over homozygotes, so-called overdominance of fitness. [. . .]

Unconditional overdominance is not the only form of balancing selection that will maintain stable heterozygosity. Selection operating in opposite directions in the two sexes or in gametic and zygotic stages, and a number of cases in which the fitnesses of genotypes are functions of their relative frequencies in the population may also lead to a stable equilibrium. A variety of situations in which fitnesses vary in space or time can also lead to stable equilibria, even though the heterozygote is not superior to the homozygotes in any of the individual spatial or temporal modes. [. . .]

Genetic variation is removed from populations by both random and deterministic forces. Every population eventually loses variation because every population is finite in size and therefore subject to random fluctuation in allele frequencies. Such random fluctuation, if unopposed by the occurrence of new variation, leads ineluctably to the fixation of one allele in the population, although the loss of heterozygosity may be extremely slow if population size is very large. But most populations of organisms, especially free-living terrestrial species, undergo periodic severe reductions in population size either regularly with the change of seasons, or in episodes of colonization, or aperiodically when a particular conjunction of adverse environmental factors creates a demographic catastrophe. At such times the loss of heterozygosity may be great, as only a few individuals survive to reproduce.

[1] Heterozygosity is a general measure of genetic variation, not simply the presence of heterozygotes in a population. However, populations with high heterozygosity usually contain many heterozygous individuals.

Irrespective of population size, newly arisen alleles are in constant danger of loss from the Mendelian mechanism itself. Even if every individual in a bisexual population left exactly two offspring, so that the population was stable in size and there was no opportunity for differential reproduction among genotypes, a heterozygote for a new mutation would have a 25 percent chance of failing to pass on that mutation to either of its offspring. If we add to this the fact that there is always some variation in family size, the probability of loss becomes even greater. If, for example, family size has a Poisson distribution, the probability of loss of a new mutant in one generation is $1/e \cong .37$.

Random fluctuations in selection intensity may also reduce genetic variation. It is true that some kinds of variation in selection will act as balancing selection, even in the absence of overdominance, yet other patterns of fluctuation will hasten homozygosity by driving allele frequency close enough to zero or one for random loss to be important, although the average selection for or against the allele may be very small or even zero.

The strongest force reducing genetic variation is, of course, selection against recessive or partly dominant deleterious genes. The effect of such selection on the species as a whole depends upon whether the same allele is favored in all populations. If there is one wild type over the whole of the species range, then the result will be uniformity over all populations. If there is local adaptation, however, with different alleles selected in different localities, the species will be polytypic but individual populations will be homozygous, as in *Acris crepitans*. There is, in this case, storage of variation in the species as a whole, rather than in the heterozygosity of individual populations, although migration between such populations will result in their heterozygosity as well.

Another not uncommon form of selection that reduces heterozygosity is selection against heterozygotes. When the heterozygote is less fit than the homozygote, there is an unstable equilibrium and the allele frequency is driven to zero or one, depending upon which side of the unstable point it finds itself on at the time the selection process begins. In effect this means that the introduction of new mutations is strongly resisted. Like selection for locally adapted genotypes, selection against heterozygotes could lead to polytypy. Indeed selection for locally adapted alleles and selection against heterozygotes are the two forms of natural selection postulated for the process that converts locally differentiated populations into those ecologically and reproductively isolated populations we call species. Maternal-fetal incompatibility is an example of selection against heterozygotes, so the maintenance of the apparently stable polymorphism for human blood groups is rather paradoxical.

Classical and balance theories

That natural selection could be both the preserver and the destroyer of intrapopulation variation would have been a surprise to Darwin. For him, natural selection was the converter of intrapopulation variation into temporal and spatial differentiation. Lacking a correct theory of genetics, and especially one that included segregation of alleles from hybrids, Darwin could not have imagined forms of selection that would actually stabilize inherited variation. Without Mendelism, the theory of natural selection is inevitably one that predicts that a more fit type will completely replace the less fit types if the trait is at all heritable. Classical Darwinism saw evolution as the passage from one more or less uniform state to another and in this sense was no different from pre-Darwinian ideas of evolution. What Darwin added was the realization that the variation present in populations was the source of the eventual variation between species, but the ontogeny of the variation itself was a mystery about which Darwin changed his mind in the course of his life. The lack of a satisfactory explanation for the origin and replenishment of variation that was constantly being reduced by 'survival of the fittest' was a serious flaw in evolutionary theory, a flaw that was not repaired until the reappearance of Mendelism.[2] Classical Darwinian theory held that heritable variations arose from a variety of sources and that natural selection sorted through these variations, rejecting all but the most fit type. Natural selection was seen as *antithetical* to variation—'Many are called, but few are chosen.'

Those who support the classical theory of population structure are the direct inheritors of this pre-Mendelian tradition. For them, variation arising from mutation is constantly being removed by the purifying force of directional selection and, to some extent, by random genetic drift. Although it seems strange to say, the classical theory owes virtually nothing to Mendelism. The fact is that almost the entire theoretical apparatus of random genetic drift and directional selection can be derived from a *haploid* model of the genome and that the introduction of diploidy and sexual recombination makes no qualitative change and only trivial quantitative changes in the predictions of evolution under these forces. For random drift, the introduction of diploidy simply alters by a factor of 2 the rate of loss of genetic variation in a population except in the totally unrealistic case of no variance in family size. For selection, so long as the heterozygote has a fitness somewhere within the range spanned by the homozygotes, the process of allelic frequency change can be adequately represented by the model of 'genic' selection, and the genetic load (or variance in fitness) in a population at equilibrium between mutation and selection is virtually the same for both models.

[2] On this repair, see Fisher (Chapter 4).

[. . .] A search through textbooks and technical literature on the theory of random drift and directional selection shows that a haploid model is often used for simplicity of derivation, diploidy being added later as a refinement.

The balance theory, as its name suggests, emphasizes that aspect of natural selection which would have been foreign to Darwin and which is the unique contribution of Mendelism to evolutionary theory, the possibility that natural selection preserves and even increases the heritable variation within populations. In the absence of special symbiotic relations between genotypes, natural selection can stabilize variation only when there are heterozygotes and segregation of genes. Only then can there be heterosis, but also the stabilization of variation by temporally fluctuating selection is critically dependent upon a diploid sexual model. The balance school sees the maintenance of variation within populations and adaptive evolution as manifestations of the same selective forces, and therefore it regards adaptive evolution as immanent in the population variation at all times. Because the alleles that are segregating in a population are maintained in equilibrium by natural selection, they are the very alleles that will form the basis of adaptive phyletic change or speciation.

The reverse is true for the classical theory, in which variation is present *faute de mieux* and in which the genetic basis for further evolution is either lacking or extremely rare most of the time in the history of a population because natural selection is efficiently sweeping out any variation that might otherwise accumulate. [. . .]

But why this continued juxtaposition of classical and balance theories? Have not the evident facts[3] of vast quantities of polymorphism and heterozygosity firmly established the balance theory? In the face of the evidence given [. . .] how can the classicists hold their ground? The answer is, Easily.

The neoclassical theory

If we take it as given that balancing selection is rare and that natural selection is nearly always directional and 'purifying,' how can we explain the observed polymorphism for electrophoretic variants at so many loci? We can do so by claiming that the variation is only apparent and not real. That is, we can suppose that the substitution of a single amino acid, although detectable in an electrophoresis apparatus, is in most cases not detectable by the organism. If it makes no difference to the physiological function of an enzyme whether

[3] Lewontin wrote when huge quantities of new studies of genetic variation were coming out. The (then) new work used gel electrophoresis. In most cases, the research was finding substantial genetic variation in natural populations. Elsewhere in the book, Lewontin reviewed the findings. However, the factual details are omitted from this extract, except for where Lewontin makes use of rough figures in the final section of this extract.

it has a glutamine or a glutamic acid residue, for instance, on the surface of the folded molecule far away from the enzyme's active site, then the variations detected by electrophoresis or by any method that is sensitive to amino acid substitutions may be completely indifferent to the action of natural selection. They are 'genetic junk,' revealed by the superior technology of the laboratory but redundant physiologically. From the standpoint of natural selection they are *neutral mutations*.

The suggestion that most, if not all, of the molecular variation in natural populations is selectively neutral has unfortunately led to widespread use of the terms 'neutral mutation theory' and 'neutralists' to describe the theory and its proponents (see almost any discussion of the problem of genic heterozygosity since 1968). But these rubrics put the emphasis in just the wrong place and obscure both the logic of the position and the historical continuity of this theory with the classical position. It is not claimed that nearly all mutations are neutral or that evolution proceeds without natural selection, chiefly by the random fixation of neutral mutations. Both these statements are patently untrue and both are foreign to the spirit of the proposition that is being made. On the contrary, the claim is that many mutations are subject to natural selection, but these are almost exclusively deleterious and are removed from the population. A second common class is the group of redundant or neutral mutations, and it is these that will be found segregating when refined physicochemical techniques are employed. In addition the theory allows for the rare favorable mutation, which will be fixed by natural selection, since after all adaptive evolution does occur. But it supposes this event to be uncommon. Finally, it also allows that occasional heterotic mutants might arise but that these do not represent a significant proportion of all the loci in the genome.

Thus the so-called neutral mutation theory is, in reality, the classical Darwin–Muller hypothesis about population structure and evolution, brought up-to-date. It asserts that when natural selection occurs it is almost always purifying, but that there is a class of subliminal mutations which are irrelevant to adaptation and natural selection. This latter class, predictable from molecular genetics and enzymology, is what is observed, they claim, when the tools of electrophoresis and immunology are applied to individual and species differences.

The neoclassical theory cannot be refuted by erecting a neutralist strawman and refuting that. So, for example, the demonstration that single amino acid substitutions can in some instances make big differences in physiology is irrelevant. The range of effects of single amino acid substitutions can be illustrated by human hemoglobin. Of 59 variant α and β hemoglobin chains listed by Harris, 43 are without known physiological effects at least in the heterozygous state in which they are found, 5 cause methemoglobinemias because they are near the site of the heme iron and are therefore mildly pathological, and 11 cause instability of the hemoglobin molecule that results in various degrees of hemolytic anemia. Although most of this last group result from the substitution of noncharged by noncharged amino acids on the inside of the three-dimensional structure, two are caused by charge changes at position 6 on the outside of the molecule. One, a substitution of lysine (+) for the normal glutamic acid (−) causes the benign hemoglobin C disease, but the other, a substitution of valine (0), is the famous hemoglobin S, causing sickle-cell anemia.

We cannot make anything of the relative proportion of pathological (16) to asymptomatic (43) substitutions, since the former are detected from the pathology they cause, whereas the latter turn up in routine electrophoretic screening. We do not know, for example, how many neutral substitutions on the surface or in the interior of the molecule go undetected because of their lack of detectable physiological effect. Nor is it certain, conversely, that the 43 surface charge changes are absolutely neutral. Most have never been seen in homozygous condition. The point is that single amino acid substitutions, charged and uncharged, on the surface or the interior of the molecule, run the gamut of effects from apparently neutral to severely pathological, and this range is in no way contradictory to the neoclassical theory.

Second, the neoclassical theory is not refuted by occasional observations of overdominance for fitness, because the theory does not deny that cases exist but only that they are common and explain a significant proportion of natural variation. So it is no use trotting out that tired old Bucephalus, sickle-cell anemia, as a proof that single-locus heterosis can exist. Anyone who has taught genetics for a number of years is tired of sickle-cell anemia and embarrassed by the fact that it is the only authenticated case of overdominance available. 'If balancing selection is so common,' the neoclassicists say, 'why do you always end up talking about sickle-cell anemia?'

Finally, the neoclassical theory cannot be disposed of by pointing to the elephant's trunk and the camel's hump. The theory does not deny adaptive evolution but only that the vast quantity of molecular variation within populations and, consequently, much of the molecular evolution among species, has anything to do with that adaptive process. [. . .]

The neoclassical argument is made up of two complementary parts. First, it attempts to show that the balance theory is untenable because it involves internal contradictions, and then attempts to show that the neoclassical theory is compatible with the data. Both parts of the argument are essential. To show only that the neoclassical view is compatible with the facts, to show that it is a sufficient theory, is not enough, for then the two theories stand side by side with no way to choose between them. It is a cornerstone of the neoclassical argument that the balance view is irreconcilable with all the important known facts.

This two-sided argument is applied to two different sets of facts, the amount of heterozygosity in populations and the rate of substitution of alleles in evolution. The neoclassicists maintain that the amount of allelic variation and the rate of amino acid substitution in proteins during evolution are both too large to be accounted for by selection but can be satisfactorily explained by assuming that the genetic variation for amino acid substitutions is neutral and that the differences in amino acid composition of most proteins, are the result of random fixation of these neutral alleles during evolution. [. . .]

Evidence from total heterozygosity

Let us consider the argument that the amount of variation within populations is too great for balancing selection. If we consider a locus with two alleles and with fitnesses.[4]

AA	*Aa*	*aa*
$1 - s$	1	$1 - t$

then at gene frequency equilibrium the heterozygosity is

$$H = \frac{2st}{(s + t)^2} \qquad (1)$$

and the mean fitness of the population is

$$\overline{W} = 1 - \frac{st}{(s + t)} \qquad (2)$$

or, substituting equation (1) into (2),

$$\overline{W} = 1 - H\left(\frac{s + t}{2}\right) = 1 - H\bar{s} \qquad (3)$$

where $(s + t)/2 = \bar{s}$ is the average fitness advantage of the heterozygote.

As a low estimate, the heterozygosity per locus is 10 percent and the proportion of polymorphic loci is 30 percent. If we take a conservative estimate of the size of the genome, which will favor the balance theory, there are 10,000 genes coding for enzymes and proteins in Drosophila.

[4] For non-mathematicians it may help to know that Lewontin is considering a genetic locus with heterozygous advantage. (The values of s and t are between 0 and 1, making the homozygotes worse off than the heterozygote.) He derives a formula for the fitness of an average individual in this case (equation (3)). Nothing crazy happens here. But he then multiplies up to take account of the number of gene loci that could be heterozygous, and derives an absurdity.

Then there are 3000 loci segregating, with an average heterozygosity per segregating locus of 0.10/0.30 = 33 percent. If the effects on fitness of the various loci are independent, then the mean fitness of the population will be the product of the fitnesses at the separate loci and, for the genome as a whole,

$$\overline{W} = (1 - 0.33\bar{s})^{3000} \cong e^{-1000\bar{s}}$$

if $\bar{s}$ is small. Suppose that the overdominance were only 10 percent at each locus. Then

$$\overline{W} = e^{-100} \cong 10^{-43}.$$

Relative to the fitness of a completely heterozygous individual, the average fitness of the population is thus 10^{-43}. In a population that is neither increasing nor decreasing rapidly over long periods, the average reproductive rate per individual must be around 1, so the reproductive ability of a complete heterozygote would be 10^{43}, an absurdity. One can hardly imagine a Drosophila female, no matter how many loci she was heterozygous for, laying 10^{43} eggs.

We could ask, reciprocally, how large a value of $\bar{s}$ could be postulated and still allow a reasonable fitness for the hypothetical fittest genotype. If we were generous and allowed that genotype a fitness of 100, relative to the mean, then $\bar{s} \cong 0.005$. But if the average heterosis at a locus is so low, less than 0.5 percent, then the normal fluctuation of the environment and random genetic drift become vastly more important than the average selection coefficient. A locus with an average heterosis of 0.005 is likely to experience long sequences of generations when there is no heterosis at all. The balance hypothesis is in serious difficulties if it must rely on postulated average selection of such magnitude.

[*The Genetic Basis of Evolutionary Change* (New York: Columbia UP, 1974).]

MOTOO KIMURA

13 Recent development of the neutral theory viewed from the Wrightian tradition of theoretical population genetics

After some historical reflections, the paper defines the neutral theory of molecular evolution and contrasts it with the traditional neo-Darwinian theory of evolution. The neutral theory is quantitative and predicts the rate of evolution. Molecular evolution is clock-like, and conservative—that is, functionally more important regions of DNA evolve at a slower rate. The evolutionary rate of pseudogenes can be used to estimate the total human mutation rate, at about 250 mutations per offspring. Selectionist explanations for the molecular clock are unsatisfactory. [Editor's summary.]

The late Professor Sewall Wright was my idol when I was young. Soon after graduating from Kyoto University, I read Wright's 1931 classic 'Evolution in Mendelian populations' and his subsequent papers on random genetic drift and the distribution of gene frequencies.[1] These papers impressed me deeply and, in fact, inspired me to become a theoretical population geneticist. Without this foundation I would never have been able to propose the neutral theory or to incorporate new knowledge from molecular genetics into the framework of population genetics.

When the neutral theory was proposed, the only available data consisted of amino acid sequences of a few proteins in related organisms, such as hemoglobin molecules in some vertebrate species. Also, genetic variability at the molecular level could be inferred only from electrophoretic data on enzyme polymorphisms for a few species such as fruit flies and humans. Resolution of the ensuing controversy regarding the pros and cons of the neutral theory of molecular evolution was much limited by nonavailability of DNA data. The situation has changed dramatically with the emergence of DNA sequence data, which resulted from the development of rapid DNA sequencing techniques. I am glad to note that, following this development, strong evidence for the neutral theory has accumulated steadily with the passage of time, particularly during the past decade.

In this paper I shall review recent developments of molecular evolutionary studies from the standpoint of the neutral theory. I shall also discuss some neutralist views on evolution in general.

According to the neutral theory, the great majority of evolutionary mutant substitutions at the molecular level are caused by random fixation, through sampling drift, of selectively neutral (i.e., selectively equivalent) mutants under continued mutation pressure. This view is in sharp contrast to the traditional neo-Darwinian (i.e., the synthetic) theory of evolution, which claims that the spreading of mutants within the species in the course of evolution can occur only with the help of positive natural selection.

The neutral theory also asserts that most intraspecific variability at the molecular level (including protein and DNA polymorphisms) is selectively neutral, and it is maintained in the species by the balance between mutational input and random extinction. In other words, the neutral theory regards protein and DNA polymorphisms as a transient phase of molecular evolution

[1] Kimura does not cite here Chapter 5 in Section A, but that chapter illustrates Wright's perspective. The papers that Kimura does cite are mainly included in: S. Wright, *Evolution: Selected Papers*, ed. W. B. Provine (Chicago: University of Chicago Press, 1986).

and rejects the notion that the majority of such polymorphisms are adaptive and actively maintained in the species by some form of balancing selection.

The neutral theory differs from traditional theories of evolution in that it is quantitative—namely, we can derive simple formulae for such quantities as the rate of evolution and the amount of intraspecific variability—and in that we can check the validity of the formulae by comparing theoretical predictions with actual data.

First, let us consider the cumulative process in which neutral mutants are substituted sequentially at a given locus or site through random genetic drift under continued input of new mutants. Then we have for the rate of evolution per generation the formula

$$k_g = v_o, \tag{1}$$

where k_g represents the long-term average per generation of the number of mutants that spread through the population and v_o is the rate of production of neutral mutants per locus (or site) per generation. This formula is based on the well-known property that, for neutral mutations, the long-term rate of mutant substitution is equal to the mutation rate.

If we denote by v_T the total mutation rate, and if f_o is the fraction of neutral mutations at the time of occurrence, so that $f_o = v_o / v_T$, then Eq. 1 may be rewritten as

$$k_g = v_T f_o. \tag{2}$$

Advantageous mutations may occur, but the neutral theory assumes that they are so rare that they may be neglected in our quantitative consideration. Thus, $(1 - f_o)$ represents the fraction of definitely deleterious mutants that are eliminated from the population without contributing to either evolution or polymorphism, even though the selective disadvantages involved may be very small in the ordinary sense. The above formulation has a remarkable simplicity in that the evolutionary rate (on the long-term basis) is independent of population size and environmental conditions of each organism. [. . .]

The first definitive evidence supporting the neutral theory was the discovery that synonymous base substitutions, which do not cause amino acid changes, almost always occur at much higher rate than nonsynonymous—that is, amino-acid-altering—substitutions. It was also found that evolutionary base substitutions at other 'silent' sites, such as introns, occur at comparably high rates. These observations suggest that molecular changes that are less likely to be subjected to natural selection occur more frequently in evolution and therefore show higher evolutionary rates. This is easy to understand from the neutral theory, because such changes are more likely to turn out to be nondeleterious (i.e., selectively neutral) and therefore f_0 in Eq. 2 is larger for them.

More than a decade and a half ago, in collaboration with Ohta, I enumerated five principles that govern molecular evolution, one of which states that functionally less important molecules or parts of a molecule evolve (in terms of mutant substitutions) faster than more important ones. When this principle was proposed, accompanied by its neutralist explanation, much opposition was voiced by the neo-Darwinian establishment, but I am glad to note that it has become a part of common knowledge among molecular biologists, even if few of them seem to realize that its theoretical basis stems from the neutral theory. It is now a routine practice to search for various signals by comparing a relevant region of homologous DNA sequences of diverse organisms and to pick out a constant or 'consensus' pattern, but to disregard variable parts as unimportant.

I once predicted, on the basis of the neutral theory (i.e., using Eq. 2), that the maximal evolutionary rate is set by the mutation rate ($k_g \leq v_T$) and that the maximal rate is attained when all the mutations are selectively neutral (i.e., when $f_0 = 1$). A few years later, this prediction was dramatically vindicated by the discovery of very high evolutionary rates for pseudogenes (or 'dead' genes), which have lost their function. What is especially interesting, as revealed by the studies of pseudogenes, is that the rates of substitution are equally high in all three codon positions. The estimated rate in globin pseudogenes is about $k = 5 \times 10^{-9}$ substitutions per nucleotide site per year.

If the neutralist interpretation of the high evolutionary rate of the mouse globin pseudogene is correct, it will enable us to estimate the total mutation rate due to base substitutions per gamete per generation in humans. For this purpose, we assume that the human genome consists of 3×10^9 nucleotide sites, and we take into account the fact that the rate of molecular evolution per year is significantly lower in humans than in rodents. According to my estimate, the mouse line evolves faster than the human line with respect to amino acid replacements per year in hemoglobin α and β chains by a factor of 34.5/14.6, or approximately 2.4. Note that, from the standpoint of the neutral theory, the difference of the evolutionary rates in these lines is caused by the difference in their neutral mutation rates per year. Thus, assuming an average generation span of 20 years for the human lineage, the total mutation rate per generation is

$$V_T = (3 \times 10^9) \times (5 \times 10^{-9}) \times 20 \div 2.4,$$

or $V_T = 125.0$, which means that the total number of new mutations per generation due to base substitutions amounts to 125 per gamete, and twice as many per zygote.

In the above calculation, we assumed that the mouse line evolved 2.4 times faster than the human line. It is possible, and also likely, however, that this is an underestimate. In fact, Li and Tanimura obtained a result that the rate of synonymous substitutions in rodents is about 7 times higher than that in higher primates. If we adopt this estimate, we obtain $V_T \approx 43$, which is about $\frac{1}{3}$ of the above estimate. Still, this is a very high value when compared with the traditional estimates of the genomic mutation rate. From the consideration of genetic load,[2] the mutational load becomes intolerably high unless the great majority (say, 99.5%) of them are selectively neutral (i.e., nondeleterious).

An equally interesting example suggesting neutral evolution is the recent observation that the evolutionary rate of the eye lens protein αA-crystallin has been much enhanced in the blind mole rat, *Spalax ehrenbergi*. This animal is completely blind and is adapted to a burrowing subterranean way of life (and possibly has been for the last 25 million years, according to fossil evidence). Although this animal is completely blind, the crystallins are still expressed in the atrophied lens cells. Generally speaking, αA-crystallin is a slowly evolving protein, with an average replacement rate of about 0.3×10^{-9} per amino acid site per year in rodents and other vertebrates. In the mole rat lineage, however, this rate has increased severalfold. In this case, although the rate is much increased, its maximum estimate is still only $\frac{1}{5}$ of the observed rate of the globin pseudogene. This is quite understandable because in this animal the αA-crystallin gene is still expressed (i.e., transcribed and translated), and even if the eyes are no longer used for vision, vestigial amounts of the protein still exist in the body, so that some selective constraint should still remain (i.e., $f_o < 1$).

According to the neutral theory, mutation pressure plays a predominant role in molecular evolution. In recent years, much evidence corroborating this has been added. One of the most remarkable examples demonstrating this is the very rapid evolutionary change observed in RNA viruses such as influenza viruses, which are known to have very high mutation rates: genes of RNA viruses show evolutionary rates per year roughly a million times those of DNA organisms. It is remarkable that, in this case, synonymous substitutions also predominate over nonsynonymous substitutions, similar to what has been found in genes of DNA-containing organisms. A recent analysis of data on influenza A virus evolution also confirmed the existence of very clear, clocklike progression of base substitutions, in which the rate of synonymous substitutions (13.1×10^{-3} per site per year) is about 3.5 times that of nonsynonymous substitutions (3.6×10^{-3} per site per year). These observations can readily be explained by the neutral theory by noting that, in

[2] 'Load' is a technical concept. The genetic load of a population equals the (relative) chance that a random individual will die without reproducing because it possesses inferior genes relative to other genes in the population. Muller's paper (Chapter 56) is about loads. The question as to what is the total deleterious mutation rate is important but unanswered, and we therefore do not know what the load due to mutation is.

Eq. 2, the value of v_T is about a million times higher in the RNA genome than in the DNA genome, while the values of f_o remain roughly the same.

Compared with evolution at the phenotypic level, molecular evolution is characterized by two outstanding features. The first is the constancy of the rate—i.e., for each protein or gene region, the rate of amino acid or nucleotide substitution is approximately constant per site per year (hence the term 'molecular evolutionary clock'). The second is the 'conservative nature' of the changes—i.e., functionally less important molecules, or portions of molecules, evolve faster than more important ones.

As to the first feature (i.e., constancy of the rate), this may be explained by the neutral theory by assuming that v_T/g remains the same (constant) among diverse lineages and over time for a given protein or gene, for which f_o is assumed to be constant. In other words, the theory assumes that for a given gene, the production of neutral mutations per year is nearly constant among diverse organisms whose generation spans are very different. Note that 'mutation' here refers to changes that lead to DNA base replacements but not to lethal or 'visible' changes. These latter types of mutations, whose incidence has been known to be generation dependent, are now suspected to be largely caused or controlled by various movable genetic elements such as transposons and insertion sequences in the genome. On the other hand, it is likely that errors in DNA replication and repair are the main causes of DNA changes that are responsible for molecular evolution. Thus, the mutation rate for nucleotide substitutions may depend on the number of cell divisions in the germ lines, particularly in the male line, and this will make the molecular mutation rate roughly proportional to years. Experimental studies on this subject are much needed.

As to the second feature (i.e., the conservative nature), it can easily be understood from the neutral theory, because the less drastic or more conservative the mutational change, the more likely it is to turn out to be nondeleterious, and therefore selectively neutral. This means that for more conservative changes the values of f_o in Eq. 2 are larger. As mentioned already, it is now a common practice among molecular biologists to search for various signals by comparing homologous DNA sequences of diverse organisms and to pick out 'consensus' patterns as important while disregarding variable parts as unimportant.

From the standpoint of the neutral theory, a universally valid and exact molecular evolutionary clock would exist only if, for a given molecule, the mutation rate for neutral alleles *per year* (v_o/g) were *exactly equal* among all organisms at all times (which is rather unlikely in nature). Thus, any deviation from the exact equality of neutral mutation rate per year makes the molecular clock less exact. In other words, the variance of the evolutionary rates among different lineages for a given molecule may tend to become larger than expected from the simple Poisson distribution, as often noted in actual observations. Gillespie,[3] who criticizes the neutral theory on the basis of such observations, claims that a model of evolution, which he calls the 'episodic model', can fit the data better. His model is based on the idea that molecular evolution is episodic, with short bursts of rapid substitution being separated by long periods of no substitution. According to him, each environmental change presents a challenge to the species that may be met by amino acid substitutions caused by positive natural selection. I think that Gillespie's theory is highly unrealistic in that it assumes that the numbers of episodes in different lineages (which must experience different environments) follow the same probability distribution. It is also highly problematical to assume that natural selection acts in such a way that the number of mutant substitutions per episode follows the same probability distribution for all episodes in all lineages. In his theory, natural selection is invoked arbitrarily to fit the data, while neglecting all the effects of the mutation rate, population size, and selective constraint. If it turns out that difference of evolutionary rates among lineages is mainly caused by differences of v_T/g—i.e., mutation rate in terms of base substitutions per site per year—Gillespie's 'episodic clock' theory breaks down completely. [. . .]

In conclusion, I would like to emphasize the importance of random genetic drift as a major cause of evolution. We must be liberated, so to speak, from the selective constraint posed by the neo-Darwinian (or the synthetic) theory of evolution. Wright, in his later years, used to claim that he had never attributed any significance to random drift except as an agent to bring about shift of adaptive peaks. As shown in Provine's recent book,[4] however, Wright in his papers of the early 1930s used to attach much more weight to random drift. Personally, I was mainly influenced by Wright's earlier papers, so that he is truly the forerunner in whose footsteps I have followed. I admire him very deeply.

[*Proc. of the National Academy of Sciences USA*, 88 (1991), 5969–73.]

[3] Gillespie is one of the best known critics of neutralism. See his book: J. H. Gillespie, *The Causes of Molecular Evolution* (New York: Oxford University Press, 1991).

[4] W. B. Provine, *Sewall Wright and Evolutionary Biology* (Chicago: University of Chicago Press, 1986).

Section C

Adaptation

The papers in this section deal with three main issues: what adaptation means; whether adaptations evolve gradually; and the importance of adaptation in living things.

The use of the word adaptation in biology differs from its ordinary language use in that it refers to a static condition. If someone refers to adaptation in a non-biological context, they are probably referring to the way an individual, or an institution, changes its behaviour to suit some conditions in the environment—adapting to new surroundings, new competitors, a new job. In the biological sense, that kind of adaptation has already happened in the past—by natural selection—and the word refers to the state of being well adjusted to the environment.

It is easy to point to things that are biological adaptations—our hands, eyes, hearts, digestive systems, anything without which we should live and reproduce less effectively. However, it turns out to be difficult to give a universally satisfactory criterion of what an adaptation is. Indeed, the biologists who do research on adaptation themselves do not agree about what adaptation means. The contributors to this section argue between two ways of defining adaptation. (There are further disagreements about the meaning of adaptation, but they are not covered here.) One, used by Williams (Chapter 15) and Grafen (Chapter 16), defines adaptation in terms of design: this is the engineering concept of adaptation, according to which adaptations are properties of organisms that look well designed in an engineering sense. An eye is optically well designed for vision; skeletal and muscular systems are well designed for support and powered movement; enzymes are chemically well designed as catalysts. One objection to this concept is that, although it is easy enough to apply to big and obvious adaptations like eyes, there can also be grey-zone cases where its application is awkward and subjective: granted that eyes are adaptations, what about the number of small black spots on a butterfly's hindwing, the number of bands on a snail, or fingerprint patterns? Supporters of a design concept would have something to say here, but the possible subjectivity, or at any rate difficulty, of applying the design concept in some cases, has led other biologists to prefer a definition in terms of relative reproductive success.

Reeve and Sherman (Chapter 17) define an adaptation as the variant among a set of alternatives that causes maximum reproductive success. The scientific merit of this criterion is its objectivity: it is based on a clear measurement protocol. It relates the study of adaptation to the kind of work contained in Section B, on selection in action. Grafen argues this is overly restrictive, or even a confusion. Although participants in this debate might sometimes see two wholly contrasting ways of thinking about the deep nature of biological organization, it is important also to realize that in many cases they will closely agree on whether something is an adaptation. After all, attributes that produce high reproductive success must also suit the organism for life, and vice versa. It is a technical controversy about how to recognize and imagine adaptations; it is not a controversy about whether adaptations exist, or can be known to exist, or about how adaptations evolve. Also, it is not a verbal squabble: active research scientists are not concerned with definition in a merely verbal sense.

To some extent, the definition that a biologist prefers will be associated with the kind of research he or she does. If you have a design concept of adaptation your research is likely to be to reverse-engineer the design; if you have a reproductive success concept, you are more likely to measure reproduction. Both kinds of work are valuable, where they are practicable. The two concepts come most strongly into contrast where one kind of research is possible but not the other, such as the examples of the eye described by Grafen. Also, Reeve and Sherman's definition ties adaptation to the Darwinian mechanism by which it evolves. We can then ask—suppose it is shown in the future that Darwinism is wrong, and 'adaptations' evolve by some other means: will it then be impossible to recognize adaptations? And how did people think of adaptation before Darwinism existed? (The extract from Hume (Chapter 62) supplies a religious relation for the design criterion of adaptation. Before Darwin, the existence of adaptive design in nature was one of the main theological arguments for the existence of God. Natural selection, however, explains the existence of design without any need for supernatural factors.)

The next theme is the question as to by what kind of steps adaptations evolve. Fisher (Chapter 14) argues for one extreme view that biologists find highly provocative, but also highly difficult to refute. Fisher argues that adaptations evolve in a large number of very small steps. There is an alternative extreme view, which could be called Goldschmidtian, according to which a new adaptation evolves in one big macromutational jump. (Orr and Coyne (Chapter 18) describe Goldschmidt's view.) Fisher's argument is enough to dispose of Goldschmidtism, and biologists do not take Goldschmidt's ideas seriously now. However, as Orr and Coyne say, rejection of Goldschmidt's view does not force us to accept Fisher's. It could still be that in the evolution of real adaptations there are some large steps, perhaps in addition to some small Fisherian steps. The question can be made more testable by asking about the magnitude of the effects of the genes that control the adaptation, and part of Orr and Coyne's paper not included here was concerned with that. There are classic case studies, such as butterfly mimicry, and an increasing amount of evidence from modern molecular genetic work on so-called 'quantitative trait loci', but we lack the space for either and only the conceptual positions are set out here.

For any adaptation you can ask whether it evolved by smooth 'hill-climbing', in which each step, even if infinitely small, takes you toward the adaptation, or whether some jumps are needed. In a way, part of Darwin's contribution was to show that characteristics such as the eye, which initially seemed to need to evolve in one big jump, can really evolve in many small stages if the alternatives are investigated carefully. Nilsson and Pelger's paper (Chapter 50) illustrates the point.

How important a concept is adaptation? The papers by Cain (Chapter 19) and by Gould and Lewontin (Chapter 20) provide contrasting views. For Cain, adaptation is one of the most important concepts in biology, organisms are well—even perfectly—adapted, and critics have been too quick to invent bogus alternatives to adaptation, and to dismiss characteristics as non-adaptive before they have done the hard work it would take to show it. For Gould and Lewontin, the importance of adaptation is exaggerated; its supporters fail to consider sound alternatives and invoke adaptation uncritically, often after only shabby research. Common ground can be found, however, and there is much deep understanding of life to be gained from thinking the issues through. One conclusion that I believe should not be drawn is that good research on adaptation is impossible. Adaptation can be studied both by measurements of reproductive success and by reverse engineering. Section B contains examples of one approach. Kettlewell's (Chapter 7) work shows how moth coloration is adaptive; Karn and Penrose's (Chapter 9) how birth weight is adaptive; and Gibbs and Grant's (Chapter 11) how beak size in Darwin's finches is adaptive. The reverse engineering approach is used in Section H on two baffling features of life: Medawar (Chapter 46) reverse-engineers senescence and Maynard Smith (Chapter 48) sex. Crick's (Chapter 47) and Sniegowski *et al.*'s (Chapter 52) papers are of a similar sort.

Finally, I have included an extract from Dawkins' famous book *The selfish gene* (Chapter 21). I selected the passage in which he argues that the gene is the unit of selection. The unit of selection matters for understanding adaptation; the unit of selection is the unit that adaptations evolve for the benefit of. Biologists have variously argued that selection acts on genes, on organisms, on groups of organisms, and on other units of life. Dawkins explains the reasoning used to find out which of these units is a true unit of selection. His argument has been widely influential (though it is not universally accepted). If his argument is right, then the key to understanding adaptations is to understand which versions of a gene will be favoured by natural selection.

R. A. FISHER

14 The nature of adaptation

Fisher describes a geometric model of adaptation. He uses it to argue that large genetic changes are less likely to improve the quality of adaptation than are small genetic changes. He illustrates the point with a microscope analogy. [Editor's summary.]

In order to consider in outline the consequences to the organic world of the progressive increase of fitness of each species of organism, it is necessary to consider the abstract nature of the relationship which we term 'adaptation'. This is the more necessary since any *simple* example of adaptation, such as the lengthened neck and legs of the giraffe as an adaptation to browsing on high levels of foliage, or the conformity in average tint of an animal to its natural background, lose, by the very simplicity of statement, a great part of the meaning which the word really conveys. For the more complex the adaptation, the more numerous the different features of conformity, the more essentially adaptive the situation is recognized to be. An organism is regarded as adapted to a particular situation, or to the totality of situations which constitute its environment, only in so far as we can imagine an assemblage of slightly different situations, or environments, to which the animal would on the whole be less well adapted; and equally only in so far as we can imagine an assemblage of slightly different organic forms, which would be less well adapted to that environment. This I take to be the meaning which the word is intended to convey, apart altogether from the question whether organisms really are adapted to their environments, or whether the structures and instincts to which the term has been applied are rightly so described.

The statistical requirements of the situation, in which one thing is made to conform to another in a large number of different respects, may be illustrated geometrically. The degree of conformity may be represented by the closeness with which a point A approaches a fixed point O. In space of three dimensions we can only represent conformity in three different respects, but even with only these the general character of the situation may be represented. The possible positions representing adaptations superior to that represented by A will be enclosed by a sphere passing through A and centred at O. If A is shifted through a fixed distance, r, in any direction its translation will improve the adaptation if it is carried to a point within this sphere, but will impair it if the new position is outside. If r is very small it may be perceived that the chances of these two events are approximately equal, and the chance of an improvement tends to the limit $\frac{1}{2}$ as r tends to zero; but if r is as great as the diameter of the sphere or greater, there is no longer any chance whatever of improvement, for all points within the sphere are less than this distance from A. For any value of r between these limits the actual probability of improvement is

$$\frac{1}{2}\left(1-\frac{r}{d}\right),$$

where d is the diameter of the sphere.

The chance of improvement thus decreases steadily from its limiting value $\frac{1}{2}$ when r is zero, to zero when r equals d. Since A in our representation may signify either the organism or its environment, we should conclude that a change on either side has, when this change is extremely minute, an almost equal chance of effecting improvement or the reverse; while for greater changes the chance of improvement diminishes progressively, becoming zero, or at least negligible, for changes of a sufficiently pronounced character.

The representation in three dimensions is evidently inadequate; for even a single organ, in cases in which we know enough to appreciate the relation between structure and function, as is, broadly speaking, the case with the eye in vertebrates, often shows this conformity in many more than three respects. It is of interest therefore, that if in our geometrical problem the number of dimensions be increased, the form of the relationship between the magnitude of the change r and the probability of improvement, tends to

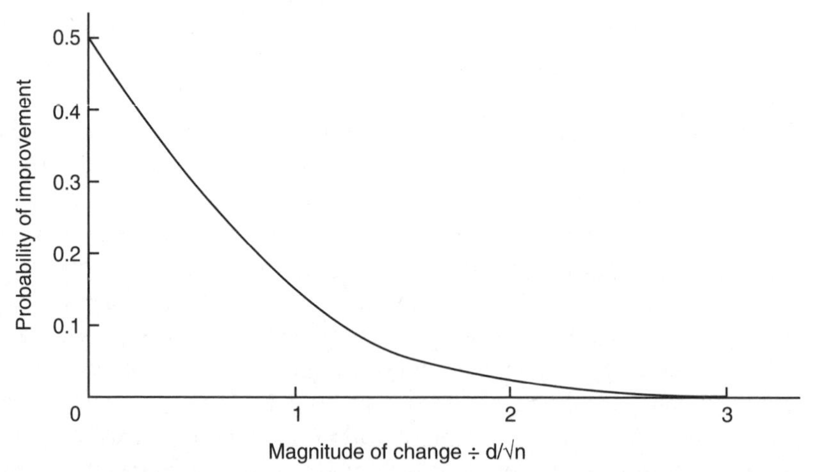

Figure 14.1 The relation between the magnitude of an undirected change and the probability of improving adaptation, where the number of dimensions (n) is large

$$p=\frac{1}{\sqrt{2\pi}}\int_{x}^{\infty}e^{-\frac{1}{2}t^{2}}\,dt,\; x=r\sqrt{n}/d.$$

a limit which is represented in Figure 14.1. The primary facts of the three dimensional problem are conserved in that the chance of improvement, for very small displacements tends to the limiting value $\frac{1}{2}$, while it falls off rapidly for increasing displacements, attaining exceedingly small values, however, when the number of dimensions is large, even while r is still small compared to d.

For any degree of adaptation there will be a standard magnitude of change, represented by $d/\sqrt{n}$, and the probability of improvement will be determined by the ratio which the particular change considered bears to this standard magnitude. The higher the adaptation the smaller will this standard be, and consequently the smaller the probability that a change of given magnitude shall effect an improvement. The situation may be expressed otherwise by supposing changes of a given magnitude to occur at random in all directions, and comparing the rates of evolutionary progress caused by two opposite selective agencies, one of which picks out and accumulates all changes which increase the adaptation, and another which similarly picks out and accumulates all which diminish it. For changes very small compared to the standard, these two agencies will be equally effective, but, even for changes of only one-tenth of the standard, the destructive selection is already 28 per cent. more effective than the selection favouring adaptation. At one half the standard it is over three and a half times as powerful, at the standard value itself, at which the probability of improvement is still, as the diagram shows, nearly one in six, the selection destroying adaptation is thirteen times as effective as that building it up, and at twice and three times the standard value the ratio has risen to the values 236 and 7,852 respectively.

The conformity of these statistical requirements with common experience will be perceived by comparison with the mechanical adaptation of an instrument, such as the microscope, when adjusted for distinct vision. If we imagine a derangement of the system by moving a little each of the lenses, either longitudinally or transversely, or by twisting through an angle, by altering the refractive index and transparency of the different components, or the curvature, or the polish of the interfaces, it is sufficiently obvious that any large derangement will have a very small probability of improving the adjustment, while in the case of alterations much less than the smallest of those intentionally effected by the maker or the operator, the chance of improvement should be almost exactly one half.

If therefore an organism be really in any high degree adapted to the place it fills in its environment, this adaptation will be constantly menaced by any undirected agencies liable to cause changes to either party in the adaptation. The case of large mutations to the organism may first be considered, since their consequences in this connexion are of an extremely simple character. A considerable number of such mutations have now been observed, and these are, I believe, without exception, either definitely pathological (most often lethal) in their effects, or with high probability to be regarded as deleterious in the wild state. This is merely what would be expected on the view, which was regarded as obvious by the older naturalists, and I believe by all who have studied wild animals, that organisms in general are, in fact, marvellously and intricately adapted, both in their internal mechanisms, and in their relations to external nature. Such large mutations occurring in the natural state would be unfavourable to survival, and as soon as the numbers affected attain a certain small proportion in the whole population, an equilibrium must be established in which the rate of elimination is equal to the rate of mutation. [. . .]

As to the physical environment, geological and climatological changes must always be slowly in progress, and these, though possibly beneficial to some few organisms, must as they continue become harmful to the greater number, for the same reasons as mutations in the organism itself will generally be harmful. For the majority of organisms, therefore, the physical environment may be regarded as constantly deteriorating, whether the climate, for example, is becoming warmer or cooler, moister or drier. [. . .] Probably more important than the changes in climate will be the evolutionary changes in progress in associated organisms. As each organism increases in fitness, so will its enemies and competitors increase in fitness; and this will have the same effect, perhaps in a much more important degree, in impairing the environment, from the point of view of each organism concerned. Against the action of Natural Selection in constantly increasing the fitness of every organism, at a rate equal to the genetic variance in fitness which that population maintains, is to be set off the very considerable item of the deterioration of its inorganic and organic environment. This at least is the conclusion which follows from the view that organisms are very highly adapted. Alternatively, we may infer that the organic world in general must tend to acquire just that level of adaptation at which the deterioration of the environment is in some species greater, though in some less, than the rate of improvement by Natural Selection, so as to maintain the general level of adaptation nearly constant.

[*The Genetical Theory of Natural Selection* (Oxford, OUP, 1930).]

G. C. WILLIAMS

15 Adaptation and natural selection

We begin by distinguishing the main function of an organ from other effects or incidental consequences of it. Some advantageous properties of an organism may be simple effects of physics or chemistry, rather than evolved adaptations. Adaptations can often be identified by analogy with human goals or designs. [Editor's summary.]

Difficulties in interpretation, especially with respect to the many supposedly group-related adaptations, may result from inappropriate criteria for distinguishing adaptations from fortuitous effects. They are also encouraged by imperfections of terminology. Any biological mechanism produces at least one effect that can properly be called its goal: vision for the eye or reproduction and dispersal for the apple. There may also be other effects, such as the apple's contribution to man's economy. In many published discussions it is not at all clear whether an author regards a particular effect as the specific function of the causal mechanism or merely as an incidental consequence. In some cases it would appear that he has not appreciated the importance of the distinction. In this book I will adhere to a terminological convention that may help to reduce this difficulty. Whenever I believe that an effect is produced as the function of an adaptation perfected by natural selection to serve that function, I will use terms appropriate to human artifice and conscious design. The designation of something as the *means* or *mechanism* for a certain *goal* or *function* or *purpose* will imply that the machinery involved was fashioned by selection for the goal attributed to it. When I do not believe that such a relationship exists I will avoid such terms and use words appropriate to fortuitous relationships such as *cause* and *effect*. [. . .]

Thus I would say that reproduction and dispersal are the goals or functions or purposes of apples and that the apple is a means or mechanism by which such goals are realized by apple trees. By contrast, the apple's contributions to Newtonian inspiration and the economy of Kalamazoo County are merely fortuitous effects and of no biological interest. [. . .]

A frequently helpful but not infallible rule is to recognize adaptation in organic systems that show a clear analogy with human implements. There are convincing analogies between bird wings and airship wings, between bridge suspensions and skeletal suspensions, between the vascularization of a leaf and the water supply of a city. In all such examples, conscious human goals have an analogy in the biological goal of survival, and similar problems are often resolved by similar mechanisms. Such analogies may forcefully occur to a physiologist at the beginning of an investigation of a structure or process and provide a continuing source of fruitful hypotheses. At other times the purpose of a mechanism may not be apparent initially, and the search for the goal becomes a motivation for further study. Adaptation is assumed in such cases, not on the basis of a demonstrable appropriateness of the means to the end but on the indirect evidence of complexity and constancy. Examples are (or were) the rectal glands of sharks, cypress 'knees,' the lateral lines of fishes, the anting of birds, the vocalization of porpoises.

The lateral line is a good illustration. This organ is a conspicuous morphological feature of the great majority of fishes. It shows a structural constancy within taxa and a high degree of histological complexity. In all these features it is analogous to clearly adaptive and demonstrably important structures. The only missing feature, to those who first concerned themselves with this organ, was a convincing story as to how it might make an efficient contribution to survival. Eventually painstaking morphological and physiological studies by many workers demonstrated that the lateral line is a sense organ related in basic mechanism to audition. The fact that man does not have this sense organ himself, and had not perfected artificial receptors in any way analogous, was a handicap in the attempt to understand the organ. Its constancy and complexity, however, and the consequent conviction that it must be useful in some way, were incentives and guides in the studies that eventually elucidated the workings of an important sensory mechanism.

I have stressed the importance of the use of such concepts as biological means and ends because I want it clearly understood that I think that such a conceptual framework is the essence of the science of biology. Much of this book, however, will constitute an attack on what I consider unwarranted uses of the concept of adaptation. This biological principle should be used only as a last resort. It should not be invoked when less onerous principles, such as those of physics and chemistry or that of unspecific cause and effect, are sufficient for a complete explanation.

For an example that I assume will not be controversial, consider a flying fish that has just left the water to undertake an aerial flight. It is clear that there is a physiological necessity for it to return to the water very soon; it cannot long survive in air. It is, moreover, a matter of common observation that an aerial glide normally terminates with a return to the sea. Is this the result of a mechanism for getting the fish back into water? Certainly not; we need not invoke the principle of adaptation here. The purely physical principle of gravitation adequately explains why the fish, having gone up, eventually comes down. The real problem is not how it manages to come down, but why it takes it so long to do so. To explain the delay in returning we would be forced to recognize a gliding mechanism of an aerodynamic perfection that must be attributed to natural selection for efficiency in gliding. Here we would be dealing with adaptation.

In this example it would be absurd to recognize an adaptation to achieve the mechanically inevitable. [. . .]

How, ultimately, does one ascertain the function of a biological mechanism? In this book I have assumed, as is customary, that functional design is something that can be intuitively comprehended by an investigator and convincingly communicated to others. Although this may often be true, I suspect that progress in teleonomy will soon demand a standardization of criteria for demonstrating adaptation, and a formal terminology for its description. [. . .]

Perhaps the main reason why biologists have not adopted a formal system for determining functional relationships is that many of the problems are so readily solved intuitively. We do not need weighty abstractions to help us decide that the eye is a visual mechanism. Also there are many helpful parallels between natural and artificial mechanisms, and it is so convenient as to be inevitable that parallel terminology be used. The close analogy between the lens of a camera and the lens of an eye make the term *lens* appropriate for both.

[*Adaptation and Natural Selection* (Princeton: Princeton UP, 1966).]

A. GRAFEN

16 Adaptation versus selection in progress

Adaptation, in the sense of a feature of an organism that has a function and evolved by natural selection (Williams, Chapter 15), should be distinguished from selection-in-progress—studies of changes in gene frequency driven by natural selection in modern populations. Reeve and Sherman (Chapter 17) defend a closer relation between the study of adaptation and of selection-in-progress. [Editor's summary.]

The distinction between adaptation and selection in progress is simple yet important. An adaptation in the sense of Williams[1] is a feature of an organism that can reasonably be said to serve a purpose and is the result of natural selection in the past. Selection in progress is gene frequencies changing now as a result of differences in design between genetically different individuals. An organism may have an adaptation even if selection is not operating on it now. I do not know whether genetic variation is currently affecting the eye in humans, and I do not need to know in order to recognize the eye as an adaptation, to study its function, and to analyze its adaptive value. On the other side, selection in progress may be modifying an existing adaptation, creating a new adaptation, or simply changing the value of a quantitative trait back and forth as generations proceed. A paleontological analogy can be made between adaptations and the bulk of

[1] Grafen refers to G. C. Williams, *Adaptation and Natural Selection* (Princeton: Princeton University Press, 1966), from which the previous extract (Chapter 15) is taken.

existing fossils, on the one hand, and between selection in progress and present-day corpses, some of which are currently being turned into fossils, on the other.

I am sure this distinction has been widely appreciated before, but I do not think it has been made explicitly in print, presumably because there was no need. Darwin was interested in selection in progress—partly because it was evidence for the mutability of species—and in adaptation. His theories of natural and sexual selection are still our only explanations for the existence of organic complexity and adaptation. Wright's[2] four-volume treatise on evolution is almost entirely concerned with selection in progress and is a fund of information about balanced polymorphisms, selection coefficients, effective population sizes, rates of gene substitution, linkage disequilibrium, dominance, and epistasis. Fisher's book, in contrast, is mainly concerned with understanding adaptation and treats selection in progress as an important but logically subsidiary topic. Its topics include mimicry, sex ratios, extravagant male characters, infanticide, and the heroic virtues. Kimura's neutral theory[3] is about what fraction of genetic variability can be attributed to selection in progress, as opposed to random drift and mutation pressure, and is all but irrelevant to the study of adaptation.

The study of adaptations begins with trying to answer the question why. Why do male red deer have antlers? Why are kingfishers brightly colored? Why are black grouse polygynous? These are the kinds of questions that have always been asked about animals, and the key to them was provided by Darwin. The comparative approach, the theory of evolutionarily stable strategies, and functional morphology are methods of studying certain kinds of adaptations. I suspect that most authors and most readers of this book[4] are interested in explaining adaptations.

The study of selection in progress is also fairly old. Animal breeders who keep track of their stocks are interested in selection in progress. The school of ecological genetics is devoted to the study of selection in progress in nature and has made many fascinating discoveries. To distinguish between different hypothesized modes of evolution—for example, the Fisher-Haldane mode of a succession of more-or-less independent gene substitutions and the Wrightian shifting balance—it is important to study selection in progress.

[2] Wright's four-volume treatise is: S. Wright, *Evolution and Genetics of Populations*, 4 vols (Chicago: University of Chicago Press, 1968–1978). Fisher's book, referred to one sentence down, is Fisher, R. A., *The Genetical Theory of Natural Selection* (Oxford: Oxford University Press, 1930; Variorum edition, 2000, published by Oxford University Press, Oxford). The contrast between the Fisher and Wright schools is a recurrent theme in this anthology.

[3] See Kimura, Chapter 13.

[4] 'This book', referred to here and at the end of the next paragraph, is the original book from which this extract was taken. The book was concerned with measurements of reproductive success in natural populations.

I believe that the authors and readers of this book are less interested in currently changing gene frequencies than in adaptations.

To illustrate the differences in the kind of study necessary to investigate these two distinct problems, I shall use as an example the spot number on the hind wing of *Maniola jurtina*, a character much studied by ecological geneticists. The present book is about measuring the reproductive success of individuals, a technique the ecological geneticists did not use. The experiments I propose will therefore be hypothetical, and I do not wish to suggest that they are superior to those in fact used.

Suppose first that we wish to discover the adaptive significance of spot number. The obvious experiment is to paint spots on or off the hind wings and to compare (for example) predation, mating success, and thermo-regulation in the groups with different numbers of spots. If we found that spottier butterflies were eaten less often but had the same mating success and temperature control, we could conclude that the function (or better, *a* function) of the spots was to avoid predation. LRS[5] is not very useful here because it is too all-encompassing a measure. We wish to know *why* the butterfly has the spots, not *how much* more successful more spotted individuals are than less spotted. To understand the adaptive significance of spot number, we want to pin down more exactly the mechanism of advantage. We can make a start by finding if spot number correlates with components of LRS, hoping to find where it is useful to look more closely for the reasons behind the advantage of spots.

If, on the other hand, we are interested in selection in progress on spot number, then we are looking for evidence of gene-frequency changes at the loci that affect spot number. It would be pointless in this case to create variation by painting spots. Ford and colleagues measured the frequency of adult morphs in successive generations and sought to exclude the other possible causes of the observed changes. With data on LRS of individuals, we could find the covariance between LRS and natural spot number. According to the 'secondary theorem of natural selection' of Robertson, the selective change in a character is equal to its genetic covariance with LRS divided by mean LRS. The genetic covariance is equal to the phenotypic covariance (the one we observe) multiplied by the heritability of the character. So by showing that the heritability was not zero, which is to say there is additive genetic variance for spot number, the covariance between spot number and LRS could be used to demonstrate that there was selection in progress at the loci affecting spot number. It would not demonstrate that spot number was part of the causal chain from genes to differential success of individuals, for it could be another, pleiotropic, effect of the genes that determine spot

[5] LRS: lifetime reproductive success or the number of offspring left in a lifetime.

number. In fact, Ford reports that the selection they detected by measuring spot numbers was probably the result of differential parasitism of the butterfly larvae by a hymenopteran. The reason for the correlation between spot number as an adult and susceptibility to parasite attack as a larva is not known.

We have examined distinct ways of studying the two distinct problems. One particularly important difference between them is the kind of variation exploited. It was simplest, though not necessary, to use artificial variation in the study of adaptation, and it was necessary to use natural variation to study selection in progress.

The distinction between adaptation and selection in progress does not mean that there are no connections between them. One obvious connection is that current adaptations are the result of selection that was in progress at some time in the past. Another connection arises in some modern theories of sexual selection and the maintenance of sexual reproduction. Hamilton and Zuk proposed that sexual selection is a defense against certain kinds of parasitism, in which females choose comparatively unparasitized males so that their offspring will in turn be comparatively resistant to parasites. This is a good case for illustrating the distinction between adaptation and selection in progress, because an adaptation in one character (female choice) is based on the continuing existence of selection in progress not in itself, but in another character (resistance to parasitism).

['On the Uses of Data on Lifetime Reproductive Success'. In T. H. Clutton-Brook (ed.), *Reproductive Success* (Chicago, University of Chicago Press, 1988), 454–71.]

H. K. REEVE AND P. W. SHERMAN

17 An operational, nonhistorical definition of adaptation

A definition of adaptation is given and three features of the definition are described. The definition—adaptations are phenotypic variants conferring the highest fitness among sets of variants—comes close to requiring research on selection-in-progress and partly contrasts with the perspective of Grafen (Chapter 16). [Editor's summary.]

We now propose a simple definition of adaptation that captures the essential research motives of evolutionary biologists interested in questions of phenotype existence: [. . .] *An adaptation is a phenotypic variant that results in the highest fitness among a specified set of variants in a given environment.* This definition undoubtedly strikes some readers as being too simplistic. Therefore its key features require discussion and defense.

First, we emphasize that adaptation is a *relative* concept, defined only in relation to explicit alternatives. This is because natural selection sorts among the phenotypic alternatives available each generation, with the individuals exhibiting the greatest relative reproductive success contributing disproportionately to subsequent generations. As Williams put it,[1] 'Selection has nothing to do with what is necessary or unnecessary, or what is adequate or inadequate, for continued survival. It deals only with an immediate better-vs.-worse within a system of alternative, and therefore competing, entities'. Natural selection is a little like a game of poker: The best hand (phenotype) wins (reproduces) regardless of whether it is a pair of twos or four aces.

A second feature of our proposed definition is that it is in a sense *genotype-free*. By this we mean that it refers only to phenotypic features and not necessarily to genotypes at specific loci. We take this approach for two reasons. First, it does not really matter for our definition precisely how genotypes are connected to phenotypes, as long as there is some connection. Second, a given phenotype might be produced by a variety of genotypes. As a consequence of the many-to-one relationship between genotypes and the phenotype, there are numerous ways the same evolutionarily stable state can be produced. For example, a selectively favored trait A might spread because allele *a* arose at locus X, or because allele *a′* arose at locus Y, and so forth. Whether a given trait resulted from locus X, Y, or some other locus is immaterial. What does matter is that the overall probability of seeing a favored trait increases with the number of loci that can generate it. Thus our definition focuses on phenotypic features that may be produced by multiple genotypes, each potentially experiencing a different history, which nonetheless may converge on a small number of identifiable stable states.

A third feature of our proposed definition is that it is in an important sense *history-free*. There is no reference to a specific historical process or evolutionary mechanism that leads to the predominance of the most adapted member(s) of the phenotype-set. We take this approach because natural selection sorts among existing variants every generation without regard to their prior states. To put this point another way: *Whatever is important about a trait's history is already recorded in the environmental context and the biological attributes of the organism.*

['Adaptation and the Goals of Evolutionary Success',
Quarterly Review of Biology, 68 (1993), 1–32.]

[1] G. C. Williams, *Adaptation and Natural Selection* (Princeton: Princeton University Press, 1966). The quote is from p. 31, which is not included in Chapter 15.

H. ALLEN ORR AND JERRY A. COYNE

18 The genetics of adaptation: a reassessment

The extract begins by noting that most modern evolutionary biologists assume adaptive evolution to proceed by a large number of small genetic steps. Orr and Coyne argue that the theoretical and empirical basis for this assumption is unconvincing. (Only the theoretical part of their argument is included in this extract.) They distinguish a strong from a weak version of 'macromutationism'. They then argue that Fisher's theory (Chapter 14) is incomplete. Adaptive evolution in large genetic steps is more plausible when all relevant factors are taken into account. [Editor's summary.]

It is a tenet of evolutionary biology that adaptations nearly always result from the substitution of many genes of small effect. This view, which we call 'micromutationism,' originated with Darwin, who clearly believed that adaptations were constructed from innumerable subtle variations:[1] 'We have many slight differences which may be called individual differences. . . . These individual differences are highly important for us, as they afford materials for natural selection'. Indeed, to Darwin, the essence of natural selection was that 'extremely slight modifications in the structure or habits of one inhabitant would often give it an advantage over others'.

Shortly after publication of *On the Origin of Species*, however, this view was challenged by T. H. Huxley:[2] 'Mr. Darwin's position might . . . have been even stronger than it is if he had not embarrassed himself with the aphorism, "Natura non facit saltum," which turns up so often in his pages. We believe . . . that Nature does make jumps now and then, and a recognition of the fact is of no small importance in disposing of many minor objections to the doctrine of transmutation.' Continuing over the next 70 years, this debate evolved into a conflict between Mendelian geneticists and biometricians, with the former emphasizing the importance of large mutations in evolution and the latter emphasizing the gradual, imperceptible transformation of species by natural selection.

This controversy was supposedly resolved during the modern synthesis, in which there was general agreement that, although adaptive evolution was based on Mendelian genes, many alleles of small effect were involved. Dobzhansky,[3] for example, argued that interracial and interspecific differences are 'caused in a majority of cases by cooperation of numerous genes, each of

[1] Both quotations from C. Darwin, *The Origin of Species* (London: John Murray, 1859).

[2] T. H. Huxley, 'The origin of species', *Westminster Review*, 17 (1860), 541–70. (Reprinted in D. L. Hull (ed.), *Darwin and His Critics* (Cambridge, Mass.: Harvard University Press, 1973).)

[3] T. Dobzhansky, *Genetics and the Origin of Species* (New York: Columbia University Press, 1937), 26.

which taken separately has only slight effects on the phenotype.' Similarly, J. Huxley[4] concluded that the 'detailed analysis of the last ten or fifteen years . . . has revealed large numbers of gene-differences with extremely small effects, down almost to the limit of detectability. It is not only possible but highly probable that among these are to be sought the chief building-blocks of evolutionary change, and that it is by means of small mutations, notably in the form of series of multiple allelic steps, each adjusted for viability and efficiency by recombinations and further small mutations, that progressive and adaptive evolution has occurred.' These views were also voiced by Muller, Wright, and Mayr.

Although a few biologists have suggested an evolutionary role for mutations of large effect, the neo-Darwinian view has largely triumphed, and the genetic basis of adaptation now receives little attention. Indeed, the question is considered so dead that few may know the evidence responsible for its demise.

Here we review this evidence. We conclude—unexpectedly—that there is little evidence for the neo-Darwinian view: its theoretical foundations and the experimental evidence supporting it are weak, and there is no doubt that mutations of large effect are sometimes important in adaptation.

We hasten to add, however, that we are not 'macromutationists' who believe that adaptations are nearly always based on major genes. The neo-Darwinian view could well be correct. It is almost certainly true, however, that some adaptations involve many genes of small effect and others involve major genes. The question we address is, *How often* does adaptation involve a major gene? We hope to encourage evolutionists to reexamine this neglected question and to provide the evidence to settle it.

What are macromutations?—The micromutational view of Darwin, Fisher, and others is clear: adaptations arise by allelic substitutions of slight effect at many loci, and no single substitution constitutes a major portion of an adaptation. There are, in contrast, at least two forms of macromutationism. The first is exemplified by the extreme saltationism of Goldschmidt: single 'systemic mutations' produce important, complex adaptations in essentially perfect form (Goldschmidt believed that systemic mutations were chromosomal rearrangements). As Charlesworth notes, this 'strong' version of macromutationism is almost certainly wrong. It is highly unlikely that a single mutation could create adaptations as complex as eyes or legs, much less new taxa differing by many adaptations.

The second form of macromutationism posits that adaptation often involves one or a few alleles of large effect. Although these alleles do not produce perfect adaptations by themselves, they are responsible for a large portion of the adaptation. This 'weak' version, which is more realistic than

[4] J. S. Huxley, *Evolution: The Modern Synthesis*, 2nd edn. (London: Allen & Unwin, 1963), 115.

Goldschmidt's view, is the form of macromutation we consider in the rest of this article. Although the term 'macromutationism' has unfortunate historical connotations, we use it for lack of a better word.

It should be noted that we are not concerned with whether most substitutions in DNA between species have small or large effects on fitness. Because of their large number, most of these substitutions must have small effects on fitness. Instead, we ask whether one or two genes may often be responsible for most of the increase in fitness during the evolutionary change of a character. In short, do most adaptations involve fixation of major genes?

The theory

The primary theoretical argument for micromutationism [. . .] is the brief discussion in the second chapter of Fisher's *The Genetical Theory of Natural Selection*.[5] Fisher offered a mechanical analogy, arguing that adaptation invovles 'conformity of parts.' Each component of an adapted organism, like that of a complex machine, must successfully interact with many other parts. It is obvious, he argued, that any random change in a finely tuned machine is likely to worsen its functioning, especially if the change is large. A random change in a reasonably well-focused microscope, for example, will almost always worsen the focus, but an extremely small adjustment is far less likely to harm the focus and may even improve it. Fisher formalized this argument with a geometrical analogy. If the arrangement of parts conferring the highest fitness is represented by the center of a sphere, the species' present position can be represented by some point on the sphere's surface. Any displacement inside the sphere improves adaptation while any movement outside worsens it. Fisher shows that a trivially small displacement in a random direction has a 50% chance of entering the sphere and improving the adaptation while much larger displacements have a negligibly small chance of entering the sphere.

To Fisher, the lesson was simple: large mutations have a very small chance of being favorable. Adaptation must therefore be based on many gene substitutions of individually small effect. As Turner argues, Fisher held an extreme form of micromutationism, believing that adaptations were based on *innumerable* loci of very small effect. This led to Fisher's belief that mutation plays little or no creative role in evolution but that natural selection shapes adaptations out of an infinite supply of very small mutations.

There are, however, several problems with Fisher's argument. First, as Kimura points out, Fisher shows only that small mutations are more likely to be favorable, not that they are more likely to be substituted. These two

[5] Chapter 14 in this Reader.

possibilities have often been confused, as in Simpson's[6] assertion that the 'chance that a mutation will be favored by selection and the chance that it will or can be integrated into a genetic system . . . are inversely related to the effect of the mutation.' But these probabilities are not identical. The substitution rate of a class of mutations depends not only on its chance of being advantageous but also on its mutation rate and its probability of fixation once it has arisen. As Kimura points out, the substitution rate, $k(x)$, for new mutations of phenotypic effect x is proportional to

$$m(x)P_a(x)\,2x, \tag{1}$$

where $m(x)$ is the rate of mutation to alleles of effect x, $P_a(x)$ is Fisher's probability that a mutation of phenotypic effect x is favorable, and $2x$ is proportional to the probability of fixation of such a mutation (assuming no dominance and that the selection coefficient of a favorable mutation is proportional to its phenotypic effect). The important point is that a mutation's probability of fixation is directly proportional to its phenotypic effect. Thus, Kimura argues, even if large mutations are less likely to be favorable than small ones, they are nevertheless fixed more easily. Taking both factors into account, Kimura concludes that mutations of more *intermediate* phenotypic effect might be the most frequent components of adaptations. (Although Kimura's argument assumes that all mutations are unique, there is no alternative theory allowing for recurrent mutation.)

Fixation of a large mutation also, of course, contributes more to the trait under selection than does fixation of a small mutation. Thus, even if large mutations were fixed less often than smaller ones, they might still account for the overwhelming majority of the response to selection. [. . .]

Second, it is important to realize that Fisher's argument does not apply in cases in which an adaptive landscape has more than a single peak. Fisher viewed adaptation as an endless slog up a single adaptive peak that slowly moves about as the environment changes. Not surprisingly, he ignored the possibility that a large phenotypic change could place its bearer near a distant but unoccupied adaptive peak. We do not know, of course, how often species are within striking distance of only one adaptive peak. There are certainly cases in which a species is suddenly thrown far from its old optimum, as when radically different habitats or distant islands are colonized. The optimum can also shift in large but less dramatic ways, as when a species suddenly encounters new predators, parasites, competitors, or diseases.

[6] G. G. Simpson, *The Meaning of Evolution* (New Haven: Yale University Press, 1949), 234. The source for Kimura is his 1983 book *The Neutral Theory of Molecular Evolution* (Cambridge: Cambridge University Press).

Third, Fisher's model assumes that all phenotypic dimensions count equally: whenever part of an organism interacts with other parts, each interaction has the same effect on fitness. A mutation affecting body size might, for example, also affect foraging efficiency, thermoregulation, fecundity, and mobility. According to Fisher, a change in any one of these characters would affect fitness as much as a change in any other. However, conformity in some dimensions would surely have greater effects on fitness than conformity in others. A change in body size, for example, might affect fitness far more through fecundity than through thermoregulation. Relaxing Fisher's assumption reduces the 'effective number of dimensions' and so increases the chance that a large mutation would be favorable.

Finally, as Maynard Smith points out, the deleterious effects of a major mutation might be dampened by the developmental system of organisms, which have probably evolved to minimize the consequences of developmental accidents.

In summary, while Fisher provides some reason for believing that small mutations are more likely to be favorable than larger ones, he did not convincingly show that adaptation must usually involve substitution of these small mutations.

[*American Naturalist*, 140 (1992), 725–42.]

A. J. CAIN

19 The perfection of animals

Cain criticizes the view that special features of small taxa may be adaptations but that general features of large taxa are non-adaptive. (For instance, the particular colour pattern of an individual butterfly species might be conceded to be an adaptation. The pentadactyl (five-digit) limb in all tetrapods (amphibians, reptiles, birds, mammals) is an example of a general character in a large taxon, and often said to be non-adaptive.) Cain traces the history of the view he opposes to Owen and Darwin. He counter-argues that selection is powerful enough to re-organize characters of large taxa if they are non-adaptive. He looks at evidence from ecological genetic studies of modern populations, from the convergent evolution of faunas such as the mammals of South America and mammals elsewhere on Earth, and from the functional anatomy of general characters, such as the hard exoskeleton of arthropods. [Editor's summary.]

It is commonly thought at the present day [. . .] that most of the particular features of any animal are adaptive to its particular mode of life, but its general plan which it shares perhaps with an enormous number of other forms cannot be adaptive to a particular mode of life, and therefore must be due to its ancestry. This is almost certainly incorrect for four reasons. First, the belief has arisen as a carry-over from a previous epoch of ideas which are not valid. Second, it is gradually being realized that if we personally cannot see any adaptive or functional significance of some feature, this is far more likely to be due to our own abysmal ignorance than to the feature being truly non-adaptive, selectively neutral or functionless. Third, everything that is known of the power of natural selection and the nature of evolution strongly suggests that there has been ample time for the complete reconstruction of the older groups to make them better adapted to their modes of life if this had been necessary; their remarkable constancy of plan combined with plasticity in pretty well every detail of that plan over hundreds of millions of years almost forces us to the conclusion that they are as they are because that is what, in competition with all the other great groups, they need to be. And last, some direct evidence is now being obtained of the highly adaptive nature of features characterizing some major groups.

This is not, of course, to say that all animals are perfectly adapted for their present modes of life. The environment is always changing, and populations cannot adapt instantaneously. You cannot change mice into men by selection in ten generations, and a major overhaul may take a long time. But if we allow a time-lag of twenty million years (for the large animals, and no doubt much less for very small ones) or even sixty million for the clear establishment of a major change, for reasons suggested below, we still have plenty of time from the Cambrian to the present day to completely remodel the older groups if it were necessary.

The thesis I wish to put forward, therefore, is that broadly speaking, the major plans of construction shown by the older groups are soundly functional and retained merely because of that. The phyla and classes are the main possible ways of living in the face of competition from each other. Their plans are adaptive for broad functional specializations; the particular features of lesser groups are, as has long been agreed, adaptive for more particular functions. [. . .]

Owen on the nature of limbs

The unfortunate personal relationship between Owen and Darwin has often been commented on, but almost nothing has been said on the far more important subject of what Darwin took over from Owen, agreed to, and reinterpreted in evolutionary terms.

Owen [. . .] begins by giving examples of the adaptive modification of the pentadactyl limb—the 'fin' of the dugong, forelimb of the mole, wing of the bat, fore and hind legs of the horse, grasping limbs of monkeys, and finally the manipulatory forelimbs and ambulatory hindlimbs of Man. [. . .]

He points out how remarkable it is that in the tetrapods generally there are never more than five digits on each limb, and that when the number is reduced, one can still see exactly which digits have been lost and which retained. [. . .] His solution is given in a particularly clear and graphic passage:[1] 'Something also I would fain add with a view to remove or allay the scruples of those who may feel offended at any expressions that seem to imply that any part or particle of a created being could be made in vain.

Those physiologists who admit no other principle to have governed the construction of living being than the exclusive and absolute adaptation of every part to its function, are apt to object to such remarks as have been offered regarding the composition of the skeleton of the whale's fin and of the chick's head, that "nothing is made in vain"; and they deem that adage a sufficient refutation of the idea that so many apparently superfluous bones and joints should exist in their particular order and collocation in subordination to another principle; conceiving, quite gratuitously in my opinion, the idea of conformity of type to be opposed to the idea of design.

But let us consider the meaning which in such discussions is commonly attached to the phrase "made in vain". Were the teleologist to analyse his belief in the principle governing organization, he would, perhaps, find it to mean, that so far as he can conceive of mechanism directly adapted to a special end, he deems every organic mechanism to have been so conceived and adapted. In a majority of instances he finds the adaptation of the organ to its function square with his notions of the perfection of a machine constructed for such an end; and in the exceptional cases, where the relation of the ascertained structure of an organ is not so to be understood, he is disposed to believe that that structure may be, nevertheless, as directly needed to perform the function, although he perceives that function to be a simple mechanical action, and might conceive a more simple mechanism for performing it. The fallacy perhaps lies in judging of created organs by the analogy of made machines; but it is certain that in the instances where that analogy fails to explain the structure of an organ, such structure does not exist "in vain" if its truer comprehension lead rational and responsible beings to a better conception of their own origin and Creator . . . the recognition of an ideal Exemplar for the Vertebrated animals proves that the knowledge of such a being as Man must have existed before Man appeared. For the Divine mind which planned the Archetype also foreknew all its modifications.'

Only the shallowest mind can believe that in a great controversy one side is mere folly. Owen was no fool, and the strength of his case was apparently great. Yet it would seem that, as can happen to the wisest men, he was wrong, not so much because of his chain of reasoning, but because of the inadequacy of the information he was basing it on. [. . .] I think there is no doubt, and it is no great criticism of him, that his mind was so taken up with

[1] R. Owen, *On the Nature of Limbs* (London, 1849), 84–6.

his anatomy and the important philosophical issues it raised that he just did not think sufficiently about the actual way in which limbs are used. What he could see was adaptive, he duly recognized as such; what he could not, instead of reflecting on the need for further information before coming to a decision, he decided must be archetypal, and he used it for edification. Owen's attitude is still a very usual one, except that what is not evidently adaptive is described as ancestral, not archetypal. [. . .]

Darwin's interpretation

The effect of Owen's arguments on Darwin is clearly set out in the section on morphology in Chapter 13 of 'On the Origin of Species'.

'What can be more curious than that the hand of man, formed for grasping, that of a mole for digging, the leg of the horse, the paddle of the porpoise, and the wing of the bat, should all be constructed on the same pattern, and should include the same bones, in the same relative pattern? . . . Nothing can be more hopeless than to attempt to explain the similarity of pattern in members of the same class by utility or by the doctrine of final causes. The hopelessness of the attempt has been expressly admitted by Owen in his most interesting work on the "Nature of Limbs". On the ordinary view of the independent creation of each being, we can only say that so it is—that it has so pleased the Creator to construct each animal and plant.' [. . .]

Darwin continues, 'The explanation is manifest on the theory of the natural selection of successive slight modifications—each modification being profitable in some way to the modified form, but often affecting by correlation of growth other parts of the organization. In changes of this nature there will be little or no tendency to modify the original pattern, or to transpose parts. . . . If we suppose that the ancient progenitor, the archetype as it may be called, of all mammals, had its limbs constructed on the existing general plan, for whatever purpose they served, we can at once perceive the plain signification of the homologous construction of the limbs throughout the whole class. . . . Nevertheless, it is conceivable that the general pattern of an organ might become so much obscured as to be finally lost, by the atrophy and ultimately by the complete abortion of certain parts, by the soldering together of other parts, and by the doubling or multiplication of others—variations which we know to be within the limits of possibility. In the paddles of the extinct gigantic sea-lizards, and in the mouths of certain suctorial crustaceans, the general pattern seems to have been thus to a certain extent obscured.'

Darwin, therefore, originated the evolutionary interpretation which has been followed ever since, that the general plan of the pentadactyl limb is not now adaptive, although it must have been in the common ancestor, but its modifications are adaptive. In the course of evolution the plan has been modified in different ways in different groups of mammals but has been retained as the substratum on which in each evolutionary line and in every life-history the modifications are imposed. In general, the plan or archetype common to all the diversely adapted members of a given group cannot itself be adaptive for any one mode of life, and is clearly there only by inheritance. In the section in Chapter 6 of the 'Origin' dealing with 'Organs of little apparent importance' he makes the very just remark that 'we are much too ignorant in regard to the whole economy of any one organic being, to say what slight modifications would be of importance or not'. But he states in the same section that 'the chief part of the organization of every being is simply due to inheritance; and consequently, though each being assuredly is well fitted for its place in nature, many structures now have no direct relation to the habits of life of each species. Thus, we can hardly believe that the webbed feet of the upland goose or of the frigate-bird are of special use to these birds; we cannot believe that the same bones in the arm of the monkey, in the foreleg of the horse, in the wing of the bat, and in the flipper of the seal, are of special use to these animals. We may safely attribute these structures to inheritance. But to the progenitor of the upland goose and of the frigate-bird, webbed feet no doubt were as useful as they now are to the most aquatic of existing birds. So we may believe . . . that the several bones in the limbs of the monkey, horse, and bat, which have been inherited from a common progenitor, were formerly of more special use to that progenitor, or its progenitors, than they now are to these animals having such widely diversified habits. . . . Hence every detail of structure in every living creature (making some little allowance for the direct action of physical conditions) may be viewed, either as having been of special use to some ancestral form, or as being now of special use to the descendants of this form—either directly, or indirectly through the complex laws of growth.' [. . .]

Insecurity of this interpretation

It is clear from the above quotations that Darwin was much too impressed by Owen's ideas. Translated directly into evolutionary terms they seemed to explain so much that was otherwise wholly obscure. [. . .] As Darwin said, most of the features of any organism would be 'simply due to inheritance'—and those that had been nearly lost would be vestigial organs. [. . .]

But there are two reasons for apparent obscurity of function; one is indeed that a function has changed or been abolished without the structure serving it having changed to correspond; the other is that we simply do not know enough to say anything. Moreover, in the passages quoted above from the 'Origin', Darwin is confusing two rather different classes of phenomena. It may perhaps be that we can see, and state positively, from the known habits of the upland goose of South America or the frigate bird that they *never* use their webbed feet for swimming and the webs are of absolutely no other use to them. One would need a pretty comprehensive study of their life-histories before saying anything so definite. But let it be allowed that this is so; then these would be examples of a present divergence of structure and function explicable on the theory of evolution by a recent change in mode of life, and valid evidence (which is what Darwin was looking for) against any theory of fixity of species.

However, where we are dealing with structures which have persisted for hundreds of millions of years in hundreds of billions of individual life-histories, and which are still so little understood from a functional point of view, it is a very rash assertion that they are merely ancestral. Such, for example, is the pentadactyl limb. Owen's arguments are so clearly unsound precisely in this matter of actual present-day function. The flipper of a seal, for example, is not used merely as a simple flat plane: it executes complicated movements during swimming involving bending both along and across the axis. It is still used to some extent for movement on land. The use of the ends of the digits, when bent, for scratching may be of great importance in dislodging settlers. Adaptations need not be one hundred per cent necessary for individual survival to be called adaptations, a point not well appreciated even recently, when adaptive structures have been 'explained' as the mere by-products of physiological processes because they were thought to be not absolutely necessary, and therefore not really adaptations. [. . .]

Darwin's conversion of Owen's idea to an explanation of so many features of animals was much too facile. It was based far more on ignorance of actual function than on positive knowledge, and we need not wonder, therefore, that he and Owen (and the 'Naturphilosophen') assessed the imperfection of animals, the degree to which they are not adapted for their mode of life, as far higher than anyone else had done for centuries. For some features he was surely right; those beetles on oceanic islands that have the elytra immovable so that the wings, which are present and apparently well-formed are not usable, are perhaps as good an example as any of a change in requirements rendering useless an important structure which persists (as yet) by simple inheritance. But to extend such an explanation to major features of great groups is not permissible without further evidence. Every fresh piece of work that bears on function at all shows us again and again functional significance where we might not have expected it and highlights our vast ignorance about almost all living things. Often when I have been putting forward this point of view, I have been asked 'What, then, is your explanation of such and such a structure?' and if I could not reply, the whole viewpoint was rejected. But this is merely to repeat the Owenian error; the interpretation of the course and nature of evolution is not to be based on what one individual happens not to know. Nevertheless, there is a correct feeling behind it that some positive evidence should be forthcoming. Since the Darwinian point of view has been accepted so generally by those who have interested themselves in evolution and systematics, and the whole subject has been merely ignored by many others, direct evidence of the functional significance of characters of major groups has not yet been systematically searched for. However, there is some, and the indirect evidence is considerable and cogent.

The power of natural selection

We owe to Fisher, Haldane and Sewall Wright the development of the mathematics of population genetics. Perhaps the most remarkable single conclusion is the enormous power of only a few per cent of selection to determine gene-frequencies. Except in very restricted circumstances, mutation pressure and random sampling errors can play only a very minor role, but migration, of course, if massive, can have an overwhelming effect. At the time when Sewall Wright pointed out that random processes might have a considerable effect in very small populations, it was widely believed that many characters were neutral or non-adaptive, and he helpfully suggested that genetic drift might be responsible for them. As I have pointed out, there was no real basis for the idea of their neutrality. Looking at skins of closely related species of bird reposing on a museum tray, one might well be at a loss to produce an explanation for some of their interspecific differences; but a stuffed bird on a museum tray is not in the best position to show what it does with its characters. Extensive field work may be necessary before their significance is realized, even if it is great. In fact, in every case which has been carefully examined, the supposed influence of random drift, postulated on the basis of insufficient knowledge, has been greatly reduced or actually disproved for the characters under consideration. This has now happened with very diverse organisms and characters. The chromosomal inversions of *Drosophila pseudoobscura* thought by Dobzhansky and Queal to show drift from population to population were shown on more careful analysis by Wright and Dobzhansky to be responding remarkably to temperature, and the beautiful work of Dobzhansky and his school since has emphasized the extraordinary complexity of selective forces which may be acting, both within the genotype and in the external environment, on a given inversion. Similarly, it was widely proclaimed that the human blood groups must be of no selective significance, and therefore could be used as markers for the study of human migration; but this conclusion was merely due to insufficient information. The *medionigra* gene in a colony near Oxford of the moth *Panaxia dominula* originally studied by Fisher and Ford was shown by them to be highly subject to selection although in a small population fluctuating greatly in size; their work has been extended by Sheppard and Sheppard and

Cook both in the original colony and in artificial ones. Sheppard has shown that the gene is associated with non-random mating, and Williamson has discussed its maintenance in the original population. What might seem to be a trivial and entirely neutral alteration in the colour pattern is associated with strong selection of more than one sort, ample to determine its frequency even in small populations.

In the case of the banding and colour varieties of the shell in the snails *Cepaea nemoralis* and *hortensis* it had been confidently asserted that it could not matter to a snail whether it had one band on its shell or two. Cain and Sheppard, Cain, and Sheppard were able to show that definite visual selection was exerted by predators in *C. nemoralis* and that some strong non-visual selection must also be acting to maintain the polymorphism in face of this visual pressure. Clarke has shown similar selection in *C. hortensis* and proposed an additional mode of visual selection. Lamotte (and other workers reviewed by him) has produced evidence of differential physiological response of the morphs to heat and cold. Cain and Currey have now shown that considerable differences from place to place in the non-visual selective forces controlling the balance of the polymorphism are likely. Clarke and Murray have been able to use a very careful survey made in 1926 by Captain C. Diver and the late Professor A. E. Boycott of morph frequencies in *Cepaea nemoralis* on the sand-dunes at Berrow (Somerset). By repeating the survey and comparing results, they have demonstrated very considerable selection in populations at first sight varying at random. This is not to say that random processes have no effect on the morph frequencies in snail populations; Goodhart has described a situation in which flooding may well have been responsible for considerable local changes. But these studies do show, as do so many other studies on genes in the wild, that merely to fail on a casual inspection to see any selective significance in a particular variation does *not* license the observer to proclaim that there is none. And more positively, they show, as indeed Fisher had done many years before on analysing the data of Nabours on the grouse locusts *Apotettix eurycephalus* and *Paratettix texanus*, that very considerable selection coefficients (even up to 50 per cent in some combinations of genes investigated by Fisher) are actually found to act in the wild. As Sheppard has pertinently remarked, we need both the mathematical models and some knowledge of what actually goes on in the wild to determine the power of selection. [. . .]

But, it will be rightly said, to show that all characters are determined by selection does not show that they are adaptive. Dobzhansky has particularly urged that very many characters are mere by-products of others that are selected for, and in themselves of no selective value or even somewhat deleterious. He reminds us that genes do not determine characters in a simple one–one correspondence; the organism is a complex and integrated whole and the alteration of the action of any gene is likely to produce all sorts of changes (pleiotropic effects) throughout the phenotype. To single out any one of these as 'the' action of the gene is incorrect. He gives as an example the three orbital bristles found in all of the more than 600 known species of the genus Drosophila, the most anterior of which is always proclinate (bent forwards) and the other two reclinate (bent backwards).[2] 'Now, why should this character be retained so tenaciously in so many species? Is it really important for the flies of this genus to have one proclinate and two reclinate orbital bristles? . . . When one considers traits in which species of insects and other organisms often differ, such as the differences between Drosophila species mentioned above, the supposition that all or even most of them are directly useful to their possessors stretches too much one's credulity . . . In fact, some Drosophila mutants have one or more of the orbital bristles missing, and the mutant flies seem to suffer no inconvenience on this account. But the processes which result in the formation of certain bristles may give rise also to the other traits, morphological and physiological, in the same organism. The proclinate or reclinate position of a bristle, though quite unimportant in itself, may be an outward visible sign of the occurrence in the organism of quite important developmental processes. The latter are not necessarily disturbed when other factors cause some particular bristle to be missing; a mutant may survive without it.' [. . .]

It will be seen from the quotations given, that Dobzhansky's argument rests entirely on his own incredulity. Now it is certainly true that nothing is known about the value of many different characters in Drosophila; but as long as one accepts that anything not understood is a mere pleiotrope, no investigations will be made. One possible function for bristles that suggests itself straight away is in relation to toilet. Drosophila must be very liable to get covered with sticky and possibly highly deleterious micro-organisms. These strong bristles near the eyes may act with others on the legs as brushes, or even have some tactile function like crude vibrissae. Dobzhansky remarks 'The usefulness of a trait must be demonstrated, it cannot just be taken for granted'. But equally, its uselessness cannot be taken for granted, and indirect evidence on the likelihood of its being selected for and actually adaptive cannot be ignored. In any case, the argument that flies in cultures do not seem to be affected by the loss of a bristle is inconclusive; a few per cent selection, enough to fix the character in a short time, would not have been noticed, or selection might be far more stringent in the wild. [. . .]

Where comparatively trivial characters have been investigated, some very definite functional significances have turned up. One might well ask why the chick of the kittiwake *Rissa tridactyla* should have a black band on the neck,

[2] This quotation, and that in the next paragraph are from: T. Dobzhansky, *American Naturalist*, 90 (1956), 337–47.

which is not found on that of other gulls. Cullen's remarkable analysis has shown that this and many other features of the kittiwake are directly related to its nesting on narrow ledges of cliffs instead of on the ground as do other gulls. The black band is shown off when the chick hides its beak as an appeasement gesture to prevent fights which might well end in both the birds involved falling from the ledge. Here, a wholly 'trivial' character might have a coefficient of selection of 50–100 per cent. [. . .]

But perhaps the most remarkable functional interpretation of a 'trivial' character is given by Manton's work on the diplopod Polyxenus, in which she has shown that a character formerly described as an 'ornament' (and what could sound more useless?) is almost literally the pivot of the animal's life. Polyxenus, is a very remarkable minute millipede which can actually walk upside-down on the ceiling of small crevices and even moult there. Manton shows that a curious Y-shaped bar of chitin on the legs enables the animal to use a very wide leg-swing in walking and develop considerable fleetness without using long legs. Speed is necessary as it has to make long journeys for its food, and short legs are an advantage in the crevices where it hides. Its gait is basically of a slow pattern, thus enabling it to have many leg-tips touching the ceiling of a crevice at once; also more secure adherence is obtained by means of special lappets at the tips. She further points out that the Y-shaped bar is also produced completely independently in some very fast-running centipedes for the same reason, namely, to strengthen the joints of a very widely-swinging leg.

To sum up this section, therefore, we can say that the theoretical power of natural selection is very great indeed, and studies in the field have shown that large coefficients are associated with what might seem very trivial characters. Where investigations have been undertaken, trivial characters have proved to be of adaptive significance in their own right. There may well be some characters which are necessary consequences of the production of others but of no selective value in themselves; but it is doubtful if any have been demonstrated to be in this state. The chances that any effect of a widespread gene of a wild-type genotype can be, or remain, neutral for long are slight indeed. Also, the evidence on which characters have been called non-adaptive is invariably wholly negative. If it is taken dogmatically that many characters *must* be non-adaptive, then of course there will be no motive to investigate them, and they will continue to be quoted as non-adaptive whether they are or not; but the positive evidence suggests an adaptive nature.

Adaptive radiation and convergence

Even where we have only a general idea of the adaptations involved and their genetic basis, we can sometimes see that adaptation is indeed affecting a very great number of characters. Adaptive radiation, the deployment of a basal stock into a large number of niches with consequent divergence of lines, is one of the most usual and pervasive of evolutionary processes, but few studies have yet appeared in which the actual adaptive nature of the divergences is investigated. [. . .] Simpson's remarkable elucidation of the evolution of mammals in South America deals with groups of a high taxonomic rank, with the advantages of a good fossil history. He shows how in the earliest Palaeocene probably not more than three stocks of primitive mammals, two eutherian and one marsupial, got into South America, which for most of the Tertiary was an island continent like Australia, and proceeded immediately to radiate into the available niches. The marsupials produced such generalized forms as the present-day American opossums and the rather shrewlike little Caenolestes, but for the most part specialized in the carnivore habit, producing weasel-like, cat-like and other types, even including a marsupial version, Thylacosmilus, of the 'true' sabre-toothed tigers of the northern world. Of the two eutherian stocks, one gave rise to the ground-sloths, tree sloths, anteaters, glyptodonts and armadillos, several of which are still extant. The other produced parallels with nearly all the large herbivores of the rest of the world—elephant-like, rodent-like, camel-like, horse-like, and others resembling big herbivores such as uintatheres which are now extinct. Much later, in the late Eocene to Oligocene, some true rodents got in and produced the great radiation of caviomorph rodents well known at the present day (capybara, agouti, guinea pig, paca, viscacha etc.) and some advanced lemuroids also arriving in this period produced the New World monkeys, which are generally agreed to have an independent origin from the Old World monkeys. Lastly, very late on, in the late Miocene to Recent, as several large islands became interpolated between South and southern North America and finally the Isthmus of Panama was completed, the fauna of the northern world invaded South America in force. Simpson points out that we have good fossil evidence of the consequences; broadly, those South American forms most like North American ones became extinct (this included all the remarkable herbivores and nearly all the marsupials) but those unlike anything coming in, and some very generalized forms, survived and in a few cases even managed to invade North America.

This fossil history of the South American mammalian fauna is of the first importance to all students of evolution. It is the only one which is reasonably sufficient for us to be sure of the course of events and which relates to an island continent virtually undisturbed (until near the end) either by largescale immigration or by considerable changes in climate. [. . .]

It is not to be expected, of course, that the resemblance of different stocks occupying the same niche in different continents should be perfect, even if they are closely related and therefore very similar to start with. Circumstances will never be exactly the same. For example, the proportions of the main classes of food available may differ and so may their characteristics; if one region of savannah is much more subject to fire than another, its woody plants, becoming fire-resistant, may also become useless as a standby for food for grazers in times of great scarcity, while their bark in the other region may remain just edible. The competitors will be different in each region, and available niches may be shared out in different ways. A mode of life open to a specialist animal in one region may be too unreliable in another to allow any form to specialize in it because of a different seasonal régime. With this in mind, the convergence actually found in the now extinct South American mammals and others is remarkable. The development of three-toed and one-toed 'horses' independently in South America and in the northern world has produced a fantastic convergence in the structure of the foot (though far less in the skull). Very small details are seen to evolve in parallel—and at widely different periods—in the true horses and in Diadiaphorus and Thoatherium, the South American convergent forms. But in fact [. . .] the single-toed 'horse' foot evolved *three* times independently; [it is found] again with astonishingly detailed similarities, in the Miocene hyracoids of East Africa. Simpson has examined the function of the sabre teeth of big carnivores and shown that, although this has been denied, they are beautifully adapted for stabbing; he points out that they have arisen three times at least, twice within the true carnivores (once in North American Eocene creodonts, once in the Oligocene to Pleistocene machairodont cats) and once in the marsupials of South America. Again, convergence extends to small details. [. . .]

Convergence, if we have (as usual) only an imperfect fossil record, may be difficult to detect. Up to now it has usually been assumed, by Darwin and others, that convergence will never be so good as to mislead us. We may allow that it is ecologically very unlikely that two widely different groups such as insects and brachiopods will show convergence, because it is so unlikely that similar selective pressures would act for long enough always in the same direction to produce it, and at least as unlikely that the intermediate forms would be able to persist in the face of competition from others more specialized whose niches would have to be traversed to reach the desired result. But, as remarked above, rather similar stocks may well converge greatly and in the absence of a good fossil record, give rise to much confusion. Convergence is now being suspected at all sorts of levels of the animal kingdom; to the examples I have previously given,[3] many more could be added, such as Kleinenberg's acceptance of a diphyletic and Slijper's of a triphyletic origin of the Cetacea. It must be made clear that the polyphyletic origin of *all* 'natural' groups is not being asserted. It may be that some groups have been so successful in their own lines and have spread so widely

[3] Not in this extract, but in A. Cain, *Proceedings of the Linnean Society of London*, 170 (1959), 185–217.

that they occupy completely a given broad niche everywhere and no others can get in. If so, then there will be no convergence on them and they may well be monophyletic, even perhaps in the strictest sense. But it does not follow at all that other groups are not capable, given the opportunity, of converging into that niche. I have pointed out before[4] some examples of animals that transgress the definition of their phyla in respect of major characters—a mollusc with no anus, a coelenterate with a so-called terminal pore to its gut, a protozoon (and there are several others) that is multicellular; and in a large number of groups the profound changes that have come about with the adoption of parasitism, such as those mentioned by Darwin as occurring in the mouthparts of some crustacea are evidence that even major features can be altered if necessary. It may be objected that parasitism involves only loss or hypertrophy, so that the changes are morphologically simple, but in parasitic crustacea at least this is not so, remarkable attachment systems and root-like feeding systems being developed. One can point also to the profound change in the life history of sessile tunicates from the free-living tadpole to the adult, which involves a gain as well as a loss, or the development of the mouthparts of true flies (Diptera) from the biting mouthparts of primitive insects. It is only the existence of intermediate forms that allows us to keep the Diptera and Thysanura (for example) in the same class; the whole tendency of our present system of 'natural' classification is to put together everything which can be included under some simple definition, a tendency we inherit from Linnaeus and others because it is convenient in cataloguing the enormous diversity of living things to have simple definitions of our groups. We ignore the great diversity in each group [. . .], manufacture a definition of the group which in some cases may be full of exceptions, and then tend to think of it as having an unalterable plan. But even if there is a constant general plan expressible in a simple definition, it does not follow that it has become incapable of alteration if the need should arise; and in some cases the diversity within major groups (in the invertebrates) is so great that we could make out classifications differently. Until a quantitative method of expressing overall differences can be produced and used to make a real map, not a highly distorted sketch-map, of the animal kingdom, our accepted classification will continue to bias our thoughts.

Adaptive features of major groups

Because of the general assumption that the major features of greater groups must be merely ancestral, there is as yet little direct evidence on their actual function. Perhaps the best available is the beautiful work of Manton, on the locomotory adaptations of arthropods. One small piece of this huge body of

[4] Ibid. 234–44.

work has been mentioned above, namely the explanation of the 'ornamental' Y-shaped bar on the legs of Polyxenus. She has herself summarized her work on Peripatus, Polyxenus and the Scutigeromorpha. Briefly, it can be said that every feature of the skeleton investigated shows a soundly functional significance in relation to speed and power of locomotion, or the ability to push hard either forwards or dorsally in burrowing (Chilopoda), and in Peripatus to squeeze through extremely narrow and irregular apertures. Now if Peripatus is so highly modified in relation to this ability in the curious mandibles, the body wall and cuticle, the nature of the muscles, the construction of the legs, and the body cavity, it requires further investigation to see whether the excretory system, for example, is not also modified for the same reason. Similarly, Manton describes the whole anatomy of Polyxenus as explicable on the assumption that its ancestors were Diplopods burrowing by pushing into soil, and secondarily it has become modified for the mode of life described above. But we might well ask, since it may be found deep in soil or under bark, whether it has retained some of the modifications for slight pushing because it needs them, in which case its anatomy will be explicable in terms of its *present* mode of life. [. . .]

The features that Manton deals with characterize orders and subclasses, and in the case of Peripatus a phylum or subphylum. All indicate the importance of adaptation and the absolute necessity of knowing the ecology of the forms concerned before coming to any conclusions. Drosophila is so convenient an animal in so many ways that it is easy to forget its unsuitability in others. It has four different ecologies in each life-history (for egg, larva, pupa and adult) hardly one of which is well understood for any species, and functional analysis may be extremely difficult in any stage. A wider survey of the animal kingdom is necessary before probabilities of adaptation can be assessed. [. . .]

Conclusions

It seems, then, that the grounds on which so much of the diversity of animals has been asserted to be non-adaptive and merely ancestral are mistaken, and based almost entirely on a lack of information. The indirect evidence available points strongly to the adaptive nature of the major plans on which animals are built, and of almost all the details. It may well be that some features are truly neutral and due only to an ancestral arrangement; the course of the recurrent laryngeal has been suggested to me as an example. But the direct evidence available, necessarily scanty, demonstrates the adaptive nature of a vast number of characters related to groups of very high as well as low rank.

[J.D. Carthy and C. L. Duddington (eds.), *Viewpoints in Biology*, vol. 3 (London: Butterworth, 1964), 36–63.]

S. J. GOULD AND R. C. LEWONTIN

20 The spandrels of San Marco and the Panglossian paradigm: a critique of the adaptationist programme

In the adaptationist research programme, features of biological organisms are studied on the assumption that they are adaptations. Like all scientific assumptions, it is unfalsifiable by the research that assumes it. However, other processes can drive evolution, such as genetic drift and the correlation of growth. Darwin allowed for non-adaptive processes in evolution. [Editor's summary.]

The adaptationist programme

We wish to question a deeply engrained habit of thinking among students of evolution. We call it the adaptationist programme, or the Panglossian paradigm. It is rooted in a notion popularized by A. R. Wallace and A. Weismann (but not, as we shall see, by Darwin) towards the end of the nineteenth century: the near omnipotence of natural selection in forging organic design and fashioning the best among possible worlds. This programme regards natural selection as so powerful and the constraints upon it so few that direct production of adaptation through its operation becomes the primary cause of nearly all organic form, function, and behaviour. Constraints upon the pervasive power of natural selection are recognized of course (phyletic inertia primarily among them, although immediate architectural constraints, as discussed in the last section, are rarely acknowledged). But they are usually dismissed as unimportant or else, and more frustratingly, simply acknowledged and then not taken to heart and invoked.

Studies under the adaptationist programme generally proceed in two steps:

(1) An organism is atomized into 'traits' and these traits are explained as structures optimally designed by natural selection for their functions. For lack of space, we must omit an extended discussion of the vital issue: 'what is a trait?' Some evolutionists may regard this as a trivial, or merely a semantic problem. It is not. Organisms are integrated entities, not collections of discrete objects. Evolutionists have often been led astray by inappropriate atomization. [. . .] Our favourite example involves the human chin. If we regard the chin as a 'thing', rather than as a product of interaction between two growth fields (alveolar and mandibular), then we are led to an interpretation of its origin (recapitulatory) exactly opposite to the one now generally favoured (neotenic).

(2) After the failure of part-by-part optimization, interaction is acknowledged via the dictum that an organism cannot optimize each part without imposing expenses on others. The notion of 'trade-off' is introduced, and organisms are interpreted as best compromises among competing demands.

Thus, interaction among parts is retained completely within the adaptationist programme. Any suboptimality of a part is explained as its contribution to the best possible design for the whole. The notion that suboptimality might represent anything other than the immediate work of natural selection is usually not entertained. As Dr Pangloss said in explaining to Candide why he suffered from venereal disease: 'It is indispensable in this best of worlds. For if Columbus, when visiting the West Indies, had not caught this disease, which poisons the source of generation, which frequently even hinders generation, and is clearly opposed to the great end of Nature, we should have neither chocolate nor cochineal.' The adaptationist programme is truly Panglossian. Our world may not be good in an abstract sense, but it is the very best we could have. Each trait plays its part and must be as it is.

At this point, some evolutionists will protest that we are caricaturing their view of adaptation. After all, do they not admit genetic drift, allometry, and a variety of reasons for non-adaptive evolution? They do, to be sure, but we make a different point. In natural history, all possible things happen sometimes; you generally do not support your favoured phenomenon by declaring rivals impossible in theory. Rather, you acknowledge the rival, but circumscribe its domain of action so narrowly that it cannot have any importance in the affairs of nature. Then, you often congratulate yourself for being such an undogmatic and ecumenical chap. We maintain that alternatives to selection for best overall design have generally been relegated to unimportance by this mode of argument. Have we not all heard the catechism about genetic drift: it can only be important in populations so small that they are likely to become extinct before playing any sustained evolutionary role.

The admission of alternatives in principle does not imply their serious consideration in daily practice. We all say that not everything is adaptive; yet, faced with an organism, we tend to break it into parts and tell adaptive stories as if trade-offs among competing, well designed parts were the only constraint upon perfection for each trait. It is an old habit. As Romanes complained about A. R. Wallace in 1900: 'Mr. Wallace does not expressly maintain the abstract impossibility of laws and causes other than those of utility and natural selection . . . Nevertheless, as he nowhere recognizes any other law or cause . . ., he practically concludes that, on inductive or empirical grounds, there *is* no such other law or cause to be entertained.'

The adaptationist programme can be traced through common styles of argument. We illustrate just a few; we trust they will be recognized by all:

(1) If one adaptive argument fails, try another. Zig-zag commissures of clams and brachiopods, once widely regarded as devices for strengthening the shell, become sieves for restricting particles above a given size. A suite of external structures (horns, antlers, tusks) once viewed as weapons against predators, become symbols of intraspecific competition among males. The eskimo face, once depicted as 'cold engineered', becomes an adaptation to

generate and withstand large masticatory forces. We do not attack these newer interpretations; they may all be right. We do wonder, though, whether the failure of one adaptive explanation should always simply inspire a search for another of the same general form, rather than a consideration of alternatives to the proposition that each part is 'for' some specific purpose.

(2) If one adaptive argument fails, assume that another must exist; a weaker version of the first argument. Costa and Bisol,[1] for example, hoped to find a correlation between genetic polymorphism and stability of environment in the deep sea, but they failed. They conclude: 'The degree of genetic polymorphism found would seem to indicate absence of correlation with the particular environmental factors which characterize the sampled area. The results suggest that the adaptive strategies of organisms belonging to different phyla are different.'

(3) In the absence of a good adaptive argument in the first place, attribute failure to imperfect understanding of where an organism lives and what it does. This is again an old argument. Consider Wallace on why all details of colour and form in land snails must be adaptive, even if different animals seem to inhabit the same environment:[2] 'The exact proportions of the various species of plants, the numbers of each kind of insect or of bird, the peculiarities of more or less exposure to sunshine or to wind at certain critical epochs, and other slight differences which to us are absolutely immaterial and unrecognizable, may be of the highest significance to these humble creatures, and be quite sufficient to require some slight adjustments of size, form, or colour, which natural selection will bring about.'

(4) Emphasize immediate utility and exclude other attributes of form. Fully half the explanatory information accompanying the full-scale Fibreglass *Tyrannosaurus* at Boston's Museum of Science reads: 'Front legs a puzzle: how *Tyrannosaurus* used its tiny front legs is a scientific puzzle; they were too short even to reach the mouth. They may have been used to help the animal rise from a lying position.' (We purposely choose an example based on public impact of science to show how widely habits of the adaptationist programme extend. We are not using glass beasts as straw men; similar arguments and relative emphases, framed in different words, appear regularly in the professional literature.) We don't doubt that *Tyrannosaurus* used its diminutive front legs for something. If they had arisen *de novo*, we would encourage the search for some immediate adaptive reason. But they are, after all, the reduced product of conventionally functional homologues in ancestors (longer limbs of allosaurs, for example). As such, we do not need an explicitly adaptive explanation for the reduction itself. It is likely to be a developmental correlate

[1] R. Costa and P. M. Bisol, *Biological Bulletin*, 155 (1978), 125–33; quote in next sentence from pp. 132–3.

[2] A. R. Wallace, *Darwinism* (London: Macmillan, 1889), 148.

of allometric fields for relative increase in head and hindlimb size. This non-adaptive hypothesis can be tested by conventional allometric methods and seems to us both more interesting and fruitful than untestable speculations based on secondary utility in the best of possible worlds. One must not confuse the fact that a structure is used in some way [. . .] with the primary evolutionary reason for its existence and conformation.

Telling stories

All this is a manifestation of the rightness of things, since if there is a volcano at Lisbon it could not be anywhere else. For it is impossible for things not to be where they are, because everything is for the best. (Dr Pangloss on the great Lisbon earthquake of 1755 in which up to 50,000 people lost their lives.)

We would not object so strenuously to the adaptationist programme if its invocation, in any particular case, could lead in principle to its rejection for want of evidence. We might still view it as restrictive and object to its status as an argument of first choice. But if it could be dismissed after failing some explicit test, then alternatives would get their chance. Unfortunately, a common procedure among evolutionists does not allow such definable rejection for two reasons. First, the rejection of one adaptive story usually leads to its replacement by another, rather than to a suspicion that a different kind of explanation might be required. Since the range of adaptive stories is as wide as our minds are fertile, new stories can always be postulated. And if a story is not immediately available, one can always plead temporary ignorance and trust that it will be forthcoming, as did Costa and Bisol, cited above. Secondly, the criteria for acceptance of a story are so loose that many pass without proper confirmation. Often, evolutionists use *consistency* with natural selection as the sole criterion and consider their work done when they concoct a plausible story. But plausible stories can always be told. The key to historical research lies in devising criteria to identify proper explanations among the substantial set of plausible pathways to any modern result.

We have criticized Barash's work on aggression in mountain bluebirds for this reason. Barash mounted a stuffed male near the nests of two pairs of bluebirds while the male was out foraging. He did this at the same nests on three occasions at 10 day intervals: the first before eggs were laid, the last two afterwards. He then counted aggressive approaches of the returning male towards both the model and the female. At time one, aggression was high towards the model and lower towards females but substantial in both nests. Aggression towards the model declined steadily for times two and three and plummeted to near zero towards females. Barash reasoned that this made evolutionary sense since males would be more sensitive to intruders before eggs were laid than afterwards (when they can have some confidence that their genes are inside). Having devised this plausible story, he considered his work as completed (1976, pp. 1099, 1100):

'The results are consistent with the expectations of evolutionary theory. Thus aggression toward an intruding male (the model) would clearly be especially advantageous early in the breeding season, when territories and nests are normally defended . . . The initial aggressive response to the mated female is also adaptive in that, given a situation suggesting a high probability of adultery (i.e. the presence of the model near the female) and assuming that replacement females are available, obtaining a new mate would enhance the fitness of males . . . The decline in male–female aggressiveness during incubation and fledgling stages could be attributed to the impossibility of being cuckolded after the eggs have been laid . . . The results are consistent with an evolutionary interpretation.'

They are indeed consistent, but what about an obvious alternative, dismissed without test by Barash? Male returns at times two and three, approaches the model, tests it a bit, recognizes it as the same phoney he saw before, and doesn't bother his female. Why not at least perform the obvious test for this alternative to a conventional adaptive story: expose a male to the model for the *first* time after the eggs are laid.

Since we criticized Barash's work, Morton *et al.* repeated it, with some variations (including the introduction of a female model), in the closely related eastern bluebird *Sialia sialis*. 'We hoped to confirm', they wrote, that Barash's conclusions represent 'a widespread evolutionary reality, at least within the genus *Sialia*. Unfortunately, we were unable to do so.' They found no 'anticuckoldry' behaviour at all: males never approached their females aggressively after testing the model at any nesting stage. Instead, females often approached the male model and, in any case, attacked female models more than males attacked male models. 'This violent response resulted in the near destruction of the female model after presentations and its complete demise on the third, as a female flew off with the model's head early in the experiment to lose it for us in the brush'.[3] Yet, instead of calling Barash's selected story into question, they merely devise one of their own to render both results in the adaptationist mode. Perhaps, they conjecture, replacement females are scarce in their species and abundant in Barash's. Since Barash's males can replace a potentially 'unfaithful' female, they can afford to be choosy and possessive. Eastern bluebird males are stuck with uncommon mates and had best be respectful. They conclude: 'If we did not support

[3] E. S. Morton, M. S. Geitgey, and S. McGrath, 'On bluebird responses to apparent female adultery', *American Naturalist*, 112 (1978), 968–71. The quote a few lines on is from D. P. Barash, 'Male response to apparent female adultery in the mountain bluebird: an evolutionary interpretation', *American Naturalist*, 110 (1976), 1097–1101.

Barash's suggestion that male bluebirds show anticuckoldry adaptations, we suggest that both studies still had "results that are consistent with the expectations of evolutionary theory", as we presume any careful study would.' But what good is a theory that cannot fail in careful study (since by 'evolutionary theory', they clearly mean the action of natural selection applied to particular cases, rather than the fact of transmutation itself).

The master's voice re-examined

Since Darwin has attained sainthood (if not divinity) among evolutionary biologists, and since all sides invoke God's allegiance, Darwin has often been depicted as a radical selectionist at heart who invoked other mechanisms only in retreat, and only as a result of his age's own lamented ignorance about the mechanisms of heredity. This view is false. Although Darwin regarded selection as the most important of evolutionary mechanisms (as do we), no argument from opponents angered him more than the common attempt to caricature and trivialize his theory by stating that it relied exclusively upon natural selection. In the last edition of the *Origin*, he wrote:[4]

As my conclusions have lately been much misrepresented, and it has been stated that I attribute the modification of species exclusively to natural selection, I may be permitted to remark that in the first edition of this work, and subsequently, I placed in a most conspicuous position—namely at the close of the Introduction—the following words: 'I am convinced that natural selection has been the main, but not the exclusive means of modification.' This has been of no avail. Great is the power of steady misinterpretation.

Romanes, whose once famous essay on Darwin's pluralism versus the panselectionism of Wallace and Weismann deserves a resurrection, noted of this passage: 'In the whole range of Darwin's writings there cannot be found a passage so strongly worded as this: it presents the only note of bitterness in all the thousands of pages which he has published.' Apparently, Romanes did not know the letter Darwin wrote to *Nature* in 1880, in which he castigated Sir Wyville Thomson for caricaturing his theory as panselectionist:

I am sorry to find that Sir Wyville Thomson does not understand the principle of natural selection . . . If he had done so, he could not have written the following sentence in the Introduction to the Voyage of the Challenger: 'The character of the abyssal fauna refuses to give the least support to the theory which refers the

[4] C. Darwin, *The Origin of Species*, 6th edn. (London: John Murray, 1859), 395. The Romanes ref. below is: G. J. Romanes, *Darwin, and After Darwin* (London: Longmans, 1890). The next Darwin quotation is from C. Darwin, 'Sir Wyville Thompson and Natural Selection', *Nature*, 23 (1880), 32.

evolution of species to extreme variation guided only by natural selection.' This is a standard of criticism not uncommonly reached by theologians and metaphysicians when they write on scientific subjects, but is something new as coming from a naturalist . . . Can Sir Wyville Thomson name any one who has said that the evolution of species depends only on natural selection? As far as concerns myself, I believe that no one has brought forward so many observations on the effects of the use and disuse of parts, as I have done in my 'Variation of Animals and Plants under Domestication'; and these observations were made for this special object. I have likewise there adduced a considerable body of facts, showing the direct action of external conditions on organisms.

We do not now regard all of Darwin's subsidiary mechanisms as significant or even valid, though many, including direct modification and correlation of growth, are very important. But we should cherish his consistent attitude of pluralism in attempting to explain Nature's complexity.

A partial typology of alternatives to the adaptationist programme

In Darwin's pluralistic spirit, we present an incomplete hierarchy of alternatives to immediate adaptation for the explanation of form, function, and behaviour.

(1) No adaptation and no selection at all. At present, population geneticists are sharply divided on the question of how much genetic polymorphism within populations and how much of the genetic differences between species is, in fact, the result of natural selection as opposed to purely random factors. Populations are finite in size and the isolated populations that form the first step in the speciation process are often founded by a very small number of individuals. As a result of this restriction in population size, frequencies of alleles change by *genetic drift*, a kind of random genetic sampling error. The stochastic process of change in gene frequency by random genetic drift, including the very strong sampling process that goes on when a new isolated population is formed from a few immigrants, has several important consequences. First, populations and species will become genetically differentiated, and even fixed for different alleles at a locus in the complete absence of any selective force at all.

Secondly, alleles can become fixed in a population *in spite of natural selection*. Even if an allele is favoured by natural selection, some proportion of population, depending upon the product of population size N and selection intensity s, will become homozygous for the less fit allele because of genetic drift. If Ns is large this random fixation for unfavourable alleles is a rare phenomenon, but if selection coefficients are on the order of the reciprocal of population size ($Ns = 1$) or smaller, fixation for deleterious alleles is common. If many genes are involved in influencing a metric character like shape, metabolism or behaviour, then the intensity of selection on each locus will be small and *Ns* per locus may be small. As a result, many of the loci may be fixed for non-optimal alleles.

Thirdly, new mutations have a small chance of being incorporated into a population, even when selectively favoured. Genetic drift causes the immediate loss of most new mutations after their introduction. With a selection intensity *s*, a new favourable mutation has a probability of only 2*s* of ever being incorporated. Thus, one cannot claim that, eventually, a new mutation of just the right sort for some adaptive argument will occur and spread. 'Eventually' becomes a very long time if only one in 1000 or one in 10 000 of the 'right' mutations that do occur ever get incorporated in a population.

(2) No adaptation and no selection on the part at issue; form of the part is a correlated consequence of selection directed elsewhere. Under this important category, Darwin ranked his 'mysterious' laws of the 'correlation of growth'. Today, we speak of pleiotropy, allometry, 'material compensation' and mechanically forced correlations in D'Arcy Thompson's sense. Here we come face to face with organisms as integrated wholes, fundamentally not decomposable into independent and separately optimized parts.

Although allometric patterns are as subject to selection as static morphology itself, some regularities in relative growth are probably not under immediate adaptive control. For example, we do not doubt that the famous 0.66 interspecific allometry of brain size in all major vertebrate groups represents a selected 'design criterion,' though its significance remains elusive. It is too repeatable across too wide a taxonomic range to represent much else than a series of creatures similarly well designed for their different sizes. But another common allometry, the 0.2 to 0.4 intraspecific scaling among homeothermic adults differing in body size, or among races within a species, probably does not require a selectionist story though many, including one of us, have tried to provide one. R. Lande has used the experiments of Falconer to show that selection upon *body size alone* yields a brain–body slope across generations of 0.35 in mice.

More compelling examples abound in the literature on selection for altering the timing of maturation. At least three times in the evolution of arthropods (mites, flies and beetles), the same complex adaptation has evolved, apparently for rapid turnover of generations in strongly *r*-selected feeders on superabundant but ephemeral fungal resources: females reproduce as larvae and grow the next generation within their bodies. Offspring eat their mother from inside and emerge from her hollow shell, only to be devoured a few days later by their own progeny. It would be foolish to seek adaptive significance in paedomorphic morphology *per se*; it is primarily a by-product of selection for rapid cycling of generations. In more interesting cases, selection for small size (as in animals of the interstitial fauna) or rapid maturation (dwarf males of many crustaceans) has occurred by progenesis, and descendant adults contain a mixture of ancestral juvenile and adult features. Many biologists have been tempted to find primary adaptive meaning for the mixture, but it probably arises as a by-product of truncated maturation, leaving some features 'behind' in the larval state, while allowing others, more strongly correlated with sexual maturation, to retain the adult configuration of ancestors.

(3) The decoupling of selection and adaptation.

(i) Selection without adaptation. Lewontin has presented the following hypothetical example:[5] 'A mutation which doubles the fecundity of individuals will sweep through a population rapidly. If there has been no change in efficiency of resource utilization, the individuals will leave no more offspring than before, but simply lay twice as many eggs, the excess dying because of resource limitation. In what sense are the individuals or the population as a whole better adapted than before? Indeed, if a predator on immature stages is led to switch to the species now that immatures are more plentiful, the population size may actually decrease as a consequence, yet natural selection at all times will favour individuals with higher fecundity.'

(ii) Adaptation without selection. Many sedentary marine organisms, sponges and corals in particular, are well adapted to the flow régimes in which they live. A wide spectrum of 'good design' may be purely phenotypic in origin, largely induced by the current itself. (We may be sure of this in numerous cases, when genetically identical individuals of a colony assume different shapes in different microhabitats.) Larger patterns of geographic variation are often adaptive and purely phenotypic as well. [. . .] Many hemimetabolous aquatic insects reach smaller adult size with reduced fecundity when they grow at temperatures above and below their optima. Coherent, climatically correlated patterns in geographic distribution for these insects—so often taken as *a priori* signs of genetic adaptation—may simply reflect this phenotypic plasticity.

'Adaptation'—the good fit of organisms to their environment—can occur at three hierarchical levels with different causes. It is unfortunate that our language has focused on the common result and called all three phenomena 'adaptation': the differences in process have been obscured and evolutionists have often been misled to extend the Darwinian mode to the other two levels as well. First, we have what physiologists call 'adaptation': the phenotypic plasticity that permits organisms to mould their form to prevailing circumstances during ontogeny. Human 'adaptations' to high altitude fall into this category (while others, like resistance of sickling heterozygotes to malaria, are genetic and Darwinian). Physiological adaptations are not heritable, though the capacity to develop them presumably is. Secondly, we have a 'heritable' form of non-Darwinian adaptation in humans (and, in rudimentary

[5] R. C. Lewontin, *Behavioral Science*, 24 (1979), 5–14.

ways, in a few other advanced social species): cultural adaptation (with heritability imposed by learning). Much confused thinking in human sociobiology arises from a failure to distinguish this mode from Darwinian adaptation based on genetic variation. Finally, we have adaptation arising from the conventional Darwinian mechanism of selection upon genetic variation. The mere existence of a good fit between organism and environment is insufficient evidence for inferring the action of natural selection.

(4) Adaptation and selection but no selective basis for differences among adaptations. Species of related organisms, or subpopulations within a species, often develop different adaptations as solutions to the same problem. When 'multiple adaptive peaks' are occupied, we usually have no basis for asserting that one solution is better than another. The solution followed in any spot is a result of history; the first steps went in one direction, though others would have led to adequate prosperity as well. Every naturalist has his favourite illustration. In the West Indian land snail *Cerion*, for example, populations living on rocky and windy coasts almost always develop white, thick and relatively squat shells for conventional adaptive reasons. We can identify at least two different developmental pathways to whiteness from the mottling of early whorls in all *Cerion*, two paths to thickened shells and three styles of allometry leading to squat shells. All 12 combinations can be identified in Bahamian populations, but would it be fruitful to ask why—in the sense of optimal design rather than historical contingency—*Cerion* from eastern Long Island evolved one solution, and *Cerion* from Acklins Island another?

(5) Adaptation and selection, but the adaptation is a secondary utilization of parts present for reasons of architecture, development or history. We have already discussed this neglected subject [. . .].[1] If blushing turns out to be an adaptation affected by sexual selection in humans, it will not help us to understand why blood is red. The immediate utility of an organic structure often says nothing at all about the reason for its being. [. . .]

[*Proc. of the Royal Society of London*, ser. B, 205 (1979), 581–98.]

RICHARD DAWKINS

21 The selfish gene

What entities do adaptations evolve for the benefit of? What units does natural selection act on? Larger genetic units, such as whole chromosomes, do not last long over evolutionary time; shorter units do. The length of DNA that lasts long enough for natural selection to adjust its frequency can act as a unit of selection: the unit is defined as a gene. Other kinds of entity, such as individual organisms, are too short-lived to act as units of selection. The argument

[1] In a section of the paper not included here.

that genes are units of selection does not overlook the way in which genes interact during the development of individual organisms. [Editor's summary.]

In the title of this book[1] the word gene means not a single cistron but something more subtle. My definition will not be to everyone's taste, but there is no universally agreed definition of a gene. Even if there were, there is nothing sacred about definitions. We can define a word how we like for our own purposes, provided we do so clearly and unambiguously. The definition I want to use comes from G. C. Williams. A gene is defined as any portion of chromosomal material that potentially lasts for enough generations to serve as a unit of natural selection. In the words of the previous chapter, a gene is a replicator with high copying-fidelity. Copying-fidelity is another way of saying longevity-in-the-form-of-copies and I shall abbreviate this simply to longevity. The definition will take some justifying.

On any definition, a gene has to be a portion of a chromosome. The question is, how big a portion—how much of the ticker tape? Imagine any sequence of adjacent code-letters on the tape. Call the sequence a *genetic unit*. It might be a sequence of only ten letters within one cistron; it might be a sequence of eight cistrons; it might start and end in mid-cistron. It will overlap with other genetic units. It will include smaller units, and it will form part of larger units. No matter how long or short it is, for the purposes of the present argument, this is what we are calling a genetic unit. It is just a length of chromosome, not physically differentiated from the rest of the chromosome in any way.

Now comes the important point. The shorter a genetic unit is, the longer—in generations—it is likely to live. In particular, the less likely it is to be split by any one crossing-over. Suppose a whole chromosome is, on average, likely to undergo one cross-over every time a sperm or egg is made by meiotic division, and this cross-over can happen anywhere along its length. If we consider a very large genetic unit, say half the length of the chromosome, there is a 50 per cent chance that the unit will be split at each meiosis. If the genetic unit we are considering is only 1 per cent of the length of the chromosome, we can assume that it has only a 1 per cent chance of being split in any one meiotic division. This means that the unit can expect to survive for a large number of generations in the individual's descendants. A single cistron is likely to be much less than 1 per cent of the length of a chromosome. Even a group of several neighbouring cistrons can expect to live many generations before being broken up by crossing over.

The average life-expectancy of a genetic unit can conveniently be expressed in generations, which can in turn be translated into years. If we take a whole chromosome as our presumptive genetic unit, its life story lasts for only one generation. Suppose it is your chromosome number 8a,

[1] The extract is from a book entitled *The selfish gene.*

inherited from your father. It was created inside one of your father's testicles, shortly before you were conceived. It had never existed before in the whole history of the world. It was created by the meiotic shuffling process, forged by the coming together of pieces of chromosome from your paternal grandmother and your paternal grandfather. It was placed inside one particular sperm, and it was unique. The sperm was one of several millions, a vast armada of tiny vessels, and together they sailed into your mother. This particular sperm (unless you are a non-identical twin) was the only one of the flotilla which found harbour in one of your mother's eggs—that is why you exist. The genetic unit we are considering, your chromosome number 8a, set about replicating itself along with all the rest of your genetic material. Now it exists, in duplicate form, all over your body. But when you in your turn come to have children, this chromosome will be destroyed when you manufacture eggs (or sperms). Bits of it will be interchanged with bits of your maternal chromosome number 8b. In any one sex cell, a new chromosome number 8 will be created, perhaps 'better' than the old one, perhaps 'worse', but, barring a rather improbable coincidence, definitely different, definitely unique. The life-span of a chromosome is one generation.

What about the life-span of a smaller genetic unit, say 1/100 of the length of your chromosome 8a? This unit too came from your father, but it very probably was not originally assembled in him. Following the earlier reasoning, there is a 99 per cent chance that he received it intact from one of his two parents. Suppose it was from his mother, your paternal grandmother. Again, there is a 99 per cent chance that she inherited it intact from one of her parents. Eventually, if we trace the ancestry of a small genetic unit back far enough, we will come to its original creator. At some stage it must have been created for the first time inside a testicle or an ovary of one of your ancestors.

Let me repeat the rather special sense in which I am using the word 'create'. The smaller sub-units which make up the genetic unit we are considering may well have existed long before. Our genetic unit was created at a particular moment only in the sense that the particular *arrangement* of sub-units by which it is defined did not exist before that moment. The moment of creation may have occurred quite recently, say in one of your grandparents. But if we consider a very small genetic unit, it may have been first assembled in a much more distant ancestor, perhaps an ape-like pre-human ancestor. Moreover, a small genetic unit inside you may go on just as far into the future, passing intact through a long line of your descendants.

Remember too that an individual's descendants constitute not a single line but a branching line. Whichever of your ancestors it was who 'created' a particular short length of your chromosome 8a, he or she very likely has many other descendants besides you. One of your genetic units may also be present in your second cousin. It may be present in me, and in the Prime

Minister, and in your dog, for we all share ancestors if we go back far enough. Also the same small unit might be assembled several times independently by chance: if the unit is small, the coincidence is not too improbable. But even a close relative is unlikely to share a whole chromosome with you. The smaller a genetic unit is, the more likely it is that another individual shares it—the more likely it is to be represented many times over in the world, in the form of copies. [. . .]

I am using the word gene to mean a genetic unit that is small enough to last for a large number of generations and to be distributed around in the form of many copies. This is not a rigid all-or-nothing definition, but a kind of fading-out definition, like the definition of 'big' or 'old'. The more likely a length of chromosome is to be split by crossing-over, or altered by mutations of various kinds, the less it qualifies to be called a gene in the sense in which I am using the term. A cistron presumably qualifies, but so also do larger units. A dozen cistrons may be so close to each other on a chromosome that for our purposes they constitute a single long-lived genetic unit. The butterfly mimicry cluster is a good example. As the cistrons leave one body and enter the next, as they board sperm or egg for the journey into the next generation, they are likely to find that the little vessel contains their close neighbours of the previous voyage, old shipmates with whom they sailed on the long odyssey from the bodies of distant ancestors. Neighbouring cistrons on the same chromosome form a tightly-knit troupe of travelling companions who seldom fail to get on board the same vessel when meiosis time comes around.

To be strict, this book should be called not *The Selfish Cistron* nor *The Selfish Chromosome*, but *The slightly selfish big bit of chromosome and the even more selfish little bit of chromosome*. To say the least this is not a catchy title so, defining a gene as a little bit of chromosome which potentially lasts for many generations, I call the book *The Selfish Gene*.

[. . .] Selfishness is to be expected in any entity that deserves the title of a basic unit of natural selection. We saw that some people regard the species as the unit of natural selection, others the population or group within the species, and yet others the individual. I said that I preferred to think of the gene as the fundamental unit of natural selection, and therefore the fundamental unit of self-interest. What I have now done is to *define* the gene in such a way that I cannot really help being right!

Natural selection in its most general form means the differential survival of entities. Some entities live and others die but, in order for this selective death to have any impact on the world, an additional condition must be met. Each entity must exist in the form of lots of copies, and at least some of the entities must be *potentially* capable of surviving—in the form of copies—for a significant period of evolutionary time. Small genetic units have these properties: individuals, groups, and species do not. It was the great achievement of Gregor Mendel to show that hereditary units can be treated in practice as indivisible and independent particles. Nowadays we know that this is a little too simple. Even a cistron is occasionally divisible and any two genes on the same chromosome are not wholly independent. What I have done is to define a gene as a unit which, to a high degree, *approaches* the ideal of indivisible particulateness. A gene is not indivisible, but it is seldom divided. It is either definitely present or definitely absent in the body of any given individual. A gene travels intact from grandparent to grandchild, passing straight through the intermediate generation without being merged with other genes. If genes continually blended with each other, natural selection as we now understand it would be impossible. Incidentally, this was proved in Darwin's lifetime, and it caused Darwin great worry since in those days it was assumed that heredity was a blending process. Mendel's discovery had already been published, and it could have rescued Darwin, but alas he never knew about it: nobody seems to have read it until years after Darwin and Mendel had both died. Mendel perhaps did not realize the significance of his findings, otherwise he might have written to Darwin.

Another aspect of the particulateness of the gene is that it does not grow senile; it is no more likely to die when it is a million years old than when it is only a hundred. It leaps from body to body down the generations, manipulating body after body in its own way and for its own ends, abandoning a succession of mortal bodies before they sink in senility and death.

The genes are the immortals, or rather, they are defined as genetic entities that come close to deserving the title. We, the individual survival machines in the world, can expect to live a few more decades. But the genes in the world have an expectation of life that must be measured not in decades but in thousands and millions of years.

In sexually reproducing species, the individual is too large and too temporary a genetic unit to qualify as a significant unit of natural selection. The group of individuals is an even larger unit. Genetically speaking, individuals and groups are like clouds in the sky or dust-storms in the desert. They are temporary aggregations or federations. They are not stable through evolutionary time. Populations may last a long while, but they are constantly blending with other populations and so losing their identity. They are also subject to evolutionary change from within. A population is not a discrete enough entity to be a unit of natural selection, not stable and unitary enough to be 'selected' in preference to another population.

An individual body seems discrete enough while it lasts, but alas, how long is that? Each individual is unique. You cannot get evolution by selecting between entities when there is only one copy of each entity! Sexual reproduction is not replication. Just as a population is contaminated by other populations, so an individual's posterity is contaminated by that of his sexual partner. Your children are only half you, your grandchildren only a quarter you. In a few generations the most you can hope for is a large number of descendants, each of whom bears only a tiny portion of you—a few genes—even if a few do bear your surname as well.

Individuals are not stable things, they are fleeting. Chromosomes too are shuffled into oblivion, like hands of cards soon after they are dealt. But the cards themselves survive the shuffling. The cards are the genes. The genes are not destroyed by crossing-over, they merely change partners and march on. Of course they march on. That is their business. They are the replicators and we are their survival machines. When we have served our purpose we are cast aside. But genes are denizens of geological time: genes are forever.

Genes, like diamonds, are forever, but not quite in the same way as diamonds. It is an individual diamond crystal that lasts, as an unaltered pattern of atoms. DNA molecules don't have that kind of permanence. The life of any one physical DNA molecule is quite short—perhaps a matter of months, certainly not more than one lifetime. But a DNA molecule could theoretically live on in the form of *copies* of itself for a hundred million years. Moreover, just like the ancient replicators in the primeval soup, copies of a particular gene may be distributed all over the world. The difference is that the modern versions are all neatly packaged inside the bodies of survival machines.

What I am doing is emphasizing the potential near-immortality of a gene, in the form of copies, as its defining property. To define a gene as a single cistron is good for some purposes, but for the purposes of evolutionary theory it needs to be enlarged. The extent of the enlargement is determined by the purpose of the definition. We want to find the practical unit of natural selection. To do this we begin by identifying the properties that a successful unit of natural selection must have. In the terms of the last chapter, these are longevity, fecundity, and copying-fidelity. We then simply define a 'gene' as the largest entity which, at least potentially, has these properties. The gene is a long-lived replicator, existing in the form of many duplicate copies. It is not infinitely long-lived. Even a diamond is not literally everlasting, and even a cistron can be cut in two by crossing-over. The gene is defined as a piece of chromosome which is sufficiently short for it to last, potentially, for *long enough* for it to function as a significant unit of natural selection.

Exactly how long is 'long enough'? There is no hard and fast answer. It will depend on how severe the natural selection 'pressure' is. That is, on how much more likely a 'bad' genetic unit is to die than its 'good' allele. This is a matter of quantitative detail which will vary from example to example. The largest practical unit of natural selection—the gene—will usually be found to lie somewhere on the scale between cistron and chromosome.

It is its potential immortality that makes a gene a good candidate as the basic unit of natural selection. But now the time has come to stress the word 'potential'. A gene *can* live for a million years, but many new genes do not even make it past their first generation. The few new ones that succeed do so partly because they are lucky, but mainly because they have what it takes, and that means they are good at making survival machines. They have an effect on the embryonic development of each successive body in which they find themselves, such that that body is a little bit more likely to live and reproduce than it would have been under the influence of the rival gene or allele. For example, a 'good' gene might ensure its survival by tending to endow the successive bodies in which it finds itself with long legs, which help those bodies to escape from predators. This is a particular example, not a universal one. Long legs, after all, are not always an asset. To a mole they would be a handicap. Rather than bog ourselves down in details, can we think of any *universal* qualities that we would expect to find in all good (i.e. long-lived) genes? Conversely, what are the properties that instantly mark a gene out as a 'bad', short-lived one? There might be several such universal properties, but there is one that is particularly relevant to this book: at the gene level, altruism must be bad and selfishness good. This follows inexorably from our definitions of altruism and selfishness. Genes are competing directly with their alleles for survival, since their alleles in the gene pool are rivals for their slot on the chromosomes of future generations. Any gene that behaves in such a way as to increase its own survival chances in the gene pool at the expense of its alleles will, by definition, tautologously, tend to survive. The gene is the basic unit of selfishness.

The main message of this chapter has now been stated. But I have glossed over some complications and hidden assumptions. The first complication has already been briefly mentioned. However independent and free genes may be in their journey through the generations, they are very much *not* free and independent agents in their control of embryonic development. They collaborate and interact in inextricably complex ways, both with each other, and with their external environment. Expressions like 'gene for long legs' or 'gene for altruistic behaviour' are convenient figures of speech, but it is important to understand what they mean. There is no gene which single-handedly builds a leg, long or short. Building a leg is a multigene cooperative enterprise. Influences from the external environment too are indispensable: after all, legs are actually made of food! But there may well be a single gene which, *other things being equal*, tends to make legs longer than they would have been under the influence of the gene's allele.

As an analogy, think of the influence of a fertilizer, say nitrate, on the growth of wheat. Everybody knows that wheat plants grow bigger in the presence of nitrate than in its absence. But nobody would be so foolish as to claim that, on its own, nitrate can make a wheat plant. Seed, soil, sun, water, and various minerals are obviously all necessary as well. But if all these other factors are held constant, and even if they are allowed to vary within limits, addition of nitrate will make the wheat plants grow bigger. So it is with single genes in the development of an embryo. Embryonic development is controlled by an interlocking web of relationships so complex that we had best not contemplate it. No one factor, genetic or environmental, can be considered as the single 'cause' of any part of a baby. All parts of a baby have a near infinite number of antecedent causes. But a *difference* between one baby and another, for example a difference in length of leg, might easily be traced to one or a few simple antecedent differences, either in environment or in genes. It is *differences* that matter in the competitive struggle to survive; and it is genetically-controlled differences that matter in evolution.

As far as a gene is concerned, its alleles are its deadly rivals, but other genes are just a part of its environment, comparable to temperature, food, predators, or companions. The effect of the gene depends on its environment, and this includes other genes. Sometimes a gene has one effect in the presence of a particular other gene, and a completely different effect in the presence of another set of companion genes. The whole set of genes in a body constitutes a kind of genetic climate or background, modifying and influencing the effects of any particular gene.

[*The Selfish Gene* (Oxford: Oxford University Press, 1976), 28–37.]

Section D

Speciation and biodiversity

You do not have to look around for long to see that life is organized into discrete species. There are robins, dandelions, sparrows, and humans, each of which has a formal Linnaean binomial such as *Homo sapiens*. Some species are easier to recognize than others, and most species vary from place to place; the existence of discrete species is therefore clearest at a particular site rather than across their ranges as a whole. However, the existence of species is not in doubt, and the theory of evolution aims to give an account of why life has this form. Ernst Mayr has been as influential a thinker on this subject as anyone else in the past century and I have included two of his contributions (Chapters 22 and 23). The second and longer piece (Chapter 23) discusses a series of possible species' criteria, and favours the 'biological species concept'. The biological species concept conceives of species in terms of interbreeding: it has a species as a group of interbreeding organisms. The concept may apply better to some living things, such as vertebrates and insects, than to others, such as some plants and microbes. The concept therefore tends to enjoy most support among zoologists (Mayr is himself a distinguished ornithologist), though this is not an absolute rule.

Mayr has also argued that the nature of biological species has general implications for the way we should think about variation and classification. In Chapter 22 he contrasts typological thinking, or essentialism, in which category membership is defined by reference to some 'type' attributes, with population thinking in which membership is defined by interbreeding rather than any attributes an individual may possess. In population thinking, all the variants within a group are equivalent, none being a better member of the group than any other. The argument has been taken further by Ghiselin and by Hull, who have argued that biological species are not classes, in which classification is by defining attributes, but are more like individuals: you are who you are because of a birth event and the continuity afterwards; much the same is true of species in evolution. The contrast between typological and population thinking is one of the great conceptual tensions within evolutionary biology, with some biologists preferring one mode of thought, others the other. Indeed, a single biologist may be more of a 'typologist' with regard to some questions, such as the existence or not of 'body plans',[1] and more

[1] The idea that major groups such as arthropods or vertebrates have a distinct body design. The differences between major groups would then differ in kind from the differences between species. Some hint of this issue can be found towards the end of the extract from Cain in Section C (Chapter 19).

of a population thinker with regard to other questions, such as intraspecific racial variation.

All modern life almost certainly shares a unique common ancestor, and there are now an unknown large number of (perhaps 30 million) species on Earth. (Crick's 'frozen accident' interpretation of the universal genetic code is one piece of evidence for the single common ancestor—see Section H, (Chapter 47).) On many occasions one species must have split into two. How does this happen? We have two main ideas. One is that speciation occurs as an incidental consequence, or by-product, when two subpopulations of an ancestral species evolve apart. Initially we have one species, and it becomes geographically subdivided into a number of subpopulations. One subpopulation in one place may evolve one set of adaptations to its local conditions; another subpopulation may evolve a different set of adaptations to the different local conditions where it lives. The populations are evolutionarily diverging. Later, members of the two divergent populations may interbreed for some reason (perhaps their ranges change and they meet up again). They produce hybrid offspring. The parents from the two populations have slightly different sets of genes and the two sets of genes may not interact well together in the hybrid offspring. Also, the two populations may have evolved different mating behaviour, and then mate preferentially with their own kind, given the choice. These mating preferences, and the lowered fitness of hybrid offspring because of incompatibilities between the parental genes, are two steps towards speciation. In this theory, speciation has just 'dropped out' as an incidental consequence while two populations have been evolving independently of each other.

The second theory is 'reinforcement'. In this theory, speciation is actively favoured by natural selection rather than being an incidental consequence of unrelated evolutionary change. Suppose that hybrid offspring between the members of two populations (or of two near-species) have lower fitness than the offspring of pure crosses within one of those populations (or near-species). Natural selection then favours more discriminating mating, to avoid mating with a partner of the other type and producing hybrid offspring. As (over evolutionary time) individuals come to mate more with members of their own population, speciation is occurring.

The 'by-product' theory of speciation is well supported, and widely accepted by biologists. It is the standard theory of speciation. Reinforcement is altogether more controversial. Some biologists doubt whether it occurs to any significant extent at all. I begin the extracts on this subject with Darwin (Chapter 24), who originated the by-product theory. Darwin doubted whether natural selection directly favoured speciation, and he exchanged many letters with Wallace on the topic. Wallace sometimes seems to be supporting something like the theory of reinforcement, and Darwin to be criticizing it. However, probably neither of them had the full theory of reinforcement, as it is now understood, in mind. I have not included any extracts from the Darwin–Wallace correspondence. Dobzhansky is the first definite supporter (though not the first proposer) of the theory of reinforcement, from the 1930s onwards. I include an extract from the 1970 edition of a book that originally appeared in 1936 (Chapter 25). Dobzhansky clearly argues that reinforcement is a common process.

But in the 1970s and 1980s, several criticisms were made of the theory of reinforcement. By the late 1980s, the theory had few supporters, but it may then have begun something of a revival in the 1990s. One reason was a finding made by Coyne and Orr. My extract from Coyne and Orr (Chapter 27) mentions that finding in a section about reinforcement. But they, unlike Dobzhansky, implicitly assume that their readers will be sceptical. They suggest that we should be finding out more about the importance of the process; they do not treat it as a normal part of speciation.

The by-product theory has been more consistently successful. Rice and Hostert (Chapter 26) gathered together a large number of experimental results to show that speciation happens as an automatic consequence when natural selection favours different characteristics in separate (or nearly separate) populations. They particularly looked at evidence of preferential mating. Members of the separate populations seem to evolve a preference for mating with their own type, even though this preference has not been directly selected for in the experiment. It is a remarkably strong result, remarkable not least for the way in which biologists have overlooked it. It is a solid rock on which we can stand when seeking to go further into the question. (I have included only a part, perhaps less than a half, of Rice and Hostert's paper; the original also discusses reinforcement, among other topics.) Coyne and Orr (Chapter 27) look at another part of the by-product theory—the reduction of hybrid fitness due to genetic incompatibilities between the parents. There are masses of evidence for this process, and it is a second solid rock to stand on.

However, not everyone accepts that genetic incompatibilities explain why hybrids are adaptively inferior. An alternative is that hybrids are poorly adapted ecologically. For instance, if hybrids are intermediate in form between the two parental species, the hybrids may not have the right adaptations to exploit the local resources. This ecological theory is defended by Schluter (Chapter 28). Currently, I suspect it is a minority view. It is the topic of active research and makes an interesting 'compare and contrast' with the genetic 'Dobzhansky-Muller' theory favoured by Coyne and Orr.

Finally, I include a brief introduction by Grant (Chapter 29) to hybrid speciation in plants. Hybrid speciation is one of the best confirmed mechanisms for the origin of new species; it has even been possible to replicate the process in the laboratory, as Grant describes. Hybrid speciation is a distinctively botanical contribution to evolutionary biology: it has been worked out exclusively in research on plants, by plant scientists, and is probably much more frequent in plant evolution than in animals or microbes. However, we know little about microbial speciation—a matter touched on in Ochman *et al.*'s paper in Section F (Chapter 36) and Schopf's paper in Section G (Chapter 43).

E. MAYR

22 Typological versus population thinking

Rather imperceptibly a new way of thinking began to spread through biology soon after the beginning of the nineteenth century. It is now most often referred to as population thinking. What its roots were is not at all clear, but the emphasis of animal and plant breeders on the distinct properties of individuals was clearly influential. The other major influence seems to have come from systematics. Naturalists and collectors realized increasingly often that there are individual differences in a group of human beings. Population thinking, despite its immense importance, spread rather slowly, except in those branches of biology that deal with natural populations.

In systematics it became a way of life in the second half of the nineteenth century, particularly in the systematics of the better-known groups of animals, such as birds, mammals, fishes, butterflies, carabid beetles, and land snails. Collectors were urged to gather large samples at many localities, and the variation within populations was studied as assiduously as differences between localities. From systematics, population thinking spread, through the Russian school, to population genetics and to evolutionary biology. By and large it was an empirical approach with little explicit recognition of the rather revolutionary change in conceptualization on which it rested. So far as I know, the following essay, excerpted from a paper originally published in 1959, was the first presentation of the contrast between essentialist and population thinking, the first full articulation of this revolutionary change in the philosophy of biology. [Author's summary.]

The year of publication of Darwin's *Origin of Species*, 1859, is rightly considered the year in which the modern science of evolution was born. It must not be forgotten, however, that preceding this zero year of history there was a long prehistory. Yet, despite the existence in 1859 of a widespread belief in evolution, much published evidence on its course, and numerous speculations on its causation, the impact of Darwin's publication was so immense that it ushered in a completely new era.

It seems to me that the significance of the scientific contribution made by Darwin is threefold:

1. He presented an overwhelming mass of evidence demonstrating the occurrence of evolution.

2. He proposed a logical and biologically well-substantiated mechanism that might account for evolutionary change, namely, natural selection. [. . .]
3. He replaced typological thinking by population thinking.

The first two contributions of Darwin's are generally known and sufficiently stressed in the scientific literature. Equally important but almost consistently overlooked is the fact that Darwin introduced into the scientific literature a new way of thinking, 'population thinking.' What is this population thinking and how does it differ from typological thinking, the then prevailing mode of thinking? Typological thinking no doubt had its roots in the earliest efforts of primitive man to classify the bewildering diversity of nature into categories. The *eidos* of Plato is the formal philosophical codification of this form of thinking. According to it, there are a limited number of fixed, unchangeable 'ideas' underlying the observed variability, with the *eidos* (idea) being the only thing that is fixed and real, while the observed variability has no more reality than the shadows of an object on a cave wall, as it is stated in Plato's allegory. The discontinuities between these natural 'ideas' (types), it was believed, account for the frequency of gaps in nature. Most of the great philosophers of the seventeenth, eighteenth, and nineteenth centuries were influenced by the idealistic philosophy of Plato, and the thinking of this school dominated the thinking of the period. Since there is no gradation between types, gradual evolution is basically a logical impossibility for the typologist. Evolution, if it occurs at all, has to proceed in steps or jumps.

The assumptions of population thinking are diametrically opposed to those of the typologist. The populationist stresses the uniqueness of everything in the organic world. What is true for the human species—that no two individuals are alike—is equally true for all other species of animals and plants. Indeed, even the same individual changes continuously throughout its life-time and when placed into different environments. All organisms and organic phenomena are composed of unique features and can be described collectively only in statistical terms. Individuals, or any kind of organic entities, form populations of which we can determine only the arithmetic mean and the statistics of variation. Averages are merely statistical abstractions; only the individuals of which the populations are composed have reality. The ultimate conclusions of the population thinker and of the typologist are precisely the opposite. For the typologist, the type (*eidos*) is real and the variation an illusion, while for the populationist the type (average) is an abstraction and only the variation is real. No two ways of looking at nature could be more different.

The importance of clearly differentiating these two basic philosophies and concepts of nature cannot be overemphasized. Virtually every controversy in the field of evolutionary theory, and there are few fields of science with as many controversies, was a controversy between a typologist and a populationist. Let me take two topics, race and natural selection, to illustrate the great difference in interpretation that results when the two philosophies are applied to the same data.

Race[1]

The typologist stresses that every representative of a race has the typical characteristics of that race and differs from all representatives of all other races by the characteristics 'typical' for the given race. All racist theories are built on this foundation. Essentially, it asserts that every representative of a race conforms to the type and is separated from the representatives of any other race by a distinct gap. The populationist also recognizes races but in totally different terms. Race for him is based on the simple fact that no two individuals are the same in sexually reproducing organisms and that consequently no two aggregates of individuals can be the same. If the average difference between two groups of individuals is sufficiently great to be recognizable on sight, we refer to such groups of individuals as different races. Race, thus described, is a universal phenomenon of nature occurring not only in man but in two thirds of all species of animals and plants.

Two points are especially important as far as the views of the populationist on race are concerned. First, he regards races as potentially overlapping population curves. For instance, the smallest individual of a large-sized race is usually smaller than the largest individual of a small-sized race. In a comparison of races the same overlap will be found for nearly all examined characters. Second, nearly every character varies to a greater or lesser extent independently of the others. Every individual will score in some traits above, in others below the average for the population. An individual that will show in all of its characters the precise mean value for the population as a whole does not exist. In other words, the ideal type does not exist.

Natural selection

A full comprehension of the difference between population and typological thinking is even more necessary as a basis for a meaningful discussion of the most important and most controversial evolutionary theory, namely, Darwin's theory of evolution through natural selection. For the typologist everything in nature is either 'good' or 'bad,' 'useful' or 'detrimental.' Natural selection is an all-or-none phenomenon. It either selects or rejects, with

[1] Livingstone (Chapter 57) also applies population thinking to dismiss the idea of races in humans.

rejection being by far more obvious and conspicuous. Evolution to him consists of the testing of newly arisen 'types.' Every new type is put through a screening test and is either kept or, more probably, rejected. Evolution is defined as the preservation of superior types and the rejection of inferior ones, 'survival of the fittest' as Spencer put it. Since it can be shown rather easily in any thorough analysis that natural selection does not operate in this described fashion, the typologist comes by necessity to the conclusions: (1) that natural selection does not work, and (2) that some other forces must be in operation to account for evolutionary progress.

The populationist, on the other hand, does not interpret natural selection as an all-or-none phenomenon. Every individual has thousands or tens of thousands of traits in which it may be under a given set of conditions selectively superior or inferior in comparison with the mean of the population. The greater the number of superior traits an individual has, the greater the probability that it will not only survive but also reproduce. But this is merely a probability, because under certain environmental conditions and temporary circumstances, even a 'superior' individual may fail to survive or reproduce. This statistical view of natural selection permits an operational definition of 'selective superiority' in terms of the contribution to the gene pool of the next generation.

[*Evolution and the Diversity of Life* (Cambridge, Mass.: Harvard UP), 26–9.]

E. MAYR

23 Species concepts and their application

We should distinguish three sorts of species concept: typological concepts; concepts used by local naturalists; and concepts based on interbreeding. Biological species concepts incorporate both reproductive isolation from other groups and interbreeding within a group. Biological species concepts can be difficult to apply when: (1) morphological data is used to assess interbreeding; (2) information is lacking; (3) speciation is incomplete; (4) reproduction is asexual. [Editor's summary.]

[. . .] A study of all the species definitions published in recent years indicates that they are based on three theoretical concepts, neither more nor less. An understanding of these three concepts is a prerequisite for the investigation of the problem of speciation.

(1) *The Typological Species Concept.* This is the simplest and most widely held species concept. Species here means 'a different thing,' something that 'looks different' (from the Latin *specere*, to look at, to regard), 'a different kind.' This is the concept the mineralogist has in mind when he speaks of 'species of minerals' or the physicist who speaks of 'nuclear species.'

This simple concept of everyday life was made the basis of the *eidos* in Plato's philosophy. Different authors have stressed different aspects of Plato's *eidos*, some its independence of perception, others its transcendent reality, and still others its eternity and immutability. All these concepts take for granted that there is an unchanging essence, an *eidos*, which alone has objective reality. Objects, on the other hand, are for Plato and his adherents merely varying manifestations ('shadows') of the *eidos*. The individuals of a natural species, being merely shadows of the same 'type,' do not stand in any special relation to each other. Variation, under this concept, is due to the imperfections in the visible manifestations of the 'idea' implicit in each species.

There are, however, limits to the amount of variation that can be ascribed to the varying manifestations of a single *eidos*. Where it transgresses these limits, more than one *eidos* must be involved. Degree of morphological difference, thus, determines species status. The two aspects of the typological species concept, subjectivity and definition by degree of difference, depend on each other and are logical correlates. The typological species concept, translated into practical taxonomy, is the morphologically defined species.

In recent years most systematists have found this typological-morphological concept inadequate and have rejected it. They have pointed out that this concept treats the individuals of a species like an aggregation of inanimate objects, a singularly inappropriate treatment for a reproductive community. They have also called attention to the fact that the morphological species criterion is highly misleading in cases of polymorphic diversity within species or of morphologically extremely similar species. Where the taxonomist applies morphological criteria, he uses them as secondary indications of reproductive isolation.

(2) *The Nondimensional Species Concept*. This concept is based on the relation of two coexisting natural populations in a nondimensional system, that is, at a single locality and at the same time (sympatric and synchronous). This is the species concept of the local naturalist. If one studies the birds, the mammals, the butterflies, or the snails near his home town, he finds each species clearly defined and sharply separated from all other species. This is sometimes better appreciated by primitive natives than by modern civilized man. Some 30 years ago I spent several months with a tribe of superb woodsmen and hunters in the Arfak Mountains of New Guinea. They had 136 different vernacular names for the 137 species of birds that occurred in the area, confusing only two species. It is not, of course, pure coincidence that these primitive woodsmen arrive at the same conclusion as the museum taxonomists, but an indication that both groups of observers deal with the same, non-arbitrary discontinuities of nature.

This striking discontinuity between sympatric populations is the basis of the species concept in biology. The two taxonomists who, more than anyone else, were responsible for the acceptance of species in biology were local naturalists, John Ray in England and Carolus Linnaeus in southern Sweden. But anyone can test the reality of these discontinuities for himself, even where the morphological differences are slight. In eastern North America, for instance, there are four rather similar species of the genus *Catharus*, the Veery (*C. fuscescens*), the Hermit Thrush (*C. guttatus*), the Olive-backed or Swainson's Thrush (*C. ustulatus*), and the Gray-cheeked Thrush (*C. minimus*). These four species are sufficiently similar visually that they confuse not only the human observer, but also silent males of the other species. The species-specific songs and call notes, however, permit easy species discrimination. Rarely more than two species breed in the same area and the overlapping species $f + g$, $g + u$ and $u + m$ usually differ considerably in their foraging habits and niche preference, so that competition is minimized with each other and with two other thrushes, the Robin (*Turdus migratorius*) and the Wood Thrush (*Hylocichla mustelina*), with which they share their geographic range. In connection with their different foraging and migratory habits the four species differ from each other (and from other thrushes) in the relative length of wing and leg elements and in the shape of the bill. The rather extraordinary number of small differences between these at first sight very similar species has been worked out in detail. Most importantly, no hybrids or intermediates among these four species have ever been found. Each is a separate genetic, behavioral, and ecological system, separated from the others by a complete biological discontinuity, a gap.

Indeed the most characteristic attribute of a species in such a nondimensional system is that it is separated by a gap from other units in this system. The gap that surrounds a species is the core of the species concept. The term 'species' signifies a very definite mutual relation between sympatric populations, between units in a nondimensional system, namely that of reproductive isolation. The great advantage of the criterion of interbreeding between two populations in a nondimensional system is that its presence or absence can be determined unequivocally. Reproductive isolation thus supplies an objective yardstick, a completely nonarbitrary criterion, for the determination of species status of a population. The word 'species' indicates a relationship, like the word 'brother.' Being a brother is not an inherent property of an individual, as hardness is the property of a stone. An individual is a brother only with respect to someone else. A population is a species only with respect to other populations. To be a different species is not a matter of difference but of distinctness.

(3) *The Interbreeding-population Concept*. The concept of the multidimensional species is a collective concept. It considers species as groups of populations that actually or potentially interbreed with each other. Such populations, in order to retain their identity, cannot coexist at the same place and at the same time. The multidimensional-species concept thus deals

with allopatric and allochronic populations, populations distributed in the dimensions of space and time, and classifies them on the basis of mutual interbreeding.

This concept has the weakness of all collective concepts, that of practical difficulties of delimitation: which discontinuous populations shall be judged 'potentially' interbreeding? Even though the multidimensional concept comes much closer to reality than the nondimensional concept, it is evident that it lacks the latter's objectivity.

Species definitions

When the term species is applied to inanimate objects, as in 'species of minerals,' it is based on the typological species concept. When the term is used in biology, it is based to a greater or lesser degree on the two other concepts, the nondimensional ('reproductive gap') and the multidimensional ('unlimited gene exchange'). Parts of these two concepts have been incorporated into nearly all species definitions in biology in the last 100 years. Most of the definitions proposed in the last 25 years have avoided all reference to morphological distinctness. For instance, I defined species[1] as 'groups of actually or potentially interbreeding natural populations which are reproductively isolated from other such groups,' and Dobzhansky defined the species as 'the largest and most inclusive . . . reproductive community of sexual and cross-fertilizing individuals which share in a common gene pool.'

Definitions that stress this dual biological significance of species, reproductive isolation and community of gene pools, are usually referred to as 'biological' species definitions. This designation has been questioned on the grounds that this is not an exclusive terminology, since many of the other species definitions also refer to living species and their biological attributes. This cannot be denied and yet the designation 'biological species' would seem best for the modern concept for three reasons. First, it has never been used for any other species concept or definition; no confusion can arise as to the intent of an author who uses this terminology. Second, this terminology emphasizes that the underlying concept is based on the biological meaning of the species, that is, to serve as a protective device for a well-integrated, co-adapted set of gene complexes. Third, alternative terminologies are even more ambiguous. [. . .] The term 'biological species concept' for a concept emphasizing interbreeding within the population system and reproductive isolation against others is now so widely adopted and so uniformly used that it could hardly lead to misunderstanding.

[1] Quotes respectively from E. Mayr, 'Speciation phenomena in birds', *American Naturalist*, 74 (1940), 249–58, and T. Dobzhansky, *American Naturalist*, 84 (1950), 401–18.

If we wanted to single out the aspects most frequently stressed in recent discussions of the biological species concept, we would list these three:

(1) Species are defined by distinctness rather than by difference;
(2) Species consist of populations rather than of unconnected individuals; and
(3) Species are more unequivocally defined by their relation to non-conspecific populations ('isolation') than by the relation of conspecific individuals to each other. The decisive criterion is not the fertility of individuals but the reproductive isolation of populations.

The typological species concept treats species as random aggregates of individuals that have in common 'the essential properties of the type of the species' and that 'agree with the diagnosis.' This static concept ignores the fact that species are reproductive communities. The individuals of a species of animals recognize each other as potential mates and seek each other for the purpose of reproduction. A multitude of devices insure intraspecific reproduction in all organisms. The species is also an ecological unit that, regardless of the individuals composing it, interacts as a unit with other species with which it shares the environment. The species, finally, is a genetic unit consisting of a large, intercommunicating gene pool, whereas the individual is merely a temporary vessel holding a small portion of the contents of the gene pool for a short period of time. [. . .]

Difficulties in the application of the biological species concept

The general adoption of the biological species concept has done away with a bewildering variety of 'standards' followed by the taxonomists of the past. One taxonomist would call every polymorph variant a species, a second would call every morphologically different population a species, and a third would call every geographically isolated population a species. This lack of a universally accepted standard confused not only the general biologists who wanted to use the work of the taxonomist, but the taxonomists themselves. Agreement on a single yardstick, the biologically defined category species, to be applied by everybody, has been a great advance toward mutual understanding.

Yet not all difficulties were eliminated by the discovery of this yardstick. Some taxonomists confused themselves and the issue by failing to understand that there is a difference between the species as a category and the species as a taxon. The species as category is characterized by the biological species concept. The practicing taxonomist, however, deals with taxa, with populations and groups of populations, which he has to assign to one category or another, for instance either to the category species or to the category sub-species. The nonarbitrary criterion of the category species, biologically defined, is that of the interbreeding or noninterbreeding. When confronted with the task of having to assign a taxon to the correct category, the occurrence or potentiality of interbreeding is usually only inferred. This, as Simpson has stressed, poses in most cases only a pseudo problem. Whether a given taxon deserves to be placed in the category species is a matter of the total available evidence.

The evidence that the definition is met in a given case with a sufficient degree of probability is a different matter [from the validity of the concept]. The evidence is usually morphological, but to conclude that one therefore is using or should use a morphological concept of the category (not taxon) species is either a confusion in thought or an unjustified relapse into typology. The evidence is to be judged in the light of known consequences of the genetical situation stated in the definition [of the category].[2]

Taxonomy is not alone in encountering difficulties when trying to assign concrete phenomena to categories. Most of the universally accepted concepts of our daily life encounter similar difficulties. The transition in category from subspecies to species is paralleled by the transitions from child to adult, from spring to summer, from day to night. Do we abandon these categories because there are borderline cases and transitions? Do we abandon the concept tree because there are dwarf willows, giant cactuses, and strangler figs? Such conflicts are encountered whenever one is confronted with the task of assigning phenomena to categories. [. . .]

Lack of information

The Ranking of Variant Individuals. Whether certain morphologically rather distinct individuals belong to the same species or not is a routine problem of taxonomy. [. . .] It is important to emphasize the difficulties, caused by sexual dimorphism, age differences, genetic polymorphism, and nongenetic habit differences, which face the student of insects, of parasites, and indeed of any group of living animals, because some paleontologists seem to believe that it is only in work with fossils that one has to cope with the difficulty of having to draw inferences from morphological types.

No one will deny that the application of the biological species concept to fossil specimens is a difficult task. Yet, in principle, it does not differ from the task of the neontologist who only rarely can study natural populations but is usually forced to classify preserved specimens. The task of the paleontologist is clarified if he remembers that fossils are the remains of formerly living organisms that, when they were alive, were members of genetically defined

[2] G. G. Simpson, *Principles of Animal Taxonomy* (New York: Columbia University Press, 1961), 150.

populations exactly as the species living today are. Morphological criteria are used by the paleontologist as inferences on the natural populations that left the fossil remains. There is no justification for abandoning the biological approach merely because it is sometimes difficult to decide whether or not several morphological types in a sample are conspecific. No one makes the absurd demand that the paleontologist test the reproductive isolation of the species he recognizes. Yet by proper consideration of all the available morphological, ecological, stratigraphic, and distributional evidence it can usually be inferred with high probability whether certain specimens when living were or were not members of the same population. [. . .]

The Ranking of Populations. The criterion of species status, 'sympatric coexistence without interbreeding,' raises practical problems also where two populations occur in contiguous geographic areas but in very different habitats. Where the evergreen rain forest of central Africa comes in contact with open-country vegetation, one may find the forest drongo *Dicrurus ludwigii* within 50 meters of the very similar savanna drongo *D. adsimilis*, but not on the same tree. Indeed they never interbreed. The same is true of other closely related species pairs wherever habitats meet along a sharp border. Even though such species replace each other spatially they must nevertheless be considered sympatric. The potential mates are within cruising range of each other during the breeding season, and could freely interbreed if they were not kept apart by specific isolating mechanisms. The terms 'sympatric' or 'coexistence' in species definitions must be conceived broadly, to include populations the individuals of which are within cruising range of each other during the breeding season, even though the habitats in which they occur do not overlap in space.

Incompleteness of speciation

Evolution is a gradual process and, in general, so is the multiplication of species (except by polyploidy). As a consequence one finds many populations in nature that have progressed only part of the way toward species status. They may have acquired some of the attributes of distinct species and lack others. One or another of the three most characteristic properties of species—reproductive isolation, ecological difference, and morphological distinguishability—is in such cases only incompletely developed. The application of the species concept to such incompletely speciated populations raises considerable difficulties. The various situations usually encountered can be classified under six headings.

(1) *Evolutionary continuity in space and time.* Species that are widespread in space or time may have terminal populations that behave toward each other like distinct species even though they are connected by an unbroken chain of interbreeding populations. [. . .] For instance, when Leopard Frogs (*Rana*

pipiens) from the northern United States are crossed with frogs from southern Florida or from Texas most of the embryos die during development.

Intermediacy of populations between successive species would be the normal situation in paleontology if all populations had left a fossil record. Actually the breaks in the fossil record are so frequent that it has been possible in only a few cases to piece together unbroken lineages connecting good species. The evolution from *Micraster leskei* through *M. cortestudinarium* to *M. coranguinum* is one such case. In other cases, cited in the literature, the differences in the lineages are so slight that neontologists would be inclined to consider the consecutive forms merely subspecies of a single polytypic species. Even though the number of cases causing real difficulties to the taxonomist is very small, it cannot be denied that an objective delimitation of species in a multidimensional system is an impossibility.

(2) *Acquisition of reproductive isolation without equivalent morphological change*. This group of cases raises a difficulty more practical than fundamental. When the reconstruction of the genotype in an isolated population has resulted in the acquisition of reproductive isolation, such a population must be considered a biological species, regardless of how little it may have changed morphologically.

(3) *Morphological differentiation without acquisition of reproductive isolation*. The acquisition of isolating mechanisms in isolated populations sometimes lags far behind morphological divergence. Such populations will be as different morphologically as good species and yet interbreed indiscriminately where they come in contact. The West Indian snail genus *Cerion* illustrates this situation particularly well. Whenever reproductive isolation and morphological differentiation do not coincide, the decision as to species status must be based on a broad evaluation of the particular case. The solution is generally a rather unsatisfactory compromise.

(4) *Reproductive isolation based on habitat isolation*. Numerous cases have been described in the literature in which natural populations acted toward each other like good species (in areas of contact) as long as their habitats were undisturbed. Yet the reproductive isolation broke down as soon as the characteristics of these habitats were changed, usually by the interference of man. The toads *Bufo americanus* and *B. fowleri* in North America, and the flycatchers *Terpsiphone rufiventer* and *T. viridis* are well-known examples. Prior to the habitat disturbance no one would have questioned the status of these species, but afterward they behaved like conspecific populations.

(5) *The incompleteness of isolating mechanisms*. Very few isolating mechanisms are all-or-none devices. They are built up step by step (except in polyploidy) and most isolating mechanisms of an incipient species will be imperfect and incomplete. Species level is reached when the process of speciation has become irreversible, even if some of the (component) isolating mechanisms have not yet reached perfection. To determine whether or not

an incipient species has reached the point of irreversibility is often impossible.

(6) *Attainment of different levels of speciation in different local populations.* The perfecting of isolating mechanisms may proceed at different rates in different populations of a polytypic species. Two widely overlapping species may, as a consequence, be completely distinct at certain localities but may freely hybridize at others. [. . .] The compromise solution that the practicing taxonomist often adopts, other things being equal, is to compare the sizes of the areas of undisturbed sympatry and of hybridization. Whichever is the larger determines species status.

The species is a population separated from others by a discontinuity, but not every discontinuity entitles the isolated population to species rank. If we designate as an *isolate* any more or less isolated population or array of populations, we can distinguish in sexually reproducing organisms between geographical, ecological, and reproductive isolates, of which only the last are species. The unspoken assumption made by certain authors, that the three kinds of isolates coincide, is not supported by the known facts and has led to unwarranted conclusions regarding the pathways of speciation.

The six types of phenomena described in the preceding paragraphs are consequences of the gradual nature of the ordinary process of speciation. Determination of species status of a given population is difficult or impossible in many of these cases.

The difficulties posed by asexuality

The criterion of interbreeding among natural populations, the ultimate test of conspecificity in the higher animals, is unavailable in uniparentally reproducing organisms. It is evident that the absence of this criterion provides the most formidable and most fundamental obstacle to the application of the biological species concept. What should the evolutionist consider the 'unit of evolution' in such organisms?

Asexuality in existing organisms is almost certainly a secondary phenomenon. All existing asexual organisms seem to be derived from sexual forms. Asexually reproducing lines have, sooner or later, one of three fates: they are lost by extinction, or they mutate, or they exchange genes with some other line by some process of recombination. Indeed clandestine sexuality appears to be rather common among so-called asexual organisms. The expression 'uniparental reproduction' is being used increasingly, instead of 'asexual reproduction,' to overcome this and other difficulties. Many biologists, for instance, are reluctant to refer to parthenogenesis as asexual reproduction.

It is too early for a definitive proposal concerning the application of the species concept to asexually or uniparentally reproducing organisms. If mutation and survival were random among the descendants of an asexual individual, one would expect a complete morphological (and genetic) continuum. Yet discontinuities have been found in most carefully studied groups of asexual organisms and this has made taxonomic subdivision possible. For this phenomenon I have advanced the explanation[3] 'that the existing types are the survivors among a great number of produced forms, that the surviving types are clustered around a limited number of adaptive peaks, and that ecological factors have given the former continuum a taxonomic structure.' Each adaptive peak is occupied by a different 'kind' of organism and if each 'kind' is sufficiently different from other kinds it will be legitimate to call such a cluster of genotypes a species.

Various proposals have been made to resolve the difficulty that asexuality raises for the biological species concept. Some authors have gone so far as to abandon the biological species concept altogether and return to the morphological species for sexual and asexual organisms. I can see nothing that would recommend this solution. It exaggerates the importance of asexuality, which is both secondary and limited in its extent, and reintroduces the subjectivity and arbitrariness of the morphological species. [. . .]

Is the biological species concept invalidated by the difficulties in its application that have been listed?

One can confidently answer this question: 'No!' Almost any concept is occasionally difficult to apply, without thereby being invalidated. The advantages of the biological species are far greater than its shortcomings. Difficulties are rather infrequent in most groups of animals and are well circumscribed where they do occur. Such difficulties are least frequent in nondimensional situations where (except in paleontology) most species studies are done. Indeed the biological species concept, even where it has to be based on inference, nearly always permits the delimitation of a sounder taxonomic species than does the morphological concept.

The importance of a nonarbitrary definition of species

Whoever, like Darwin, denies that species are nonarbitrarily defined units of nature not only evades the issue, but fails to find and solve some of the most interesting problems of biology. These problems will be apparent only to the student who attempts to determine species status of natural populations. The correct classification of the many different kinds of varieties, of polymorphism, of polytypic species, of biological races, would all be meaningless, indeed would be ignored, but for an interest in arranging natural populations and phenotypes into biological species. Application of the concept has led to advances in the sorting of fossil specimens. Even though the evidence is

[3] E. Mayr, 'Difficulties and Importance of the Biological Species', in E. Mayr (ed.), *The Species Problem* (American Association for the Advancement of Science, Publication No. 5, 1957), 371–88.

largely morphological, an interpretation of fossil specimens based on biological concepts forces the paleontologist to make clear-cut decisions: morphologically different specimens found in the same exposure (the same sample) must be either different species or intrapopulation variants (excepting the relatively rare instances of secondary deposits).

It was not possible to state the problem of the multiplication of species with precision until the biological species concept had been developed. Only after the naturalists had insisted on the sharp definition of local species was there a problem of the bridging of the gap between species. And only then did the problem arise whether or not the species is a unit of evolution, and what sort of unit.

[*Animal Species and Evolution* (Cambridge, Mass.: Harvard UP, 1963), ch. 2.]

CHARLES DARWIN

24 The sterility of hybrids

Darwin is here mainly concerned to argue against what could now be called a 'creationist' view of species: the view that different species have somehow been specially endowed with an inability to interbreed. He argues that various facts about sterility do not fit this view. Instead they make sense if sterility evolves as an incidental consequence when two species evolve other differences. This extract finishes as Darwin notes a puzzle, that domestic varieties do not appear to have evolved sterility even though they have evolved differences in other characters. [Editor's summary.]

The view generally entertained by naturalists is that species, when intercrossed, have been specially endowed with the quality of sterility, in order to prevent the confusion of all organic forms. This view certainly seems at first probable, for species within the same country could hardly have kept distinct had they been capable of crossing freely. The importance of the fact that hybrids are very generally sterile, has, I think, been much underrated by some late writers. On the theory of natural selection the case is especially important, inasmuch as the sterility of hybrids could not possibly be of any advantage to them, and therefore could not have been acquired by the continued preservation of successive profitable degrees of sterility. I hope, however, to be able to show that sterility is not a specially acquired or endowed quality, but is incidental on other acquired differences. [. . .]

Laws governing the sterility of first crosses and of hybrids

We will now consider a little more in detail the circumstances and rules governing the sterility of first crosses and of hybrids. Our chief object will be to see whether or not the rules indicate that species have specially been endowed with this quality, in order to prevent their crossing and blending together in utter confusion. The following rules and conclusions are chiefly drawn up from Gärtner's admirable work on the hybridisation of plants. I have taken much pains to ascertain how far the rules apply to animals, and considering how scanty our knowledge is in regard to hybrid animals, I have been surprised to find how generally the same rules apply to both kingdoms.

It has been already remarked, that the degree of fertility, both of first crosses and of hybrids, graduates from zero to perfect fertility. It is surprising in how many curious ways this gradation can be shown to exist; but only the barest outline of the facts can here be given. When pollen from a plant of one family is placed on the stigma of a plant of a distinct family, it exerts no more influence than so much inorganic dust. From this absolute zero of fertility, the pollen of different species of the same genus applied to the stigma of some one species, yields a perfect gradation in the number of seeds produced, up to nearly complete or even quite complete fertility; and, as we have seen, in certain abnormal cases, even to an excess of fertility, beyond that which the plant's own pollen will produce. So in hybrids themselves, there are some which never have produced, and probably never would produce, even with the pollen of either pure parent, a single fertile seed: but in some of these cases a first trace of fertility may be detected, by the pollen of one of the pure parent-species causing the flower of the hybrid to wither earlier than it otherwise would have done; and the early withering of the flower is well known to be a sign of incipient fertilisation. From this extreme degree of sterility we have self-fertilised hybrids producing a greater and greater number of seeds up to perfect fertility.

Hybrids from two species which are very difficult to cross, and which rarely produce any offspring, are generally very sterile; but the parallelism between the difficulty of making a first cross, and the sterility of the hybrids thus produced—two classes of facts which are generally confounded together—is by no means strict. There are many cases, in which two pure species can be united with unusual facility, and produce numerous hybrid-offspring, yet these hybrids are remarkably sterile. On the other hand, there are species which can be crossed very rarely, or with extreme difficulty, but the hybrids, when at last produced, are very fertile.[1] Even within the limits of the same genus, for instance in Dianthus, these two opposite cases occur.

The fertility, both of first crosses and of hybrids, is more easily affected by unfavourable conditions, than is the fertility of pure species. But the degree of fertility is likewise innately variable; for it is not always the same when the same two species are crossed under the same circumstances, but depends in

[1] This is approximately to say that two species may have little prezygotic, but large postzygotic, isolation; or large prezygotic, but little postzygotic, isolation. However, 'species which can be crossed very rarely' might have postzygotic isolation at an early stage. For instance, the zygotes may fail to develop.

part upon the constitution of the individuals which happen to have been chosen for the experiment. So it is with hybrids, for their degree of fertility is often found to differ greatly in the several individuals raised from seed out of the same capsule and exposed to exactly the same conditions. [. . .]

By a reciprocal cross between two species, I mean the case, for instance, of a stallion-horse being first crossed with a female ass, and then a male-ass with a mare: these two species may then be said to have been reciprocally crossed. There is often the widest possible difference in the facility of making reciprocal crosses. Such cases are highly important, for they prove that the capacity in any two species to cross is often completely independent of their systematic affinity, or of any recognisable difference in their whole organisation. On the other hand, these cases clearly show that the capacity for crossing is connected with constitutional differences imperceptible by us, and confined to the reproductive system. [. . .]

Now do these complex and singular rules indicate that species have been endowed with sterility simply to prevent their becoming confounded in nature? I think not. For why should the sterility be so extremely different in degree, when various species are crossed, all of which we must suppose it would be equally important to keep from blending together? Why should the degree of sterility be innately variable in the individuals of the same species? Why should some species cross with facility, and yet produce very sterile hybrids; and other species cross with extreme difficulty, and yet produce fairly fertile hybrids? Why should there often be so great a difference in the result of a reciprocal cross between the same two species? Why, it may even be asked, has the production of hybrids been permitted? To grant to species the special power of producing hybrids, and then to stop their further propagation by different degrees of sterility, not strictly related to the facility of the first union between their parents, seems to be a strange arrangement.

The foregoing rules and facts, on the other hand, appear to me clearly to indicate that the sterility both of first crosses and of hybrids is simply incidental or dependent on unknown differences, chiefly in the reproductive systems, of the species which are crossed. The differences being of so peculiar and limited a nature, that, in reciprocal crosses between two species the male sexual element of the one will often freely act on the female sexual element of the other, but not in a reversed direction. It will be advisable to explain a little more fully by an example what I mean by sterility being incidental on other differences, and not a specially endowed quality. As the capacity of one plant to be grafted or budded on another is so entirely unimportant for its welfare in a state of nature, I presume that no one will suppose that this capacity is a *specially* endowed quality, but will admit that it is incidental on differences in the laws of growth of the two plants. We can sometimes see the reason why one tree will not take on another, from differences in their rate of growth, in the hardness of their wood, in the period of the flow or nature of their sap, &c.; but in a multitude of cases we can assign no reason whatever. [. . .]

Fertility of varieties when crossed, and of their mongrel offspring

It may be urged, as a most forcible argument, that there must be some essential distinction between species and varieties, and that there must be some error in all the foregoing remarks, inasmuch as varieties, however much they may differ from each other in external appearance, cross with perfect facility, and yield perfectly fertile offspring. I fully admit that this is almost invariably the case. But if we look to varieties produced under nature, we are immediately involved in hopeless difficulties; for if two hitherto reputed varieties be found in any degree sterile together, they are at once ranked by most naturalists as species. For instance, the blue and red pimpernel, the primrose and cowslip, which are considered by many of our best botanists as varieties, are said by Gärtner not to be quite fertile when crossed, and he consequently ranks them as undoubted species. If we thus argue in a circle, the fertility of all varieties produced under nature will assuredly have to be granted.

If we turn to varieties, produced, or supposed to have been produced, under domestication, we are still involved in doubt. For when it is stated, for instance, that the German Spitz dog unites more easily than other dogs with foxes, or that certain South American indigenous domestic dogs do not readily cross with European dogs, the explanation which will occur to everyone, and probably the true one, is that these dogs have descended from several aboriginally distinct species. Nevertheless the perfect fertility of so many domestic varieties, differing widely from each other in appearance, for instance of the pigeon or of the cabbage, is a remarkable fact; more especially when we reflect how many species there are, which, though resembling each other most closely, are utterly sterile when intercrossed. Several considerations, however, render the fertility of domestic varieties less remarkable than at first appears.[2] [. . .]

[*The Origin of Species* (London: John Murray, 1859), 264–82.]

[2] Darwin then discusses several interesting points. However, this is one topic for which a modern view differs from Darwin's. We now have extensive experimental evidence of reproductive isolation between populations that have evolved apart. See Rice and Hostert's extract (Chapter 26). Darwin did mention some (unconvincing) evidence of the sort we now have in abundance, and from convincingly controlled experiments.

25

THEODOSIUS DOBZHANSKY

Reproductive isolation as a product of genetic divergence and natural selection

Two theories of speciation have been suggested: new species may evolve (1) as a by-product of evolutionary divergence between populations, and (2) by natural selection (that is, by 'reinforcement'). The extract looks at evidence for (2) from experiments, then for (1) from crosses between biogeographically separate populations, and then for (2) from higher ethological isolation between sympatric than between allopatric populations (that is, character displacement). Theory (2) applies mainly to prezygotic isolation. [Editor's summary.]

To state that races are incipient species is not tantamount to saying that every race is a future species. Race differentiation is reversible; race divergence may be superseded by convergence. This is, in fact, what is happening to the human species. To become species, races must evolve reproductive isolation. The question naturally presents itself, what causes bring about the development of reproductive isolating mechanisms? Two hypothetical answers have been proposed. First, reproductive isolation is a by-product of the accumulation of genetic differences between the diverging races. The same genes that make the races diverge in morphological and physiological traits render them reproductively isolated. Second, the isolation is built up by natural selection, when and if the gene exchange between the diverging populations generates recombination products of low fitness. The establishment of reproductive isolation is a special kind of genetic divergence. These two hypotheses are not mutually exclusive. Needless disputes have arisen because they were mistakenly treated as alternatives.

The first hypothesis has long been implicit in the thinking of systematists but its genetic formulation is due to Muller. The gene pool of a population is an integrated system of genes; evolutionary changes are not mere additions or subtractions of unrelated gene elements. The initial advantage of most mutations that arise and become established in a species is slight. As the accumulation of gene differences continues, genes that at one time might have been easily dispensed with become essential constituents of the genotype. In the course of evolution, the functions of a gene in the development may undergo changes. If in two or more races or species the gene functions diverge, the gene systems may no longer be compatible in hybrids. In Muller's opinion, all isolating mechanisms may arise in this manner:

> Which kind of character becomes affected earliest, and to what degree . . . will depend in part upon its general complexity (which is correlated with the number of genes affecting it), in part on the nicety or instability of the equilibria of processes necessary for its proper functioning, and in part on the accidental circumstances that determined just which incompatible mutations happened to become established first.

The second hypothesis was, according to Grant, suggested as far back as 1889 by A. R. Wallace,[1] and later by Fisher and Dobzhansky. It starts from the same premise as that of Muller (see above), namely, that the genotype of a species is an integrated system adapted to the ecological niches in which the species lives. Gene recombination in the offspring of species hybrids may lead to the formation of discordant gene patterns that decrease the reproductive potentials of both interbreeding populations. Suppose that incipient species, A and B, are in contact in a certain territory. Mutations arise in either or in both species that make their carriers averse to mating with the other species. The nonmutant individuals of A that cross to B will produce a progeny inferior to the pure species. Since the mutants breed only or mostly within the species, their progeny will be superior in fitness to that of the nonmutants. Consequently, natural selection will favor the spread and establishment of the mutant condition.

Sturtevant and Bruce Wallace have pointed out one of the possible causes that might initiate such a process. Suppose that the gene arrangement *ABCDEFGH* in a chromosome is modified in one race to *AFEDCBGH* and in another race to *ABGFEDCH*. Heterozygotes carrying the ancestral and either of the modified arrangements will produce few or no inviable offspring. In a hybrid carrying the two modified arrangements, crossing over in the section *CDEF* will give chromosomes *AFEDCH* and *ABGFEDCBGH*. Such chromosomes may be inviable. Hence prevention of the interbreeding of carriers of *AFEDCBGH* and *ABGFEDCH* will have a selective advantage.

There is good experimental evidence that selection can build up reproductive isolation. Koopman made use of the observation that the ethological isolation between *Drosophila pseudoobscura* and *D. persimilis* is weaker at a low (16°c) than at a higher (25°c) temperature. He placed in population cages equal numbers of females and males of the two species, marked by two different recessive mutants. The offspring of matings within and between species are distinguishable by inspection; if the two species are *aaBB* and *AAbb*, the hybrids are wild-type, *AaBb*. In every generation, the hybrids were destroyed, and the populations were continued with equal numbers of the pure species. Therefore, the flies that mated with representatives of their own species had their progenies included among the parents of the next generation, whereas those mating with the other species suffered 'genetic death'. After only five generations of selection, the proportions of hybrids among the offspring fell to a fraction of the former value.[2] [. . .]

[1] Wallace probably did not have the theory of reinforcement, though he did have several components of it. His writings are disentangled by H. Cronin, *The Ant and the Peacock*, pp. 416–25 (Cambridge: Cambridge University Press, 1991).

[2] Dobzhansky here reviews several further experiments of the same sort. However, their relevance is challenged by Rice and Hostert (Chapter 26).

In order that natural selection may promote reproductive isolation, there must be a challenge of loss of fitness owing to gene flow between populations. Reduced viability or fertility of hybrid offspring provides such a challenge. This is another way of saying that postmating isolating mechanisms may act as stimuli for the development of premating isolation. Postmating isolating mechanisms (i.e. hybrid inviability, sterility, breakdown, or combinations of these) are, then, consequences of differential adaptedness of races or species to the conditions of life in their respective distribution areas. They are by-products of genetic divergence [. . .]

Rick has studied the wild tomato (*Lycopersicon peruvianum*), which grows in stream valleys along the coast of Peru. Intercrosses of northern strains with southern ones give few viable seeds. Nevertheless, geographically intermediate populations form a 'compatibility bridge', and the genetic unity of the species is maintained. Rick states:

> It is not difficult to understand how such barriers might arise gradually in races that have been isolated for long periods of time. Different reaction norms for rates of embryo and endosperm development, osmotic values, and other developmental characteristics might have become fixed by selection while races were adapting to the new environments into which they were migrating.

Similar situations have been observed by, among others, Grant in Gilia, Stebbins in Elymus, Kruckeberg in Streptanthus, Vickery in species of monkey flowers (Mimulus), and Levin and Kerster in Phlox.

Hoenigsberg and Koref-Santibañez found differences in courtship patterns between some laboratory strains of *Drosophila melanogaster*, which result in a preference for homogamic matings. Similar preferences exist, as pointed out previously, in Transitional populations of *D. paulistorum* and in geographic strains of *D. birchii*. It is unlikely that these rudiments of ethological isolation were built by natural selection for their function as isolating mechanisms; on the other hand, they are genetic raw materials from which reproductive isolation may be compounded by natural selection. It is appropriate to mention at this point that not all genetic divergence leads to changed mating preferences. Robertson selected a population of *D. melanogaster* for adaptedness to a modified diet. Although the adaptation involved multiple, polygenic gene differences, no trace of ethological isolation between the original and the changed strains was found.

There is ample, though of necessity indirect, evidence that selection builds isolating mechanisms in nature. At least the premating isolating mechanisms between closely related species should be enhanced in the geographic areas where hybridization is most likely to occur. The observations of Ehrman on sympatric and allopatric strains of *Drosophila paulistorum* are among the most elegant verifications of this prediction. The ethological isolation is greater among sympatric than among allopatric strains of the same pairs of semispecies.

In a series of papers Grant has supplied a demonstration that mechanisms preventing the hybridization of species of Gilia arise by selection under conditions of sympatry. Five related species occur in the foothills and valleys of California and are often found growing side by side. Four other species occur in coastal localities in North and South America; they are completely allopatric with respect to one another, and largely so with respect to the five inland species. Experimentally obtained species hybrids are highly sterile in all combinations tried. Yet the allopatric species can be crossed quite easily, giving 18.1 hybrid seeds per flower on the average. In contrast, the sympatric species are separated by crossability barriers and yield only 0.2 seeds per flower when cross-pollinated artificially.

Mating call differences are important isolating factors between related species of anuran amphibians. Littlejohn has analyzed the mating calls of allopatric and sympatric populations of two species of Australian frogs:

> Whereas mating calls of remote allopatric populations of *Hyla ewingi* and *H. verreauxi* are very similar, those of the sympatric populations are quite distinct. . . . It is suggested that the marked differences between sympatric populations have resulted from the direct action of selection for increased reproductive efficiency, i.e., the slight differences present in the allopatric populations have been reinforced in the sympatric populations.

Some evidence of sympatric reinforcement of species differences in mating calls has also been recorded in *Microhyla olivacea* and *M. carolinensis* and in *Hyla regilla* and *H. californiae*. No such reinforcement was found, however, in *Pseudacris clarki* and *P. nigrita*, or in *P. ornata* and *P. streckeri*. [. . .]

The examples just cited of the reinforcement of premating isolating mechanisms may be viewed as instances of character displacement, which Brown and Wilson[3] defined as 'the situation in which, when two species of animals overlap geographically, the differences between them are accentuated in the zone of sympatry and weakened or lost entirely in the parts of their ranges outside this zone.' Habitat, temporal, and ethological isolations are particularly likely to arise in this manner. [. . .]

Instances of the lack or the weakness of premating isolating mechanisms, where the populations of related species are wholly or largely allopatric, are perhaps as significant as their presence where the species are sympatric. Thus Zaslavsky found that hybrids between the ladybird beetle species (or semispecies) *Chilocorus bipustulatus* and *Ch. geminus* are sterile, but detected no ethological isolation at all, at least under experimental conditions. Hybrid belts, formed where the geographic areas of two species come into contact,

[3] Brown, W. L. and Wilson, E. O., 'Character displacement', *Systematic Zoology*, 5 (1958), 49–64.

have been studied to find whether premating isolating mechanisms may be formed there. The most thoroughly investigated cases are those of two species of crows (Corvus) in Europe, and of grackles (Quiscalus) in the United States. The evidence is ambiguous.

Whether postmating isolating mechanisms can be reinforced by natural selection is also an open problem. If the progeny of hybrids is inferior in fitness, it would seem advantageous to the species concerned to prevent hybridization, either by premating isolation or, failing that, by such postmating mechanisms as inviability or sterility of F_1 hybrids. Group selection could, theoretically, bring such a result about. However, because the efficiency of group selection is low relative to the selection of individual genotypes, it is doubtful that isolating mechanisms frequently arise in this way.

[*The Genetic Basis of Evolutionary Change* (New York: Columbia University Press, 1970), 376–82.]

WILLIAM R. RICE AND ELLEN E. HOSTERT

26 Laboratory experiments on speciation: what have we learned in 40 years?

Experimental studies of speciation provide extensive evidence that prezygotic isolation tends to evolve between two populations that are kept apart for several generations, evolving adaptations to different environmental conditions. However, experimental evidence for the process of reinforcement is either flawed in design or negative. [Editor's summary.]

Beginning in the 1950s and continuing to the present, many researchers have set out to duplicate all or part of the speciation process under controlled laboratory conditions. Here we attempt to integrate these studies and ask what we can conclude about the major models of speciation. Speciation via polyploidy, which appears to be common in plants, and other chromosomal mechanisms are not discussed here. The extensive body of purely theoretical work on speciation is deemphasized. Instead we focus on inferences deduced from experimental studies. Throughout we define species via the biological species concept, that is, 'groups of interbreeding natural populations that are reproductively isolated from other such groups'.[1]

To define and integrate the major models of speciation, we begin with the 'basic allopatry' or geographical model of speciation. In this model, a species range becomes dissected into two parts by a physical barrier (mountain range, river, etc.), which prevents gene flow between them. The populations

[1] E. Mayr, *Animal Species and Evolution* (Cambridge, Mass.: Harvard University Press, 1963), 19, which is included in Chapter 23 above.

are presumed to evolve independently because of the allopatry induced by their physical isolation. Genetic divergence accrues as a result of adaptation to the prevailing environmental conditions and by means of sampling drift. Prezygotic (i.e., positive assortative mating[2] that reduces the production of hybrids) and postzygotic (i.e., reduced viability and/or fertility of hybrids) reproductive isolation develop between the physically isolated populations as an incidental by-product of genetic differences that gradually accrue between them. Once pre- and/or postzygotic isolation is complete, speciation has occurred.

The three[3] other major modes of speciation include the reinforcement, divergence-with-gene-flow, and bottleneck models. All of these can be expressed as simple modifications of the basic allopatry model.

In the *reinforcement model*, it is presumed that the physical barrier breaks down before complete reproductive isolation has evolved in allopatry. Heterotypic matings between previously separated subpopulations are presumed to produce low-fitness hybrid offspring, and this selects for positive assortative mating. If this selection is successful and leads to complete prezygotic isolation, then the speciation that began in allopatry is completed despite renewed gene flow between subpopulations. [. . .]

The basic allopatry model

The major prediction to be tested concerning the basic allopatry model is that sampling drift and/or adaptation to different environments can lead to genetic differentiation that produces incidental reproductive isolation. Substantial experimental evidence bears on this prediction.

Sampling Drift.—One simple way to determine the potential for sampling drift to generate reproductive isolation among isolated populations is to look for pre- and postzygotic isolation among inbred lines. We have found no reports of hybrid inviability or sterility in crosses between different inbred

[2] *Assortative mating* means 'like mates with like': if the population contains tall and short individuals then tall males mate with tall females, short males with short females. *Positive assortative mating* means the same thing, but can be contrasted with *negative assortative mating*, or disassortative mating, in which unlike types mate (tall males with short females). Later on, the expressions *homotypic* and *heterotypic matings* are used; their meanings are similar—homotypic matings occur with (positive) assortative mating. *Prezygotic isolation* means that hybrids between two types are not formed, usually because of non-interbreeding; (positive) assortative mating is a form of prezygotic isolation: *postzygotic isolation* means hybrids are formed but fail to reproduce. *Allopatry* means other-place; *sympatry* means same-place. One other technical term that appears in the paper is *gene flow*: it means interbreeding, and is particularly used in a spatial context: when individuals from different places interbreed, their genes 'flow' through the species' range. Gene flow is often produced by migration and interbreeding.

[3] Only material on the first is included here; the divergence-with-gene-flow and the bottleneck models are excluded.

lines of *Drosophila* species; however, prezygotic isolation has been observed. [. . .] Overall studies of mating among inbred strains suggest that sampling drift can both contribute to or detract from isolation among populations.

Divergent Selection and Prezygotic Isolation.—Besides sampling drift, genetic differentiation in response to divergent selection among allopatric populations can lead to reproductive isolation as a correlated response[4] via incidental pleiotropy or genetic hitchhiking, that is, sampling error-induced linkage disequilibrium between alleles affecting the divergently selected character(s) and alleles affecting positive assortative mating. In practice, it is usually impossible to differentiate between pleiotropy and genetic hitchhiking, thus we pool these two causative factors and refer to them collectively as 'pleiotropy/hitchhiking.'

Many experimental studies have looked for isolation as a correlated response to divergent selection. For example, Burnet and Connolly[5] divided a founder stock of *D. melanogaster* into three groups. The first and second were selected for increased and decreased locomotor activity, respectively, and the third was an unselected control. After 112 generations, the selected groups manifest markedly divergent locomotor activity, in the selected directions, whereas the controls remained unchanged. When the lines selected for increased or decreased activity were tested for nonrandom mating, a 50% excess of homotypic mating was observed (i.e., the percentage of homotypic matings was about 75 instead of the random-mating expectation of 50). [. . .]

When we surveyed 14 studies from the literature in which divergent selection was applied to allopatric populations and then a measure was taken for the development of prezygotic isolation, we were surprised to find such a large excess of positive results (10 positive to 4 negative; part A of Table 26.1). While allowing for the fact that negative results are less likely to be published, it still remains clear that it is not unusual to find prezygotic isolation as a fortuitous by-product of adaptation to divergent selection regimes.

One issue in studies such as those outlined above is the degree to which isolation is the result of sampling drift that occurred while the populations were being selected in allopatry, versus isolation, which is a by-product of genetic difference built up because of divergent selection among populations.

[4] Imagine a species of bird evolving different sized beaks in two populations. The beaks are adapted to different food in the two places the populations occupy. If beak size also influences mate recognition then prezygotic isolation arises by *pleiotropy*: the same character influences both adaptation to diet and mate recognition. Alternatively, beak size may only evolve in relation to diet but the characters influencing mate recognition may be genetically closely linked to the genes influencing beak size. Then evolution in the beak size genes will drag along the mate recognition genes, and prezygotic isolation evolves by *hitch-hiking*.

[5] B. Burnet and K. Connolly, in J. H. F. van Abeelen (ed.), *The Genetics of Behaviour* (Amsterdam: North-Holland, 1974), 201–58.

In most of the studies that we surveyed, it was impossible to tease these two factors apart, but in two studies it was possible (part B of Table 26.1).

Kilias *et al.*[6] collected two base populations from different geographical localities in Greece. Each of these was split into two allopatric populations, one of which was reared under cold-dry-dark conditions, the other under warm-moist-light conditions. After 5 years of adaptation under allopatry, divergently selected populations derived from the same or different original base populations showed prezygotic isolation (about a 50% excess of homotypic matings relative to the random mating expectation) but parallel-selected populations experiencing the same environmental conditions showed no isolation. If sampling drift were a major factor leading to prezygotic isolation, then prezygotic isolation should have accrued between allopatric populations experiencing both divergent and parallel selection.

Because isolation was found only among divergently selected populations, this study supports the idea that pleiotropy of the selected variation itself, or tightly linked variation, was responsible for the development of prezygotic isolation. [. . .]

Divergent Selection and Postzygotic Isolation.—Two major contexts for postzygotic isolation exist. The first is unconditional and occurs when hybrids between divergently selected lines have lowered viability and/or fertility under benign conditions. The second is environment-dependent and occurs whenever hybrids have an intermediate phenotype that is selectively inferior in specific environmental contexts.

It is commonplace for hybrids (from the F_1 and many offspring from back-crosses, the F_2, F_3, etc.) between divergently selected lines to have an intermediate phenotype, and this will lead to environment-dependent postzygotic isolation whenever populations in different habitats or regions become differentiated because of divergent selection (see below). This type of isolation, though intuitively obvious, is rarely measured in laboratory studies, owing to the difficulty in duplicating divergent, multifarious natural selection. Most laboratory studies measure viability and fecundity only under benign conditions, and therefore will overlook environment-dependent postzygotic isolation and cause this form of isolation to be unappreciated, despite its potential importance in nature. [. . .]

It is difficult to make any generalizations concerning the relative frequency with which this form of isolation develops in laboratory studies. All we can observe is that in three of the four cases in which unconditional postzygotic isolation was sought, it was found.

Overall, laboratory studies strongly support the conclusion that prezygotic and environment-dependent postzygotic reproductive isolation can readily

[6] G. Kilias, S. N. Alahiotis, and M. Pelecanos, *Evolution*, 34 (1980), 730–7.

Table 26.1: Prezygotic isolation experiments grouped by method.[7]

STUDY	PREZYGOTIC REPRODUCTIVE ISOLATION?
Part A: divergent selection in allopatry	
Koref-Santibanez and Waddington 1958	No
Ehrman 1964, 1969	Yes/No, inconsistent across samples
del Solar 1966	Yes
Kessler 1966	Yes, but asymmetrical
Barker and Cummins 1969	No
Grant and Mettler 1969	Yes
Burnet and Connolly 1974	Yes
Soans *et al.* 1974	Yes
Hurd and Eisenberg 1975	Yes
van Dijken and Scharloo 1979	No
de Oliveira and Cordeiro 1980	Yes
Kilias *et al.* 1980	Yes
Koepfer 1987	Yes, but asymmetrical
Dodd 1989	Yes
Part B: parallel selection in allopatry	
Kilias *et al.* 1980	No
Dodd 1989	No
Part C: divergent selection with hybrid inviability in sympatry (destroy hybrids experiments)	
Koopman 1950	Yes
Wallace 1953	Yes, but transient
Knight *et al.* 1956	Yes
Kessler 1966	Yes
Paterniani 1969	Yes
Ehrman 1971, 1973, 1979	Yes, but complex pattern across years
Barker and Karlsson 1974	Yes
Crossley 1974	Yes
Dobzhansky *et al.* 1976	Yes
Part D: divergent selection with hybrid viability in sympatry	
Thoday and Gibson 1962	Yes
Grant and Mettler 1969	No
References (18 experiments) cited in Thoday and Gibson 1970 and Scharloo 1971	No, 18 of 18 experiments
Spiess and Wilke 1984	No
Part E: divergent selection with hybrid viability in sympatry and with isolation via pleiotropy	
Coyne and Grant 1972	Yes, in one of two replicates
Soans *et al.* 1974	Yes
Hurd and Eisenberg 1975	Yes
Rice 1985	Yes
Rice and Salt 1988, 1990	Yes

[7] See Rice and Hostert's original paper for the references.

develop as a fortuitous by-product of pleiotropy or hitchhiking associated with genes that adapt populations to different environmental conditions. Limited support for unconditional postzygotic isolation is also present. Drift alone may play a role in the development of such coincidental reproductive isolation, but the experimental evidence for this is quite meager.

The reinforcement model

The observational basis for suspecting that reinforcement is an important speciation mechanism is remarkably compelling: it is common to observe stronger levels of prezygotic isolation in areas where a pair of closely related species have overlapping ranges, compared with the same comparison when the species are sampled from nonoverlapping portions of their ranges. We found so many published records of this pattern [. . .] that there seems little doubt the pattern is general. Several examples include crickets, frogs, fruit flies, damselflies, and fish.

These observations certainly are consistent with the idea that prezygotic isolation has evolved to prevent the production of low-fitness hybrid offspring. The logical jump between the observational data and the conclusion that the reinforcement model of speciation is in fact operating is made tenuous for three reasons: (1) there are other biological explanations for the observed pattern, (2) there are strong theoretical objections to the reinforcement models, and (3) no repeatable laboratory experiments have been able to duplicate even the early stages of the reinforcement model.

One group of laboratory studies that has been used to support the reinforcement model are the numerous 'destroy-the-hybrids' experiments, typically carried out with *Drosophila* species (part C of Table 26.1). Many variations of the experimental design exist, but the basic protocol is to collect equal numbers of male and female virgins from each of two genetically marked strains. These are held separately until sexually mature and then mixed in a common mating chamber and finally allowed to produce offspring. Through the use of genetic markers, offspring can be classified as being derived from homotypic or heterotypic matings, and from the former a new set of males and females is collected and treated as described above. Repeated cycles of the protocol generate strong, multigenerational selection for homotypic mating. Almost all of the experiments of this kind that we have located in the literature report the evolution of increased prezygotic isolation between the selected strains.

These studies clearly indicate that most *Drosophila* laboratory populations have the requisite additive genetic variation for the evolution of homotypic mating. However, because all of the hybrids are destroyed each generation, these studies do not truly test the reinforcement model. The protocol therefore simulates the case in which speciation *already has been completed* via postzygotic isolation and asks if prezygotic isolation will follow. The key 'ingredient' missing is gene flow between the strains.

What happens when gene flow is permitted? [. . .]

A second, more extensive group of experiments testing the reinforcement model (part D of Table 26.1) are the large collection of 'disruptive-selection experiments' that accrued after the remarkable results of Thoday and Gibson.[8] Thoday and Gibson's protocol applied strong disruptive selection to an arbitrary character, bristle number, and then asked if prezygotic isolation would develop to prevent the production of the low-fitness offspring that resulted from heterotypic matings between the extreme types. Remarkably, after only 12 generations of strong disruptive selection on bristle number, complete prezygotic isolation was observed between the high-bristle-number and low-bristle-number selected lines. This experimental outcome would appear to provide strong experimental support for the reinforcement model.

Thoday and Gibson's results were so striking that laboratories around the world set out to repeat them. [. . .] All attempts to repeat the experimental outcome with new stocks have failed. The one thing that is repeatable about Thoday and Gibson's experimental protocol is that it does not lead to prezygotic isolation. [. . .]

Overall, the 'destroy-the-hybrids' experiments do not provide support for the reinforcement model because they prevent gene flow between the selected populations by imposing complete postzygotic isolation. When gene flow is permitted, as occurred in most attempts to repeat Thoday and Gibson's work, prezygotic isolation did not evolve. The available laboratory evidence therefore provides no support for the reinforcement model of speciation. [. . .]

[*Evolution*, 47 (1993), 1637–53.]

JERRY A. COYNE AND H. ALLEN ORR

27 The evolutionary genetics of speciation

Epistatic interactions between genes are the distinctive feature of speciation. A table reviewing genetic research on speciation shows that the fitness of hybrids between two species is usually influenced by many gene loci. Recent research suggests that reinforcement can occur, though its importance remains unknown. Postmating, but prezygotic, isolation is another recent research topic. The main theory of postzygotic isolation is the Dobzhansky–Muller model, which makes several testable predictions, such as asymmetries in the allelic

[8] K. Mather, in M. Ashburner *et al.* (eds.), *The Genetics and Biology of Drosophila*, vol. 3c (London: Academic Press, 1983).

combinations that influence hybrid fitness, and the snowball effect. Haldane's rule describes an intermediate stage in speciation. Evidence exists in support of two genetic explanations for Haldane's rule. [Editor's summary.]

Studying reproductive isolation

WHAT IS NOVEL ABOUT SPECIATION?

Some of our colleagues have suggested that speciation is not a distinct field of study because—as a by-product of conventional evolutionary forces like selection and drift—the origin of species is simply an epiphenomenon of normal population-genetic processes. But even if speciation is an epiphenomenon, it does not follow that the mathematics or genetics of speciation can be inferred from traditional models of evolution in single lineages. Under the BSC,[1] the origin of species involves reproductive isolation, a character that is unique because it requires the joint consideration of two species and usually an interaction between the genomes of two species. The distinctive feature of the genetics of speciation is therefore epistasis.[2] This is necessarily true for all forms of postzygotic isolation, in which an allele that yields a normal phenotype in its own species causes hybrid inviability or sterility on the genetic background of another (see below). Epistasis also occurs in many forms of prezygotic isolation. Sexual isolation, for example, usually requires the coevolution of male traits and female preferences, so that the fitness of a male trait depends on whether the choosing female is conspecific or heterospecific.

These ubiquitous (and complex) interactions between the genomes of two species guarantee that the mathematics of speciation will differ from that describing evolutionary change within species and that speciation may well show emergent properties not seen in traditional models. Indeed, such properties have already been seen for postzygotic isolation (e.g., the snowball effect; see below).

[1] BSC stands for 'biological species concept', which Coyne and Orr had defined in part of their chapter not included here. The biological species concept is explained by Mayr in Chapter 23. Footnote 2 to Chapter 26, by Rice and Hostert, defines a number of terms also used in this extract, including sympatric, allopatric, and prezygotic and postzygotic isolation.

[2] Epistasis means that the effect of a gene depends on which other genes are present in the same body as it. Here we are interested in epistasis for fitness, between genes coming from the same species or from two different species (or near-species). For example, a gene that is normally found only in chimpanzees will work well with other chimp genes in a chimp body. But, if one such chimp gene were substituted for one human gene in an otherwise human body, that chimp gene might no longer work well. Its interactions with other human genes would probably snarl-up, unlike its normal interactions with chimp genes. Then the chimp gene has high fitness in a chimp body with chimp genes, but low fitness in a human body with human genes: that's epistasis, for fitness.

Two motives usually underlie genetic analyses of speciation. First, [. . .] we would like to understand the genetic basis of cladogenesis. That is, we would like to know the number of genes involved in reproductive isolation, the distribution of their phenotypic effects, and their location in the genome. Second, we expect genetic analyses of reproductive isolation to shed light on the process of speciation, as different evolutionary processes should leave different genetic signatures. The observation of more genes causing hybrid male than hybrid female sterility has suggested, for example, that these critical substitutions were driven by sexual selection.

WHAT TRAITS SHOULD WE STUDY?

Because speciation is complete when reproductive isolation stops gene flow in sympatry, the genetics of speciation properly involves the study of only those isolating mechanisms evolving up to that moment. The further evolution of reproductive isolation, while interesting, is irrelevant to speciation. Although widely recognized, this point is understandably often ignored in practice. If speciation is allopatric and several isolating mechanisms evolve simultaneously, it is hard to know which will be important in preventing gene flow when the taxa become sympatric. *Drosophila simulans* and *D. mauritiana*, for example, are allopatric, and in the laboratory show sexual isolation, sterility of F_1 hybrid males, and inviability of both male and female backcross hybrids. We have no idea which of these factors would be most important in preventing gene exchange in sympatry, or if other unstudied factors—like ecological differences—would play a role.

It seems likely, in fact, that several isolating mechanisms evolve simultaneously in allopatry and act together to both prevent gene flow in sympatry and allow coexistence. (Although reproductive isolation is sufficient for speciation, different species must coexist in sympatry in order to be seen.) There are two reasons why multiple isolating mechanisms seem likely. In theory, no single isolating mechanism except for distinct ecological niches or some types of temporal divergence can at the same time completely prevent gene flow and allow coexistence in sympatry. Two species solely isolated by hybrid sterility, for example, cannot coexist: One will become extinct through excessive hybridization or ecological competition. Species subject only to sexual isolation are ecologically unstable because they occupy identical niches. Second, direct observation often shows that complete reproductive isolation in nature often involves several isolating mechanisms. Schluter, for example, describes several species pairs having incomplete prezygotic isolation. When hybrids are formed, however, they are ecologically unsuited for the parental habitats, and do not thrive.

We know little about the temporal order in which reproductive isolating mechanisms appear. [. . .]

A SUMMARY OF GENETIC STUDIES

Because there are relatively few studies of the genetics of speciation, we have summarized them all in Table 27.1. We used two criteria for including a study in this table. First, the character studied must be known to cause reproductive isolation between species in either nature or the laboratory or be plausibly involved in such isolation. Second, the genetic analysis must have been fairly

Table 27.1: Summary of existing genetic analyses of reproductive isolation between closely related species

SPECIES PAIR	TRAIT	NUMBER OF GENES
D. heteroneura/D. silvestris	Head shape	9
D. melanogaster/D. simulans	Hybrid inviability	≥9
	Female pheromones	≥5
D. mauritiana/D. simulans	Hybrid male sterility	≥15
	Hybrid female sterility	≥4
	Hybrid inviability	≥5
	Male sexual isolation	≥2
	Female sexual isolation	≥3
	Genital morphology	≥9
	Shortened copulation	≥3
D. mauritiana/D. sechellia	Female pheromones	≥6
D. simulans/D. sechellia	Hybrid male sterility	≥6
	Hybrid inviability	≥2
	Female sexual isolation	≥2
D. mojavensis/D. arizonae	Hybrid male sterility	≥3
	Male sexual isolation	≥2
	Female sexual isolation	≥2
D. pseudoobscura/D. persimilis	Hybrid male sterility	≥9
	Hybrid female sterility	≥3
	Sexual isolation	≥3
D. pseudoobscura USA/Bogota	Hybrid male sterility	≥5
D. buzatti/D. koepferae	Hybrid male inviability	≥4
	Hybrid male sterility	≥7
D. subobscura/D. madeirensis	Hybrid male sterility	≥6
D. virilis/D. littoralis	Hybrid female viability	≥5
D. virilis/D. lummei	Male courtship song	≥4
	Hybrid male sterility	≥6
D. hydei/D. neohydei	Hybrid male sterility	≥5
	Hybrid female sterility	≥2
	Hybrid inviability	≥4

(Continued)

Table 27.1—continued

SPECIES PAIR	TRAIT	NUMBER OF GENES
D. montana/D. texana	Hybrid female inviability	≥2
D. virilis/D. texana	Hybrid male sterility	≥3
Ostrina nubialis, Z and E races	Female pheromones	1
	Male perception of pheromones	2
Laupala paranigra/L. kohalensis	Song pulse rate	≥8
Spodoptera latifascia/S. descoinsi	Pheromone blend	1
Xiphophorus helleri/X. maculatus	Hybrid inviability	2
M. lewisii/M. cardinalis	8 floral traits	(see note)[3]
M. guttatus/M. micranthus	Bud growth rate	8
	Duration of bud development	10
Mimulus, four taxa	(see note)[3]	(see note)[3]
M. guttatus pops.	Hybrid inviability	2 (system 1) ≥2 (system 2)
M. guttatus/M. cupriphilus	Flower size	3–7
Helianthus annuus/H. petiolarus	Pollen viability	≥14

rigorous, with one of three methods used: (1) classical genetic analyses, in which species differing in molecular or morphological mutant markers are crossed and the segregation of reproductive isolation with the markers examined. Here we included only those studies in which markers were distributed among all major chromosomes; (2) simple Mendelian analyses in which segregation ratios in backcrosses or F_2s indicated that an isolating mechanism was due to changes at a single locus; or (3) biometric analyses, in which measurement of character means and variances in backcrosses or F_2s yielded a rough estimate of gene number. Table 27.1 also gives the actual or the minimum number of genes involved for each isolating mechanism. The footnotes give more detail about the type of genetic analysis, whether each pair of species was sympatric or allopatric, and a brief summary and critique of the results.[3]

Several caveats are necessary. First, despite our attempts to comprehensively comb the literature, we have surely missed some studies. Second, the quality of the analyses is uneven: Some classical genetic studies involved detailed mapping experiments with many markers, while others relied on only one marker per chromosome. Third, in most cases the number of genes

[3] The table footnotes have been excluded in this extract.

causing a single reproductive isolating mechanism is an underestimate. This may reflect a limited number of markers, failure to test all possible interactions between chromosomes, or the use of biometric approaches, which nearly always underestimate true gene number. Finally, data are given for single isolating mechanisms, but several mechanisms may often operate together to impede gene flow in nature.

The most striking feature of Table 27.1 is the imbalance of both species and isolating mechanisms. Roughly 75% of all the studies involve *Drosophila*, with only seven other pairs of taxa, mostly plants from the genus *Mimulus*. Moreover, nearly two thirds of the *Drosophila* work is on hybrid sterility and inviability. There is no published genetic study of ecological isolation (but see below). We obviously must extend such studies to other groups and other forms of reproductive isolation.

For convenience, we discuss prezygotic and postzygotic isolation separately.

Prezygotic isolation

[. . .]

REINFORCEMENT

One of the greatest controversies in speciation concerns reinforcement: The process whereby two allopatric populations that have evolved some postzygotic isolation in allopatry undergo selection for increased sexual isolation when they later become sympatric. Reinforcement was introduced and popularized by Dobzhansky,[4] who apparently considered it the necessary last step of speciation. Its wide popularity may have reflected its seductive assumption of a creative (and not an incidental) role for natural selection in speciation. While theoretical studies of reinforcement appeared only recently, two analyses of *Drosophila* supported the idea: In species pairs with overlapping ranges, sexual isolation was stronger when populations derived from areas of sympatry than from allopatry.

In the 1980s, however, several critiques eroded the popularity of reinforcement. First, Templeton pointed out that the pattern of stronger sexual isolation among sympatric than allopatric populations could be caused not by reinforcement but by differential fusion, in which species could persist in sympatry only if they had evolved sufficiently strong sexual isolation in allopatry. Thus stronger isolation in sympatry might not reflect direct selection, but the loss by fusion of weakly isolated populations. Moreover, it became clear that some of the data offered in support of reinforcement were flawed. Finally, the first serious theoretical treatment of reinforcement showed that even under favorable conditions (e.g., complete sterility of hybrids), extinction of populations occurred more often than reinforcement.

[4] Chapter 25 is from a later discussion of reinforcement by Dobzhansky.

Recently, however, a combination of empirical and theoretical work has resurrected the popularity of reinforcement. In an analysis of 171 pairs of *Drosophila* species, Coyne and Orr found that recently diverged pairs show far more sexual isolation when sympatric than allopatric. [. . .] Although in 1989 there were no theoretical studies showing that reinforcement was feasible, two such investigations have appeared recently. Liou and Price and Kelly and Noor showed that reinforcement can occur frequently even if hybrids have only moderate postzygotic isolation. [. . .]

Newer data also support reinforcement. These include a reanalysis of the earlier literature, finding many more possible examples, and two new studies of species pairs with partially overlapping ranges. [. . .]

The data and theory reviewed above strongly suggest that reinforcement of sexual isolation can occur. This conclusion represents one of the most radical changes of views about speciation over the past decade. Future work must determine whether reinforcement is rare or ubiquitous across animal taxa and whether it exists in plants.[5] [. . .]

POLLINATOR ISOLATION

Pollinator isolation probably represents a common form of reproductive isolation in plants. A variant of this is the isolation of insect-pollinated plants from self-compatible species, a mechanism described in *Mimulus*. The few data at hand (Table 27.1) show that the difference between outcrossing and inbreeding is due to several genes, but that differences in flower shape, color, or nectar reward that attract different pollinators may be due to one or a few major genes. This latter observation, if common, might suggest a rapid form of speciation.

POSTMATING, PREZYGOTIC ISOLATION

Biologists have begun to appreciate that sexual selection is not limited to obvious behavioral and morphological traits that act before copulation, but can include cryptic characters that act between copulation and fertilization. Such selection can lead to female control of sperm usage, male–male sperm competition within multiply inseminated females, and the mediation of such competition by the female. Selection acting between copulation and fertilization has also been invoked to explain the striking diversity of male genitalia among animal species, bizarre conformations of female reproductive tracts, and postcopulation courtship behavior by males. Moreover,

[5] At this point in the original, Coyne and Orr discuss ecological isolation. The extract from Schluter (Chapter 28) in this anthology concerns ecological isolation (though for postzygotic isolation).

the relatively rapid evolution of proteins involved in reproduction is also consistent with sexual selection.

Just as sexual selection acting on male plumage or courtship behavior can pleiotropically produce sexual isolation, so postcopulation, prezygotic sexual selection can produce cryptic sexual isolation detectable only after fertilization. Such isolation may take the form of either blocked heterospecific fertilization, such as the insemination reaction of *Drosophila*, or the preferential use of conspecific sperm when a female is sequentially inseminated by heterospecific and conspecific males. This latter phenomenon has recently been described in grasshoppers, crickets, flour beetles and *Drosophila*. [. . .]

Postzygotic isolation

Postzygotic isolation occurs when hybrids are unfit. Evolutionists have, historically, pointed to three types of genetic differences as causes of these fitness problems: species may have different chromosome arrangements, different ploidy levels, or different alleles that do not function properly when brought together in hybrids. Although each of these modes of speciation has enjoyed its advocates, it is now clear that the latter two are by far the most important.[6] [. . .]

THE DOBZHANSKY–MULLER MODEL

Understanding the evolution of postzygotic isolation is difficult, because the phenotypes we are hoping to explain—the inviability and sterility of hybrids—seem maladaptive. The difficulty is best seen by considering the simplest possible model for the evolution of postzygotic isolation: change at a single gene. One species has genotype *AA*, the other *aa*, while *Aa* hybrids are completely sterile. Regardless of whether the common ancestor was *AA* or *aa*, fixation of the alternative allele cannot occur because the first mutant individual has genotype *Aa* and so is sterile. Using the metaphor of adaptive landscapes, it is hard to see how two related species can come to reside on different adaptive peaks unless one lineage passed through an adaptive valley.

This problem was finally solved by Bateson, Dobzhansky, and Muller, who noted that, if postzygotic isolation is based on incompatibilities between two or more genes, hybrid sterility and inviability can evolve unimpeded by natural selection. If, for example, the ancestral species had genotype *aabb*, a new mutation at one locus (allele *A*) could be fixed by selection or drift in one

[6] Coyne and Orr did not discuss ploidy levels in their original, referring readers to other papers. Chapter 29 in this anthology looks at the topic. Coyne and Orr did discuss chromosomal rearrangements, but negatively, to argue their unimportance; that section of their paper is not included here—we go straight to the positive material.

isolated population as the *Aabb* and the *AAbb* genotypes are perfectly fit. Similarly, a new allele (*B*) at the other locus could be fixed in a different population as the *aaBb* and *aaBB* genotypes are perfectly fit. But while each population is fit, it is entirely possible that when these populations come into contact, the resulting *AaBb* hybrids would be sterile or inviable. The *A* and *B* alleles have never been tested together within a genome, and so may not function properly when brought together in hybrids.

Alleles showing this pattern of epistasis are called complementary genes. Such genes need not, of course, have drastic effects on hybrid fitness: any particular incompatibility might lower hybrid fitness by only a small amount. It should also be noted that the Dobzhansky–Muller model is agnostic about the evolutionary causes of substitutions that ultimately produce hybrid sterility or inviability: Purely adaptive or purely neutral evolution within populations can give rise to complementary genes and thus to postzygotic isolation.

The Dobzhansky–Muller model is the basis for almost all modern work in the genetics of postzygotic isolation. There is now overwhelming evidence that hybrid sterility and inviability do indeed result from such between-locus incompatibilities. Curiously, theoretical studies of the Dobzhansky–Muller model have been slow in coming. Recent analyses, however, predict that the evolution of postzygotic isolation should show several regularities.

PATTERNS IN THE GENETICS OF POSTZYGOTIC ISOLATION

Long before any formal studies of the Dobzhansky–Muller model, Muller predicted that the alleles causing postzygotic isolation must act asymmetrically. To see this, consider the two-locus case sketched above. Although the *A* and *B* alleles might be incompatible in hybrids, their allelomorphs *a* and *b* must be compatible. This is because the *aabb* genotype must represent an ancestral state in the evolution of the two species.

There is now good evidence that genic incompatibilities do in fact act asymmetrically. The best data come from Wu and Beckenbach's study of male sterility in *D. pseudoobscura–D. persimilis* hybrids. When an X-linked region from one species caused sterility on introgression into the other species' genome, they found that the reciprocal introgression had no such effect. This observation has now been confirmed in many additional *Drosophila* hybridizations.

The second pattern expected under the Dobzhansky–Muller model was pointed out more recently. If hybrid sterility and inviability are caused by the accumulation of complementary genes, the severity of postzygotic isolation, as well as the number of genes involved, should snowball much faster than linearly with time. This follows from the fact that any new substitution in one species is potentially incompatible with the alleles from the other species at all of those genes that have previously diverged. (In our discussion above, the *B* substitution is potentially incompatible with the previous *A* substitution in the other species.) Later substitutions are therefore more likely to cause hybrid incompatibilities than earlier ones. Consequently, the cumulative number of hybrid incompatibilities increases much faster than linearly with the number of substitutions *K*. If all incompatibilities involve pairs of loci, the expected number of Dobzhansky–Muller incompatibilities increases as K^2 or (assuming a rough molecular clock) as the square of the time since species diverged. If incompatibilities sometimes involve interactions between more than two loci (see below), the number of hybrid incompatibilities will rise even faster. This snowballing effect requires that we interpret genetic studies of postzygotic isolation with caution. Because the genetics of hybrid sterility and inviability will quickly grow complicated as species diverge, it is easy to overestimate the number of genes required to cause strong reproductive isolation.

The limited data we possess are consistent with a snowballing effect, although they do not prove it. The simplest prediction of the snowballing hypothesis is that the number of mapped genes causing hybrid sterility or inviability should increase quickly with molecular genetic distance between species. But given the enormous difficulties inherent in accurately mapping and counting speciation genes, it may be some time before such direct contrasts are possible.

Third, while we have assumed that hybrid incompatibilities involve pairs of genes, analysis of the Dobzhansky–Muller model shows that more complex hybrid incompatibilities, involving interactions among three or more of genes, should be common. The reason is not intuitively obvious but is easily demonstrated mathematically. Certain paths to the evolution of new species are barred because they would require passing through intermediate genotypes that are sterile or inviable. It is easy to show, however, that the proportion of all imaginable paths to speciation allowed by selection increases with the complexity of hybrid incompatibilities. Thus, for the same reason that two-gene speciation is easier than single-gene speciation so three-gene is easier than two-gene speciations, and so on.

The evidence for complex incompatibilities is now overwhelming. They have been described in the *D. obscura* group, the *D. virilis* group, the *D. repleta* group, and the *D. melanogaster* group. Indeed, such interactions could prove more common than the two-locus interactions discussed at length by Bateson, Dobzhansky, and Muller.

Theoretical analysis of the Dobzhansky–Muller model also has yielded results that contradict popular ideas about the effect of population subdivision on speciation. Many evolutionists have maintained, for example, that speciation is most likely in taxa subdivided into small populations. This is, however, demonstrably untrue, at least for postzygotic isolation. Orr and Orr showed that if the substitutions ultimately causing Dobzhansky–Muller incompatibilities are driven by natural selection—as seems likely—the waiting time to speciation grows longer as a species of a given size is splintered into ever smaller populations. If, on the other hand, the substitutions causing hybrid problems are originally neutral, population subdivision has little effect on the time to speciation. In no case is the accumulation of hybrid incompatibilities greatly accelerated by small population size. Unfortunately, we have little empirical data bearing on this issue. Although it might seem that the effect of population size on speciation rates could be estimated by comparing the rate of evolution of postzygotic isolation on islands versus continents, this comparison is confounded by the likelihood of stronger selection in novel island habitats, which itself might drive rapid speciation.

Finally, much of the theoretical work on the evolution of postzygotic isolation has been devoted to explaining one of the most striking patterns characterizing the evolution of hybrid sterility and inviability—Haldane's rule. Because this large and confusing literature has recently been reviewed elsewhere, we will not attempt a thorough discussion here. Instead, we briefly consider the data from nature and sketch the leading hypotheses offered to explain them.

HALDANE'S RULE

In 1922, Haldane noted that, if only one hybrid sex is sterile or inviable, it is nearly always the heterogametic (XY) sex. More recent and far more extensive reviews show that Haldane's rule is obeyed in all animal groups that have been surveyed, e.g., *Drosophila*, mammals, Orthoptera, birds, and Lepidoptera (the latter two groups have heterogametic females). Indeed, it is likely that Haldane's rule characterizes postzygotic isolation in all animals having chromosomal sex determination. Moreover, these surveys show that Haldane's rule is consistently obeyed. In *Drosophila*, for instance, out of 114 species crosses producing sterile hybrids of one sex only, in 112 it is the males. Comparative work in *Drosophila* also shows that Haldane's rule represents an early stage in the evolution of postzygotic isolation: Hybrid male sterility or inviability arises quite quickly, while female effects appear only much later.

For obvious reasons, Haldane's rule has received a great deal of attention. It represents one of the strongest patterns in evolutionary biology and perhaps the only pattern characterizing speciation. In addition, the rule implies that there is some fundamental similarity in the genetic events causing speciation in all animals.

Although many hypotheses have been offered to explain Haldane's rule, most have been falsified. We will not consider these failed explanations here. Instead, we briefly review the three explanations of Haldane's rule that remain feasible. There is strong evidence for two of these and suggestive evidence for the third. In a field that has historically been rife with disagreement, a surprisingly good consensus has emerged that some combination of these hypotheses explains Haldane's rule.

The first hypothesis, the dominance theory, posits that Haldane's rule reflects that the recessivity of X-linked genes causes hybrid problems. This idea was first suggested by Muller, and his verbal theory was later formalized. The mathematical work shows that heterogametic hybrids suffer greater sterility and inviability than homogametic hybrids whenever the alleles causing hybrid incompatibilities are, on average, partially recessive ($\bar{d} < 1/2$). The reason is straightforward. Although XY hybrids suffer the full hemizygous effect of all X-linked alleles causing hybrid problems (dominant and recessive), XX hybrids suffer twice as many X-linked incompatibilities (as they carry twice as many Xs). These two forces balance when $\bar{d} = 1/2$. But if $\bar{d} < 1/2$, the expression of recessives in XY hybrids outweighs the greater number of incompatibilities in XX hybrids, and Haldane's rule results. Obviously, the dominance theory can account not only for Haldane's rule, but also for the well-known large effect of the X chromosome on hybrid sterility and inviability.

There is now strong evidence that dominance explains Haldane's rule for hybrid inviability. In particular, the dominance theory predicts that, in *Drosophila* hybridizations obeying Haldane's rule for inviability, hybrid females who are forced to be homozygous for their X chromosome should be as inviable as F_1 hybrid males. (Such unbalanced females possess an F_1 malelike genotype in which all recessive X-linked genes are fully expressed.) In both of the species crosses in which this test has been performed, unbalanced females are, as expected, completely inviable. Similarly, there is evidence from haplodiploid species that hybrid backcross males (who are haploid) suffer more severe inviability than their diploid sisters.

There is also weaker indirect evidence that the alleles causing hybrid sterility act as partial recessives: Hollocher and Wu and True *et al.* found that while most heterozygous introgressions from one *Drosophila* species into another are reasonably fertile, many homozygous introgressions are sterile. It therefore seems likely that dominance contributes to Haldane's rule for both hybrid inviability and hybrid sterility. Last, it is worth noting that the dominance theory—unlike several alternatives—should hold in all animal taxa, regardless of which sex is heterogametic.

The second hypothesis posits that Haldane's rule reflects the faster evolution of genes ultimately causing hybrid male than female sterility. Wu and his colleagues offer two explanations for this faster-male evolution: (1) In hybrids, spermatogenesis may be disrupted far more easily than oogenesis, and (2) sexual selection may cause genes expressed in males to evolve faster than those expressed in females. While there is now good evidence for faster-male evolution (see below), this theory cannot be the sole explanation of Haldane's rule. First, it cannot explain Haldane's rule for sterility in those taxa having heterogametic females, e.g., birds and butterflies. After all, spermatogenesis and sexual selection involve males per se, while Haldane's rule extends to all heterogametic hybrids, male or not. Second, the faster-male theory cannot account for Haldane's rule for inviability in any taxa. Because there is strong evidence that genes causing lethality are almost always expressed in both sexes, it seems unlikely that hybrid male lethals can evolve faster than female lethals. Finally, it is not obvious that sexual selection would inevitably lead to faster substitution of male than female alleles. One can easily imagine, for instance, forms of sexual selection in which each substitution affecting female preference is matched by a substitution affecting expression of a male character. If sexual selection causes faster evolution of male sterility, one may need to consider processes like male–male competition in addition to male–female coevolution.

Despite these caveats, there is now good evidence—at least in *Drosophila*—that alleles causing sterility of hybrid males accumulate much faster than those affecting females. (Unfortunately, both of these studies analyzed the same species pair; analogous data from other species are badly needed.) While we cannot be sure of the mechanism involved, it certainly appears that faster-male evolution plays an important role in Haldane's rule for sterility in taxa with heterogametic males.

The last hypothesis, the faster-X theory, posits that Haldane's rule reflects the more rapid divergence of X-linked than autosomal loci. Charlesworth *et al.* showed that, if the alleles ultimately causing postzygotic isolation were originally fixed by natural selection, X-linked genes will evolve faster than autosomal genes if favorable mutations are on average partially recessive ($\bar{h} < 1/2$). (It must be emphasized that this theory requires only that the favorable effects of mutations on their normal conspecific genetic background be partially recessive; nothing is assumed about the dominance of alleles in hybrids. Conversely, the dominance theory requires only that the alleles causing hybrid problems act as partial recessives in hybrids; nothing is assumed about the dominance of these alleles on their normal conspecific genetic background.) Under various cases, this faster evolution of X-linked genes can indirectly give rise to Haldane's rule.

There is some evidence that X-linked hybrid steriles and lethals do in fact evolve faster than their autosomal analogs. In their genome-wide survey of speciation genes in the *D. simulans–D. mauritiana* hybridization, True *et al.* found a significantly higher density of hybrid male steriles on the X chromosome than on the autosomes. Hollocher and Wu, however, found no such difference in a much smaller experiment. Thus, while faster-X evolution may contribute to Haldane's rule, the evidence for it is considerably weaker than that for both the dominance and faster-male theories. [. . .]

In sum, there is now strong evidence for both the dominance and faster-male theories of Haldane's rule. Future work must include better estimates of the dominance of hybrid steriles, better tests of the faster-X theory, and, most important, genetic analyses of Haldane's rule in taxa having heterogametic females. Although both the dominance and faster-male theories make clear predictions about the genetics of postzygotic isolation in these groups, we have virtually no direct genetic data from these critical taxa. Last, it is important to determine if Haldane's rule extends beyond the animal kingdom, particularly to those species of plants having heteromorphic sex chromosomes. [. . .]

Conclusions

The study of speciation has grown increasingly respectable over the past decade. The reasons are obvious: A number of important questions have been resolved by experiment and a number of patterns explained by theory. This progress reflects several fundamental but rarely recognized changes in our approach to speciation. First, the field has grown increasingly genetical. As a consequence, a large body of grand but notoriously slippery questions (How important are peak shifts in speciation? Is sympatric speciation common?) have been replaced with a collection of simpler questions (Is the Dobzhansky–Muller model correct? What is the cause of Haldane's rule?). While it would be fatuous to claim that these new questions are more important than the old, there is no doubt that they are more tractable. Second, the connection between theory and experiment has grown increasingly close. While speciation once seemed riddled with amorphous and untestable verbal theories, the past decade of work has produced a body of mathematical theory, yielding clear and testable predictions about the basis of reproductive isolation. Last, but most important, many of these predictions have been tested.

Despite this progress, many questions about speciation remain unanswered. Throughout this chapter we have tried to highlight those questions that seem to us both important and tractable. Most fall into two broad sets. The first concerns speciation in taxa that have been relatively ignored: Does reinforcement occur in plants? Do prezygotic and postzygotic isolation evolve at approximately the same rate in most taxa as in *Drosophila*? Do plants with heteromorphic sex chromosomes obey Haldane's rule? Do hybrid male steriles still evolve faster than female steriles in taxa having heterogametic females? How distinct are sympatric asexual taxa?

The second set of questions concerns the evolution of prezygotic isolation, which has received less attention than postzygotic isolation. How complex is the genetic basis of sexual isolation? How common is reinforcement? Why is sexual isolation so often asymmetric? What is the connection between adaptive radiation and sexual isolation?

It may seem that trading yesterday's grand verbal speculations for today's smaller, more technical studies risks a permanent neglect of the larger questions about speciation. We believe, however, that more focused pursuits of tractable questions will ultimately produce better answers to these bigger questions. Just as no mature theory of population genetics was possible until we understood the facts of inheritance, so no mature view of speciation seems possible until we understand the origins and mechanics of reproductive isolation.

[In R. A. Singh and C. Krimbas (eds.), *Evolutionary Genetics* (New York: Cambridge University Press, 2000), 532–61.]

DOLPH SCHLUTER

28 Ecological basis of postmating isolation

If hybrids differ from the parental species, the hybrids may be poorly adapted to exploit the local ecological resources. This is an ecological mechanism for low hybrid fitness, distinct from genetic mechanisms. The relative importance of ecological and genetic influences on hybrid fitness is controversial. Evidence for ecological influences comes from sympatric speciation, and from cases in which hybrid fitness changes when the ecological conditions change. [Editor's summary.]

More direct evidence for ecological speciation would be provided by a demonstration that ecological mechanisms directly reduce the fitness of hybrids between sympatric species that occasionally interbreed yet lack strong genetic mechanisms of postmating isolation. By 'ecological' mechanisms I mean postmating isolation stemming from traits in the hybrid that are disadvantageous in natural environments. Ecological postmating isolation may arise because an intermediate phenotype renders the hybrid less efficient at capturing prey in the wild, or because intermediate defences leave the hybrid susceptible to predation and parasitism. In contrast, 'genetic' mechanisms of postmating isolation (Dobzhansky's 'discordant gene patterns') result from the breakup of favourable gene combinations in the parent species (epistasis) and from interactions between parental alleles leading to underdominance.[1] The chief practical difference is that ecological

[1] This kind of genetic explanation for low hybrid fitness is the topic of Coyne and Orr in Chapter 27. 'Underdominance' here means that heterozygotes have lower fitness than either homozygote. When two species cross, the hybrids may be heterozygous at a locus at which the two parental species are homozygous (with different homozygotes in each species).

mechanisms of postmating isolation are detectable in the wild but usually vanish in the laboratory (i.e. if food is uniform and predators are absent), whereas genetic mechanisms are largely independent of environment and should be detectable in most laboratory environments as well as in the wild.

Genetic mechanisms of postmating isolation between the youngest species of an adaptive radiation are frequently weak or lacking, as indicated by high viability and fertility of hybrids in laboratory settings. Examples include Hawaiian and many other *Drosophila*, Hawaiian silverswords and indeed many perennial flowering plants, some East African cichlid fishes, postglacial fishes, and *Heliconius* butterflies. Most studies have examined only the F1 hybrids, whereas genetic mechanisms of postmating isolation are typically most pronounced in F2 and backcross hybrids. However, in some of these examples fitness is also high in backcrosses and/or the F2 generation, such as *Heliconius* butterflies, three-spine sticklebacks, and Lake Victoria cichlids. The persistence of such species implies that some form of selection against intermediates takes place in the wild, possibly natural selection stemming from environment.

Evidence of speciation in sympatry likewise implies some form of divergent natural selection against intermediates, possibly ecological (but not necessarily: it may arise instead from divergent sexual selection). The most compelling examples of sympatric speciation are the cichlids from Barombi Mbo and other crater lakes in Cameroon, West Africa.[2] Phylogenetic studies using mtDNA indicate that the species within a crater basin constitute a monophyletic group: all are more closely related to species within the crater basin than any is to candidate ancestral cichlid species outside the basin. The only alternative to sympatric speciation is that the species in the crater formed from a series of separate invasions to the crater by an ancestral form, but that all trace of multiple ancestry was erased by mtDNA gene flow among species within the crater. mtDNA gene flow is common between young sympatric fish species, and has more than once led researchers astray. Nevertheless, the possibility seems far-fetched in the crater lake cichlids given that no candidate immediate ancestors for the newest species inside the crater can be found outside of it. The mechanism of selection driving sympatric speciation in cichlids has not been discovered.

Direct measures of ecological selection against hybrids are few. F1 hybrids between the limnetic and benthic species of three-spine stickleback have a

[2] Sympatric speciation means that a new species evolves within the geographic range of the parental species. Contrast it with allopatric speciation, in which a new species evolves in a subpopulation, geographically outside the range of the parental species. Allopatric speciation is the standard model of speciation, as seen for instance in Chapters 26 and 27 by Rice and Hostert and Coyne and Orr, respectively. The genetic mechanisms of speciation they discuss and that Schluter is here arguing against do not work in sympatry.

high fitness in the laboratory but an intermediate phenotype that compromises their ability to acquire food from the two main habitats of native lakes. Their ability to seize and retain small, evasive zooplankton in open water is inferior to that of the limnetic species. Their rate of intake in the littoral zone is lower than that of the benthic species, mainly because the F1 hybrids do not take the larger prey. These feeding differences translate into slower growth of F1 hybrids relative to either of its parent species when in the habitat of that parent.

Ecological selection also implies that the strength of postmating isolation should change as environments change, perhaps to the point of increasing hybrid fitness and causing the collapse of the parental species pair. Something like this happened in a natural experiment on a Galápagos island. Grant and Grant[3] recorded the fate of offspring of crosses between *Geospiza fuliginosa* (small ground finch) and *G. fortis* (medium ground finch) over 20 years on the island of Daphne Major. *G. fuliginosa* is an uncommon but regular immigrant to the island, and many adults hybridize with *G. fortis*. The offspring are intermediate in beak size between the parent species and consume mainly small, soft seeds also eaten by *G. fuliginosa*. Small seeds are typically much less abundant than the large, hard seeds eaten by *G. fortis*, and hybrid survival is correspondingly poor. However, food conditions were dramatically changed in the years after record rains associated with an El Niño event, and this elevated hybrid survival to a level not less than pure *G. fortis*. Hybrids suffered no reduction in fertility through this period (they mated mainly to *G. fortis*). The pair of species essentially ceased to be biological species on Daphne. *G. fuliginosa* persists only because of recurrent immigration from another island. [. . .]

A tendency for hybrids to do worse than the parents in the habitat of those parents is not universal. For example, F1 and F2 hybrids between the irises *Iris fulva* and *I. hexagona* grow as well or better than the parent species in the native habitats. However, the species are characterized by strong genetic incompatibilities (hybrid *Iris* have low pollen fertility and poor germination). Clearly, ecological factors may play a role even when genetic incompatibilities are present, but their role in the origin of these incompatibilities is unclear. In contrast, the ecological component of hybrid fitness may yield more direct evidence of the role of divergent natural selection between environments in the origin of reproductive isolation when genetic incompatibilities have not yet built up.

These examples suggest that when genetic mechanisms of postmating isolation are weak, hybrids in nature may often be selected against

[3] The Grants' research on Darwin's finches is the topic of Chapter 11. However, that extract does not discuss the research on hybrid fitness that is being looked at here.

because they fall between the niches of their parents. They are best adapted to intermediate environments that do not exist. The significance of such mechanisms to speciation is twofold. First, any environmental agent that preferentially removes hybrids helps forestall the collapse of sympatric species by hybridization. Second, such mechanisms favour further divergence between species and may have contributed to their origin.

[*The Ecology of Adaptive Radiation* (Oxford: Oxford University Press, 2000), 199–203.]

V. GRANT

29 Hybrid speciation

New plant species can arise following hybridization between two existing plant species. Some early cases of hybrid speciation were found to be associated with increases in chromosome numbers. The increases may have allowed meiosis to proceed as normal in hybrids. Experiments have re-created hybrids, from existing members of the species that ancestrally gave rise to them; these experiments can reveal the genetic events in the origin of species. [Editor's summary.]

One of the historical problems of plant evolution, a problem which was discussed by successive authors through the eighteenth and nineteenth centuries, is the role of hybridization in species formation. Is natural hybridization a mechanism for the production of new plant species? [. . .]

By hybrid speciation we mean the origin of a new species directly from a natural hybrid. This definition brings the problem into focus. It is clear that the same sexual process which produced the natural hybrid will also bring about the breakup of its gene combination by segregation in later generations. Therefore an essential part of the mechanism of hybrid speciation is the stabilization of the breeding behavior of the hybrids. [. . .]

In the pre-Mendelian period the hybrid origin of new plant species was proposed by Linnaeus in 1744 and 1760, William Herbert in 1820, Charles Naudin in 1863, and Anton Kerner in 1891. The line of thought was continued in the early post-Mendelian period by Lotsy, Hayata, and others. [. . .]

With the rise of modern genetics in the early decades of this century, the old problem of hybrid speciation could at last be stated correctly. The problem could be seen to involve sterility barriers and segregation as well as morphological and physiological traits. In order to solve this problem it was necessary to find a mechanism by which a new, internally isolated, constant type could arise from species hybrid without loss of sexuality.

The first major breakthrough was Winge's hypothesis of amphiploidy[1] to use a much later term, which postulates that a new constant species can arise from a hybrid between two preexisting species following chromosome number doubling. Winge's hypothesis was soon confirmed experimentally by other workers in *Nicotiana*, *Raphanus-Brassica*, *Galeopsis*, and other plant groups.

Amphiploidy is a mode of speciation which involves unidirectional increases in number of chromosome sets, whereas many plant species have obviously evolved on the diploid level, and the question therefore remained whether hybrid speciation could take place without change in ploidy.

Winge himself returned to this question in his later work on *Tragopogon* and *Erophila*, and Lamprecht tackled it in *Phaseolus*. The stated objective of the experiments in *Tragopogon*, *Erophila*, and *Phaseolus* was to determine whether new fertile types or microspecies could be produced by hybridization without the accompaniment of amphiploid doubling. These experiments paved the way for studies by other workers on a process of hybrid speciation, alternative to amphiploidy. [. . .]

Chromosomal sterility stems from segmental rearrangements between the parental genomes which upset the course of meiosis, visibly or invisibly, in the hybrid. Either the homologous chromosomes do not pair normally in bivalents and do not separate properly to the poles at anaphase, or they pair but segregate to yield daughter nuclei carrying deficiencies and duplications. In either case a proportion of the meiotic products are unbalanced and do not develop into functional spores and gametes. This proportion of inviable spores and, hence, the degree of sterility rises rapidly with increase in number of heterozygous rearrangements.

Let us suppose that the chromosomally sterile hybrid undergoes doubling of chromosome number. Then, disregarding other possible complicating factors for the present, it will become meiotically normal and gametically fertile.

This recovery of fertility on the new polyploid level is most clear-cut in the first case mentioned above where the chromosomes do not form bivalents regularly in the diploid hybrid. The structurally well-differentiated genomes of the parental species can be symbolized as *A* and *B* respectively. The genomic constitution of the two diploid species is *AA* and *BB*; and that of their hybrid is *AB*. The chromosomes belonging to the *A* set have no homologous partners to pair with at meiosis in the hybrid, and neither do the

[1] Amphiploidy is a change (often doubling) of genome size by (or partly by) combining the genomes of two different species. (Compare with autopolyploidy in which genome size is multiplied up within one species.) Chapter 37 describes genomic inferences about polyploidy in the history of angiosperms.

B chromosomes. But the situation is entirely different in the allotetraploid (or amphidiploid) derivative of this hybrid with the genomic constitution to *AABB*. Now there exists a homologous pair of each chromosome type in each genome. Consequently meiosis and gamete formation can proceed normally.

In the second case mentioned earlier the homeologous chromosomes in the hybrid do form bivalents but segregate to produce genically unbalanced meiotic products. The diploid parents possess different subgenomes, A_s and A_t. Their F_1 hybrid has the genomic constitution A_sA_t and has chromosomal sterility associated with visibly normal meiosis. In this case also the tetraploid derivative, $A_sA_sA_tA_t$, is likely to recover fertility, or at least semifertility, as a result of what Darlington called differential affinity. Each A_s chromosome and each A_t chromosome has a completely homologous partner in the tetraploid. These homologous chromosomes pair preferentially in the tetraploid and pass to opposite poles to yield segmentally and genically balanced products of meiosis.

The tetraploid derivative of the chromosomally sterile hybrid, whether it has the genomic constitution *AABB* or $A_sA_sA_tA_t$, is not only fertile itself, but is also reproductively isolated from its diploid parents. Crosses between tetraploid and diploid plants frequently run into incompatibility barriers, and the hybrids, if any arise, are triploid and hence usually sterile.

The classical cases of *Primula kewensis, Raphanobrassica*, and *Galeopsis tetrahit* illustrate the various types of genomic constitution in experimental amphiploids.

The F_1 hybrid of *P. floribunda* ($2n = 18$) × *P. verticillata* ($2n = 18$), widely known by the name *P. kewensis*, is a diploid perennial herb like its parents. It showed normal bivalent pairing but was highly sterile. In three different years the otherwise sterile hybrid plant spontaneously gave rise to fertile branches, which, when studied cytologically, turned out to be tetraploid ($2n = 36$) and to have predominantly bivalent pairing.

Evidently, preferential pairing of completely homologous chromosomes occurred in the tetraploid branches. These fertile shoots arose by somatic doubling in a bud or sector of a bud. They produced seeds which developed into fertile and fairly uniform F_2 progeny.

A contrasting condition was found in *Raphanobrassica*, the amphiploid derivative of *Raphanus sativus* × *Brassica oleracea*. The parental species are both diploid with $2n = 18$ chromosomes. The F_1 hybrid exhibits complete failure of chromosome pairing and is highly sterile. It produces some unreduced diploid gametes. Union of these gave rise to tetraploid ($2n = 36$) plants in F_2, which had normal meiosis with regular bivalent formation and were mostly quite fertile. They yielded a morphologically uniform F_3 generation. The new, fertile, true-breeding line is isolated by sterility barriers from the parental diploid species.

Galeopsis tetrahit ($2n = 32$), unlike *Primula kewensis* and *Raphanobrassica*, is a naturally occurring tetraploid species in northern Europe and Asia. Müntzing proved that this annual herb is an amphiploid derived from two related diploid species in Europe, *G. pubescens* ($2n = 16$) and *G. speciosa* ($2n = 16$).

The artificial F_1 hybrid of *Galeopsis pubescens* × *speciosa* is meiotically irregular and fairly sterile with 8% good pollen and 5 to 8 bivalents. The F_1 hybrids, despite their sterility, produced an F_2 generation consisting of some sterile diploids and one sterile triploid plant. The latter arose from the union of one unreduced gamete and a reduced gamete in F_1. The triploid when backcrossed to *G. pubescens* yielded a single seed which gave a tetraploid plant in F_3. The tetraploid probably arose from the fertilization of an unreduced $3n$ egg by a normal l*n* sperm. The tetraploid F_3 plant had 16 bivalents at metaphase of meiosis and produced 70% good pollen. It yielded fertile tetraploid F_4 progeny.

The artificial allotetraploid resembled natural *Galeopsis tetrahit* in morphology as well as in chromosome number. Furthermore, it is isolated by an incompatibility barrier from *G. pubescens* and *G. speciosa*, as wild *G. tetrahit* is. It could be considered to be a synthetic form of *G. tetrahit*. To test this assumption, Müntzing crossed the artificial allotetraploid with natural *G. tetrahit*. The artificial and natural tetraploids crossed easily to produce F_{IS} which were fertile with good chromosome pairing. This was the first experimental resynthesis of a naturally occurring amphiploid species.

[*Plant Speciation*, 2nd edn. (New York: Columbia UP, 1981), ch. 19.]

Section E

Macroevolution

Macroevolution means evolution on the grand scale. It is contrasted with microevolution, the study of evolution over short time periods, such as that of a human lifetime or less. Microevolution mainly refers to changes in gene frequency within a population; the research in Section B was microevolutionary. Macroevolutionary events are more likely to take millions, probably tens of millions and maybe hundreds of millions of years. Macroevolution refers to things like the origin of major groups or mass extinctions. The studies in Section G on 'The history of life' are all macroevolutionary: they look at the origin of life, prokaryotic evolution over two billion years, the main trends in angiosperm evolution, and the radiations of life that occurred in the Cambrian and in the Tertiary (following the Cretaceous mass extinction).

Speciation (discussed in Section D) is the traditional dividing line between micro- and macroevolution. Speciation can just about be studied on the human timescale, easily in some cases such as instantaneous hybrid speciation in plants and less easily in other cases when speciation takes over a million years. Microevolution has traditionally been the sphere of the geneticist and ecologist; macroevolution of the palaeontologist. But in recent years things have changed. Genetic, and particularly molecular genetic, methods have been increasingly used to study macroevolution. This has led to some creative clashes, when molecular and fossil evidence have been used to answer the same question. An early example concerned human evolution (Chapter 53). More recently, molecular evidence has challenged fossil times for the origin of several large groups of animals (Chapter 44). The new subject of genomics (Section F) is another way in which new molecular evidence is yielding macroevolutionary insights.

I should emphasize that micro- and macroevolution are only two ends of a continuum. They are terms of convenience only. They do not imply that different processes drive evolution in the short and long term—as if microevolutionary processes somehow come to a stop beyond speciation. However, traditional population genetic evidence (Section B) differs from fossil evidence, and even from modern genomic evidence (Section F). In consequence, different kinds of hypotheses have tended to be tested in micro- and in macroevolutionary research.

The extracts I have selected here fall into three categories. The first concerns the rate of evolution. The classic research on this topic was done on fossils, by the palaeontologist G. G. Simpson. Simpson is often thought of as the man who introduced the Modern Synthesis into palaeontology and vice versa, particularly via two classic books, *The tempo and mode of evolution* (1944) and *The major features of evolution* (1953)—the latter being an up-date of the former. (In section G, Schopf's paper (Chapter 43) is explicitly inspired by the former book and Cooper and Fortey (Chapter 44) begin with a look at Simpson's ideas.) Simpson supported a pluralistic conclusion about the rates of evolution. The fossil record, according to Simpson, showed a wide variety of rates.

More recently, there has been extensive discussion of Eldredge and Gould's theory of punctuated equilibrium, originally published in 1972; Erwin and Anstey (Chapter 30) introduce the ideas, and give a modern review of the many factual studies inspired by those ideas. Eldredge and Gould criticized the idea that evolution has a constant rate through time; Erwin and Anstey refer to this as 'gradualism'. Eldredge and Gould suggested instead that evolution has a high rate at times of speciation, when a lineage splits, but a low (or zero) rate between splits: this is what is meant by a 'punctuated' pattern of evolution. As Erwin and Anstey describe, Eldredge and Gould argued from Mayr's 'peripheral isolate' theory of speciation. This is a special case of the allopatric theory discussed in Chapters 26 and 27. A peripheral isolate is a small isolated population at the edge of the main species range: Mayr suggested that speciation may often occur in these populations. If so, fossils of the speciating population would be unlikely to be preserved (the speciating population was small), and not at the same place as the ancestral population (the speciating population was elsewhere); hence the punctuated pattern in the fossil record of speciation. Erwin and Anstey's review of 58 factual studies done between 1972 and 1994 reveals a variety of observations, but a good core of support for Eldredge and Gould's basic claim.

The next three extracts (Chapters 31–3) concern the nature of homology. Homology can be defined theoretically without any problem. A homology is any identifiable attribute of organisms that is shared between species and was present in those species' common ancestor. However, we cannot observe ancestors directly, which means that homologies have to be inferred from modern organisms. That is when the problems start. Homology has proved one of the more enigmatic of evolutionary concepts. It seems to have a clear-cut meaning, but it rapidly becomes confusing when you try to apply it to real evidence. De Beer's paper (Chapter 31) works through one possible criterion after another and shows that none of them are adequate. Homology is undoubtedly a genuine and important concept; the problem is to spell out exactly what it means in practice. The problem has grown more fascinating as our understanding of the genetics of morphological development has improved recently. Attributes, such as the eyes of insects and of vertebrates, which every zoologist knows not to be homologous, turned out to be controlled by the same, or a very similar, gene. Dawkins (Chapter 32) introduces the observations, and Dickinson (Chapter 33) interestingly clarifies how to think about homology in these cases.

Homologies have two main applications. One is in the inference of phylogenetic relations. Homologies are used to work out the ancestral relationships among species. The other is in research on the history of particular features of living things—features such as organs, or molecules, or behaviour patterns. This second kind of research is illustrated by the classic example of the mammalian ear bones, derived from jaw bones in reptiles, described by De Beer. To trace the transformations of an organ, as it is modified down an evolutionary lineage, is to study the homologous forms of the organ. The readings on lens proteins (Chapter 40), eyes (Chapter 50), and language (Chapter 59) in Sections F, H, and I, are transformational studies of this sort, though the lens proteins of different species turn out not to be homologous.

The meaning of homology given here (an attribute that is present in two species and that is descended from their common ancestor) is, I believe, fairly orthodox among most, though not all, biologists. A habit has grown up among some molecular biologists of using homology to mean similarity, regardless of whether the similarity is due to descent from a common ancestor. They thus talk about the '% homology' between two molecules, meaning the percentage of amino acid, or nucleotide, sites that are the same in the two molecules. They are often criticized for their unorthodox (even uneducated) usage, and the opening of Dickinson's paper alludes to the issue.

My third macroevolutionary topic here is the relation between individual development (from egg to adult) and evolution. The relation has interested biologists since Darwin's time, and has enjoyed a revival in the kind of modern research informally called 'evo-devo'. One concept shaken up by modern evo-devo is homology itself, and the two extracts by Dawkins (Chapter 32) and Dickinson (Chapter 33) already discussed also illustrate research on evo-devo. The other two extracts concern an early idea that failed to deliver but that retains a big influence. This is the infamous theory of recapitulation described by Haeckel (Chapter 34).

The extract from Haeckel incidentally gives some feel, even at a distance of 100 years, for his infectious enthusiasm, enabling him in his time to be a successful science popularizer and propagandist. Haeckel is so closely associated with the principle of recapitulation that his name is almost interchangeable with it. Recapitulation is the idea that an individual, during its development, 'climbs up its family tree'. As the extract makes clear, it was thought to reflect the workings of the hereditary mechanism, and recapitulation was another of the victims of the Mendelian revolution. Zoological critics also pointed to cases in which early developmental stages were evolutionarily more modified than the adult stages; Garstang's poems

(Chapter 35) illustrate these critics' viewpoint.[1] Haeckel was aware of the counter-examples, as Chapter 34 shows. However, there is a world of difference between an identified exceptional process within one theoretical scheme, and the first-among-equals process of another. The main interest of Haeckel's system was in phylogenetic inference—recapitulation provided a sort of molecular clock in the biology of one century ago—but it is no longer used in that way. Although recapitulation's theoretical basis and empirical generalizations have both collapsed, Haeckel and his followers may still have identified a phenomenon of sufficient generality that biologists should be trying to explain it; currently, however, they are not.

Evo-devo is not far behind some other extracts in this book. Evolutionary genomics (Section F) is closely related to evo-devo. Tracing the evolution of the genes that control development, such as the Hox genes, can equally be classified as evolutionary genomic or evo-devo research. Moreover, developmental change may have been particularly important in human evolution. Section I (Chapters 53 and 54) includes a classic 1975 paper by King and Wilson, who argued that humans evolved mainly by changes in regulatory genes. (A regulatory gene is not the same as a gene controlling development. For instance, some regulatory genes control cellular processes. However, developmental genes are included within the general category of regulatory genes.) King and Wilson's hypothesis still retains its scientific force. Carroll (Chapter 39) looks at how genomic evidence may be about to test whether King and Wilson were right.

On another topic, Section H (Chapter 51) has an extract from Gerhart and Kirschner's *Cells, embryos, and evolution*, a book about the evolution of cells and development. However, the particular extract I chose is about innovation, rather than the relation between development and evolution, and fitted better in Section G than here in Section E.

DOUGLAS H. ERWIN AND ROBERT L. ANSTEY

30 Speciation in the fossil record

In 1972, Eldredge and Gould suggested that evolution proceeds by punctuated equilibrium rather than phyletic gradualism. This extract begins by looking at the historical and conceptual background to Eldredge and Gould's proposal. It then looks at developments in

[1] Garstang was a professional zoologist, but lived in an age that placed less emphasis on the steady grind of dry publications than we do now. He wrote scientific papers, but also liked to put his speculative and theoretical ideas about zoology into light verse. He also wrote some more serious poetry, in the style that literary critics call Georgian. He composed many such poems for his friends, among whom they circulated informally. His zoological verse became more widely known after A. C. Hardy gathered much of it in the slim posthumous volume *Larval forms*.

the theory of punctuated equilibrium after 1972, and at how the theory can be tested. Forty studies have been done, providing evidence in some cases of punctuated equilibrium, in other cases of gradualism, and in other cases of a mix of the two. [Editor's summary.]

The processes and patterns of speciation present some of the most intractable problems of evolutionary biology. The process of speciation generally requires too long a period of time to be directly observable by biologists, who can only make inferences from the populational and intrapopulational events they observe, and who must reconstruct past events from the attributes of species inferred to be closely related. Paleobiologists are likewise in an unenviable position: the fossil record is generally too coarse in temporal scale and too limited in geographic coverage to provide a detailed history of speciation events. Fossils also do not preserve all the character states that define species; distinctions between fossil species often reflect subjective judgments by taxonomists. [. . .]

As is often the case in paleobiology, evidence from the fossil record constrains the range of the mechanisms applicable to particular evolutionary case histories but does not specify mechanisms precisely. The simplifying effect of this deficiency is that paleobiology has not needed to cope with such difficult issues as the proper biological definition of species, genetic and developmental constraints, and the relative frequency of particular speciation mechanisms. Such issues, however, ultimately must be addressed in the debate over speciation: the definition of species, for example, plays a crucial role in evaluating the importance of alternative speciation mechanisms. [. . .] While such debates fall outside the purview of paleobiology, they at any rate do not discourage paleobiologists from likewise developing a pluralism of viewpoints on speciation.

Paleontologists have, over the last two centuries, developed their own empirical protocols for recognizing fossil morphospecies. These protocols were generally typological and invariant prior to the Modern Synthesis of the 1930s and 40s, becoming populational and incorporating variation afterward. Paleontological protocols have always attempted to mimic contemporary biological concepts. Their chief departure was the recognition (under the aegis of the Modern Synthesis) of the so-called evolutionary species or chronospecies, an evolving lineage incorporating temporal variation. The present debate about speciation in the fossil record is a debate about the evolutionary phenomenology of fossil morphospecies.[1] [. . .]

Paleontologists, taking their cue from Darwin's *Origin of Species*, historically settled into an automatic discussion of the gaps in the fossil record when describing speciation. Species were regarded as the incidental by-products of

[1] A morphospecies is a species that is recognized by measurements of morphology. Compare the (paleontologically impractical) interbreeding criterion of a species: Chapter 23 above.

evolution, and speciation was viewed as a long-term continuous process rather than an episodic one. The incompleteness of the fossil record was deployed as the primary explanation for an inability to document gradual change from one morphospecies into another. Typology[2] partially contributed to this problem, but the eventual study of variation in fossils did little to relieve the emphasis on gaps in the fossil record: even variable morphospecies failed to display gradual, directional change. Today many fossil species are still known from only a very few specimens, from very few localities, and from very few beds. Mere observation of evolutionary tempo depends upon retrieval, and therefore upon abundance in the fossil record. The best known fossil species—i.e., those that are abundant, widely distributed, and stratigraphically long-ranging—seemed to display little net morphological change over their durations. One of the early objections voiced to the *Origin of Species* was that of the English paleontologist John Phillips, who in 1860 observed that most fossil species appeared, persisted through one to three formations, and then disappeared. [. . .]

Despite the anecdotal (but taxonomically practical) knowledge that many fossil species persisted unchanged, a textbook[3] widely used for several decades still attributed such observations to the incompleteness of the record, and caricatured speciation as a long-term gradual process that could only be glimpsed in the fossil record through the episodic windows provided by preservation and sedimentation. Under the aegis of this dominant textbook, paleontologists continued to name species, but rarely examined speciation *per se* in the fossil record. Rather, they directed evolutionary research to trends in taxa above the species level. [. . .]

Stratigraphic paleontologists (who enjoyed significant economic support from the petroleum industry) favored index or guide fossils in their investigations: geographically widespread species that displayed little or no change over geologically brief durations. The few studies that had purported to demonstrate gradualism at the species level in the fossil record, thereby 'proving' that evolution was paleontologically verified, were falsified by the late 1960s. Workers using less coarse sampling designs, and who were more critical in their analysis of morphology, were responsible for these reappraisals. The concept of chronospecies remained a theoretical construct, but with few substantive examples.

[2] See Chapter 23 on typology. The particular problem here is this. Suppose we have a range of shades of grey, from white to black. We define 'white' typologically as all white plus all light grey shades that are sufficiently similar to 'type' white, where 'sufficiently similar' takes us about halfway up the range; we define black as black plus all the darker shades sufficiently similar to black. The smooth continuum of shades will then show a punctuational jump from white to black. Likewise if species are typologically defined.

[3] R. C. Moore, G. C. Lalicker, and A. G. Fischer, *Invertebrate Fossils* (New York: McGraw-Hill, 1952).

By the late 1960s, with improved time resolution in the stratigraphic record and discovery of new sections filling in many gaps, the view of overwhelming incompleteness in the fossil record was becoming less plausible. [. . .]

Punctuated equilibria: a revolution?

The theory of punctuated equilibria (irreverently known as 'evolution by jerks')[4] was born as a straightforward attempt to translate into the fossil record Mayr's hypothesis of peripheral isolates. Eldredge and Gould suggested that peripheral isolates—that is, small populations separated from the main body of the species—would be essentially invisible to the fossil record until (and unless, since many 'proto-species' would simply disappear) population size and geographic distribution increased sufficiently to raise the likelihood of preservation. From this perspective, the discontinuities in the fossil record were not artifacts; they were real.

In effect, Eldredge and Gould argued that the prevailing paradigm (gradualism punctuated by stratigraphic gaps) was wrong. If the dominant mode of speciation involves peripheral isolates, as Mayr contended, then the sudden appearance of new species in the record is to be expected *even if the record is highly complete*. Stratigraphic or taphonomic incompleteness would merely accentuate the pattern. In their view, once the population expanded, morphologic stasis[5] would ensue [. . .]. Stasis was not promoted out of theoretical concerns, but was based upon an empirical assessment of the actual patterns displayed by known fossil morphospecies.

In their initial proposal of punctuated equilibria, Eldredge and Gould arguably went beyond Mayr's concept of speciation (which did not include stasis). Speciation in their view should no longer be regarded as a long-term process (coextensive with evolution itself) but as an *event* (in the geological sense of 'event'). Recognition of speciation as an event effectively decoupled intraspecific evolution from macroevolution, creating two levels of a new evolutionary hierarchy. The claim of morphologic stasis during the duration of a species, while arguably based on the fossil record, implicitly called into question the significance of intraspecific adaptation and variation, except as expressed through the generation of peripheral isolates. It was this claim which was to cause the greatest controversy.

Considerable confusion was generated by the changing nature of the

[4] N. Eldredge, *Evolution*, 25 (1971), 156–67. N. Eldredge and S. J. Gould in T.J.M. Schopf (ed.), *Models in Paleobiology* (San Francisco: Freeman, Cooper, 1972), 82–115. The irreverent source is J. R. G. Turner, in D. M. Raup and D. Jablonski (eds.), *Patterns and Processes in the History of Life* (Berlin: Springer Verlag, 1986), 183–209.

[5] Stasis is used, in the theory of punctuated equilibrium, to refer to absence of evolutionary change—that is, constancy in form.

claims for punctuated equilibria between 1972 and the mid 1980s. Ruse[6] recognizes three phases. The first phase is that just described. The second phase linked punctuated equilibria with macromutationist ideas (in the sense of Goldschmidt). This phase is exemplified by Gould's 1980 paper entitled 'Is a new and general theory of evolution emerging?'[7] which proposed evolution as a three-tier hierarchy and which explicitly called into question the efficacy of natural selection. Gould later claimed that this line of argument was developed independently of the theory of punctuated equilibria, a claim which may well be true; nonetheless, many readers missed the distinction. The third phase was marked by a retreat from what Ruse (in our view, fairly) characterizes as the extremism of the second phase and a decline in anti-selectionist and macromutationist views. This latest phase also engendered support for stasis from quantitative genetic models, along with concomitant claims that punctuated equilibria had been fully predicted by neo-Darwinism.

The theory of punctuated equilibria involved several assumptions and made several testable claims about the nature of speciation. The first of these assumptions was that reproductive isolation was generally linked to morphologic change. Thus the morphologic transitions observed in the fossil record were taken as a valid proxy for speciation. Second, Eldredge and Gould relied on a particular model of isolation; yet the pattern they described could well be produced by other forms of isolation as well. Indeed, many biological criticisms of the theory focused more on Mayr's model than on empirical patterns from the fossil record. Third, Eldredge and Gould claimed that speciation occurred rapidly. As paleontologists, they meant this in a geological sense, involving tens of thousands of years. Many neobiologists, however, misinterpreted this as a claim for biological rapidity, involving only a few generations. This third claim is often difficult to test because: (1) there are no intermediates available in beds below the base of a punctuated species; (2) the diastem immediately below the base of such a species usually accounts for more time than the beds above or below the diastem; and (3) chronostratigraphic resolution in fossiliferous sedimentary rocks is rarely more precise than ±500,000 years. Fourth and finally, Eldredge and Gould coupled geologically rapid speciation to a claim of long-term morphologic stasis (which is much easier to document). [. . .] We can use these four major assertions as a framework in which to consider research on speciation in the fossil record over the past two decades.

The first assumption is among the most troubling. Clearly, if the definition of speciation is the acquisition of reproductive isolation, then paleobiologists

[6] M. Ruse, in A. Somit and S. A. Peterson (eds.), *Dynamics of Evolution* (Ithaca, NY: Cornell University Press, 1989), 139–67.

[7] *Paleobiology*, 6 (1980), 119–30.

cannot observe speciation directly. [. . .] With the exception of the groundbreaking work of Jackson and Cheetham[8] and similar studies comparing species in fossil and living material, it is difficult to address the linkage between reproductive isolation and morphological change [. . .] Studies of Darwin's finches in the Galapagos and *Drosophila* in Hawaii demonstrate that the assumption of a linkage between reproductive isolation and morphologic change might in fact be reasonable. Nonetheless, it is this aspect of punctuated equilibria which has received the greatest challenge from population geneticists. Conversely, numerous population geneticists have developed models of genetic change consistent with the tenets of punctuated equilibria. Futuyma[9] observed that much of this disagreement stems from a division between Fisherian and Wrightian traditions in evolutionary genetics. The former emphasizes gene action and the significance of selection, while the latter places greater importance on epistasis and genetic integration. The legacy of Sewall Wright is thus the tradition underlying punctuated equilibria. The genetic constraints favored by the Wrightian approach provide a compelling argument for the probability of stasis. [. . .]

The second assumption, the actual geography of small isolated populations that gave rise to new species, is equally intractable. Taphonomic studies of fossil preservation, now numerous, emphasize that the preservation of even a well-skeletonized organism is indeed a rare event. Preserved fossil populations represent, at best, only submicroscopic percentages of their once-living populations. Sedimentation and preservation of sedimentary beds is a highly episodic process. Populations small enough to produce the geologically rapid morphologic changes called for in punctuations have, therefore, a zero probability of preservation. The taphonomic and stratigraphic window through which paleobiologists might hope to see the biogeography of micro-demes—be they sympatric, allopatric, peripatric, or whatever—usually is represented by a single bedding plane [. . .]

On the other hand, the fossil morphospecies whose patterns of change have been tracked through multiple time horizons all represent enormously large once-living populations. To be common enough to be sampled in fossil populations in densely spaced strata over a significant time duration means, in most instances, that population densities were comparable to those of the Ordovician bryozoan *Parvohallopora*, whose colony fragments likely numbered in the hundreds per square meter in thickets dispersed over 600 km of epeiric seaway; these population densities were maintained over a time interval exceeding seven million years. Enormous quantities of fossil material

[8] J. B. C. Jackson, and A. H. Cheetham, *Science*, 248 (1990), 579–83.

[9] D. J. Futuyma, *American Naturalist*, 134 (1989), 318–21, and in D. Otte and J. A. Endler (eds) *Speciation and its Consequences*, (Sunderland, Mass.: Sinauer, 1989), 557–78.

are available for the Jurassic bivalve *Gryphaea*, for planktonic forams and radiolaria, and for Permian fusulines. [. . .]

The third assumption inherent in the punctuated equilibria theory involves the rapidity of speciation. The distribution of morphologic change through time and space remains a compelling question, regardless of one's views on the linkage between morphologic change and the onset of reproductive isolation. Moreover, paleobiologists have the only long-duration data applicable to the question. These data constitute a rich body of empirical knowledge on the temporal dynamics of morphological evolution. Since 1972 numerous paleontological studies have explicitly investigated the tempo of morphological change in the fossil record, with varying results. Many of the observed patterns, such as the maintenance of long-term sustained gradualism, or the even more problematic oscillations between stasis and gradualism, may not yet have an adequate theoretical basis. Calculated selection coefficients seem to be too weak to justify conventional natural selection as a cause for these long-term patterns. Problems of temporal scaling are also involved, as discussed by Gingerich.[10] The results so far ostensibly support a pluralism of tempos and modes in the historical record of speciation. No single pattern appears dominant, and their variety suggests that new research agendas are needed to investigate the conditions under which any one of them might be predicted. The results shown in Table 30.1,[11]

Table 30.1: Varieties of tempos and modes of speciation

	TEMPO		
MODE	GRADUALISM	GRADUALISM AND STASIS	PUNCTUATION AND STASIS
Non-branding	Ordovician bivalves: Bretsky & Bretsky 1977 Permian fusulines: Ozawa 1975 Permian bryozoans: Pachut & Cuffey 1991 Jurassic bivalves: Fortey 1988 Palepgene primates: Godinot 1985 Neogene radiolarians: Baker 1983 Neogene forams:	Ordovician bryozoans: Brown& Daly 1985 Ordovician trilobites: Cisne *et al.* 1982 Sheldon 1987; 1993 Jurassic ammonites: Raup & Crick 1981; 1982 Cretaceous bivalves: Geary 1987 Cenozoic ostracods: Benson 1983 Neogene radiolarians: Kellogg 1985; 1983	Devonian corals: Pandolfi & Burke 1989 Devonian brachiopods: Isaacson & Perry 1979 Jurassic bivalves: Johnson 1985 Cretaceus echinoids: Smith & Paul 1985 Eocene condylarths: West 1979 Neogene echinoids: McNamara 1990

[10] P. D. Gingerich, *Science*, 222 (1983), 159–61, and *American Journal of Science*, 293-A (1993), 453–78. What rate do we see when evolution reverses its direction? If the interval between measurements is short we see a high rate, but if it is long the observed net rate is slower because the faster spurts between tend to cancel. Chapter 11 is an example of such a reversal.

[11] See the original paper for the references. Branching here refers to speciation, or a split in a lineage. The studies are divided into three groups: those in which no speciational splits were seen, those with a few splits, and those with many splits.

	TEMPO		
MODE	GRADUALISM	GRADUALISM AND STASIS	PUNCTUATION AND STASIS
	Malmgren & Kennett 1981 Arnold 1983 Banner & Lowry 1985 Neogene bivalves: Hayami & Ozawa 1975 Neogene gastropods: Geary 1990 Neogene echinoids: McNamara 1990 Neogene rodents: Chaline & Laurin 1986	Sachs & Hasson 1979 Neogene forams: Scott 1982 Malmgren *et al.* 1983 Malmgren & Berggren 1987 Wei & Kennett 1988 Neogene fish: Bell *et al.* 1985	
Limited Branching:	Jurassic ammonites: Callmon 1985 Eocene primates: Gingerich 1976; 1985	Ordovician trilobites: Portey 1985 Cretaceous ammonites: Reyment 1975	Devonian brachiopods: Goldman & Mitchell 1990 Cretaceous ostracods: Reyment 1982 Cenozoic ostracods: Whately 1985 Eocene ostracods: Reyment 1985 Neogene diatoms: Sorhannus 1990a, b Neogene radiolarians: Kellogg & Hays 1975
Multiple Branching:	Eocene condylarths: Gingerich 1976; 1985 Eocene primates: Bown & Rose 1987		Cambrian trilobites: Robison 1975 Ordovician trilobites: Fortey 1974 Henry & Clarkson 1975 Devonian stromatoporoids Fagerstrom 1978 Devonian brachiopods: Fagerstrom 1978 Devonian trilobites: Eldredge & Gould 1972 Jurassic bivalves: Hallam 1982 Cenozoic bivalves: Stanley & Yang 1987 Neogene radiolaria: Lazarus 1983; 1986 Neogene bryozoans: Cheetham 1987 Neogene molluscs: Eldredge & Gould 1972 Williamson 1981 Kelley 1983 Neogene ungulates: Vrba 1980

Note: From non-anecdotal paleontological studies published between 1972 and [1995]. Position in the table generally reflects the conclusions of the authors cited, and not subsequent debates.

representing studies of 58 lineages published during the last 22 years, collectively provide a very different result from the initial predictions of Eldredge and Gould. The greatest departure may be in the preponderance of studies illustrating *both* stasis and gradualism in the history of a single lineage.

The fourth issue raised by the theory of punctuated equilibria concerns morphologic stasis. A principal claim of Eldredge and Gould is that 'stasis is data'. Many illustrations of both gradual change (gradualism) and stasis are overly simplistic. In several instances, the same data have even been interpreted as gradualism by one set of authors and stasis by another. In neither tempo should one expect (in continuously varying characters) absolute stasis, nor a directional trend without minor reversals. Distinguishing between ecophenotypic[12] effects, selection, and random walks is difficult in most cases. Sequences of short duration that can not be distinguished statistically from the null hypothesis of a random walk represent neither stasis nor gradualism; the tempo in those instances is simply unresolved. [. . .] Multiple forms of both gradualism and stasis have obscured the once-simple distinction between the two, and the issue of temporal scale confounds the issue even more.

Characters with nominal and ordinal states—for example, those with discrete states, four toes v. five toes—have a tough time displaying gradualism (except as shifting frequencies, a condition rarely ascertainable from the fossil record). These are the characters most commonly used in cladistic analysis, and this style of cladistics assumes punctuation at the character level. Such characters may, however, represent phenotypic thresholds that come to expression following the buildup of genotypic change. The underlying genetic changes could thus well be gradualistic; only the morphologic expression is punctuated.

Most of the studies in Table 30.1 have been based upon continuously varying characters—that is, shapes, angles, etc. These are intrinsically the characters most likely to display temporal variation. In such characters some gradualism is to be expected. There is simply no such thing as complete stasis in features, such as shapes, measured on a continuous scale. Such

[12] Ecophenotypic effects means that the form of the phenotype is influenced by the environmental conditions. Its relevance here is that there can be an ecophenotypic switch between very different phenotypes, depending on the environment. The phenotypes may be different enough to look like different species (and there are examples where they have been so classified), but if you take eggs laid by one female and rear some in one set of conditions they grow up as 'species 1' and rear the rest in another set of conditions they grow up as 'species 2'. Switches of this kind have been demonstrated in snails and other species. The environmental conditions are things like temperature and mineral composition of the water. A punctuational jump between forms may therefore not be evolutionary. It would be bogus evidence for the theory of punctuated equilibrium, which is properly a theory about evolutionary change.

characters are also strongly subject to both ontogenetic and ecophenotypic variation. [. . .]

Stratigraphic position in a local section may not accurately reflect phylogeny [. . .]. One of the problems involved in using cladistics to investigate tempo and mode in speciation is that whole-organism or 'phenetic tempo' may become methodologically unobservable. Naturally, the primitive states lingering within any lineage are much more likely to display stasis than are the derived states, and phenetic approaches will automatically find more stasis. Conversely, recognition of 'cladistic tempo' may be based on only a small number of derived states. Finally, the multivariate conventions for recognizing phenetic morphospecies may incorporate perceptual biases that may be difficult to detect casually, thus intrinsically biasing the results. Canonical multivariate techniques, for example, are designed to minimize within-group variation and maximize across-group differences. It would be unfair to delineate species by such a technique and then 'discover' that their canonical variates display stasis. [. . .] Both cladists and pheneticists should become aware that their standard manipulations of data might render biased conclusions regarding the tempo of morphological evolution.

Current status

An interesting departure in Table 30.1 from conventional expectations is that fully one-fourth of the studies we assessed describe single lineages that sequentially or simultaneously display both gradualism and stasis—more than half of which represent metazoans. Several other studies could be relocated to this column as well. For example, the Eocene mammalian condylarths that display multiple branching and gradations in one region display no branching and stasis in an adjacent region. The Permian fusulines studied by Ozawa evolved gradually in four characters, but not in five others; therefore they might also be moved to the center column. Similarly, the short-term micro-evolution in Neogene bivalves and in Permian bryozoans might easily be interpreted as just short segments of oscillatory stasis or random walks (therefore they might land in any column). If some of the punctuated patterns fail to hold up—such as those observed in Cenozoic molluscs by Williamson, which might be attributable to ecophenotypic responses to increased salinity during a lacustrine regression[13]—then the gradation-and-stasis column might come to hold as many empirical examples as does the punctuated pattern. This possibility is not so surprising, because all of the taxa displaying gradualism in the particular characters studied by

[13] P. G. Williamson, *Nature*, 293 (1981), 437–43. P. W. Kat and G. M. Davis, *Nature*, 304 (1983), 660–1.

the researchers in this table must simultaneously display stasis in others, namely those characters that define and delineate their lineage. Even the most famous example of natural selection and gradualism[14]—the peppered moth, *Biston betularia*—displays gradualism only in its protective coloration, and ostensibly stasis in all other features. Caution is also required in proclaiming sequential alternations of gradualism and stasis, because varying sedimentation rates could generate condensed and diluted intervals in succession that falsely appear as rapid gradualism and stasis.

The study by Brown and Daly,[15] while not reaching any explicit conclusions regarding tempo and mode of evolution, is a highly informative example. In three species of the Ordovician bryozoan *Parvohallopora*, the characters defining each species lineage were invariant over a duration exceeding many millions of years, and thus represent stasis. Many other characters were measured, however. Using a statistical technique that reduces nonheritable variation, all were plotted stratigraphically and tested for linear trends. Both stasis and sustained trends appear in some characters in some stratigraphic sections, but not in other characters or in the same character in other sections. Some sections display short-term reversing clines in some characters. Therefore these lineages display stasis, oscillations or random walks, and local but not regional gradualism in a mosaic of biometric characters. If the study had included fewer lineages, fewer characters, fewer sections, sampling over a more limited interval, or if it had failed to reduce ecophenotypic variation, then the conclusions might well have been different, as they might also have been if only selected results had been published.

Brown and Daly and Jackson and Cheetham made efforts to account for the effects of heritable versus nonheritable variation. Hallam similarly removed the effects of ontogenetic variation from the evolution of the Jurassic bivalve *Gryphaea*. No explicit applications of the technique of fluctuating asymmetry have yet been made in the analysis of tempo and mode. [. . .]

More than one-fourth of the studies in Table 30.1 are based on protozoans, which may not form species comparable to those in outbreeding metazoans. Protozoan lineages, rather, may alternate between sexual and asexual reproduction for unknown durations. The onset of environmental stress might trigger more widespread sexual reproduction, possibly producing the alternations between stasis and gradualism seen in some lineages. Many of these alternations, however, surely reflect unrecognized hiatuses in deep-sea sediments. The increase in the size of the initial chamber in Permian fusulines might likewise reflect an extremely sluggish decrease in outbreeding during

[14] The subject of Chapters 7 and 8, above.

[15] D. G. Brown and E. J. Daly, in C. Nielsen and G. Larwood (eds.), *Bryozoa: Ordovician to Recent* (Fredensborg, Denmark: Olsen and Olsen, 1985), 51–84.

the Permian (the initial chamber is normally larger in asexually produced generations). Overall, however, the tempos and modes of protozoan evolution show patterns that have many parallels among metazoans.

The asymmetry of Table 30.1 suggests that nonbranching lineages are more likely to display either gradualism or gradualism-and-stasis, and that highly branched lineages are more likely to display stasis. Therefore the factors that can increase the frequency of branching might also lead to more stasis. The factors that produce less branching could lead to more gradualism. [. . .]

Stratigraphic resolution and acuity

Prior to 1980 paleontologists generally assumed that the fossil record was sufficiently complete to study both ecological and evolutionary processes. In other words, there were no inherent or systematic biases known in the stratigraphic record that should limit the scope of the questions paleontologists could address. Thus during the 1970s paleoecologists attempted to describe ecological processes within fossil communities without asking whether or not the record was sufficient to resolve time to the scale of tens or hundreds of years. Other paleontologists, too, reasonably assumed that there was no bar to analyzing the process of speciation and that observed patterns of divergence accurately recorded speciation events.

These assumptions came to a screeching halt in 1980. Papers by Schindel and Sadler[16] demonstrated that in most circumstances marine paleontologists simply could not resolve time finely enough to observe ecological events over a long duration, and raised substantive questions about the ability to resolve fine-scale evolutionary processes. In brief, the difficulty is that deposits which record brief intervals of time (event beds, for example) are too widely spaced in the record to provide continuous information on ecological and speciation events. Yet other deposits are so condensed and/or bioturbated that fine-scale temporal information has been lost. Interestingly, such is not the case with terrestrial megafloral deposits, which are often limited to the leaves deposited in a single year immediately around the locality [. . .].

Additional complexity is added by taphonomic processes. For example, time-averaging produces assemblages which combine fossils over a long span of time, removing seasonal and yearly fluctuations and potentially reflecting more stable, long-term community structure. Large storms may rework thousands of years of sediment into a single time-averaged event bed. Longer-term time averaging, produced for example by condensed sections,

[16] D. E. Schindel, *Paleobiology*, 6 (1980), 408–26, and 8 (1982), 340–53. P. M. Sadler, *Journal of Geology*, 89 (1981), 569–84.

combines specimens that lived at very different times. Burrowing may partially or completely rework the stratigraphic order of fossils, further destroying the utility of the fossil record for some types of studies.

For speciation studies, time averaging and other taphonomic processes generally limit paleontologists' abilities to reliably discern microevolutionary patterns. As much as we all may want to believe in the evidence from the fossil record, recent studies in taphonomy suggest that we must never overlook the null hypothesis that any particular sequence is insufficient to record ecological and short-term evolutionary processes. Certainly there are stratigraphic intervals where sedimentation rate is high enough, preservation is good enough, and sampling is sufficient that intraspecific events can be chronicled. This, however, is not the null hypothesis, and such circumstances must be carefully documented by the investigator. As to the hope that we might actually be able to follow the process of speciation in the fossil record, that largely depends upon the rate at which speciation occurs in a particular setting.

['Speciation in the Fossil Record', in *New Approaches to Speciation in the Fossil Record*, ed. Douglas H. Erwin and Robert L. Anstey (New York: Columbia University Press, 1995), 11–30.]

GAVIN DE BEER

31 Homology: an unsolved problem

Some organs in different species appear to be fundamentally similar, and are referred to as homologies. Since Darwin, homologies have been explained by inheritance from a common ancestor. Anatomists have traced homologies, such as those between the leaves of different plant species, between leaves and flowers, between the ear bones of mammals and certain jaw bones of reptiles, between the laryngeal nerve of mammals and the fourth cranial nerve of fish. In the case of the laryngeal nerve, homology explains why it takes a detour round the top of the heart as it passes from brain to larynx. Foolproof criteria for homology have been hard to find. Homologous organs do not always develop in the same way, from the same embryonic cells, and they may not be coded for by the same genes in different species. [Editor's summary.]

The concept of homology

The term *homology* is derived from the Greek *homologia* which means 'agreement', and is applied to corresponding organs and structures of plants and of animals which show 'agreement' in their fundamental plan of structure, as for example the leaf of an oak tree with the leaf of an ash tree, or the right forelimb of a dog with the right forelimb of a horse. Richard Owen introduced the term into biological language in 1843 to express similarities in basic structure found between organs of animals which he considered to be more fundamentally similar than others.

The basis of such similarity and its fundamental nature was for Owen, as for other anatomists of the Transcendental School who considered ideas that grouped facts to be more important than the facts themselves, that such organs corresponded to their representatives in a hypothetical 'archetype', a primeval pattern which was regarded as a sort of blueprint on which groups of similar animals had been created. This concept was pre-Darwinian and pre-evolutionary. The way to define an archetype was to make an abstraction of all the similarities that could be found in common in a group of animals, paying no attention to the variations which individuals and populations showed.[1] [. . .]

As it turned out, Owen was right in basing homology and homologous organs, or homologues, on their structure regardless of their function. An organ is homologous with another because of what it *is*, not because of what it *does*. Homologous organs are the 'same' organs however modified in detailed form and in the function that they carry out. The forelimb of a horse is homologous with the wing of a bat, although the former serves for locomotion on land and the latter for flight in the air. Homology is therefore to be distinguished sharply from *analogy*, the term applied by Owen to structures that perform similar functions but do not correspond to the same representative in the archetype. The wings of an insect serve the same function as the wings of a bird and are analogous to them, not homologous with them. The entire science of comparative anatomy is concerned with the recognition of homologous organs in different groups of organisms, plants and animals, and their distinction from analogous organs.

Mention must also be made of Étienne Geoffroy Saint-Hilaire (1772–1844) whose obsession with unity of type led him to believe that *all* animals were built on the same plan of structure, a view in the tradition from Aristotle to Owen, which was shattered by Cuvier (1769–1832) who contended that there were four plans of structure in animals. Geoffroy Saint-Hilaire did, however, put forward a criterion in comparative anatomy: 'the only general principle that can be applied is given by the position, the relations, and the dependences of the parts, that is to say by what I name and include under the term *connections*.' This [. . .] is still the way in which a comparative anatomist studies the morphology of organs to satisfy himself that they are, or are not, what is called homologous.

Darwin's bombshell of evolution, which burst in 1859, had a profound effect on the concept of the explanation of homology, but without touching the criteria by which it is established. At one stroke, it was obvious that

[1] Archetypes are an instance of what Mayr discusses as 'typological thinking' in Chapter 22.

metaphysical 'archetypes' do not exist, and that homology between organs is based on their correspondence with representatives in a common ancestor of the organisms being compared, from which they were descended in evolution. 'What can be more curious,' asked Darwin, 'than that the hand of a man, formed for grasping, that of a mole for digging, the leg of the horse, the paddle of the porpoise, and the wing of the bat, should be all constructed on the same pattern, and should include similar bones, in the same relative positions?' In the 6th edition of the *Origin of Species* (1872) he went on to quote Sir William Flower: 'We may call this conformity to type, without getting much nearer to an explanation of the phenomenon, but is it not powerfully suggestive of true relationship, of inheritance from a common ancestor?'

In other words, it is homologous organs that provide evidence of affinity between organisms that have undergone descent with modification from a common ancestor, i.e. evolution. Furthermore, since evolution is the explanation of the 'agreement' between homologous organs, their study, if they are hard parts susceptible of fossilization, is not restricted to the morphology of living organisms, but the entire range of palaeontology is available for it. So, provided with a cast-iron explanation in terms of affinity, of inheritance in evolution from a common ancestor, it looked as if the concept of homology was at last soundly based and presented no more problems of principle; however, as will be seen below, it unfortunately does.

Homology in plants: leaves and flowers

The leaf of a land plant is a lateral appendage of the stem, morphologically different from the stem, with, typically, a bud in the axil between the leaf-base and the stem. The leaf contains plastids with chlorophyll and is therefore green; a foliage leaf is exposed to sunlight with the energy of which the chloroplast performs the chemical reactions of photosynthesis. Foliage leaves can differ widely in detailed shape, from the needles of conifers to the stalked undivided blades of lilies, the indented leaf of the oak, the subdivided compound leaf made up of leaflets of the pea. The whole leaf, or a leaflet, can be modified into a tendril of a climbing plant as in the vine, ending in adhesive discs as in Virginia creeper. In the fly-catching sundew, the leaf bears tentacles that secrete a sticky substance that catches the fly, digests it, and then absorbs it. Leaves can also be modified into scales and bracts, but the most interesting modification is into floral leaves.

The flower of an angiosperm typically consists of four concentric whorls of elements. The frond or foliage leaf of a fern shows in its simplest form that it is a sporophyll: it forms and bears spores on its under surface. The innermost whorl of the elements of a flower is formed by the carpellary leaves, the carpels, which usually grow together to form an enclosed chamber, the ovary, surmounted by its style and stigma; but the carpels betray their sporophyll nature by the fact that they produce spores. These spores which develop into embryo-sacs, are contained within the ovules or future seed-coats. [. . .] The spores produced by the carpels are sedentary macrospores, which is why the carpels are regarded as the female elements in the flower.

The second whorl of floral elements consists of the stamens, thin stalked structures ending in anthers which produce pollen-sacs containing the pollen grains which are microspores, adapted to travel and dispersal to find the macrospores, which is why the stamens are regarded as male sporophylls. The third whorl is made up of petals, which show clear similarity to the structure of foliage leaves in spite of the fact that they may be of different colours. [. . .]

The evolutionary derivation of the parts of a flower from the unspecialized leaves of an ancestor is supported by the facts that in some Cycads, the most primitive gymnosperms living, the carpels are simple sporophylls, like foliage leaves bearing ovules, and that in the Magnoliaceae, the most primitive living angiosperms, the stamens are often broad sporophylls, bearing their spores (pollen grains) on their under surface.

Homology in animals: the ear ossicles

In reptiles the hinge between the upper and lower jaws is the joint and articulation between two bones: the quadrate of the upper jaw and the articular of the lower jaw (Figure 31.1(a)). The quadrate abuts against the side

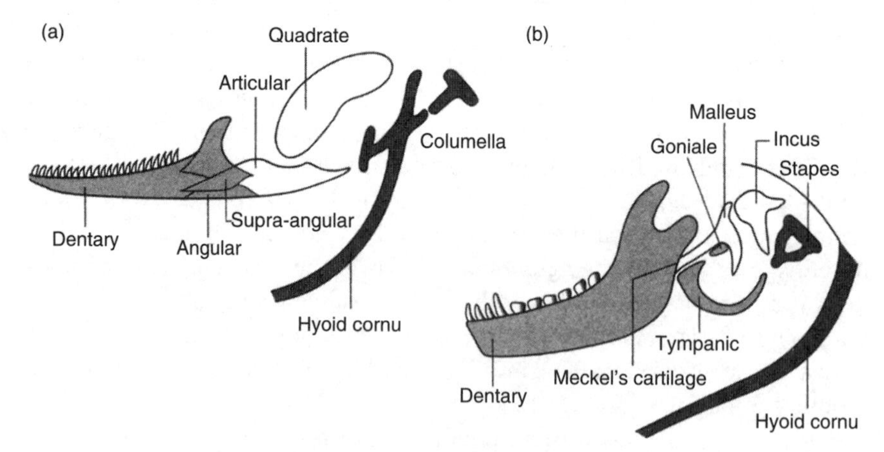

Figure 31.1: Homology in vertebrates illustrated by the hinge of the lower jaw in reptiles (a); and the ear-ossicles (stapes) in mammals (b). (After E. S. Goodrich (1930) Studies on structure of vertebrates, *Macmillan.)*

of the auditory capsule by its otic process. Both quadrate and articular are cartilage-bones, preformed in cartilage which then becomes ossified. The reptilian lower jaw also contains a number of membrane-bones, ossifications without cartilaginous precursors, such as the dentary in front, which bears the teeth, the angular and supra-angular behind situated laterally to the articular, the pre-articular and coronoid on the inner side of the jaw. Some fossil reptiles show even more bones.

The jaws are part of the 1st visceral or mandibular arch which is separated from the 2nd or hyoid arch by the tympanic cavity, derived from the 1st visceral pouch, and connected with the throat by the eustachian tube. In the hyoidarch, the uppermost skeletal element is the columella auris, cartilage-bone, a rod conveying vibrations of sound from the tympanic membrane on which sound waves impinge, to the fenestra ovalis of the auditory capsule where the vibrations are imparted to the lymph fluid which stimulates the sense organs of hearing. As the tympanic cavity lies between the 1st (or mandibular) arch and the 2nd (or hyoid) arch, the quadrate and articular bones project into the tympanic cavity from in front, and the columella auris from behind, and the latter is able to vibrate in an open space instead of in thick tissue.

In mammals (Fig. 31.1(b)) the conditions at first sight seem to be very different, because the lower jaw consists of a single bone, the dentary, from which an uprising extension articulates with the fossa of a membrane-bone of the brain case, the squamosal. The hinge of the lower jaw in mammals is therefore different from that in reptiles. When the question is asked what has happened in mammals to the old hinge bones of the reptiles, the answer is sensational. These bones have become inserted between the columella auris and the tympanic membrane and are known as the incus and malleus respectively, while the columella, now called the stapes, continues to fit into the fenestra ovalis, receiving the vibrations from the incus which in turn receives them from the malleus impinging on the tympanic membrane. The leverage which these bones can exert on one another makes the transmission of vibrations more sensitive. So there is a chain of three ear ossicles in mammals, and between two of them, the incus and the malleus, is the old hinge joint of the lower jaw of reptiles.

The other bones of the reptilian lower jaw have also changed their functions and their names. The angular in mammals has become the tympanic bone which surrounds and protects the tympanic cavity; the pre-articular (also called goniale) becomes attached to the front of the malleus; the coronoid and supra-angular disappear.

The important point to notice in these changes is the perfect morphological correspondence between the conditions in reptiles and in mammals. All the elements that are cartilage-bones in the former are so also in the latter: the same is true of the membrane-bones and their relative positions correspond exactly. This correspondence also extends to minute details. The columella in reptiles is frequently pierced by a hole through which the stapedial artery passes; this is constant for the stapes of mammals, and is the reason why it is called the 'stirrup'. The lateral head vein runs back medially to the quadrate in reptiles and to the incus in mammals. The facial nerve passes out of the brain case and runs backwards on the median side of the quadrate in reptiles and of the incus in mammals. The nerve passes above the tympanic cavity on the outer side of the stapedial artery and gives off a branch, the chorda tympani, which runs forwards above the tympanic cavity and then down on the median side of the lower jaw elements, articular or malleus, in exactly the same way in reptiles and in mammals. [. . .] What makes this study even more significant is that the results of comparative anatomy are confirmed by those of palaeontology, for there are fossil reptiles that show advances towards the mammalian condition, and the superseding of the quadrate-articular hinge of the lower jaw by the squamosal-dentary articulation. All this evolution took place without any functional discontinuity. It is a sobering thought that every man carries in his ear ossicles the homologue of the lower jaw hinge of his reptilian ancestors. [. . .]

The courses taken by certain nerves and blood vessels in adult mammals are determined by the structure of their embryos which repeat the embryonic conditions of the ancestors' embryos. The recurrent laryngeal nerve is an example of how the topology of homologous structures determines some curious anomalies in adult anatomy (Fig. 31.2). The recurrent laryngeal nerve is a branch of the vagus nerve which in fishes has four branchial branches, each of which passes down behind visceral pouches 3, 4, 5, and 6, and runs forwards ventrally but on the median side of the arterial arches that also run down behind those visceral pouches which, in fishes, are pierced as gill-slits.

In mammals these arterial arches are reduced in number by the disappearance of arches 1, 2, and 5. The 3rd or carotid, the 4th or systemic aorta, and the 6th or pulmonary persist. The systemic aorta persists only on the left side where there is still the old connection between the aorta and the pulmonary artery by means of the ductus arteriosus, which is of great importance to the embryo when still in the uterus where respiration is carried out by the placenta. At birth respiration immediately becomes pulmonary, and the ductus arteriosus closes up and becomes nothing but a ligament. But the old 4th branchial branch of the vagus, now called the recurrent laryngeal nerve still loops round the remains of the ductus arteriosus, remnant of the old 6th arterial arch.

In early stages of development, the heart lies far forward, in the neck, and the laryngeal nerve does not have far to go to innervate the larynx. But as development proceeds, the heart and the arterial arches are drawn back into the thorax. This is why the recurrent laryngeal nerve on the left side, after running backwards and looping round the ductus arteriosus, then runs

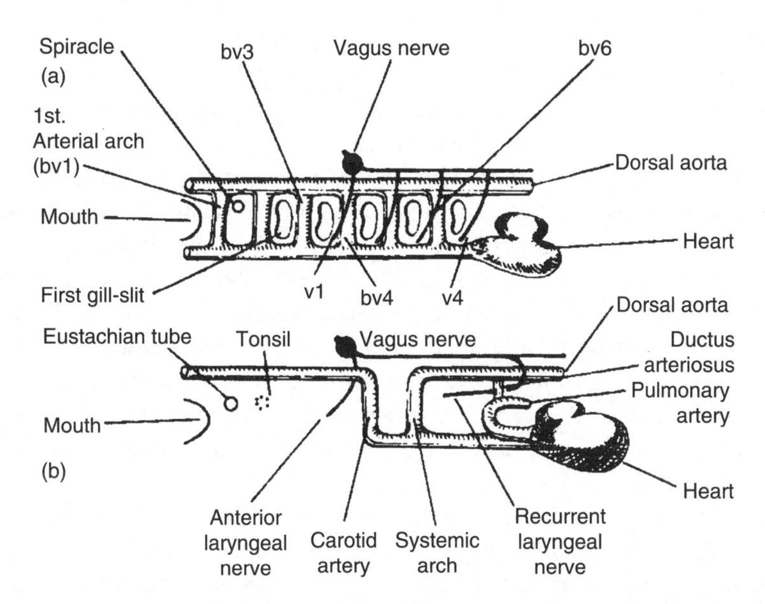

Figure 31.2: Morphology of the arterial arches and the vagus nerve in (a) dogfish; (b) rabbit. bv 1, 3, 4, 6, blood-vessels running in the 1st, 3rd, 4th, and 6th visceral arches; v, vagus nerve; vl, 4, 1st, 4th branch of vagus. (After de Beer 1966.)

forwards again to innervate the muscles of the larynx. In man, this course of the nerve is several inches longer than it need have been in the adult if it went straight to the larynx from the point where the nerve emerges from the skull. In the giraffe its course must be several feet longer. The explanation is the homology between the mammalian ductus arteriosus and the 6th arterial arch of the fish, which is respected in descendant forms, resulting in apparently anomalous conditions. [. . .]

Homology and embryology

Since every organ and structure in any organism has come into existence only as a result of embryonic development, it is natural to look to embryology for evidence on homologous structures. At late stages of development, when morphological relations between structures are established, such studies may yield valuable results, as in the case of the ear ossicles mentioned above. But at very early stages, such research leads to disappointment. [. . .]

Attention must now be paid to the germ layers. It was discovered a hundred and fifty years ago by C. Pander and K. E. von Baer that the fertilized eggs of all animals above the jelly-fish give rise to layers of tissue, three in number: ectoderm, endoderm, and mesoderm, which become folded up in different ways. It was then found that in general, ectoderm gives rise to epidermis, nervous system, sense organs, and nephridia; endoderm to the alimentary canal and its derivatives (in vertebrates: thyroid, lungs, liver, pancreas, appendix, urinary bladder); mesoderm to dermis, connective tissue, cartilage, bone, muscles, germ cells, coelomoducts or genital ducts, and also to kidneys where nephridia have been lost.

Very soon, this generalization became a dogma, and it was held that homologous organs *must* always arise from the 'correct' germ layer. This position was first shaken when experiments involving extirpation of the neural crest (from which nerve cells arise) in newt embryos also resulted in absence of cartilages of the jaws and other visceral arches. It was morphological heresy to think that cartilage could arise from ectoderm. The orthodox view was that no valid conclusions could be drawn from experimentally mutilated embryos. It therefore became necessary to demonstrate the facts from the study of embryos on which no experiments had been performed, and this is what I did in 1947.

In newt eggs, ectodermal tissues arise from the upper superficial part of the egg which is black, because of the presence of innumerable small melanin granules, which persist for a long time in the cells derived from it, and indicate their ectodermal origin. On the other hand, endodermal and mesodermal cells contain small globules of yolk which betray their origin. By means of these natural indicators I was able to show that not only the cartilages of the jaws and visceral arches consist of cells containing the tell-tale melanin granules, but also the osteoblasts of the dermal bones of the skull (frontal, parietal), and the odontoblasts in the papillae which secrete the dentine that composes the body of the teeth. Enamel had always been regarded as an ectodermal product, formed from the stomodaeal epidermis which grows in and lines the front of the mouth cavity. But enamel can be formed from ectodermal stomodaeal cells (with melanin granules), or from endodermal cells (with yolk globules), according to where the tooth rudiments are, for they act as enamel organizers. [. . .]

Before leaving embryology, there is a further aspect of the subject that is worth consideration. It is sometimes called sexual homology, and it refers to the correspondence between organs of the genital system that have undergone different development in the two sexes. For instance, the testis corresponds to the ovary; the scrotum containing the testes of the male corresponds to the labia majora of the vulva of the female, and the correspondence is made even more obvious in abnormal cases where the ovaries undergo 'descent' like testes, and pass by the canal of Nuck into the labia. The penis corresponds to the clitoris which, although diminutive, also contains erectile tissue. Part of the prostate corresponds to the uterus, and this fact can be made use of in certain pathological cases where enlargement of the prostate can be treated by sex hormones.

Are these corresponding organs homologous? Not in the strict sense, since it is not possible to refer them to a single representative in a common ancestor, which in vertebrates was certainly not hermaphrodite. They are the result of divergent embryonic development consequent on sexually dimorphic differentiation, due in part to genes and in part to sex hormones. The rudiments from which they have developed are homologous.

Homology and genetics

The converse is no less instructive. In *Drosophila* there is a gene, 'eyeless', which deprives its possessor of eyes.[2] It is a recessive character, which is important because it means that when its effect is produced, the fly has inherited the 'eyeless' allele from both parents, and no normal eye-controlling allele is present. If a stock of individuals pure (homozygous) for the 'eyeless' gene is inbred for many generations, there is high mortality as would be expected from the adverse effects of natural selection acting on a gene with such lethal effects. But eventually, flies appear in the offspring possessing normal eyes. It can easily be shown that the 'eyeless' gene has not changed, because when one of these phenotypically eye-possessing but genotypically homozygous 'eyeless' flies is mated with the original wild stock, i.e., the 'eyeless' gene is put back into the original gene complex, the virulent effects of the 'eyeless' gene reappear. What has happened during the inbreeding is that all the other pairs of alleles making up the gene complex have been reshuffled until a gene complex has been produced that prevents the phenotypic manifestation of the 'eyeless' allele. Other genes must therefore deputize for the absent normal gene that controls the formation of eyes. But why should they, and by what mechanism? Nobody can deny that the restored eyes that develop in genetically 'eyeless' stocks are homologous with the original normal eyes. Therefore, *homologous structures need not be controlled by identical genes*, and *homology of phenotypes does not imply similarity of genotypes*.

[*Homology: An Unsolved Problem*, Oxford Biology Readers (Oxford: OUP, 1971).]

RICHARD DAWKINS

32 The ey gene

The fruitfly gene symbolized by ey *functions in the development of eyes. The way geneticists refer to this gene as 'eyeless' is instructively odd. The* ey *gene can produce eyes*

[2] De Beer here discusses research done in the 1920s by T. H. Morgan on the 'eyeless' gene of the fruitfly (T. H. Morgan, 'The variability of eyeless'. *Publications of the Carnegie Institution*, 399 (1929), 139). This same gene is the subject of modern work discussed in the following two extracts (Chapters 32 and 33). Dawkins (Chapter 32) discusses what is meant by calling the gene 'eyeless'.

in abnormal parts of the fruitfly body. The ey *gene is similar to a gene that functions in eye development in mice. That mouse gene can even induce eye development in fruitflies.* [Editor's summary.]

An intriguing set of experimental results [was] recently reported by a group of workers in Switzerland associated with Professor Walter Gehring. I shall briefly explain what they found [. . .] Before I begin, I need to apologize for a maddeningly silly convention adopted by geneticists over the naming of genes. The gene called *eyeless* in the fruitfly *Drosophila* actually makes eyes! (Wonderful, isn't it?) The reason for this wantonly confusing piece of terminological contrariness is actually quite simple, and even rather interesting. We recognize what a gene does by noticing what happens when it goes wrong. There is a gene which, when it goes wrong (mutates), causes flies to have no eyes. The position on the chromosome of this gene is therefore named the *eyeless* locus ('locus' is the Latin for 'place' and it is used by geneticists to mean a slot on a chromosome where alternative forms of a gene sit). But usually when we speak of the locus named *eyeless* we are actually talking about the normal, undamaged form of the gene at that locus. Hence the paradox that the *eyeless* gene makes eyes. It is like calling a loudspeaker a 'silence device' because you have discovered that, when you take the loudspeaker out of a radio, the radio is silent. I shall have none of it. I am tempted to rename the gene *eyemaker*, but this would be confusing too. I shall certainly not call it *eyeless* and shall adopt the recognized abbreviation *ey*.

Now, it is a general fact that although all of an animal's genes are present in all its cells, only a minority of those genes are actually turned on or 'expressed' in any given part of the body. This is why livers are different from kidneys, even though both contain the same complete set of genes. In the adult *Drosophila, ey* usually expresses itself only in the head, which is why the eyes develop there. George Halder, Patrick Callaerts and Walter Gehring discovered an experimental manipulation that led to *ey*'s being expressed in other parts of the body. By doctoring *Drosophila* larvae in cunning ways, they succeeded in making *ey* express itself in the antennae, the wings and the legs. Amazingly, the treated adult flies grew up with fully formed compound eyes on their wings, legs, antennae and elsewhere. Though slightly smaller than ordinary eyes, these 'ectopic' eyes are proper compound eyes with plenty of properly formed ommatidia. They even work. At least, we don't know that the flies actually see anything through them, but electrical recording from the nerves at the base of the ommatidia shows that they are sensitive to light.

That is remarkable fact number one. Fact number two is even more remarkable. There is a gene in mice called *small eye* and one in humans called *aniridia*. These, too, are named using the geneticists' negative convention: mutational damage to these genes causes reduction or absence of eyes or parts of eyes. Rebecca Quiring and Uwe Waldorf, working in the same Swiss laboratory, found that these particular mammal genes are almost identical, in their DNA sequences, to the *ey* gene in *Drosophila*. This means that the same gene has come down from remote ancestors to modern animals as distant from each other as mammals and insects. Moreover, in both these major branches of the animal kingdom the gene seems to have a lot to do with eyes. Remarkable fact number three is almost too startling. Halder, Callaerts and Gehring succeeded in introducing the mouse gene into *Drosophila* embryos. *Mirabile dictu*, the mouse gene induced ectopic eyes in *Drosophila*. [. . .] It is an insect compound eye that has been induced, not a mouse eye. The mouse gene has simply switched on the eyemaking developmental machinery of *Drosophila*. Genes with pretty much the same DNA sequence as *ey* have been found also in molluscs, marine worms called nemertines, and sea-squirts. *Ey* may very well be universal among animals, and it may turn out to be a general rule that a version of the gene taken from a donor in one part of the animal kingdom can induce eyes to develop in recipients in an exceedingly remote part of the animal kingdom.

[*Climbing Mount Improbable* (London: Viking, 1996).]

W. J. DICKINSON

33 Molecules and morphology: where's the homology?

To show that two organs, such as the eyes of insects and of mammals, are homologous, it is not enough to show that similar genes are expressed during eye development in both cases. The genetic homology may correspond to anatomical homology at some level other than that of eyes. [Editor's summary.]

A few years ago, molecular biologists were chastised for sloppy and confusing use of the term 'homology'. Many treated homology as an objective observation rather than an inference, and as a quantitative trait ('percentage homology') rather than a relationship of common evolutionary origin that either does or does not exist. There is another source of confusion that threatens to become increasingly troublesome as the fascinating molecular homologies that lie at the heart of developmental mechanisms are unraveled: there is not necessarily a simple relationship between homology of molecules (or even pathways) and homology of the anatomical features in whose development those components participate. In other words, some recent suggestions notwithstanding, molecular similarities in the developmental mechanisms that produce specific organs are not, by themselves, strong evidence for homology of those organs.

The central point of this article is that questions of homology can be examined at multiple levels and that homology between a pair of structures can simultaneously be present at some levels but absent at others. The term 'levels of homology' refers to a nested series of progressively more ancient and inclusive ('deeper') relationships. The classic textbook example of homology, the vertebrate forelimb, conveniently illustrates the point. Considered only as forelimbs, the wings of birds and bats are homologous; considered as wings, they are not. In other words, the last common ancestor of these two groups had forelimbs but not wings. Note that this conclusion is partly based on evidence other than that derived from the direct comparisons of wings: the comparative anatomy of other vertebrate forelimbs; the fossil record; and other anatomical comparisons that reveal, for example, the relationship of bats to other mammals have also been considered (at least implicitly). If doubt remained, sequence data could be collected to confirm that bats descend from wingless mammals.

Now, suppose the molecular mechanisms controlling development in birds and bats are examined. Given the known conservation of mechanisms in vertebrates, homologous molecules and conserved pathways would certainly be found operating in the development of wings in both groups. However, such similarities would not be interpreted as supporting the 'surprising' conclusion that the two sorts of wings are, after all, homologous. Instead, the molecular similarities would be recognized as reflecting homology at a deeper level (forelimbs). In other studies, the danger arises when evidence bearing on homology is less extensive or decisive (or less well known to the average molecular or developmental biologist) than in this example.

Turning to real molecular examples, the evolving interpretation of comparisons between homeobox genes and clusters in different organisms is instructive. When these were discovered in vertebrates following their initial characterization in insects, early speculations centered on the possibility that insects and vertebrates share a conserved mechanism of segmentation, even though this contradicted the conventional view that the last common ancestor of arthropods and vertebrates was not segmented. However, the discovery of homeobox clusters in unsegmented creatures like *Caenorhabditis elegans* undermined these speculations, particularly since analyses of expression patterns in the worm confirmed that there is no relationship between homeobox genes and reiterated cell lineages that might be regarded as a primitive form of segmentation. Homeobox genes are involved in other divergent processes such as limb development in vertebrates and gut differentiation in insects. Thus, the focus of interpreting homeobox gene function shifted progressively from segmentation to anterior-posterior polarity and to axial patterning in general. Homology at even deeper levels, such as positional information *per se* or simply transcriptional regulation, may be most relevant to some homeobox gene comparisons.

A cautious initial interpretation of similarities among insects and vertebrates would have considered all of these possibilities and recognized the need for additional information to distinguish between them. As in the example of bird and bat wings mentioned above, more detailed analyses of the relevant systems would not, in isolation, have resolved the question. The progressive interpretation summarized above depended on information about additional species (e.g. *C. elegans*) and other contexts of expression within species (e.g. limbs and guts); in turn, those comparisons depend, at least implicitly, on additional data of various kinds (such as that relevant to phylogeny).

A second example highlights the classic problem of convergence, with the deceptive twist that truly homologous molecules may be involved in processes that are only analogous. Products of the *hedgehog* gene in *Drosophila* and of an avian homolog serve strikingly similar functions in wing development. Quite properly, their roles in that context are recognized as analogous, not homologous. Again, *hedgehog* homologs play comparable roles in intercellular signaling in various other developmental contexts in both insects and vertebrates. Undoubtedly, there is deep and interesting homology here but the wing is not the level at which it should be sought.

The probability of encountering such convergence is greatly increased by three well-established features of molecular evolution: (1) even within a single species the same molecule can assume functions in quite different developmental pathways; (2) gene duplication generates paralogous[1] gene families whose members can encompass an even wider range of roles; (3) domain shuffling[2] generates molecules with clear homology in some regions but potentially with quite different overall functions. All of these can be subsumed within the idea of 'levels of homology' adopted here. Molecules with multiple functions could presumably be traced back to some primordial function while paralogous sequences, whether entire molecules or domains, could be traced to a molecule (and function) that existed before some relevant duplication occurred. In either case, the molecular biologist's job (not necessarily simple) would be to determine the context in which an ancestral molecule functioned at the point where paths merge when traced backwards from the current examples under consideration. Such an analysis would identify the level at which the contemporary functions and contexts could usefully be said to be homologous.

[1] On gene duplication and paralogy, see Chapter 38, footnote 7.

[2] Domain shuffling. Many proteins are made up of several recognizable components, called domains. A domain has a particular biochemical skill, or function, which may become, or cease to be, useful in a particular protein. Domains are sometimes added to, or subtracted from, particular proteins during evolution, in a process called domain shuffling. After a domain is added to a protein, part of that protein will be homologous with part of some other protein (where the domain was duplicated from) that it was previously unrelated to. The human genome paper also discusses domain shuffling (Chapter 38, footnote 8), and Gerhart and Kirschner (Chapter 51) describe an example.

Molecular similarities have sometimes not been interpreted in an appropriately cautious manner. Based on comparisons of function and expression of the *orthodenticle* gene in *Drosophila* and of homologs in vertebrates, Finkelstein and Boncinelli suggest that, contrary to prevailing opinion, head specialization may have occurred before the ancestral lineages separated. However, the facts permit hypotheses similar to those proposed for interpreting analyses of homeobox genes mentioned above: these *orthodenticle* homologs could be deeply conserved components involved in axial patterning (or another aspect of positional information) not specifically related to cephalization.

Defects caused by *eyeless* in *Drosophila* and a homolog, *Small eye*, in mice have prompted speculation that arthropod and vertebrate eyes are homologous despite fundamental differences in organization. This situation may be comparable to that of the *hedgehog* gene in wing development. The roles of these genes in eye development should be termed homologous only if other evidence suggests that an antecedent of both *eyeless* and *Small eye* functioned in the development of an eye in a common ancestor of arthropods and vertebrates.

Kispert *et al.* suggest homology between the vertebrate notochord and the insect hindgut because the *Brachyury* (*T*) gene and a *T*-related gene (*Trg*), respectively, are required for normal development of those organs. In this case, the molecular homology is confined to a DNA-binding region. This region could have combined with other domains to generate molecules with distinct functions either before or after the separation of vertebrate and arthropod lineages.

Finally, homologous transcription factors seem to play similar roles in regulating some genes in the liver of mammals and the fat body of *Drosophila*, leading to speculations about the homology of these organs. As with other developmental regulators, these factors belong to a limited number of families and typically function in a variety of contexts. Again, coincidental similarities between analogous systems are to be expected. This case is also confused by the seemingly interchangeable use of the terms 'homology' and 'analogy' in the discussion. [. . .]

The issues raised in this article have not always been given adequate attention. It is noteworthy that the majority of 'surprising' anatomical homologies thus far proposed on the basis of molecular data involve comparisons between insects and vertebrates. This certainly reflects the intense effort devoted to molecular analyses of development in these particular systems. As other groups receive more attention, the incidence of convergent examples will surely increase, reinforcing the importance of caution and precision in the interpretation of molecular similarities.

In no case am I arguing that suggested inferences about organ level homology are definitely wrong; I claim only that the molecular evidence alone is weak and that some authors have been vague or ambiguous with respect to the level of homology suggested. It must be recognized that molecular similarities could reflect homology at any of several levels, that other data must be evaluated to decide which level is most likely in a particular case, and that the level under discussion must be carefully specified in reports of hypotheses and conclusions. Anatomical homology will become a useless concept if it is inferred in all organs in which homologous molecules are found to have similar functions.

[*Trends in Genetics*, 11 (1995), 119–20.]

E. HAECKEL

34 The fundamental law of organic evolution

The fundamental law of evolution states that 'ontogeny is a recapitualtion of phylogeny'. That is, individual development from egg to adult is a condensed re-enactment of the evolutionary history of the species to which the individual belongs. The law has exceptions that can be analysed. The law can be used to trace the ancestry of human beings. Comparative embryology shows similarities between early human fetal stages and those of other species, such as dogs—which some theologians have tried to deny. (As explained further in the editorial introduction to this section, Haeckel's ideas are now thought fundamentally unsound. However, he identified facts that still need to be understood, and he was highly influential. The confident style of the extract gives some feel for why he was so influential.) [Editor's summary.]

The story of the evolution of man, as it has hitherto been expounded to medical students, has usually been confined to embryology—or, more correctly, *ontogeny*—or the science of the development of the individual human organism. But this is really only the first part of our task, the first half of the story of the evolution of man in that wider sense in which we understand it here. We must add as the second half—as another and not less important and interesting branch of the science of the evolution of the human stem—phylogeny: this may be described as the science of the evolution of the various animal forms from which the human organism has been developed in the course of countless ages. Everybody now knows of the great scientific activity that was occasioned by the publication of Darwin's *Origin of Species* in 1859. The chief direct consequence of this publication was to provoke a fresh enquiry into the origin of the human race, and this has proved beyond question our gradual evolution from the lower species. We give the name of 'Phylogeny' to the science which describes this ascent of man from the lower ranks of the animal world. The chief source that it draws upon for facts is 'Ontogeny,' or embryology, the science of the development of the individual organism. Moreover, it derives a good deal of support from paleontology, or the science of fossil remains, and even more from comparative anatomy, or morphology.

These two branches of our science—on the one side ontogeny or embryology, and on the other phylogeny, or the science of race-evolution—are most vitally connected. The one cannot be understood without the other. It is only when the two branches fully co-operate and supplement each other that 'Biogeny' (or the science of the genesis of life in the widest sense) attains to the rank of a philosophic science. The connection between them is not external and superficial, but profound, intrinsic, and causal. This is a discovery made by recent research, and it is most clearly and correctly expressed in the comprehensive law which I have called 'the fundamental law of organic evolution,' or 'the fundamental law of biogeny.' This general law, to which we shall find ourselves constantly recurring, and on the recognition of which depends one's whole insight into the story of evolution, may be briefly expressed in the phrase: 'The history of the fœtus is a recapitulation of the history of the race'; or, in other words, 'Ontogeny is a recapitulation of phylogeny.' It may be more fully stated as follows: The series of forms through which the individual organism passes during its development from the ovum to the complete bodily structure is a brief, condensed repetition of the long series of forms which the animal ancestors of the said organism, or the ancestral forms of the species, have passed through from the earliest period of organic life down to the present day.

The causal character of the relation which connects embryology with stem-history is due to the action of heredity and adaptation. When we have rightly understood these, and recognised their great importance in the formation of organisms, we can go a step further and say: Phylogenesis is the mechanical cause of ontogenesis. In other words, the development of the stem, or race, is the cause, in accordance with the physiological laws of heredity and adaptation, of all the changes which appear in a condensed form in the evolution of the fœtus.

The chain of manifold animal forms which represent the ancestry of each higher organism, or even of man, according to the theory of descent, always form a connected whole. We may designate this uninterrupted series of forms with the letters of the alphabet: A, B, C, D, E, etc., to Z. In apparent contradiction to what I have said, the story of the development of the individual, or the ontogeny of most organisms, only offers to the observer a part of these forms; so that the defective series of embryonic forms would run: A, B, D, F, H, K, M, etc.; or, in other cases, B, D, H, L, M, N, etc. Here, then, as a rule, several of the evolutionary forms of the original series have fallen out. Moreover, we often find—to continue with our illustration from the alphabet—one or other of the original letters of the ancestral series represented by corresponding letters from a different alphabet. Thus, instead of the Roman B and D, we often have the Greek B and Δ. In this case the text of the biogenetic law has been corrupted, just as it had been abbreviated in the preceding case. But, in spite of all this, the series of ancestral forms remains the same, and we are in a position to discover its original complexion.

In reality, there is always a certain parallel between the two evolutionary series. But it is obscured from the fact that in the embryonic succession much is wanting that certainly existed in the earlier ancestral succession. If the parallel of the two series were complete, and if this great fundamental law affirming the causal nexus between ontogeny and phylogeny in the proper sense of the word were directly demonstrable, we should only have to determine, by means of the microscope and the dissecting knife, the series of forms through which the fertilised ovum passes in its development; we should then have before us a complete picture of the remarkable series of forms which our animal ancestors have successively assumed from the dawn of organic life down to the appearance of man. But such a repetition of the ancestral history by the individual in its embryonic life is very rarely complete. We do not often find our full alphabet. In most cases the correspondence is very imperfect, being greatly distorted and falsified by causes which we will consider later. We are thus, for the most part, unable to determine in detail, from the study of its embryology, all the different shapes which an organism's ancestors have presented; we usually—and especially in the case of the human fœtus—encounter many gaps. It is true that we can fill up most of these gaps satisfactorily with the help of comparative anatomy, but we cannot do so from direct embryological observation. Hence it is important that we find a large number of lower animal forms to be still represented in the course of man's embryonic development. In these cases we may draw our conclusions with the utmost security as to the nature of the ancestral form from the features of the form which the embryo momentarily assumes.

To give a few examples, we can infer from the fact that the human ovum is a simple cell that the first ancestor of our species was a tiny unicellular being, something like the amœba. In the same way, we know, from the fact that the human fœtus consists, at the first, of two simple cell-layers (the *gastrula*), that the *gastræa*, a form with two such layers, was certainly in the line of our ancestry. A later human embryonic form (the *chordula*) points just as clearly to a worm-like ancestor (the *prochordonia*), the nearest living relation of which is found among the actual ascidia. To this succeeds a most important embryonic stage (*acrania*), in which our headless fœtus presents, in the main, the structure of the amphioxus. But we can only indirectly and approximately, with the aid of comparative anatomy and ontogeny, conjecture what lower forms enter into the chain of our ancestry between the gastræa and the chordula, and between this and the amphioxus. In the course of the historical development (by means of heredity in a condensed form) many intermediate structures have gradually fallen out, which must certainly have been represented in our ancestry. But, in spite of these many, and sometimes very appreciable, gaps, there is no contradiction between the two successions. In fact, it is the chief purpose of this work to prove the real harmony and the original parallelism of the two. I hope to show, on a substantial basis of facts, that we can draw most important conclusions as to our genealogical tree from the actual and easily-demonstrable series of embryonic changes. We shall then be in a position to form a general idea of the wealth of animal forms which have figured in the direct line of our ancestry in the lengthy history of organic life.

In this phylogenetic appreciation of the facts of embryology we must, of course, take particular care to distinguish sharply and clearly between the primitive, palingenetic (or ancestral) evolutionary processes and those due to cenogenesis. By *palingenetic* processes, or embryonic *recapitulations*, we understand all those phenomena in the development of the individual which are transmitted from one generation to another by heredity, and which, on that account, allow us to draw direct inferences as to corresponding structures in the development of the species. On the other hand, we give the name of *cenogenetic* processes, or embryonic *variations*, to all those phenomena in the fœtal development that cannot be traced to inheritance from earlier species, but are due to the adaptation of the fœtus, or the infant-form, to certain conditions of its embryonic development. These cenogenetic phenomena are foreign or later additions; they allow us to draw no direct inference whatever as to corresponding processes in our ancestral history, but rather hinder us from doing so.

This careful discrimination between the primary or palingenetic processes and the secondary or cenogenetic is of great importance for the purposes of the scientific history of a species, which has to draw conclusions from the available facts of embryology, comparative anatomy, and paleontology, as to the processes in the formation of the species in the remote past. It is of the same importance to the student of evolution as the careful distinction between genuine and spurious texts in the works of an ancient writer, or the purging of the real text from interpolations and alterations, is for the student of philology. It is true that this distinction has not yet been fully appreciated by many scientists. For my part, I regard it as the first condition for forming any just idea of the evolutionary process, and I believe that we must, in accordance with it, divide embryology into two sections—palingenesis, or the science of repetitive forms; and cenogenesis, or the science of supervening structures.

To give at once a few examples from the science of man's origin in illustration of this important distinction, I may instance the following processes in the embryology of man, and of all the higher vertebrates, as *palingenetic:* the formation of the two primary germinal layers and of the primitive gut, the undivided structure of the dorsal nerve-tube, the appearance of a simple axial rod between the medullary tube and the gut, the temporary formation of the gill-clefts and arches, the primitive kidneys, and so on. All these, and many other important structures, have clearly been transmitted by a steady heredity from the early ancestors of the mammal, and are, therefore, direct indications of the presence of similar structures in the history of the stem. On the other hand, this is certainly not the case with the following embryonic changes, which we must describe as cenogenetic processes: the formation of the yolk-sac, the allantois, the placenta, the amnion, the serolemma, and the chorion—or, generally speaking, the various fœtal membranes and the corresponding changes in the blood vessels. [. . .] All these and many other phenomena are certainly not traceable to similar structures in any earlier and completely-developed ancestral form, but have arisen simply by adaptation to the peculiar conditions of embryonic life (within the fœtal membranes). In view of these facts, we may now give the following more precise expression to our chief law of biogeny:—The evolution of the fœtus (or *ontogenesis*) is a condensed and abbreviated recapitulation of the evolution of the stem (or *phylogenesis*); and this recapitulation is the more complete in proportion as the original development is preserved by a constant heredity; on the other hand, it becomes less complete in proportion as a varying adaptation to new conditions increases the disturbing factors in the development.

The cenogenetic alterations or distortions of the original palingenetic course of development take the form, as a rule, of a gradual displacement of the phenomena, which is slowly effected by adaptation to the changed conditions of embryonic existence during the course of thousands of years. This displacement may take place as regards either the locality or the time of a phenomenon. [. . .]

Variations in locality, affect, in the first place, the cells, or elementary parts of which the organs are composed; but they also affect the organs themselves. Thus, for instance, the sexual glands in the human embryo, and most of the higher animals, arise out of the middle germinal layer.[1] On the other hand, the comparative embryology of the lower animals shows us that originally they did not arise from this, but from one of the primary germinal layers. However, the germ-cells have gradually changed their position, and passed over at so early a period from their original situation into the middle layer that they now seem really to arise from it. [. . .]

Variation in time, is not less instructive. It consists in the fact that the series of forms in which the organs successively appear is different in embryology from what the stem history leads us to expect. [. . .] This may appear either as an acceleration or a delay in the rise of an organ. As cases of ontogenetic acceleration we may instance, in the embryonic development of man, the early appearance of the heart, the gill-clefts, the brain, the eyes, etc. These

[1] On germinal layers in development, see the section on homology and embryology in Chapter 31 (De Beer).

organs clearly arise much earlier, in comparison with others, than was originally the case with our ancestors. We find the reverse of this in the retarded formation of the gut, the ventral cavity, and the sexual organs. These are clear instances of ontogenetic retardation. [. . .]

Before I pass from the subject I must speak further of this, one of the most brilliant achievements of the human mind in modern times. [. . .]

The facts of embryology have so great and obvious a significance in this connection that even in recent years dualist and teleological philosophers have tried to rid themselves of them by simply denying them. This was done, for instance, as regards the fact that man is developed from an egg, and that this egg or ovum is a simple cell, as in the case of other animals. When I had explained this pregnant fact and its significance in my *Natural History of Creation*, it was described in many of the theological journals as a dishonest invention of my own. The fact that the embryos of man and the dog are, at a certain stage of their development, almost indistinguishable, was also denied. When we examine the human embryo in the third or fourth week of its development, we find it to be quite different in shape and structure from the full-grown human being, but almost identical with that of the ape, the dog, the hare, and other mammals, at the same stage of ontogeny. We find a bean-shaped body of very simple construction, with a tail below and a pair of fins at the sides, something like those of a fish, but very different from the limbs of man and the mammals. Nearly the whole front half of the body is taken up by a shapeless head without face, at the sides of which we find gill-clefts and arches as in the fish. At this stage of its development the human embryo does not differ in any essential detail from that of the ape, dog, horse, ox, etc., at a corresponding period. This important fact can easily be verified at any moment by a comparison of the embryos of man, the dog, hare, etc. Nevertheless, the theologians and dualist philosophers pronounced it to be a materialistic invention; even scientists, to whom the facts should be known, have sought to deny them.

[*The Evolution of Man*, 2 vols. (London: Watt, 1905).]

W. GARSTANG

35 Three poems

These three poems implicitly criticize the theory of recapitulation, in which embryonic stages are ancestral stages (Chapter 34). The first poem suggests that an adult feature of snails originated as an adaptation in a kind of larva called a veliger. The second poem questions whether the number of legs in developing isopods corresponds to an ancestral number. The third poem looks at how a new adult stage may evolve if a formerly larval stage becomes sexually mature. [Editor's summary.]

The Ballad of the Veliger or How the Gastropod got its Twist[1]

The Veliger's a lively tar, the liveliest afloat,
A whirling wheel on either side propels his little boat;
But when the danger signal warns his bustling submarine,
He stops the engine, shuts the port, and drops below unseen.

He's witnessed several changes in pelagic motor-craft;
The first he sailed was just a tub, with a tiny cabin aft.
An Archi-mollusk fashioned it, according to his kind,
He'd always stowed his gills and things in a mantle-sac behind.

Young Archi-mollusks went to sea with nothing but a velum—
A sort of autocyling hoop, instead of pram—to wheel 'em;
And, spinning round, they one by one acquired parental features,
A shell above, a foot below—the queerest little creatures.

But when by chance they brushed against their neighbours in the briny,
Coelenterates with stinging threads and Arthropods so spiny,
By one weak spot betrayed, alas, they fell an easy prey—
Their soft preoral lobes in front could not be tucked away!

Their feet, you see, amidships, next the cuddy-hole abaft,
Drew in at once, and left their heads exposed to every shaft.
So Archi-mollusks dwindled, and the race was sinking fast,
When by the merest accident salvation came at last.

A fleet of fry turned out one day, eventful in the sequel,
Whose left and right retractors on the two sides were unequal:
Their starboard halliards fixed astern alone supplied the head,
While those set aport were spread abeam and served the back instead.

Predaceous foes, still drifting by in numbers unabated,
Were baffled now by tactics which their dining plans frustrated.
Their prey upon alarm collapsed, but promptly turned about,
With the tender morsal safe within and the horny foot without!

This manoeuvre (*vide* Lamarck) speeded up with repetition,
Until the parts affected gained a rhythmical condition,
And torsion, needing now no more a stimulating stab,
Will take its predetermined course in a watchglass in the lab.

[1] The poem includes a hypothesis about a classic zoological puzzle: torsion in gastropods (that is, snails and relatives). Many snails have twisted shells but this is not what is meant here by torsion. Torsion is a twist in the internal body parts, or viscera, which in gastropods are twisted through 180°. They form something of a U-shaped loop such that the anus is above the head. There are many hypotheses about the evolutionary origin of torsion, and Garstang here suggests it arose as a larval adaptation: veligers are one of the larval forms found among gastropods.

In this way, then, the Veliger, triumphantly askew,
Acquired his cabin for'ard, holding all his sailing crew—
A Trochosphere in armour cased, with a foot to work the hatch,
And double screws to drive ahead with smartness and despatch.

But when the first new Veligers came home again to shore,
And settled down as Gastropods with mantle-sac afore,
The Archi-mollusk sought a cleft his shame and grief to hide,
Crunched horribly his horny teeth, gave up the ghost, and died.

Isopod Phylogeny

Sing a song of six legs, a new phyletic stage!
Four and twenty Isopods cradled in a cage:
When the cage was opened, out they ran to play—
Wasn't it a jolly thing to have a jolly day!

Mother rocked the cradle between her stegopods:
The youngsters ran about her seven pereiopods:
When they found she'd one pair more than they themselves,
They called a hasty conference on oöstegal shelves.

MacBride[2] was in his garden settling pedigrees,
There came a baby Woodlouse and climbed upon his knees,
And said: 'Sir, if our six legs have such an ancient air,
Shall we be less ancestral when we've grown our mother's pair?'

The Axolotl and the Ammocoete[3]

Amblystoma's a giant newt who rears in swampy waters,
As other newts are wont to do, a lot of fishy daughters:
These *Axolotls*, having gills, pursue a life aquatic,
But, when they should transform to newts, are naughty and erratic.

[2] MacBride was a well-known supporter of the theory of recapitulation in the UK at that time. His books included *A Textbook of Embryology*. Garstang added the following note: 'cf. *A textbook of embryology*, 1914: "They (i.e. the larvae of *Portunion*) resemble the young of normal Isopods when they leave the brood-pouch, and not even the most determined opponent of the recapitulation theory could deny their ancestral significance", p. 219.'

[3] Commentary on this poem can be found in S. J. Gould, *Ontogeny and Phylogeny* (Cambridge, Mass.: Harvard University Press, 1977), 178–9. In the final verse, the ammocoete is a larval stage of cyclostome fish, somewhat resembling a lancelet (*Amphioxus*). The lancelet is often thought to be related to ancestral chordates. Perhaps it originated from the larval stage of an existing chordate? The idea is not accepted now, but Garstang's reasoning is entertaining as ever. Also, his general point remains important—that adults can evolve by the modification of larvae.

They change upon compulsion, if the water grows too foul,
For then they have to use their lungs, and go ashore to prowl:
But when a lake's attractive, nicely aired, and full of food,
They cling to youth perpetual, and rear a tadpole brood.

And newts Perennibranchiate have gone from bad to worse:
They think aquatic life is bliss, terrestrial a curse.
They do not even contemplate a change to suit the weather,
But live as tadpoles, breed as tadpoles, tadpoles altogether!

Now look at *Ammocoetes* there, reclining in the mud,
Preparing thyroid-extract to secure his tiny food:
If just a touch of sunshine more should make his gonads grow,
The Lancelet's claims to ancestry would get a nasty blow!

[*Larval Forms* (Oxford: Basil Blackwell, 1951).]

Section F

Evolutionary genomics

The genome of an individual organism is its complete set of DNA, and it is present in most of the cells in the body. The genome is made up of coding regions that code for genes and non-coding DNA. Non-coding DNA does not seem to code for anything and its function (if any) is uncertain. 'Genomics' is a fast-growing area of biology at present. Genomes are being sequenced at an astonishing rate, and genomics refers to all the scientific, and extra-scientific, activity associated with the new sequence data. Commentators have identified medical promises, and social threats, in genomics, but so far the biggest beneficiary of the work has been in our understanding of evolution.

Genomic sequences are being used to reconstruct the phylogeny, or tree, of life. Biologists have been at work on this reconstruction for almost 150 years, using anatomical evidence from living and fossil species. Molecular evidence (of which genomic sequences are the latest version) supplements the anatomical evidence, and allows us to tackle a number of phylogenetic problems that proved insoluble with anatomical evidence. Molecular clocks are also being used to estimate the dates of evolutionary events in the past. In Chapter 41, this section, Benner *et al.* look at research with molecular clocks. Molecular clocks and molecular phylogenetics also underlie several papers in the sections on the history of life (G) and on human evolution (I).

In the work on the tree of life, new genomic evidence is being applied to a long-standing problem. Genomics is also creating new areas of enquiry in itself, about questions that have not been asked (or that were unanswerable and ignored) before about the year 2000: research on the evolution of the genome. This section contains several examples of this kind of research. In Chapter 36 Ochman *et al.* look at the evolution of bacterial genomes. Bacteria have long been known to transfer genes 'horizontally' or 'laterally' between individuals of the same or different species. (Elsewhere in life, genes are transferred only 'vertically', from parent to offspring.) After all, that is the way antibiotic-resistance moves among bacteria, and antibiotic-resistance has been known about for half a century. However, the DNA sequences of different bacterial species have characteristic features, which allow new ways of identifying horizontal gene transfer. Ochman *et al.* make quantitative estimates of how much DNA in a bacterial genome originated by horizontal import.

Vision *et al.* (Chapter 37) reconstruct how the genome of flowering plants has been built up over evolutionary time. A genome can expand by duplications in all, or part, of an ancestral genome. The time when such duplications occurred can be estimated from sequence data, using a molecular clock. Vision *et al.* identify several rounds of duplication in the history of flowering plants, and tentatively associate some of them with evolutionary events such as the origin of major taxa. The particular dates that Vision *et al.* give may not hold up in future research. Molecular clocks are liable to error, and there are several ways of using them. However, the style of research performed by Vision *et al.*, in which clocks are used to date duplications, is starting to flourish in the new genomic era. Chapter 41 by Benner *et al.* contains another example of this kind of work. These authors estimate the time of origin of the genes associated with alcohol metabolism in yeast and fruitflies, together with the time of origin of fruit in plants. Whether this three-way coincidence will hold up in future research, and whether the date is correct, are both uncertain. But it illustrates the kind of interesting inference that is becoming possible.

Over time, we should be able to reconstruct the history of the human genome. I include an extract (Chapter 38) from the 2001 paper announcing the draft sequence of the human genome. That paper contained a series of novel evolutionary analyses. One of them compared the human gene-set with other species, and inferred when in the tree of life our various genes originated. Clearly, biologists will be able to provide a much thicker description of this history as the genomes of more species are sequenced. We should also in time be able to identify the genetic changes that occurred in the origin of humans from their ape relatives. Carroll (Chapter 39) discusses this topic, prospectively. Chapters 38 and 39 are closely related to human evolution. Their inclusion here, and the inclusion of Chapters 53–55 in Section I on human evolution, was more or less arbitrary.

I also include an extract from Raff (Chapter 40). He describes an amazing miniature from the gallery of genomic discoveries. The genetic basis of crystallin genes could hardly have been guessed in advance; it is another genomic contribution to the concept of homology (discussed in Section E—see that section's editorial introduction).

HOWARD OCHMAN, JEFFREY G. LAWRENCE AND EDUARDO A. GROISMAN

36 Lateral gene transfer and the nature of bacterial innovation

Unlike eukaryotes, which evolve principally through the modification of existing genetic information, bacteria have obtained a significant proportion of their genetic diversity through the acquisition of sequences from distantly related organisms. Horizontal gene transfer produces extremely dynamic genomes in which substantial amounts of DNA are introduced into and deleted from the chromosome. These lateral transfers have effectively changed the ecological and pathogenic character of bacterial species. [Authors' summary.]

Given that they are single-celled organisms and that their genome sizes vary by little more than an order of magnitude in length, bacteria display extraordinary variation in their metabolic properties, cellular structures and lifestyles. Even within relatively narrow taxonomic groups, such as the enteric bacteria,[1] the phenotypic diversity among species is remarkable. Although the enteric bacteria as a whole share a myriad of traits denoting their common ancestry, each species also possesses a unique set of physiological characteristics that define its particular ecological niche.

Several mechanisms could be responsible for the differences evident among bacterial species. Point mutations leading to the modification, inactivation or differential regulation of existing genes have certainly contributed to the diversification of microorganisms on an evolutionary timescale; however, it is difficult to account for the ability of bacteria to exploit new environments by the accumulation of point mutations alone. In fact, none of the phenotypic traits that are typically used to distinguish the enteric bacteria *Escherichia coli* from its pathogenic sister species *Salmonella enterica* can be attributed to the point mutational evolution of genes common to both. Instead, there is growing evidence that lateral gene transfer has played an integral role in the evolution of bacterial genomes, and in the diversification and speciation of the enterics and other bacteria.

The significance of lateral gene transfer for bacterial evolution was not recognized until the 1950s, when multidrug resistance patterns emerged on a worldwide scale. The facility with which certain bacteria developed resistance to the same spectrum of antibiotics indicated that these traits were being transferred among taxa, rather than being generated *de novo* by each lineage. Although the widespread impact of lateral gene transfer on bacterial evolution was not appreciated until much later, these early studies of rapid evolution by gene acquisition encompass four issues relevant to all current studies. First, how is it possible to detect and to identify cases of lateral gene transfer? Second, where do these genes come from, and by what mechanisms are they transferred? Third, what types of, and how many, traits have been introduced through lateral gene transfer? And fourth, at what relative rates are different classes of genes mobilized among genomes?

[1] Enteric bacteria are bacteria that inhabit the intestines of their hosts. *E. coli* for instance lives in the intestines of humans and other mammals.

Detecting lateral gene transfer

How is it possible to establish whether a new trait, or a specific genetic region, is the result of horizontal processes?[2] Naturally, it would be most satisfying to actually observe the conversion of a deficient strain in the presence of an appropriate donor—and all the more convincing to establish that genetic material had indeed been transferred and the manner in which it was acquired. But outside experimental settings—that is, for most cases of lateral gene transfer in the recent or evolutionary history of a bacterial species—actual transfer events are only rarely observed, and unambiguous evidence of their occurrence must be derived from other sources.

Lateral gene transfer creates an unusually high degree of similarity between the donor and the recipient strains for the character in question. Furthermore, because each transfer event involves the introduction of DNA into a single lineage, the acquired trait will be limited to the descendents of the recipient strain and absent from closely related taxa, thereby producing a scattered phylogenetic distribution. However, lateral gene transfer need not be invoked to explain the sporadic occurrence of certain phenotypic traits, such as the ability to withstand particular antibiotics, because these properties can originate through point mutations in existing genes and therefore may evolve independently in divergent lineages. Thus, additional information is needed to discriminate between convergent evolution and lateral gene transfer. Clearly, the strongest evidence for (or against) lateral gene transfer derives from a molecular genetic analysis of the DNA sequences themselves.

DNA sequence information has been used in diverse ways to identify cases of lateral gene transfer, but the underlying basis of most applications is to discover features indicating that the evolutionary history of genes within a particular region differs from that of ancestral (vertically transmitted) genes. Similar to distinctive phenotypic properties, DNA segments gained through lateral gene transfer often display a restricted phylogenetic distribution among related strains or species. In addition, these species-specific regions may show unduly high levels of DNA or protein sequence similarity to genes from taxa inferred to be very divergent by other criteria. The significance of aberrant phylogenies can be evaluated by phylogenetic congruency tests or other means.

Although gene comparisons and their phylogenetic distributions are useful for detecting lateral transfer, the DNA sequences of genes themselves provide the best clues to their origin and ancestry within a genome. Bacterial species display a wide degree of variation in their overall G+C content, but the genes

[2] Horizontal gene transfer and lateral gene transfer are two ways of saying the same thing: gene transfer from one cell, or organism, to another, rather than 'vertical' gene transfer by reproduction from parent to offspring.

in a particular species' genome are fairly similar with respect to their base compositions, patterns of codon usage and frequencies of di- and trinucleotides. Consequently, sequences that are new to a bacterial genome, in other words, those introduced through horizontal transfer, retain the sequence characteristics of the donor genome and thus can be distinguished from ancestral DNA.

It is not surprising that genomic regions often manifest several attributes that denote their acquisition through lateral gene transfer. For example, a large number of *S. enterica* genes that are not present in *E. coli* (or any other enteric species) have base compositions that differ significantly from the overall 52% G+C content of the entire chromosome. Within *S. enterica*, certain serovars (that is, lineages that exhibit a distinct composition of flagellar and/or lipopolysaccharide surface antigens) may contain more than a megabase of DNA not present in other serovars, as assessed by a genomic subtraction procedure. The base compositions of these anonymous serovar-specific sequences suggest that at least half were gained through horizontal transfer. In addition to information obtained from the sequences of the genes themselves, the regions adjacent to genes identified as being horizontally transferred often contain vestiges of the sequences affecting their integration, such as remnants of translocatable elements, transfer origins of plasmids or known attachment sites of phage integrases, which further attest to their foreign origin in the genome.[3]

Genetic exchange within and between bacterial species also acts upon homologous sequences,[4] and numerous techniques have been developed to detect such events from sequence data. However, the action of mismatch correction systems greatly reduces the efficiency of homologous recombination when donor and recipient sequences contain nonidentical bases. As a result, homologous recombination is most successful in integrating DNA into the chromosome when the donor and recipient are relatively closely related. Because this type of genetic exchange principally affects the variation in existing genes, rather than introducing unique traits to the genome, its role

[3] Translocatable elements are 'jumping genes' that can re-locate from one region of the DNA to another. Plasmids are stretches (usually circles) of DNA, separate from the main chromosome(s) in the cell. Plasmids replicate independently of the chromosome and can sometimes transfer between cells. Phages are viruses that infect bacteria; they sometimes pick up bacterial genes when they copy themselves between host cells. All three of these genetic elements are potential agents of horizontal gene transfer.

[4] Homologous sequences are sequences of DNA that are more or less the same. See the opening sentences of Chapter 33 by Dickinson on this meaning. The methods for recognizing horizontal gene transfer discussed so far are for the case in which the genes being transferred are clearly different from anything in the receiving DNA. However, horizontal gene transfer may also occur between very similar DNA molecules.

in the ecological and physiological diversification of bacteria is apt to be negligible.

The scope of lateral gene transfer

The analysis of individual genes has uncovered numerous cases of lateral gene transfer; however, the ability to recognize horizontally acquired regions on the basis of their sequence characteristics makes it possible to assess the total proportion of foreign genes within a genome without resorting to gene phylogenies, sequence alignments or homology searches. Early attempts to establish the extent of laterally transferred sequences in a genome were limited to the very few microorganisms for which there was sufficient sequence information to get an unbiased sample of genes. Using this approach, it was originally estimated that between 10% and 16% of the *E. coli* chromosome arose through lateral gene transfer—a range similar to the amount of unique DNA in the *E. coli* chromosome inferred from early alignments of the *E. coli* and *S. enterica* genetic maps.

The availability of complete genomic sequences provides an opportunity to measure and compare the cumulative amount of laterally transferred sequences in diverse bacterial genomes (Fig. 36.1). Potentially foreign genes are identified by their atypical nucleotide compositions, or patterns of codon usage bias; after correcting for genes whose atypical features are due to amino-acid composition, the remaining genes are likely to have been introduced relatively recently by lateral gene transfer. Wide variation has been observed in the size and organization of 19 genomes analysed, and the amount of horizontally acquired DNA—represented as those open reading frames (ORFs)[5] whose sequence characteristics depart from the prevalent features of their resident genome—ranges from virtually none in some organisms with small genome sizes, such as *Rickettsia prowazekii, Borrelia burgdorferi* and *Mycoplasma genitalium*, to nearly 17% in *Synechocystis* PCC6803. In all cases, very ancient horizontal transfer events, such as those disseminating transfer RNA synthetases, would not be detected using these methods.

In some species, a substantial proportion of horizontally transferred genes can be attributed to plasmid-, phage- or transposon-related sequences (Fig. 36.1, black bars). As observed in *E. coli*, a significant fraction of acquired DNA in *Synechocystis* PCC6803, *Helicobacter pylori* and *Archaeoglobus fulgidus* is physically associated with mobile DNA, which probably mediated the integration of these sequences into the chromosome. Although analyses based

[5] ORFs, or open reading frames, are stretches of the coding DNA that consist of chains of codons uninterrupted by a stop codon. They potentially code for a polypeptide. Informally, 'ORF' can be understood as 'gene'.

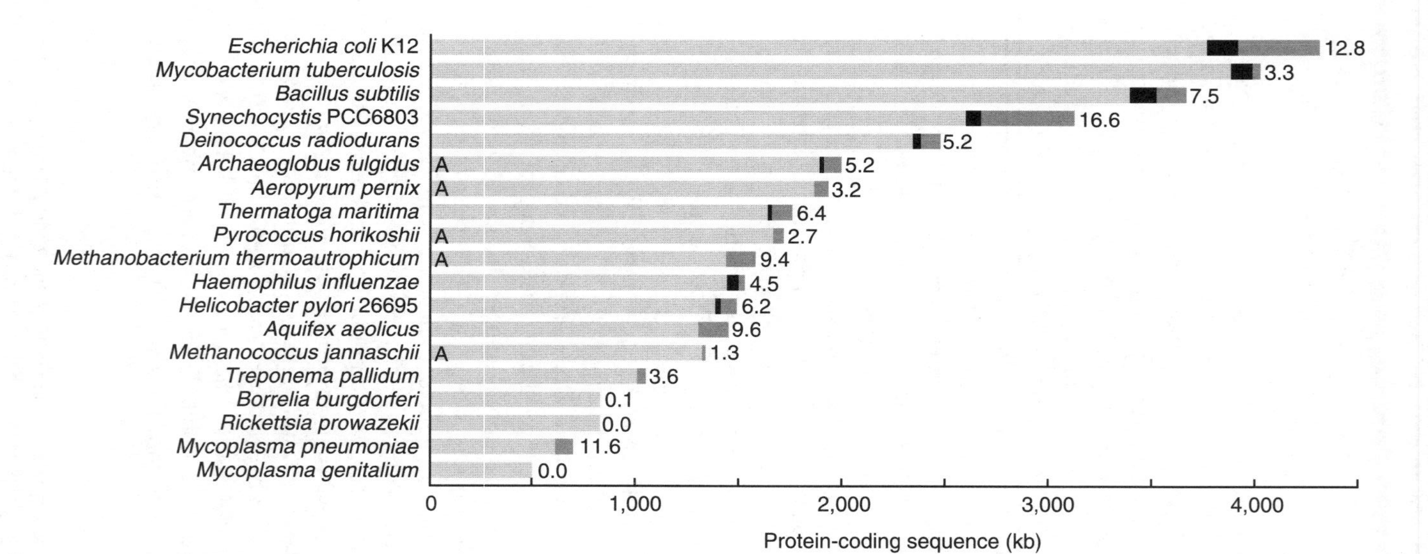

Figure 36.1: Distribution of horizontally acquired (foreign) DNA in sequenced bacterial genomes. Lengths of bars denote the amount of protein-coding DNA. For each bar, the native DNA is light grey; foreign DNA identifiable as mobile elements, including transposons and bacteriophages, is black, and other foreign DNA is dark grey. The percentage of foreign DNA is noted to the right of each bar. 'A' denotes an Archaeal genome.

on sequence features indicate that large portions of bacterial genomes are attributable to lateral transfer, these methods can underestimate the actual number of transferred genes because sequences acquired from organisms of similar base compositions and codon usage patterns, as well as ancient transfer events, will escape detection.

Comparisons of completely sequenced genomes verify that bacteria have experienced significant amounts of lateral gene transfer, resulting in chromosomes that are mosaics of ancestral and horizontally acquired sequences. The hyperthermophilic Eubacteria *Aquifex aeolicus* and *Thermotoga maritima* each contain a large number of genes that are most similar in their protein sequences and, in some cases, in their arrangements, to homologues in thermophilic Archaea. Twenty-four per cent of *Thermotoga*'s 1,877 ORFs and about 16% of *Aquifex*'s 1,512 ORFs, display their highest match (as detected by BLAST)[6] to an Archaeal protein, whereas mesophiles, such as *E. coli*, *B. subtilis* and *Synechocystis*, have much lower proportions of genes that are most similar to Archaeal homologues.

Because the detection of lateral transfer by such systematic, gene-by-gene analyses depends upon the occurrence, phylogenetic positions and evolutionary rates of homologues in the currently available databases, the specific genes, as well as the overall amount of transferred DNA, recognized by this approach could differ from those resolved by examining atypical sequence characteristics. Moreover, the utility of BLAST similarity scores to infer gene ancestry depends upon the number of acknowledged homologues, the evolutionary relationships of organisms represented in the databases and the persistance of genes in a genome; hence, values derived from whole-genome comparisons will be refined as additional genomes are completed. However, both methods reveal the potential for bacterial genomes to incorporate large numbers of unique sequences, often from very divergent organisms.

How and what sequences are acquired

In contrast to the evolution of new traits through the modification of existing sequences, the origin of new abilities through lateral gene transfer has three requirements. First, there needs to be a means for the donor DNA to be delivered into the recipient cell. Second, the acquired sequences must be incorporated into the recipient's genome (or become associated with an autonomous replicating element). And third, the incorporated genes

[6] BLAST is a computerized system, available on-line, for finding other sequences that are similar to a given sequence. For instance, if you have sequenced one gene in one species, you can use BLAST to find other genes in other species that may be evolutionarily related to that gene.

must be expressed in a manner that befits the recipient microorganism. The first two steps are largely indiscriminate with respect to the specific genes or functional properties encoded by the transferred regions, and can occur through three mechanisms: transformation, transduction and conjugation. [. . .]

Through these mechanisms, virtually any sequence—even those originating in eukaryotes or Archaea—can be transferred to, and between, bacteria. The relatively small sizes of bacterial genomes imply, however, that either the rate of transfer or the maintenance of transferred sequences is very low, or that the maintenance of horizontally transferred sequences is offset by the loss of resident sequences. Bacterial genomes do not contain arbitrary assortments of acquired genes, and the plethora of documented transfer events has provided an abundance of information about the origins and functions of the acquired sequences. Naturally, these studies are biased towards identifying genes that impart both new and consequential functions upon the recipient organism; however, they illustrate both the unprecedented range of mechanisms by which traits are disseminated among bacteria and the impact of lateral gene transfer on bacterial evolution.

Traits introduced through lateral gene transfer

ANTIBIOTIC RESISTANCE

Antimicrobial resistance genes allow a microorganism to expand its ecological niche, allowing its proliferation in the presence of certain noxious compounds. From this standpoint, it is not surprising that antibiotic resistance genes are associated with highly mobile genetic elements, because the benefit to a microorganism derived from antibiotic resistance is transient, owing to the temporal and spatial heterogeneity of antibiotic-bearing environments. Plasmids are readily mobilizable between taxa and represent the most common method of acquiring antibiotic resistance determinants. As plasmids are rarely integrated into the chromosome, the acquired traits must confer an advantage sufficient to overcome not only inactivation by mutation, but also elimination by segregation.

Transposable elements can also promote the transfer of resistance genes between bacterial genomes. When a resistant determinant, or any gene conferring a selectable phenotype, is flanked by two insertion sequences, it may be mobilized as a complex transposon; for example, two copies of the insertion sequence IS*10* flank a tetracycline resistant determinant and regulatory gene to form transposon Tn*10*. Likewise, two IS*50* elements flank a three-gene operon that confers resistance to kanamycin, bleomycin and streptomycin forming transposon Tn*5*, which can integrate into the chromosomes of phylogenetically diverse bacterial species. [. . .]

VIRULENCE ATTRIBUTES

Unlike the acquisition of antibiotic resistance, adoption of a pathogenic lifestyle usually involves a fundamental change in a microorganism's ecology. The sporadic phylogenetic distribution of pathogenic organisms has long suggested that bacterial virulence results from the presence (and perhaps acquisition) of genes that are absent from avirulent forms. Evidence for this view has taken several forms, ranging from the discovery of large 'virulence' plasmids in pathogenic *Shigella* and *Yersinia* to the ability to confer pathogenic properties upon laboratory strains of *E. coli* by the experimental introduction of genes from other species.

Recent studies have discovered that horizontally acquired 'pathogenicity islands' are major contributors to the virulent nature of many pathogenic bacteria. These chromosomally encoded regions typically contain large clusters of virulence genes and can, upon incorporation, transform a benign organism into a pathogen. Many pathogenicity islands are situated at tRNA and tRNA-like loci, which appear to be common sites for the integration of foreign sequences. [. . .]

Certain bacteriophages encode virulence determinants within their genomes, and lysogenization by such a bacteriophage results in the 'conversion' of a strain to a pathogenic variant. For example, the genes encoding exotoxin A in *Streptococcus pyogenes*, Shiga toxin in enterohaemorrhagic strains of *E. coli*, and the SopE GTP/GDP exchange factor necessary for host cell invasion by *S. enterica* are carried by converting bacteriophages. [. . .]

METABOLIC PROPERTIES

Lateral gene transfer has played a significant role in moulding bacterial genomes by mobilizing other physiological traits, which have, in effect, allowed recipient organisms to explore new environments. Although not surprising in retrospect, early alignments of the linkage maps of *E. coli* and *S. enterica* showed that many biochemical properties restricted to only one of these species, such as lactose fermentation by *E. coli* or citrate utilization by *S. enterica*, were encoded on chromosomal regions unique to each of these species. Subsequent molecular genetic analyses have established that, in many cases, species-specific traits can be attributed either to the acquisition of genes through lateral gene transfer or to the loss of ancestral genes from one lineage. In this way, most ecological innovation in Eubacteria and Archaea is fundamentally different from diversification in multicellular eukaryotes.

The acquisition of new metabolic traits by horizontal transfer emphasizes the importance of natural selection as the arbiter of lateral gene exchange—the duration of acquired sequences will be fleeting if the genes do not contribute a useful or meaningful function upon introduction to the recipient cell. Therefore, we would expect that the successful mobilization of complex metabolic traits requires the physical clustering of genes, such that all necessary genes will be transferred in a single step. As a result, horizontal inheritance will select for gene clusters and for operons, which can be expressed in recipient cells by a host promoter at the site of insertion. In this manner, *E. coli*, by acquiring the *lac* operon, gained the ability to use the milk sugar lactose as a carbon source and to explore a new niche, the mammalian colon, where it established a commensal relationship.

Whole-genome comparisons have uncovered sets of genes that are restricted to organisms that have independently adapted to a common lifestyle, such as Archaeal and Bacterial hyperthermophiles or the intracellular pathogens *Rickettsia* and *Chlamydia*. The distribution of such sequences in phylogenetically divergent but ecologically similar microorganisms has been ascribed to lateral gene transfer, and it is tempting to suggest that these genes, for which functions are presently unknown, contribute to the metabolic and physiological requirements of these unique environments.

Rate of sequence acquisition

The rate of DNA acquisition by the *E. coli* chromosome was measured indirectly by examining the amelioration of atypical sequence characteristics (for example, nucleotide composition) towards the equilibrium values displayed by this genome. Such methods allow the estimation of the time of arrival for each segment of foreign DNA detected in the genome and, hence, the rate of successful horizontal genetic transfer. Taking into account the inevitable deletion of genes—bacterial genomes are not growing ever larger in size—horizontal transfer has been estimated to have introduced successfully $\sim$ 16 kb per million years into the *E. coli* genome.

As evident from Fig. 36.1, there is broad variation in the amount of acquired DNA among bacterial genomes. Although some organisms may have only a limited capacity to transfer and exchange DNA, bacterial lifestyles can also contribute to the lower rates of gene acquisition in genomes containing small amounts of foreign DNA. For example, the intracellular habitat of *R. prowazekii* and *M. genitalium* probably shields the organism from exposure to potential gene donors and the opportunity to acquire foreign sequences.

The impact of acquired DNA

Lateral gene transfer provides a venue for bacterial diversification by the reassortment of existing capabilities. Yet, while the emergence of new phenotypic properties through lateral gene transfer furnishes several advantages, it also presents several problems to an organism. Newly acquired sequences, especially those conferring traits essential to only a portion of the bacterial life cycle, are most useful when they are appropriately and coordinately regulated with the rest of the genome. In *Salmonella*, the expression of several independently acquired virulence genes is under the control of a single regulatory system—the PhoP/PhoQ two-component system—that was already performing essential functions in the genome before the acquisition of these genes. Although the precise manner by which each of these genes is regulated has yet to be resolved, these findings suggest that the physiological capabilities and adaptation of very divergent bacteria rely on a common set of universally distributed regulatory signals.

Despite wide diversity in the structure, organization and contents of bacterial genomes, there is relatively narrow variation in genome sizes. Evidence from experimental studies shows that bacterial genomes are prone towards deleting non-essential DNA, and the small, reduced genomes of host-dependent bacteria, such as *Mycoplasma, Chlamydia* and *Rickettsia*, attest to the tendency for bacteria to delete the more expendable sequences from their genomes.

Because bacterial genomes can maintain only a finite amount of information against mutation and loss, chromosomal deletions will serve to eliminate genes that fail to provide a meaningful function, that is, the bulk of acquired DNA as well as superfluous ancestral sequences. Hence, bacterial genomes are sampling rather than accumulating sequences, counterbalancing gene acquisition with gene loss. As a result, lateral gene transfer can redefine the ecological niche of a microorganism, which will, in effect, promote bacterial speciation.

A potential result of rampant interspecific recombination is the blurring of species boundaries, and the failure of any one gene to reflect the evolutionary history of the organism as a whole. Yet, as more bacterial genomes are examined, robust consensus phylogenies are being constructed from infrequently transferred genes, which provide the benchmark for gauging the scope and impact of lateral gene transfer. Rather than diminishing the utility of molecular phylogeny, the intermingling of genes and the resulting phylogenetic incongruities document the process of gene-transfer-mediated organismal diversification.

[*Nature*, 405 (2000), 299–304.]

TODD J. VISION, DANIEL G. BROWN, AND STEVEN D. TANKSLEY

37 The origins of genomic duplications in *Arabidopsis*[1]

Large segmental duplications cover much of the Arabidopsis thaliana *genome. Little is known about their origins. We show that they are primarily due to at least four different*

[1] *Arabidopsis* is a small weed and has become a model system for genetic research on flowering plants. This paper was written when the *Arabidopsis* genome was sequenced, in 2000.

large-scale duplication events that occurred 100 to 200 million years ago, a formative period in the diversification of the angiosperms. A better understanding of the complex structural history of angiosperm genomes is necessary to make full use of Arabidopsis *as a genetic model for other plant species.* [Authors' summary.]

A. thaliana has one of the smallest angiosperm genomes. It is a well-behaved diploid with only five haploid chromosomes. Despite this, much of the genome is internally duplicated. It has been hypothesized that the duplicated blocks originated in a single polyploidy event and have since been scrambled by chromosomal rearrangements.[2] This hypothesis predicts that each region of the *Arabidopsis* genome should be present in exactly two copies. Recent comparative mapping results suggest that some regions are present in three or more copies, but it is not clear how prevalent such regions are. Here, we use the nearly complete genome sequence of *Arabidopsis* to study the evolutionary origins of duplicated blocks on a genome-wide scale. [. . .]

Duplicated blocks were identified by the presence of neighboring genes with high sequence similarity to neighboring genes elsewhere in the geneome. We considered only protein-coding genes because little conservation exists between noncoding duplicated regions in *Arabidopsis*. We used BLAST[3] to identify genes with high sequence similarity. Our data set contained 20,269 composite open reading frames (cORFs),[4] of which 2796 represented tandem arrays of related genes or the same gene present on overlapping clones. After removing low-quality matches, there were matches between 18,569 pairs of cORFs; 64% of the cORFs had at least one match. To identify duplicated blocks, we considered the proximity and transcriptional orientation of matches in both segments. We allowed singleton (nonmatching) genes within duplicated blocks because gene loss is known to follow duplication and because some genes may be transposed from their original positions. We also did not prohibit inversions within duplicated blocks because small-scale inversions are not uncommon in eukaryotic genomes. To ensure that spurious duplicated blocks would not be identified,

[2] A single polyploidy event would be one in which the whole genome was duplicated in the ancestral past. The modern genome would not be expected to look like two side-by-side duplicates of one genomic pattern, because chromosomes are continually rearranged over evolutionary time and the two wholescale duplicates would have been mixed up since the original polyploidization. The extract mentions several kinds of chromosomal rearrangement: whole chromosomes may fuse, or subdivide; within a chromosome, a region may be inverted or translocated (that is, swapped) with a region of another chromosome. Polyploidy is sometimes associated with hybrid speciation in plants, as Grant discusses in Chapter 29.

[3] BLAST is computer software, functional on-line, used to compare DNA sequences to find regions of similarity. BLAST is used here to find potential duplicates of each gene, within the *Arabidopsis* genome.

[4] 'Open reading frame', informally and approximately, means much the same as 'gene'. See Chapter 36, footnote 5.

we chose a conservative set of parameter values based on the outcome of randomization tests.

We identified 103 duplicated blocks containing seven or more matching cORFs. Candidates with fewer than seven genes are much more abundant in real than randomized data, suggesting that many smaller blocks may be present. There are duplications between all chromosomes except chromosome 2 with itself. Over 81% of cORFs fall within the bounds of at least one block. However, only 28% of these are actually present in duplicate. Interestingly, pericentromeric genes account for much of the genome that is not covered by blocks. Nearly 25% of all cORFs fall within two or more blocks, and one region, near ATAP22 on chromosome 4 falls within five blocks. Such extensive overlap among blocks provides prima facie evidence for multiple duplication events.

The number of independent duplication events can be inferred from patterns of sequence divergence between duplicated genes. A single polyploidization event will produce a unimodal distribution of divergence estimates with homogeneity among blocks. Many small independent events can also result in unimodality but with heterogeneity among blocks. A limited number of asynchronous, independent duplication events will produce a multimodal distribution.

The median estimated amino acid divergence (d_A) was between 0.325 and 0.725 amino acid substitutions per site (Δaa/site) for all but nine blocks (Fig. 37.1). Excluding these nine, there is still significant among-block heterogeneity in d_A. Thus, we reject the hypothesis they belong to a single age class. The best-fit mixture of normal distributions to these 94 medians is trimodal. Each block can be assigned to one of the three age classes (labeled, C, D, and E in progressing order of age) with greater than 50% posterior probability. The remaining blocks can be assigned, ad hoc, to two younger age classes (A and B) and one older age class (F). Some spatial overlap remains between duplicated blocks within age classes, but this may be an artifact of erroneous age class assignments and spatially overextended blocks.

The blocks in age class C collectively bound 48% of the cORFs in the genome. Adding the number of duplicated pairs of cORFs to singleton cORFs we estimate that more than 9000 cORFs were duplicated at this time. This is far larger than chromosome 1, the largest of the present complement, which we estimate to contain fewer than 6000 genes. Thus, it is likely that age class C represents either a whole-genome polyploidy event or the near-simultaneous duplication of multiple chromosomes.

The progressively older blocks in age classes D, E, and F bound 39, 11, and 3% of the cORFs in the genome, respectively. The true extent of these more ancient duplications is likely to be much greater because gene loss after each duplication event tends to obscure older blocks. Thus, these older age classes may also represent very large-scale duplication events.

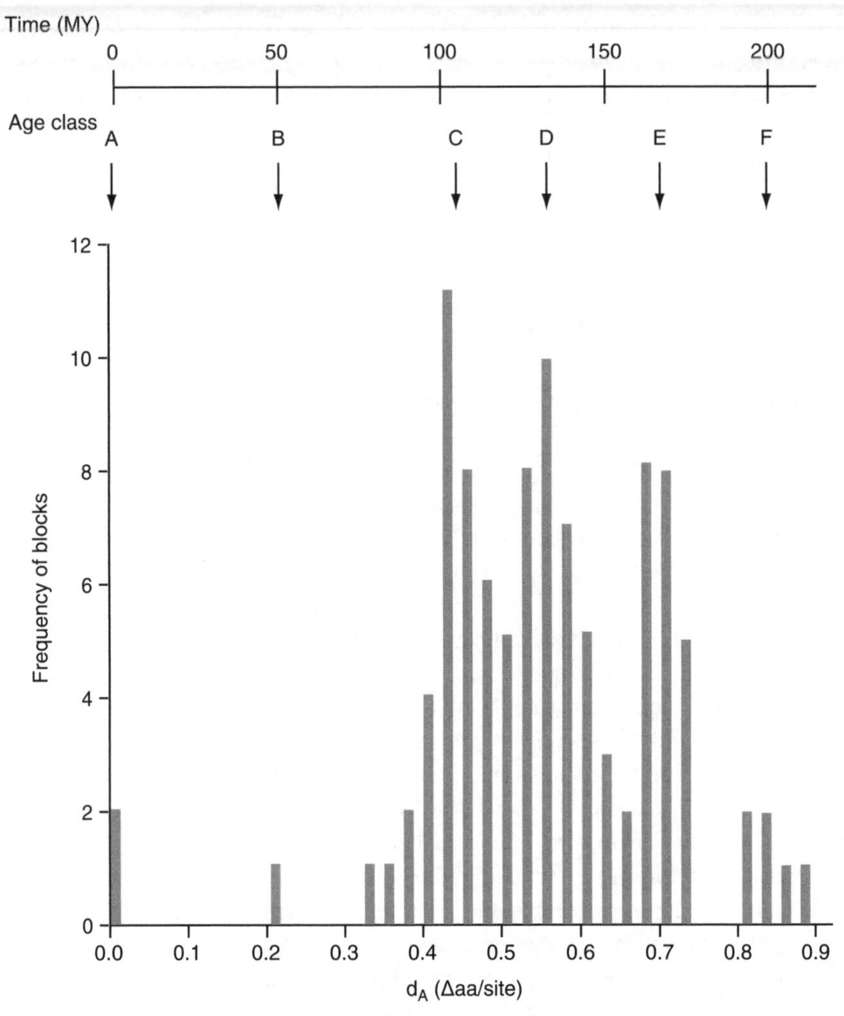

Figure 37.1: Multimodal distribution of block ages. The distribution of median d_A *values among blocks. Age classes are marked with arrows. An approximate timescale is given.*

We estimated the absolute ages of the duplicated blocks by assuming that the average extent of amino acid substitution (d_A) is linearly related to time. The average d_A values for age classes B through F (Table 37.1) yield age estimates of approximately 50, 100, 140, 170, and 200 million years ago (Mya), respectively. Thus, age classes C through F appear to date from the Mesozoic Era (65 to 245 Mya). Age class E is marginally older than the reported age of divergence between rosid and asterid eudicots, 112 to 156 Mya, whereas age class F is within the estimated time window for the divergence of monocots

Table 37.1: Features of the five age classes of duplicated blocks. d_A *is the minimum change in amino acids per between dispersed duplicated cORFs, averaged among all cORFs. Retained duplicates, ratio of presently duplicated to inferred ancestral cORFs. Block size, mean number of cORFs (including singletons) per copy.*

AGE CLASS	NO. OF BLOCKS	d_A	RETAINED DUPLICATES	BLOCK SIZE	ESTIMATED AGE (MYA)
A*	2	0†	0.90	12.5	0
B	1	0.21†	0.38	14.5	50
C	35	0.45‡	0.15	149.5	100
D	36	0.57‡	0.13	128.0	140
E	23	0.71‡	0.11	60.0	170
F	6	0.84†	0.09	57.8	200

* Probable artifact of genome assembly.
† Median among all duplicated genes.
‡ Mean of best-fit normal distribution to block medians.

and dicots, 180 to 220 Mya.[5] Thus, the older duplicated blocks reported here are likely to be a common feature of diverse groups of angiosperms. Regions contained within blocks 45, 48, 85, 88, and 100 were recently found to share common ancestry with a 105 kilobase genomic sequence from tomato, an asterid. Phylogenetic analysis in that study suggested that the rosid–asterid divergence occurred before the events leading to age classes C and E (duplicated blocks 45 and 85, respectively) and was nearly contemporaneous with age class E. Thus, the divergence and duplication dates are consistent. The two blocks assigned to age class A are likely to be artifacts of erroroneous genome assembly. In both cases, the copies are nearly identical even at the nucleotide level and are restricted to individual large-insert clones. Thus, the youngest duplicated block appears to be the sole member of age class B.

If the *Arabidopsis* genome has experienced multiple large-scale duplications, then the present complement of five chromosomes suggests a history of chromosome fusions. In fact, three such fusions have occurred since *A. thaliana* diverged from its closest extant relatives. Subchromosomal rearrangements, such as inversions and translocations, are expected to cause the average size of duplicated blocks to decrease with age, as is observed for blocks C through F (Table 37.1). Inversions can, in some cases, be inferred from our data set. [. . .]

[5] The division between monocotyledons and dicotyledons is the main division within the angiosperms. The ages of the main rounds of gene duplication can be interestingly compared with the fossil ages of the main events in angiosperm evolution, discussed by Dilcher in Chapter 45.

We have estimated the number of deleted genes in each duplicated block by counting the number of singleton cORFs. The proportion of deleted genes increases with the inferred age of the duplication event (Table 37.1). A small number of blocks deviate significantly from a 1:1 distribution of singletons between the two copies, suggesting that the loss of duplicate genes between segments may sometimes be biased, as previously observed.

The 103 duplicated blocks account for only 15% of the matches in the data set. Some proportion of the remaining matches may lie within undetected duplicated blocks or may have been transposed from their original position. Still, the remaining matches are not randomly dispersed in the genome, suggesting the presence of a separate gene duplication process. Matches not in blocks are 20% more likely to occur on the same chromosome than if the distribution were proportional to size. Of those matches on the same chromosome, the average distance is 86% of that expected between two random points. Similar findings have been reported for *Caenorhabditis elegans*, a genome that lacks large-scale duplications.

Many insertion mutants[6] in *Arabidopsis* have no obvious phenotypic effect. This may be due, in part, to redundant functions among duplicated genes. One example appears to be the *shatterproof* genes SHP1 and SHP2, MADS-box regulatory factors that must be simultaneously removed before fruit nondehiscence is observed. SHP1 and SHP2, on chromosomes 3 and 2, are within duplicated block 67 in age class C.

Our analysis implies that regions of the *Arabidopsis* genome homoelogous to genomes that diverged from *Arabidopsis* before ~100 Mya will be small, generally less than ~10 centimorgans in size. This, coupled with massive gene loss in *Arabidopsis*, has likely been responsible for the difficulty in identifying regions homoeologous between *Arabidopsis* and rice. Knowledge of the duplication history of *Arabidopsis* should facilitate such mapping efforts. For example, it has recently been proposed that segments of *Arabidopsis* chromosome 4 and rice chromosome 2 are homoeologous. The *Arabidopsis* segment is within duplicated block 92 (age class D), implying that the rice segment is also homoeologous to a part of *Arabidopsis* chromosome 5.

Our understanding of plant evolution and our use of *Arabidopsis* as a genetic model for other plants will clearly depend on a deeper appreciation for the complex duplication history of this small genome.

[*Science*, 290 (2000), 2114–16.]

[6] An insertion mutation is a mutation in which a stretch of DNA (often a transposable element) is inserted into a gene, or into a region of non-coding DNA.

INTERNATIONAL HUMAN GENOME SEQUENCING CONSORTIUM[1]

38 Initial sequencing and analysis of the human genome

The paper describing the initial draft sequence of the human genome contains material on the evolution of the human genome. The extract here is of the section on the evolution of the part of the human genome that codes for proteins (the proteome). The analysis shows the fractions of human genes that date back to various stages in human history, such as the origins of cellularity, of animals, and of vertebrates. The paper analyses the genomic changes that contributed to the historical increase in living complexity, from single celled life to humans. The expansion of the genome in vertebrates was mainly by adding new genes within gene families. Some new gene families also evolved, coding for the nervous system, development, and immunity. Human proteins are more complex on average than yeast proteins. [Editor's summary.]

Comparative proteome analysis

[. . .] Here, we aim to take a global perspective on the content of the human proteome[2] by comparing it with the proteomes of yeast, worm, fly and mustard weed. Such comparisons shed useful light on the commonalities and differences among these eukaryotes. The analysis is necessarily preliminary, because of the imperfect nature of the human sequence, uncertainties in the gene and protein sets for all of the multicellular organisms considered and our incomplete knowledge of protein structures. Nonetheless, some general patterns emerge. These include insights into fundamental mechanisms that create functional diversity, including invention of protein domains, expansion of protein and domain families, evolution of new protein architectures and horizontal transfer of genes. Other mechanisms, such as alternative splicing, post-translational modification and complex regulatory networks, are also crucial in generating diversity but are much harder to discern from the primary sequence. We will not attempt to consider the effects of alternative splicing on proteins; we will consider only a single splice form from each gene in the various organisms, even when multiple splice forms are known.

FUNCTIONAL AND EVOLUTIONARY CLASSIFICATION

We began by classifying the human proteome on the basis of functional categories and evolutionary conservation. We used the InterPro annotation protocol to identify conserved biochemical and cellular processes. InterPro is

[1] The human genome was sequenced by a huge team of scientists, working in many labs, and in several countries. Hence the unusual authorship of this paper.

[2] The proteome here refers to that subset of the genome, containing all the genes that code for proteins. Some of the genome also codes for genes, such as ribosomal RNA genes, that are not translated into proteins; some of the genome is non-coding—these other bits are not included in the proteome.

a tool for combining sequence-pattern information from four databases. [. . .]

The InterPro families are partly the product of human judgement and reflect the current state of biological and evolutionary knowledge. The system is a valuable way to gain insight into large collections of proteins, but not all proteins can be classified at present. The proportions of the yeast, worm, fly and mustard weed protein sets that are assigned to at least one InterPro family is, for each organism, about 50%.

About 40% of the predicted human proteins in the IPI[3] could be assigned to InterPro entries and functional categories. On the basis of these assignments, we could compare organisms according to the number of proteins in each category. Compared with the two invertebrates, humans appear to have many proteins involved in cytoskeleton, defence and immunity, and transcription and translation. These expansions are clearly related to aspects of vertebrate physiology. Humans also have many more proteins that are classified as falling into more than one functional category (426 in human versus 80 in worm and 57 in fly, data not shown). Interestingly, 32% of these are transmembrane receptors.

We obtained further insight into the evolutionary conservation of proteins by comparing each sequence to the complete nonredundant database of protein sequences maintained at NCBI, using the BLASTP[4] computer program and then breaking down the matches according to organismal taxonomy (Figure 38.1). Overall, 74% of the proteins had significant matches to known proteins.

Such classifications are based on the presence of clearly detectable homologues in existing databases. Many of these genes have surely evolved from genes that were present in common ancestors but have since diverged substantially. Indeed, one can detect more distant relationships by using sensitive computer programs that can recognize weakly conserved features. Using PSI-BLAST, we can recognize probable nonvertebrate homologues for about 45% of the 'vertebrate-specific' set. Nonetheless, the classification is useful for gaining insights into the commonalities and differences among the proteomes of different organisms.

PROBABLE HORIZONTAL TRANSFER[5]

An interesting category is a set of 223 proteins that have significant similarity to proteins from bacteria, but no comparable similarity to proteins from yeast,

[3] IPI, Integrated protein index. Roughly speaking, a list of all the proteins coded for in the human genome.

[4] BLASTP is a relative of BLAST, a program for identifying related genes by comparing sequences (Chapter 36, note 6). NCBI is the National Center for Biotechnological Information, Maryland, USA.

[5] Horizontal gene transfer in bacterial evolution is discussed in Chapter 36.

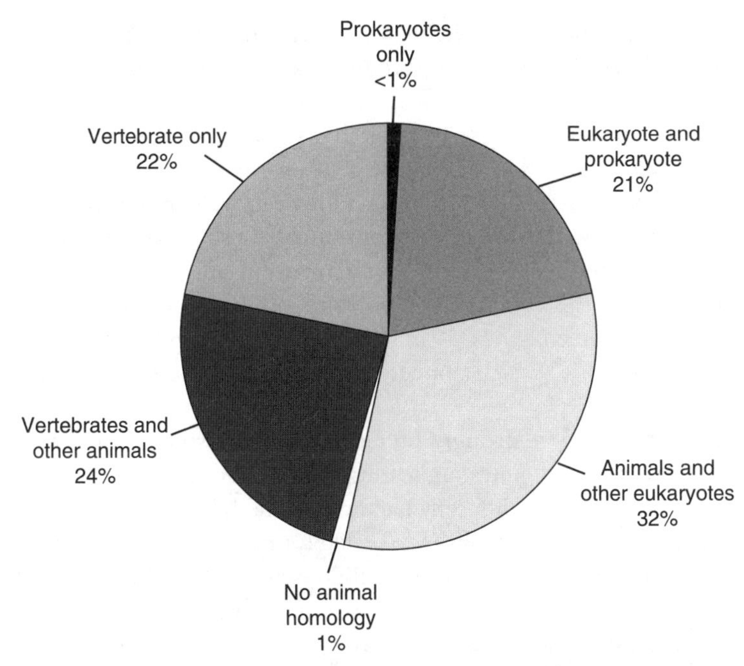

Figure 38.1: Distribution of the homologues of the predicted human proteins. For each protein, a homologue to a phylogenetic lineage was considered present if a search of the NCBI nonredundant protein sequence database, using the gapped BLASTP program, gave a random expectation (E) value of $\leq$0.001. Additional searches for probable homologues with lower sequence conservation were performed using the PSI-BLAST program, run for three iterations using the same cut-off for inclusion of sequences into the profile.

worm, fly and mustard weed, or indeed from any other (nonvertebrate) eukaryote. These sequences should not represent bacterial contamination in the draft human sequence, because [. . .].[6]

A more detailed computational analysis indicated that at least 113 of these genes are widespread among bacteria, but, among eukaryotes, appear to be present only in vertebrates. It is possible that the genes encoding these proteins were present in both early prokaryotes and eukaryotes, but were lost in each of the lineages of yeast, worm, fly, mustard weed and, possibly, from other nonvertebrate eukaryote lineages. A more parsimonious explanation is that these genes entered the vertebrate (or prevertebrate) lineage by horizontal transfer from bacteria. Many of these genes contain introns, which presumably were acquired after the putative horizontal transfer event. Similar

[6] The original paper described various technical reasons for ruling out contamination.

observations indicating probable lineage-specific horizontal gene transfers, as well as intron insertion in the acquired genes, have been made in the worm genome.

We cannot formally exclude the possibility that gene transfer occurred in the opposite direction—that is, that the genes were invented in the vertebrate lineage and then transferred to bacteria. However, we consider this less likely. Under this scenario, the broad distribution of these genes among bacteria would require extensive horizontal dissemination after their initial acquisition. In addition, the functional repertoire of these genes, which largely encode intracellular enzymes, is uncharacteristic of vertebrate-specific evolutionary innovations (which appear to be primarily extracellular proteins; see below).

We did not identify a strongly preferred bacterial source for the putative horizontally transferred genes, indicating the likelihood of multiple independent gene transfers from different bacteria. Notably, several of the probable recent acquisitions have established (or likely) roles in metabolism of xenobiotics or stress response. These include several hydrolases of different specificities, including epoxide hydrolase, and several dehydrogenases. Of particular interest is the presence of two paralogues of monoamine oxidase (MAO), an enzyme of the mitochondrial outer membrane that is central in the metabolism of neuromediators and is a target of important psychiatric drugs. This example shows that at least some of the genes thought to be horizontally transferred into the vertebrate lineage appear to be involved in important physiological functions and so probably have been fixed and maintained during evolution because of the increased selective advantage(s) they provide.

GENES SHARED WITH FLY, WORM AND YEAST

IPI.1 contains apparent homologues of 61% of the fly proteome, 43% of the worm proteome and 46% of the yeast proteome. We next considered the groups of proteins containing likely orthologues and paralogues[7] (genes that arose from intragenome duplication) in human, fly, worm and yeast.

We identified 1,308 groups of proteins, each containing at least one predicted orthologue in each species and many containing additional paralogues. The 1,308 groups contained 3,129 human proteins, 1,445 fly proteins,

[7] Orthologues and paralogues. These are two categories of homology. Thus the authors were looking for homologous genes in different species. (On homology, see Chapters 31–33, Section E.) To be exact, after a gene duplicates, there are two versions (call the A and B) of the ancestral gene. Two copies of A in two different species are called orthologues. A and B, in the same or in different species, are called paralogues.

1,503 worm proteins and 1,441 yeast proteins. These 1,308 groups represent a conserved core of proteins that are mostly responsible for the basic 'housekeeping' functions of the cell, including metabolism, DNA replication and repair, and translation.

In 564 of the 1,308 groups, one orthologue (and no additional paralogues) could be unambiguously assigned for each of human, fly, worm and yeast. These groups will be referred to as 1–1–1–1 groups. [. . .]

The 1–1–1–1 groups probably represent key functions that have not undergone duplication and elaboration in the various lineages. They include many anabolic enzymes responsible for such functions as respiratory chain and nucleotide biosynthesis. In contrast, there are few catabolic enzymes. As anabolic pathways branch less frequently than catabolic pathways, this indicates that alternative routes and displacements are more frequent in catabolic reactions. If proteins from the single-celled yeast are excluded from the analysis, there are 1,195 1–1–1 groups. The additional groups include many examples of more complex signalling proteins, such as receptor-type and src-like tyrosine kinases, likely to have arisen early in the metazoan lineage. The fact that this set comprises only a small proportion of the proteome of each of the animals indicates that, apart from a modest conserved core, there has been extensive elaboration and innovation within the protein complement. [. . .]

NEW VERTEBRATE DOMAINS AND PROTEINS

We then explored how the proteome of vertebrates (as represented by the human) differs from those of the other species considered. The 1,262 InterPro families were scanned to identify those that contain only vertebrate proteins. Only 94 (7%) of the families were 'vertebrate-specific'. These represent 70 protein families and 24 domain families.[8] Only one of the 94 families represents enzymes, which is consistent with the ancient origins of most enzymes. The single vertebrate-specific enzyme family identified was the pancreatic or eosinophil-associated ribonucleases. These enzymes evolved rapidly, possibly to combat vertebrate pathogens.

The relatively small proportion of vertebrate-specific multicopy families suggests that few new protein domains have been invented in the vertebrate lineage, and that most protein domains trace at least as far back as a common animal ancestor. This conclusion must be tempered by the fact that the InterPro classification system is incomplete; additional vertebrate-specific

[8] A domain is a region of a protein corresponding to a particular function (such as binding a substrate, or fixing the protein into a membrane). Some domains are found in more than one protein, presumably have been duplicated from one gene into another. The process is called domain shuffling, a term that is used later in this paper. Gerhart and Kirschner (Section H, Chapter 51) describe an example.

families undoubtedly exist that have not yet been recognized in the InterPro system.

The 94 vertebrate-specific families appear to reflect important physiological differences between vertebrates and other eukaryotes. Defence and immunity proteins (23 families) and proteins that function in the nervous system (17 families) are particularly enriched in this set. These data indicate the recent emergence or rapid divergence of these proteins. [. . .]

NEW ARCHITECTURES FROM OLD DOMAINS

Whereas there appears to be only modest invention at the level of new vertebrate protein domains, there appears to be substantial innovation in the creation of new vertebrate proteins. This innovation is evident at the level of domain architecture, defined as the linear arrangement of domains within a polypeptide. New architectures can be created by shuffling, adding or deleting domains, resulting in new proteins from old parts.

We quantified the number of distinct protein architectures found in yeast, worm, fly and human by using the SMART annotation resource. The human proteome set contained 1.8 times as many protein architectures as worm or fly and 5.8 times as many as yeast. This difference is most prominent in the recent evolution of novel extracellular and transmembrane architectures in the human lineage. Human extracellular proteins show the greatest innovation: the human has 2.3 times as many extracellular architectures as fly and 2.0 times as many as worm.

[. . .]

NEW PHYSIOLOGY FROM OLD PROTEINS

An important aspect of vertebrate innovation lies in the expansion of protein families. [. . .] About 60% of families are more numerous in the human than in any of the other four organisms. This shows that gene duplication has been a major evolutionary force during vertebrate evolution.

Many of the families that are expanded in human relative to fly and worm are involved in distinctive aspects of vertebrate physiology. An example is the family of immunoglobulin (IG) domains, first identified in antibodies thirty years ago. Classic (as opposed to divergent) IG domains are completely absent from the yeast and mustard weed proteomes and, although prokaryotic homologues exist, they have probably been transferred horizontally from metazoans. Most IG superfamily proteins in invertebrates are cell-surface proteins. In vertebrates, the IG repertoire includes immune functions such as those of antibodies, MHC proteins, antibody receptors and many lymphocyte cell-surface proteins. The large expansion of IG domains in vertebrates shows the versatility of a single family in evoking rapid and effective response to infection.

Two prominent families are involved in the control of development. The human genome contains 30 fibroblast growth factors (FGFs), as opposed to two FGFs each in the fly and worm. It contains 42 transforming growth factor-βs (TGFβs) compared with nine and six in the fly and worm, respectively. These growth factors are involved in organogenesis, such as that of the liver and the lung. A fly FGF protein, branchless, is involved in developing respiratory organs (tracheae) in embryos. Thus, developmental triggers of morphogenesis in vertebrates have evolved from related but simpler systems in invertebrates.

Another example is the family of intermediate filament proteins, with 127 family members. This expansion is almost entirely due to 111 keratins, which are chordate-specific intermediate filament proteins that form filaments in epithelia. The large number of human keratins suggests multiple cellular structural support roles for the many specialized epithelia of vertebrates.

Finally, the olfactory receptor genes comprise a huge gene family of about 1,000 genes and pseudogenes. The number of olfactory receptors testifies to the importance of the sense of smell in vertebrates. A total of 906 olfactory receptor genes and pseudogenes could be identified in the draft genome sequence, two-thirds of which were not previously annotated. About 80% are found in about two dozen clusters ranging from 6 to 138 genes and encompassing about 30 Mb (~1%) of the human genome. Despite the importance of smell among our vertebrate ancestors, hominids appear to have considerably less interest in this sense. About 60% of the olfactory receptors in the draft genome sequence have disrupted ORFs and appear to be pseudogenes, consistent with recent reports suggesting massive functional gene loss in the last 10 Myr.

[. . .].

CONCLUSION

Five lines of evidence point to an increase in the complexity of the proteome from the single-celled yeast to the multicellular invertebrates and to vertebrates such as the human. Specifically, the human contains greater numbers of genes, domain and protein families, paralogues, multidomain proteins with multiple functions, and domain architectures. According to these measures, the relatively greater complexity of the human proteome is a consequence not simply of its larger size, but also of large-scale protein innovation.

An important question is the extent to which the greater phenotypic complexity of vertebrates can be explained simply by two- or threefold increases in proteome complexity. The real explanation may lie in combinatorial amplification of these modest differences, by mechanisms that include alternative splicing, post-translational modification and cellular regulatory networks. The potential numbers of different proteins and protein–protein interactions are vast, and their actual numbers cannot readily be discerned from the genome sequence. Elucidating such system-level properties presents one of the great challenges for modern biology.

[*Nature*, 409 (2001), 860–921.]

SEAN B. CARROLL

39 Genetics and the making of *Homo Sapiens*

Understanding the genetic basis of the physical and behavioural traits that distinguish humans from other primates presents one of the great new challenges in biology. Of the millions of base-pair differences between humans and chimpanzees, which particular changes contributed to the evolution of human features after the separation of the Pan *and* Homo *lineages 5–7 million years ago? How can we identify the 'smoking guns' of human genetic evolution from neutral ticks of the molecular evolutionary clock? The magnitude and rate of morphological evolution in hominids suggest that many independent and incremental developmental changes have occurred that, on the basis of recent findings in model animals, are expected to be polygenic and regulatory in nature.* [Author's summary.]

[. . .]

Genetics of human evolution

GENETIC ARCHITECTURE OF TRAIT EVOLUTION

Given the dimensions of hominin[1] evolution, inferred from the fossil record and comparative anatomy, what can we expect in terms of the genetic complexity underlying trait evolution? For example, there is a long-standing tendency for events that are perceived to be relatively 'rapid' in the fossil record to be ascribed to perhaps one or a few radical mutations, including recent human evolution. Could the relative increase in brain size over 5 Myr, or its expanded cognitive function, be due to just one or a few genetic changes? The best (and, at present, the only available) guides for this question are detailed genetic studies in model organisms, which have achieved success in dissecting the genetics of complex trait formation, variation and evolution. Six essential general concepts have been established in model systems that pertain to the potential genetic architecture of human trait evolution:

[1] Hominin refers to humans plus our immediate ancestors and their relatives, who are more closely related to us than to any other living species. That is, humans plus australopithecines. It is the part of the tree of life leading to us rather than the other living great apes.

(1) Variation in continuous, quantitative traits is usually polygenic.[2] Studies of variation in model species reveal that many genes of small effect, and sometimes one or a few genes of large effect, control trait parameters. In humans, a study of variation in 20 anthropometric variables in two different ethnic human populations suggested that more than 50% of variation was polygenic.

(2) The rate of trait evolution tells us nothing about the number of genes involved. Studies of artificial selection and of interspecific divergence indicate that the intensity of selection and heritability are more important determinants of evolutionary rate than is the genetic complexity of the traits under selection. There is considerable standing variation in traits, including characters that might be thought of as highly constrained, such as limb morphology in tetrapods. In general, the observed rates of evolution under natural selection are far slower than is potentially possible. Genetic variation or genetic complexity is not the limiting factor; indeed, considerable genetic variation underlies even phenotypically invariant traits. Because the rate at which a trait emerges in the fossil record tells us nothing about genetic architecture, the temptation to invoke macromutational models for 'rapid change' must be resisted in the absence of genetic evidence.

(3) Morphological variation and divergence are associated with genes that regulate development. Comparisons of the developmental basis of body-pattern evolution in animals suggest that morphological evolution is a product of changes in the spatiotemporal deployment of regulatory genes and the evolution of genetic regulatory networks. Developmental changes in the human lineage are expected to be associated with genes that affect developmental parameters, such as those that encode transcription factors and members of signal-transduction pathways.

(4) Mutations responsible for trait variation are often in non-coding, regulatory regions. When it has been possible to localize variation in genes that underlie phenotypic variation or protein-level differences, insertions or substitutions in regulatory regions and non-coding regions are often responsible.

(5) Multiple nucleotide replacements often differentiate alleles. Fine-scale analysis of quantitative trait loci has often revealed that functional differences between alleles are due to multiple nucleotide differences. It also indicates

[2] Polygenic means influenced by more than one gene. Continuous, quantitative characters are things like height, in which individuals differ by small amounts and the population has a continuous distribution of heights. Contrast this with discrete characters, such as eye colour or gender. Thus Carroll is opposing the view just stated, that human evolution proceeds by a few radical mutations. Later in the paragraph he refers to macromutations—this is another way of talking about radical mutations. On macromutations in adaptive evolution, see Section C, Chapter 14 by Fisher and Chapter 18 by Orr and Coyne.

that non-additive interactions between sites within a locus may be key to the differentiation of alleles, and that the contribution of any individual site may be modest (and difficult to detect).

(6) There is some concordance between genes responsible for intraspecific variation and interspecies divergence. Genetic analyses of interspecies divergence is only possible under certain circumstances, when laboratory breeding can overcome species barriers and traits can be mapped. In some cases, it has been found that some of the same loci are involved in both within-species variation and between-species divergence. This raises some hope that studies of intraspecific variation in humans could lead to genes that have been important in human history.

Since human trait evolution has followed a similar, incremental course as traits studied in model systems, these six concepts suggest that we should expect a highly polygenic basis for complex traits such as brain size, craniofacial morphology and development, cortical speech and language areas, hand and digit morphology, dentition and post-cranial skeletal morphology. We should also anticipate that multiple changes in non-coding regulatory regions and in regulatory genes are of great importance. But how can we find them?

THE ARITHMETIC OF HUMAN SEQUENCE EVOLUTION

All genetic approaches to human origins are fundamentally comparative, and seek to identify genetic changes that occurred specifically in the human lineage and contributed to the differentiation of humans from our last common ancestor with either apes or other species of *Homo*. Our primary comparative reference is the genome of the chimpanzee (*Pan troglodytes*), our closest living relative, with whom we share a common ancestor that lived 5–7 Myr ago. The arithmetic that sets the problem for human evolutionary genetics is as follows: first, the most extensive comparison of chimpanzee and human genomic sequences indicates an average substitution level of ~1.2% in single-copy DNA, second, the human genome comprises $\sim 3 \times 10^9$ base-pairs; third, it is reasonable to assume that one-half of the total divergence between chimpanzees and humans occurred in the human lineage (~0.6%); and fourth, this amounts to $\sim 18 \times 10^6$ base-pair changes. In addition, there are an unknown number of gene duplications and pseudogene, transposon and repetitive element changes in each lineage. A recent small-scale survey indicated that insertions and deletions (indels) might account for another 3.4% of differences between chimpanzee and human genomes, with the bulk of that figure contributed by larger indels.[3] A good deal of genomic change might be

[3] Carroll is referring to Chapter 55 in this anthology.

the noise of neutral substitutions and the gain and loss of repetitive elements over long time spans (more than 46% of human DNA is composed of interspersed repeats), but some small fraction of the changes in genomic sequence is responsible for the hereditary differences between species. The crux of the challenge is how to identify specific changes that are biologically meaningful from the many that are not.

In the case of human evolution, there are three basic genetic issues that we would like to grasp. First, how many genes were directly involved in the origin of human anatomy, physiology and behaviour (a few, dozens, hundreds or thousands)? Second, which specific genes contributed to the emergence of particular human traits? And third, what types of change in these genes contributed to evolution (for example, gene duplications, amino-acid replacements or regulatory sequence evolution)? In the few pioneering studies that are directly addressing the genetic basis of human-chimpanzee divergence, different but somewhat complementary strategies are being pursued that are beginning to reveal the scope of human genetic evolution and, in some cases, specific genes that might have been under selection in the course of recent human evolution.

COMPARATIVE GENOMICS

The most readily detected differences between animal genomes are expansions or contractions of gene families. Although the full chimpanzee genome is not yet available, a partial comparative map indicates that there are regions of the human genome that might not be represented in chimpanzees or other apes. Such regions could be due to duplications or insertions that occurred in the hominin lineage or to deletions in the chimpanzee lineage. One gene family, dubbed *morpheus*, underwent expansion as part of a segmental duplication on human chromosome 16. This expansion is shared by other great apes, but it seems that there were human lineage-specific duplications as well.

On the basis of comparisons with other genomes, particularly the recently reported draft mouse sequence, such lineage-specific duplications are expected. In the 75 Myr or more since the divergence of the common ancestor of mice and humans, several dozen clusters of mouse-specific genes arose that are generally represented by a single gene in the human genome. The shorter divergence time between humans and apes suggests that the human-specific gene set will be smaller. It is interesting to note that a significant fraction of the mouse gene clusters encode proteins with roles in reproduction, immunity and olfaction. This indicates that sexual selection, pathogens and ecology can shape the main differences in coding content between mammals. It should also be noted that 80% of mouse genes have a 1:1 orthologue in the human genome, and that more than 99% have some homologue. These figures and synteny data suggest that there is a gene repertoire that is qualitatively nearly identical among mammals. The presence or absence of particular gene duplicates might reflect adaptively driven change, but further evidence will be necessary to determine whether positive selection has acted on genes.

THOUSANDS OF ADAPTIVE CHANGES IN THE HUMAN PROTEOME?

The first place that adaptive genetic changes have been looked for is in the coding sequences for proteins. If the 18×10^6 substitutions in the human lineage are evenly distributed throughout the genome, only a small fraction will be expected to fall within coding regions. Assuming that the average protein is ~400 amino acids in length, and that there are ~30,000 protein-coding genes, only $\sim 3.5 \times 10^7$ base pairs (or a little more than 1.5% of the genome) consists of coding regions. So, assuming neutrality and ignoring the selective removal of deleterious changes in protein sequences, ~1.5% of these 18×10^6 substitutions (or 270,000 sites) may contribute to protein evolution. A fraction of these (roughly one-quarter) are synonymous substitutions, so the total number of amino-acid replacements in the human lineage could be of the order of ~200,000. This figure is in good agreement with observed average rates of amino-acid replacement in mammals.

Various methods have been developed to detect whether amino-acid replacements could be the result of positive selection—that is, adaptive evolution. To estimate the extent of positive selection in human protein evolution, Fay *et al.* surveyed sequence-divergence data for 182 human and Old World monkey genes, and polymorphism data for a similar number of human genes. Taking into consideration the frequency of common polymorphisms (ignoring rare alleles), a greater-than-expected degree of amino-acid replacements was observed, which is evidence of selection. When extrapolated to the entire proteome, 35% of amino-acid substitutions between human and Old World monkeys were estimated to have been driven by positive selection. Applied to human–chimpanzee divergence, this would extrapolate to ~70,000 adaptive substitutions in the human lineage. This figure is substantially larger than would be expected if most mutations were neutral or nearly neutral. If it is even the correct order of magnitude, it forecasts a nightmare for the identification of key genes under selection, because this figure suggests that, on average, two or more adaptive substitutions have occurred in every human protein in the last 5 Myr.

It is possible that the figure, based on the study of less than 0.5% of the human proteome, is an overestimate of the fraction or distribution of adaptive replacements. It is clear that some proteins are under strong pressure to remain constant, whereas others, especially those involved in so-called 'molecular arms races', are under pressure to change. For example, major histocompatibility complex proteins, which interact with diverse and changing foreign substances, show clear signatures of selection. Proteins involved in reproduction that play a part in sperm competition or gemete recognition also appear to evolve faster and under some degree of positive selection. A host of human male reproductive proteins have greater-than-average ratios of amino-acid replacements. Although accelerated protein evolution can also be the consequence of relaxed constraints, the correlation of higher levels of amino-acid replacements in proteins that have a role in reproduction and immunity seems to be biologically and selectively driven.

The population genetics- and protein-sequence-based statistical estimates of adaptive evolution require three caveats regarding how much they tell us about human evolution. First, there are generally no direct functional data that either test or demonstrate whether a human protein is indeed functionally diverged from an ape orthologue. Second, the proteins for which signatures of selection have been detected generally do not affect development. And third, the proteome is just part of the whole picture of genome evolution. Non-coding sequences, including transcriptional *cis*-regulatory elements, the untranslated regions of messenger RNAs, and RNA-splicing signals, contribute considerably to evolution by affecting the time, place and level of gene expression (see above). Ever since the pioneering comparative analysis of ape and human protein-sequence divergence nearly three decades ago,[4] it has generally been anticipated that changes in gene regulation are a more important force than coding-sequence evolution in the morphological and behavioural evolution of hominins.

[*Nature*, 422 (2003), 849–57.]

R. A. RAFF

40 Co-option of eye structures and genes

At any one time, in any one species, a certain gene codes for lens crystallin. But in all vertebrates many (otherwise unrelated) genes code for crystallin, and during evolution there are rapid changes in which gene serves as the crystallin gene in a lineage. [Editor's summary.]

The origin of lens proteins presents an amazing window on gene co-option and the fluidity of gene regulation.

Eye lenses are transparent and are filled with a high concentration of a small number of proteins called crystallins. The name suggests that lens proteins somehow form crystals. They do not. Delaye and Tardieu showed

[4] See King and Wilson, Chapter 54.

that the lens is essentially a fluid of nonattracting hard spheres. The crystallins need only be stable globular proteins. That very limited functional constraint has had unexpected evolutionary consequences. One of the major vertebrate crystallins, α-crystallin is a small heat-shock protein. A study by Horwitz shows that α-crystallin functions as a molecular chaperon and prevents heat denaturation of proteins. It may well have been selected for this function as well as for its role as a crystallin. The β- and γ-crystallins are related to microbial sporulation proteins.

These relationships of crystallins to other protein families show that lens proteins had their origins via gene duplication and divergence. In itself, this is hardly astonishing. The surprises came when the study of lens crystallin biochemistry was extended to a phylogenetically broader sample, and it transpired that there were a number of taxon-specific crystallins. These taxon-specific crystallins were identified: they were familiar enzymes. A massive co-option of genes as lens crystallins was discovered and it sheds light on how genomic fluidity is exploited by selection.

Crocodiles and a few birds use ε-crystallin, which turns out to be lactate dehydrogenase. Most birds and reptiles use δ-crystallin, arginionsuccinate lyase. Other taxon-specific crystallins (ζ, η) are identical to NADPH:quinone oxidoreductase and aldehyde dehydrogenase I, respectively. Still others are related to, but not identical to, several other common enzymes, including enolase. No selective advantage can be ascribed to using any particular one of these enzymes. If one maps the distribution of ε-crystallin on a phylogenetic tree of bird orders, it suggests that losses and possibly gains have occurred within lineages of birds. Idiosyncratic distributions of crystallin enzymes, such as the use of aldehyde dehydrogenase I by elephant shrews and the use of NADPH:quinone oxidoreductase in a few groups of mammals, show that crystallins have been recruited within the past 65 million years. The repeated co-option of metabolic enzymes as crystallins represents a remarkable transience in gene usage.

[*The Shape of Life* (Chicago: University of Chicago Press, 1996).]

STEVEN A. BENNER, M. DANIEL CARACO, J. MICHAEL THOMSON, AND ERIC A. GAUCHER

41 Planetary biology—paleontological, geological, and molecular histories of life

The history of life on Earth is chronicled in the geological strata, the fossil record, and the genomes of contemporary organisms. When examined together, these records help identify metabolic and regulatory pathways, annotate protein sequences, and identify animal models to develop new drugs, among other features of scientific and biomedical interest.

Together, planetary analysis of genome and proteome databases is providing an enhanced understanding of how life interacts with the biosphere and adapts to global change. [Authors' summary.]

A key goal for biology in the postgenomic era is to use the sequences of genes and proteins to generate information about molecular, cellular, and organismal biology. In the future as in the past, much of this information will undoubtedly be obtained through biochemical, genetic, and molecular biological experiments in the laboratory. This notwithstanding, almost any approach that provides inferences, insights, or information about biology from sequence data without requiring additional experimentation will be valuable.

For this reason, considerable attention has been directed toward the fact that biomolecular sequences contain information about their historical past. The search for homologs, or protein sequences that diverged from a common ancestor, is frequently the principal tool used to annotate sequence databases. Likewise, the sequences of a set of homologous proteins suggest a tree that defines the history of the protein family. These trees can be used to infer familial relationships between the organisms that carry the proteins, define the order in which particular taxa diverged, constrain the connectivity of the deep branches joining the primary kingdoms of life, and even correlate the divergence of species with their migrations across drifting continents.

This theme has been amplified by recognizing that two other fields, geology and paleontology, also provide records of the history of life on Earth. In many respects, these records complement the record contained in molecular sequence data. For this reason, considerable effort is now being directed toward explicit connection of these three records. Here, the past is the key to the present. By understanding the history by which a protein emerged within the context of its planetary biology, we hope to better understand how it functions in contemporary life.

Joining records through dating

It is not easy, of course, to connect records that lie within rocks and bones with a record that is captured in the sequences of organic molecules, proteins, and the DNA that encodes them. One way to do so is to use historical dates as the connector. Radiochemical dating, used to date events in the geological record, offers the gold standard. Radioisotopes decay via a first-order process. The amount of isotope remaining after time t follows a simple exponential rate law, with the fraction of initial atoms remaining $f = 1 - \exp(-kt)$, where k is the rate constant for the decay process, and the half-life $\tau = \ln 2/k$. Using two isotopes of uranium in a zircon from an igneous rock, for example, precision to better than a million years is routine for a rock 500 million years (Ma) old.

This precision is envied by those who date events in the paleontological and molecular records. In paleontology, the rate of morphological change cannot follow exponential kinetics, as it does not approach an end point. Rocks that contain fossils are occasionally associated with rocks that carry datable radioisotopes, of course. These permit accurate limits to be set for dates for specific fossils in specific strata. Unfortunately, the fossil record is incomplete, meaning that we rarely (if ever) date fossils that represent branch points in a phylogenetic tree, or the first appearance of a species. This incompleteness creates large uncertainties in dates at nodes in trees, even if we have good dates for the rocks that contain the fossils.

Dating events in the molecular record is still more problematic. In the 1970s, many groups explored the possibility of constructing a molecular clock by counting replacements separating two sequences and assuming that the rate constant for amino acid replacement is invariant over time. Unfortunately, protein behaviors are too closely tied to the demands and constraints of natural selection. Amino acid replacement is faster or slower depending on how these change, making protein sequences irregular clocks at best. Collections of protein sequence families might be used to date the divergence of taxa, in the hope that this episodic rate variation averages over the collection to give an apparently time-invariant rate constant. But individual protein sequences generally serve as poor clocks, and it is difficult to correlate the molecular record for a specific protein family with the paleontological record.

To minimize the effects of selective pressure as they construct molecular clocks, most workers now examine silent sites in a gene.[1] Because they do not change the coding sequence of a protein, nucleotide substitutions at silent sites cannot alter the behavior of a protein. They are therefore most likely to be free of the selective pressures that cause protein clocks to 'tick' irregularly.

Silent sites are of many types in the standard genetic code. Some offer better clocks than others. Most useful are silent sites in codon systems that are twofold redundant. Here, exactly two codons encode the same amino acid. These codons are interconverted by transitions: a pyrimidine replacing another pyrimidine, or a purine replacing another purine. When the amino

[1] Silent sites are also called synonymous sites: they are sites in the DNA where a change in the DNA's own base sequence does not change the amino acid that is coded for. They exist because the genetic code has 64 codons, but only 20 amino acids are coded for.The approach taken in this paper concentrates on amino acids that are coded for by exactly two codons, differing by a transition. A transition is a change between *A* and *G* or between *C* and *T*. Here the simplifying assumption is made that the evolutionary changes in the DNA changes are only back and forth between the pair of codons. Other approaches more fully reconstruct the possible changes between codons, in the phylogenetic tree.

acid itself is conserved, the divergence at such sites can be modeled as an 'approach to equilibrium' kinetic process, just as radioactive decay, with the end point being the codon bias b. Here, the fraction of paired codons that are conserved $f_2 = b + (1 - b)\exp(-kt)$, where again k is the first-order rate constant and t is the time. Given an estimate of the rate constant k for these 'transition-redundant approach-to-equilibrium' processes, if k and b are time-invariant, one can estimate the time t for divergences of the two sequences. An empirical analysis suggests that codon biases and rate constants for transitions has been remarkably stable, at least in vertebrates, for hundreds of millions of years. Therefore, approach-to-equilibrium metrics provide dates for events in molecular records within phyla, especially of higher organisms. These dates are useful to time-correlate events in the molecular record with events in the paleontological and geological records.

Identifying pathways from genomic records

Simultaneous events need not, of course, be causally related, especially when simultaneity is judged using dating measurements with variances of millions of years. But observation that two events in the molecular record are nearly contemporaneous suggests, as a hypothesis, that they might be causally related. Such hypotheses are testable, often by experiment, and are useful because they focus experimental work on a subset of what would otherwise be an extremely large set of testable hypotheses.

Consider, for example, the yeast *Saccharomyces cerevisiae*, whose genome encodes ~6000 proteins. The yeast proteome has 36 million potentially interacting pairs. Some investigators in the field of systems biology are laboring to experimentally examine all of these in the laboratory, hoping to identify these interactions.

Correlating dated events in the molecular record offers a complementary approach. Gene duplications generate paralogs, which are homologous proteins within a single genome. Paralogous sequences can be aligned, their f_2 calculated, and their divergence dated. In yeast, paralog generation has occurred throughout the historical past. A prominent episode of gene duplication, however, is found with an f_2 near 0.84, corresponding to duplication events that occurred ~80 Ma, based on clock estimates that generated divergence dates in fungi. These duplications created several new sugar transporters, new glyceraldehyde-3-phosphate dehydrogenases, the nonoxidative pyruvate decarboxylase that generates acetaldehyde from pyruvate, a transporter for the thiamine vitamin that is used by this enzyme, and two alcohol dehydrogenases that interconvert acetaldehyde and alcohol.

This is not a random collection of proteins. Rather, these proteins all belong to the pathway that yeast uses to ferment glucose and produce ethanol. Correlating the times of duplication of genes in the yeast genome has identified a pathway.

The approach-to-equilibrium dating tools can be more effective at inferring possible pathways from sequence data than can other approaches, especially for recently evolved pathways. By adding the geological and paleontological records to the analysis, however, these pathways assume additional biological meaning. Fossils suggest that fermentable fruits also became prominent ~80 Ma, in the Cretaceous, during the age of the dinosaurs. Indeed, overgrazing by dinosaurs may explain why flowering plants flourished. Other genomes evidently also record episodes of duplication near this time, including those of angiosperms (which create the fruit) and fruit flies (whose larvae eat the yeast growing in fermenting fruit).

Thus, time-correlation among the three records connected by approach-to-equilibrium dates generates a planetary hypothesis about function of individual proteins in yeast, one that goes beyond a statement about a behavior ('this protein oxidizes alcohol . . .') and a pathway ('. . . acting with pyruvate decarboxylase . . .') to a statement about planetary function ('. . . allowing yeast to exploit a resource, fruits, that became available ~80 Ma'). This level of sophistication in the annotation of a gene sequence is difficult to create in any other way.

Approach-to-equilibrium dating methods are limited by the fact that silent clocks 'tick' fast, meaning that information about ancestral species is rapidly lost. In vertebrates, a typical single lineage rate constant for silent substitution is 3×10^{-9} changes/site/year. This corresponds to a half-life of about 260 million years, which provides a timescale where this tool is optimal for dating. Various approaches are conceivable to extend dates obtained from these methods back to perhaps 500 Ma in vertebrates.

Such dates prove to be interesting for biomedical research. Virtually all function peculiar to vertebrates and their associated diseases arose in the past 500 million years, including cardiovascular disease, inflammation, auto-immune diseases, pain and other neurological disorders, and certain cancers. In each, targets must be identified, animal models chosen, research directions tested, function assigned, and pathways explicated. Correlating the three records generates understanding of physiology, function, and disease that directs this effort.

Resurrecting ancient proteins from extinct organisms

The emergence of angiosperms is only one of many changes in the history of the biosphere that left a record in the genomes of the surviving organisms. Consider, for example, the Oligocene epoch, which began ~35 Ma. The dinosaurs were extinct by the start of the Oligocene, mammals had come to occupy the ecological niches that the dinosaurs had previously enjoyed, and Earth was largely a tropical rain forest. During the Oligocene, however, the planet began to cool; its mean temperature dropped by perhaps 15°C. Various explanations for the cooling have been proposed, including changes in the position of continents and seaways that separate them, changes in atmospheric carbon dioxide concentrations, cosmological events (supernovae or impacts), and changes in the planet's orbit.

For whatever reason, the succulent vegetation of the rain forest was replaced by low-nutrition, silica-containing grasses over much of the planet. The large herbivorous mammals responded. Nonruminant herbivores (the brontotheres, hyracodons, and others) came to be displaced by ruminant mammals (camels, deer, and bovids), perhaps because of the efficiency with which ruminants digest grass. Ruminants have a first stomach that holds bacteria that ferment grass. The bacteria from the first stomach are then 'eaten' in the second stomach, where their cell walls are broken by lysozymes, their nucleic acids are degraded by ribonucleases, and their proteins are digested by proteases.

Each of these protein families suffered duplications and/or rapid evolution near the time that ruminant digestion originated. To analyze these events, the sequences of ancestral proteins at nodes in their evolutionary trees are reconstructed, by inference from descendant sequences and using models for how protein sequences diverge in general. This process is much like the process used by historical linguists when reconstructing the Proto-Indo-European language from its descendant languages.

Explicit reconstructions of evolutionary intermediates assign specific amino acid replacements to specific episodes in the history of a protein family. In ancestral lysozyme genes, for example, rapid sequence evolution occurred as ruminant and ruminant-like digestion emerged. Rapid change in the sequence of a lysozyme implies rapid change in the behavior of lysozyme, which in turn suggests a change in its functional behavior. This hypothesis is inferential, of course, but can be tested. Further, it makes sense in light of a historical model. New lysozymes are expected to emerge to break open bacterial cells in the new ruminant digestion.

One way of testing such hypotheses is to resurrect the ancestral proteins and study their behavior in the laboratory. To do this, a DNA molecule encoding the ancestral protein is synthesized and expressed in an appropriate host. The ancient protein is then recovered and studied to determine whether its properties are consistent with its inferred ancestral role.

A paleobiochemical experiment was done for ruminant digestive ribonucleases, which also suffered gene duplication and an episode of rapid sequence evolution in the Oligocene. Laboratory studies on ancient ribonucleases found that the substrate specificity and stability of the emerging ruminant ribonuclease changed in this episode in a way consistent for a protein being recruited to play a new role in the digestive tract. These data added a planetary dimension to the annotation of the ribonuclease protein. Rather than saying that 'ribonuclease is involved in ruminant digestion,' we can say that digestive ribonuclease emerged near the time when ruminant digestion emerged, in animals in which ruminant digestion developed, at a time where difficult-to-digest grasses emerged, permitting their descendants to exploit a newly available resource emerging at a time of global climatic upheaval. [. . .]

Pharmacological models and biomedical research

A historical approach to contemporary biology is relevant to biomedical research, in part because it helps define the function of proteins. In general, proteins are assumed to have the same function as their homologs, a view that assumes that function is conserved when protein sequences diverge. This assumption has long been known to be poor in many proteins, including many characterized before the genomic era. When it fails, it can mislead the biomedical researcher.

A historical analysis based on reconstructed proteins can alert the scientist to the possibility of changing function. When function in a protein family changes, a signature, remains within the genome record. Many tools read these signatures, including those that exploit ancestral reconstructions as discussed above. In addition, changing function leaves many traces in the molecular record, including episodes of rapid sequence evolution between reconstructed ancestral sequences, changes in residues that appear to be more easily replaceable, increases or decreases in the amount of parallel and convergent evolution, and changes in the amount of compensatory convariation.

For example, the leptin gene in the mouse, when deleted, produced an obese rodent. The homolog for leptin was found in humans, where it was suggested to be involved in human obesity. This transfer of annotation is consistent with annotation strategies used generally in proteomics. Examining the reconstructed history of the leptin family in mammals suggested, however, an episode of rapid sequence evolution in the lineages following the divergence of rodents and primates. This suggests that in the small mammals that diverged and gave rise to apes and humans, mutated forms of leptin conferred more fitness than unmutated forms. The functional behavior of leptin must have changed in some way during this time. For the pharmacologist seeking a human antiobesity drug, this suggests (at the very least) that preclinical testing should be done in primate models rather than rodent models. [. . .]

Planetary annotation

These examples represent only pieces of a much larger puzzle, one that will have global implications when assembled. Correlation of events in the molecular, paleontological, and geological records, and the molecular dissection of historical events occurring in the past, offers a paradigm to dissect the function of the planetary proteome. Earth, after all, has only one history. It carries a finite number of species (perhaps between 1 and 2 million). When all of their genomes are sequenced, the planetary proteome will (according to current guesses) be composed of fewer than 10^5 easily recognized modules, independent units of protein sequence evolution. Consequently, one can imagine a comprehensive model of life on Earth, combining paleontology, geology, chemistry, molecular biology, structural biology, systems biology, and genomics, that captures history and function from the molecule to the planet. [. . .]

[*Science*, 296 (2002), 864–8.]

Section G

The history of life

Biologists are interested in a number of questions about the history of life. One is phylogenetic. The aim is to reconstruct the tree of life: to find out, for each taxonomic group, which other taxonomic group it shares its most recent common ancestor with. Are humans more closely related to chimpanzees, or to gorillas? Are crocodiles more closely related to birds, or to lizards? It is interesting in itself to know the relations of shared ancestry between living things. Moreover, accurate phylogenetic knowledge is needed for many further kinds of research.

A second aim is to describe the great events in the history of life. In the nineteenth century, research on fossils established that the history of life has a series of distinct phases. Three were originally distinguished (Palaeozoic, Mesozoic, Cenozoic), from about 540 million years ago to the present. Further research has added a number of earlier 'Precambrian' phases between about 4000 and 540 million years ago. The fossil fauna changes between these phases. For example, the Mesozoic was (simplifying somewhat) the age of reptiles, for vertebrate life on land, whereas the Cenozoic is the age of mammals. Biologists are interested to know the dates of these historical events, and to understand why they occurred.

This section has extracts on four topics. All four are more in the nature of case studies, than general theories. However, they illustrate several theoretical approaches. The first is on the origin of life. Maynard Smith and Szathmáry (Chapter 42) have written two books about the major transitions of evolution. Their major transitions are not for particular taxonomic (or near-taxonomic) groups, such as reptiles and mammals. The transitions are more abstract, for life forms such as replicating molecules, single cells, multicellular life, and creatures with development. In each transition, there was a change in the way in which heredity occurred. The transition, in turn, made possible the future evolution of more complex life forms (though those future events were not the reason why the transition occurred). I have extracted their discussion of the earliest transition, from the non-living to the first replicating molecular systems. They include an account of Eigen's important thinking on error limits. The final transition in Maynard Smith and Szathmáry's scheme is the origin of human language, a topic discussed by Pinker (Chapter 59) though not in the same way as Maynard Smith and Szathmáry.

Schopf (Chapter 43) summarizes the evidence for the remarkable absence of evolutionary change in early microbial life. He also makes the stimulating suggestion that evolution in the Precambrian had a different mode from the Cambrian through to the present: that evolution in the earlier period lacked the rounds of extinction and tree-like radiation of discrete species lineages that are typical of the past 600 or so million years.

Cooper and Fortey (Chapter 44) look at a kind of debate that has become quite common recently. We have two methods for reconstructing the history of life: one based on molecules and the other on fossils. In some cases, the two kinds of evidence disagree. Two examples are the Cambrian 'explosion' of about 540 million years ago, and the radiation of birds and mammals in the early Cenozoic 50–60 million years ago. The Cambrian explosion, in the fossil record, looks like the time when the main animal groups originated. But molecular evidence suggests they had a much earlier origin. Likewise, in the fossil record, mammals and birds proliferate only after the extinction of the dinosaurs; but molecular evidence suggests an earlier origin for the main mammal and bird groups. The conflict may mean that one (or both) kinds of evidence is wrong. Cooper and Fortey suggest a reconciliation in terms of a 'phylogenetic fuse'. Groups may proliferate only a long time after they originate. Molecules date the origin, and fossils date the proliferation, of each taxon; the two methods are not looking at the same events. In Section I, on human evolution, Sarich and Wilson's paper (Chapter 53) inspired the first and most famous 'fossil versus molecules' controversy, concerning the time of human origins. In that case, however, a phylogenetic fuse reconciliation is inapplicable. Molecules suggested a more recent time of human origins than was inferred from fossils.

Dilcher's paper (Chapter 45) is about the history of flowering plants (that is, angiosperms). He identifies a series of major events in the rise of flowering plants to their current dominance. He explains these events mainly in terms of coevolution with animals. Janzen (Chapter 49) also looks at angiosperm–animal coevolution, though in a different way. Not everyone agrees that coevolution with animals was so important in angiosperm evolution, but Dilcher represents one of the main (and perhaps the largest) schools of thought. It would be desirable to have a similar paper for all the main groups of living creatures. However, papers as clear and modern as Dilcher's are rare, and his work on flowering plants will have to stand, in this anthology, for a style of understanding that is applicable to all life.

JOHN MAYNARD SMITH AND EÖRS SZATHMÁRY

42 From chemistry to heredity

Some evidence suggests that life could have originated in a primitive soup; a charged surface may provide more plausible conditions. Simple test-tube systems with molecular replication have been devised. The length of a replicating molecule is limited by its accuracy of replication. If the molecule is too long relative to its copying error rate, an error threshold is crossed and the system becomes unsustainable. [Editor's summary.]

The last sentence of Darwin's *Origin of species* reads 'There is a grandeur in this view of life, with its several powers, being originally breathed by the Creator into a few forms or into one; and that, while this planet has gone cycling on according to the fixed law of gravity, from so simple a beginning endless forms most beautiful and most wonderful have been, and are being evolved'. Curiously, the phrase 'by the Creator' crept in only in the second edition: it was absent in the first. It seems that Darwin, perhaps to please his wife Emma, was willing to leave the problem of the origin of life, a problem he had little prospect of making progress with, as a matter for religious interpretation. In contrast, the problem of the origin of man was one he believed he could usefully think about.

The primitive soup

The first serious scientific attacks on the origin of life came from the Russian biochemist A. I. Oparin in 1924 and from J. B. S. Haldane in 1929. They argued that, if the primitive atmosphere lacked free oxygen, a wide range of organic compounds might be synthesized, using energy from ultraviolet light and lightning discharges. The absence of oxygen was important, because otherwise any organic compounds formed would rapidly be oxidized to carbon dioxide (CO_2) and water (H_2O). Haldane suggested that, in the absence of living organisms to feed on the organic compounds, the sea could have reached the consistency of a hot dilute soup.

In 1953, Stanley Miller, on the advice of Harold Urey, tested the idea by passing an electric discharge through a chamber containing water, methane (CH_4), and ammonia (NH_3). The results were dramatic. In these and similar experiments, a wide range of organic compounds have been produced, including many of the amino acids of which proteins are made, a variety of sugars, and various purines and pyrimidines (components of the nucleotides of which RNA and DNA are made).

Although these experiments were enormously encouraging, there were snags. Some essential molecules were present in low concentrations, or altogether absent. The sugar ribose, which forms the backbone of RNA and DNA, is produced, but the yield is low. Long-chain fatty acids, needed to form biological membranes, were absent. Perhaps more fundamental is the question of how these simple organic molecules could be strung together to form biological polymers. For example, proteins consist of strings of amino acids joined by a particular chemical bond, the peptide bond. Non-chemists can think of a protein as a poppet necklace, with different kinds of beads joined together in a precise order by a particular kind of junction. Sidney Fox found that by heating and drying a mixture of amino acids, and then dissolving the mixture in water, he could obtain strings of amino acids displaying weak catalytic activity. Unfortunately, however, the amino acids were linked together in a variety of different ways, and not only by peptide bonds. Similar difficulties arise for a second class of polymer, the nucleic acids. RNA and DNA are polymers formed of nucleotides. Even if nucleotides could be synthesized in Miller-type experiments, and this is not easy, it is not clear how they could be linked in the appropriate way. If some of the linkages were of the wrong kind, replication would be blocked.

To summarize, there is no difficulty in seeing how a wide range of organic compounds could have been formed abiotically, including many of the compounds important in modern organisms. But there is a lack of specificity in the reactions that take place, and a particular difficulty in understanding how polymers (proteins, nucleic acids), linked by specific chemical bonds, could have been formed.

The primitive pizza

A possible way out of these difficulties has recently been suggested by Günter Wächtershäuser, a somewhat unusual contributor to the field. Although he has a doctorate in chemistry, he is not working in an institute or university. He is a patent lawyer for chemical patents in Munich. (It is amusing that Einstein was also working at a patent office when he developed the special theory of relativity.) Before Wächtershäuser's first publication on the origin of life in the late 1980s, he had been strongly influenced by the philosophy of Karl Popper. This has led him to take the idea of testable hypotheses seriously, and to be disgusted by the, sadly rather common, theoretical sloppiness in this rather unconventional field.

The idea is that reactions may have taken place between ions bonded to a charged surface (for non-chemists, an ion is an electrically charged atom or molecule). For example, in the solid state common salt is not electrically charged, but in solution in water the molecules break up into positively charged sodium ions and negatively charged chlorine ions. Many important organic compounds ionize in solution: an example is the phosphate ion, PO^{3-}_{4}, which is important both in nucleic acids and in energy metabolism. Because unlike charges attract one another, ions in solution will become attached to charged surfaces. They are then free to move slowly across the surface, while maintaining a constant orientation. This would greatly increase both the speed and the specificity of the chemical reactions occurring. Because of the importance of negatively charged ions, such as the phosphate ion, in biochemistry, Wächtershäuser suggests that the relevant surface was the positively charged iron pyrites, or fool's gold.

Wächtershäuser has suggested some very specific chemical reactions that could have taken place; his ideas now need to be tested experimentally. One reason why surface bonding is important is that the molecules are held in a particular orientation, and are free to move only in a single dimension. If some of them were upside down, or all of them were free to move in three dimensions, they would never link together. Binding to a surface would also increase the local concentration of interacting molecules, and so speed up the reaction. Equally important, it would ensure that reacting molecules were held in a given orientation relative to one another, and so increase the specificity of the reactions.

The origin of replication

The first artificial replicator, not needing enzymes for its replication, was synthesized by K. von Kiedrowski in 1986. The object that was replicated was a small piece of DNA, of six base pairs: the 'units' from which new copies were made were two single-stranded molecules, each consisting of three bases linked end to end. This was an important achievement, but not yet a solution to our problem. It is a system with limited heredity, whereas we need a system with unlimited heredity. In all probability, this means a system replicating by homologous base pairing. As yet nobody has devised such a system, composed of monomers that could have arisen in the absence of living organisms and able to replicate without enzymes.

The simplest existing replicating system is shown in Figure 42.1. This is a real example of evolution in a test-tube, with no cells present. RNA molecules appear that are better and better at getting themselves replicated in this environment. It is a beautiful demonstration of the power of natural selection to generate adaptations that would never appear by chance alone. Given the right circumstances, the end point of evolution is a unique RNA molecule 235 bases long, regardless of the starting sequence. But despite this demonstration of the power of natural selection to generate unlikely adaptations, the example is in a sense cheating, if we want to explain the origin of the first replicators. The system works only because the right monomers, and more important, a complex enzyme, the Q_β replicase, are supplied. No such enzyme could have existed on the primitive Earth, before the origin of life.

One difficulty in understanding the non-enzymatic replication of a nucleic acid-like molecule lies in explaining the specificity with which the units, the monomers, are linked together. If they are linked by the wrong chemical bonds, replication will be blocked. At present, the best hope seems to lie in seeking a polymer with a chemically simpler backbone than RNA, thus reducing the number of ways in which the monomers can be linked. The extra specificity to be gained from metabolism on a surface may also help.

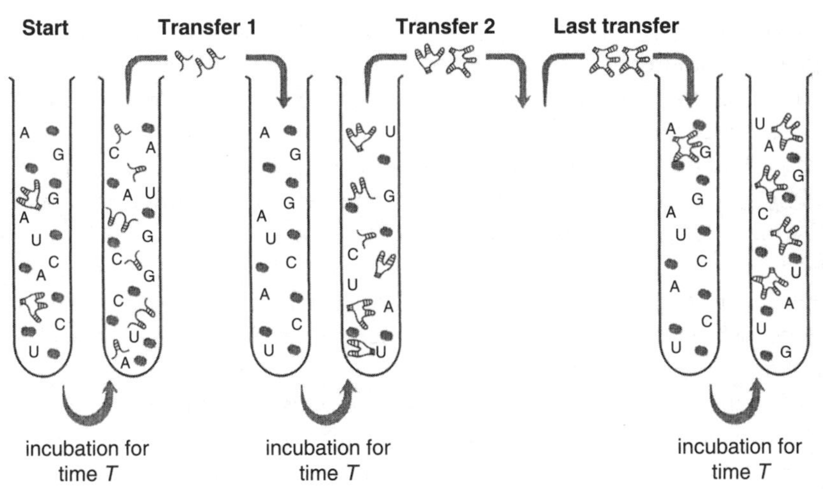

Figure 42.1: Evolution in a test-tube. The initial tube contains the four nucleotides from which RNA is synthesized, a replicase enzyme that copies RNA, and a 'primer' RNA molecule. This primer molecule is copied many times, with some errors. After incubation for time T, *a drop of solution is transferred to a new tube containing nucleotides and enzyme (but not, of course, any primer). The process of incubation and transfer can be repeated indefinitely, and evolutionary changes in the population of RNA molecules observed.*

For the present, we have to accept that the origin of molecules with unlimited heredity is an unsolved problem. But progress is rapid. Despite the last sentence of the *Origin of species*, we cannot advise creationists to put their faith in the belief that only God could create a molecule with unlimited heredity.

The accuracy of replication and the error threshold

Replication is not perfect. If it were, there would be no variation for selection to act on. But initially the problem would have been too much mutation, and not too little. Most mutations reduce fitness. Selection is therefore needed to maintain a meaningful message. The old game of Chinese whispers demonstrates that, without selection, the result is chaos. How accurate must replication be? Imagine a message—for example, a DNA molecule—that replicates to produce two copies of itself. The two copies replicate to produce four, and so on. During replication, miscopying occurs, and the erroneous copies that result are eliminated by selection. Only perfect copies survive. It is clear that, after each copying, at least one copy on average must be perfect. Otherwise selection cannot maintain the integrity of the message. This places an upper limit on the permissible mutation rate per base copied, or, equivalently, an upper limit on the length of the message, for a given mutation rate.

If the genome size, or the mutation rate per symbol, rises above this critical upper limit, the result is an accumulation of mutated messages. This is what Manfred Eigen and Peter Schuster have called the 'error threshold'. It is easy to see roughly where this upper limit lies. The requirement is that at least one perfect copy, on average, must be made at each replication. If there are n symbols, this means, approximately, that the probability of an error when replicating a symbol must be not greater than $1/n$. In other words, if the genome contains 1000 bases, the mutation rate per base, per replication, must be not greater than 1/1000.

The error rate in experiments of the kind illustrated in Fig. 42.1 is in the range 1/1000 to 1/10 000. This would permit a genome between 1000 and 10 000 bases. But this involves replication by an enzyme; if there is no enzyme, the error rate is much higher. Figure 42.2 shows an experiment by Leslie Orgel. In this experiment, an 'error' consists of a base other than C pairing with a G in the template. The error rate depends on the medium, the temperature, and so on, but very roughly the wrong base pairs with a G once in 20 times. This implies that, before there were specific enzymes, the maximum size of the genome was about 20 bases.

At first sight, this is a serious difficulty, and so it was long regarded. It presented a kind of catch-22 of the origin of life. Without a specific enzyme, the genome size is limited to about 20 bases; but with a mere 20 bases one cannot code for an enzyme, let alone the translating machinery needed to convert the base sequence into a specific protein.

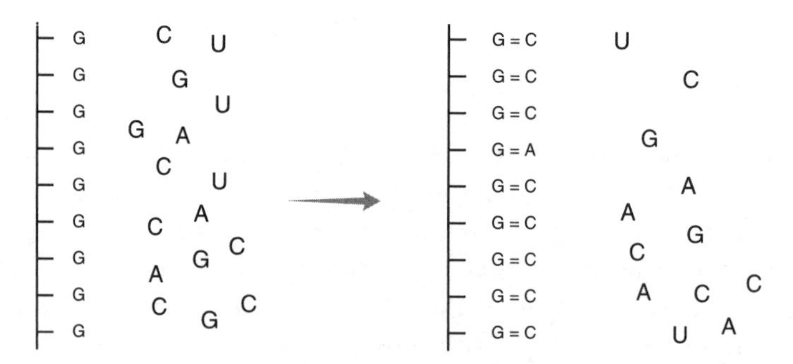

Figure 42.2: An experiment in base pairing. An RNA strand consisting entirely of a string of Gs is placed in a solution containing the four bases that constitute RNA: that is, A, C, G, and U. As expected, Cs pair with the Gs, but an occasional error occurs: in the figure, an A is shown pairing with one of the Gs. This pairing does not depend on the presence of any specific enzyme. The Cs, however, are not linked together to form a new strand: this does require an enzyme.

Escape from this dilemma came from an unexpected source. It turns out that one does not need to have proteins and translating machinery to have enzymes. RNA molecules can themselves be enzymes. The significance of this discovery is discussed in the next chapter. In essence, the first RNA molecules did not need a protein polymerase to replicate them; they replicated themselves. But before we leave the problem of replication accuracy, there is another question to answer. We said that, in the presence of an enzyme, the error rate can be reduced to 1/1000–1/10 000. This would suggest an upper limit to genome size of about 10 000 bases. Yet, even when 'nonsense DNA' has been discounted, animals and plants have genomes of 10^9 to 10^{10} bases. How can that be?

In producing this book, the compositor will no doubt make some typographical errors. A proof copy will then be sent out, and we will find most of the mistakes and ask that they be corrected. No amount of proof-reading will reduce the errors in the final version to zero, if only because new errors can happen when correcting old ones; but the number of mistakes will, we hope, be less than it was in the first version. Exactly the same is true of DNA replication in higher organisms (bacteria and upwards). After the first enzyme-catalysed replication, there are in fact two error-correcting stages, called, reasonably, proof-reading and mismatch repair. This is possible because each DNA copy consists of an original strand and a new, copied strand. Enzymes in the cell check the base pairing between new and old strands, and if there is a mismatch—that is, a non-complementary pair—alter the new strand to match the old one. After two proof-reading stages, the error rate is reduced to one in 10^9 or less. Biologists are still arguing about whether this is as low as natural selection can make it without excessive cost, or whether it is an optimal compromise between the need to reduce the frequency of harmful mutations, and the desirability of producing an occasional good one.[1]

[*The origins of life* (Oxford: Oxford University Press, 1999), 32–6.]

J. WILLIAM SCHOPF

43 Disparate rates, differing fates: tempo and mode of evolution changed from the Precambrian to the Phanerozoic

Precambrian prokaryotes show a pattern of evolution that contrasts with the relatively familiar pattern in multicellular eukaryotes for the past 550 million years. Multicellular eukaryotic evolution shows extinctions, tree-like evolutionary radiations, and bursts of

[1] In Chapter 52 Sniegowski *et al.* discuss the evolution of mutation rates, including exactly this issue.

evolutionary change. Precambrian prokaryotes have low rates of evolution and extinction; some Precambrian fossil prokaryotes are indistinguishable from prokaryotes alive on Earth today. [Editor's summary.]

When G. G. Simpson wrote *Tempo and Mode,*[1] fossil evidence of the history of life consisted solely of that known from sediments of the Phanerozoic eon, the most recent 550 million years (Ma) of geologic time. Thus, Simpson's views of the evolutionary process were based necessarily on Phanerozoic life—the familiar progression from seaweeds to flowering plants, from trilobites to humans—a history of relatively rapidly evolving, sexually reproducing plants and animals successful because of their specialized organ systems (flowers, leaves, teeth, limbs) used to partition and exploit particular environments. In short, Simpson elucidated 'normal evolution' played by the 'normal rules' of the game—speciation, specialization, extinction.

Although certainly applicable to the megascopic eukaryotes of the Phanerozoic, there is reason to question whether these well-entrenched rules apply with equal force to the earlier and very much longer Precambrian phase of microbe-dominated evolutionary history. In place of sexual multicellular plants and animals, the biota throughout much of the Precambrian was dominated by simple nonsexual prokaryotes. Rather than evolving rapidly, many Precambrian microbes evidently evolved at an astonishingly slow pace. And instead of having specialized organ systems for exploitation of specific ecologic niches, members of the most successful group of these early-evolving microorganisms—photoautotrophic cyanobacteria[2]—were ecologic generalists, able to withstand the rigors of a wide range of environments. In contrast with normal evolution, the 'primitive rules' of prokaryotic evolution appear to have been speciation, generalization, and exceptionally long-term survival.

That there is a distinction in evolutionary tempo and mode between the Phanerozoic and Precambrian histories of life is not a new idea, but it is one that has recently received additional impetus and therefore deserves careful scrutiny. However, evaluation of this generalization hinges critically on the quality and quantity of the fossil evidence available, and because active

[1] G. G. Simpson, *Tempo and Mode in Evolution* (New York: Columbia University Press, 1944). Schopf's paper was a contribution to a colloquium marking 50 years of research following Simpson's classic publication.

[2] Cyanobacteria are sometimes called blue-green algae, though they are not algae (some cyanobacteria when growing in ponds at high density cause a blue-ish green colour in the water, resembling an algal growth). Cyanobacteria can give rise to structures called stromatolites, found in the fossil record and today; fossil evidence suggests that cyanobacteria were abundant throughout the world for much of geological history; but they are now found in more localized habitats.

studies of the Precambrian fossil record have been carried out for little more than a quarter century a thorough comparison of the early history of life with that of later geologic time is not yet possible. Therefore, as a first approximation, the approach used here is to analyze the known fossil record of Precambrian cyanobacteria: well-studied, widespread, abundant, commonly distinctive, and evidently dominant members of the early prokaryotic biota. Microscopic fossils regarded as members of other prokaryotic groups are also known from the Precambrian, but their documented record is minuscule. Hence, conclusions drawn here about the early fossil record apply strictly to free-living cyanobacteria, the evolutionary history of which may or may not be representative of prokaryotes in general. To evaluate the generalization, two central questions must be addressed. First, was the tempo of Precambrian cyanobacterial evolution markedly slower than that typical of Phanerozoic eukaryotes? Second, if so, how can this difference be explained?

Tempos of evolution

In *Tempo and Mode*, Simpson coined terms for three decidedly different rate distributions in evolution, inferred from morphological comparisons of Phanerozoic and living taxa: tachytelic, for 'fast'-evolving lineages; horotelic, the standard rate distribution, typical of most Phanerozoic animals; and bradytelic, for 'slow' morphological evolution. Included among the bradytelic lineages are so-called living fossils (such as linguloid brachiopods, horseshoe crabs, coelacanth fish, crocodilians, opossums), 'groups that survive today and show relatively little change since the very remote time when they first appeared in the fossil record'.[3] Simpson's bradytely closely approximates Ruedemann's earlier developed concept of 'arrested evolution', both based on comparison of modern taxa with fossil forms that are virtually indistinguishable in morphology but are 100 Ma or more older.

HYPOBRADYTELY

Recently, a fourth term—hypobradytely—has been added to this list of rate distributions 'to refer to the exceptionally low rate of evolutionary change exhibited by cyanobacterial taxa, morphospecies that show little or no evident morphological change over many hundreds of millions of years and commonly over more than one or even two thousand million years'.[4] The

[3] Simpson (n. 1 above), 125.

[4] J. W. Schopf, in J. W. Schopf and C. Klein (eds.), *The Proterozoic Biosphere* (New York: Cambridge University Press, 1992), 556.

concept can be applied to cyanobacteria because the morphologic descriptors and patterns of cell division used to differentiate taxa at various levels of the taxonomic hierarchy are preservable in ancient sediments; as emphasized by Knoll and Golubic,[5] 'Essentially all of the salient morphological features used in the taxonomic classification of living cyanobacteria can be observed in well-preserved microfossils.' [. . .]

CAVEATS

Application of the concept of hypobradytely is not without potential pitfalls, three of which deserve particular mention. First, because of the enormous span of Precambrian time, and despite the notable paleontological progress of recent years, early biotic history is as yet very incompletely documented. In comparison with the vastly better documented record of Phanerozoic organisms—and even in geologic units of the Proterozoic (2500–550 Ma in age), by far the most studied portion of the Precambrian—the known cyanobacterial fossil record is scanty (for filamentous species amounting to ≈ 21 taxonomic occurrences per 50-Ma-long interval and, for spheroidal species, ≈ 46 occurrences). Second, assignment of some Precambrian microbial fossils (evidently $<10\%$) to the cyanobacteria can be quite uncertain, a problem that applies especially to minute morphologically simple forms (atypically small-diameter oscillatoriaceans and chroococcaceans, for example), which in the fossil state are essentially indistinguishable from various noncyanobacterial prokaryotes. Third, lack of change in the external form of morphologically simple prokaryotes may not necessarily reflect evolutionary stasis of their internal physiological machinery (the so-called Volkswagen syndrome). This last problem is especially difficult to evaluate, but it may not be of overriding importance—from early in the Proterozoic to the present, the same cyanobacterial families and many of the same (morphologically defined) genera and even species appear to have inhabited the same or closely similar environments, patterns of distribution that 'provide proxy information on physiological attributes.'[6]

Cyanobacterial hypobradytely

Are cyanobacteria hypobradytelic? Data bearing on this question fall into two classes. First, a large amount of evidence indicates that the proposition is plausible, in fact likely to be correct, but, because of a lack of accompanying relevant (chiefly environmental) information, this evidence is not fully

[5] A. H. Knoll and S. Golubic, in M. Schidlowski *et al.* (eds.) *Early Organic Evolution* (New York: Springer, 1992), 453.

[6] Ibid. 451.

compelling. Second, a small number of in-depth studies firmly support the proposition, but because of their limited number and restricted taxonomic scope these studies do not establish the proposition generally. Considered together, however, the two classes of data present a strong case.

EVIDENCE OF PLAUSIBILITY

Shown in Figure 43.1 are paired examples of morphologically comparable Proterozoic and living cyanobacteria including specimens illustrating the rather commonly cited similarity between fossil *Palaeolyngbya* and *Lyngbya*, its modern morphological counterpart. Numerous other genus- and species-level comparisons have been drawn [for example, between ≈ 850-Ma-old *Cephalophytarion grande* and the modern oscillatoriacean *Microcoleus vaginatus*; and between ≈ 2-billion year (Ga)-old *Eosynechococcus moorei* and the living chroococcacean *Gloeothece coerulea* (*Gloeobacter violaceus*)]. Examples such as these—and the fact that over the past quarter century such similarities have been noted repeatedly and regarded as biologically and taxonomically

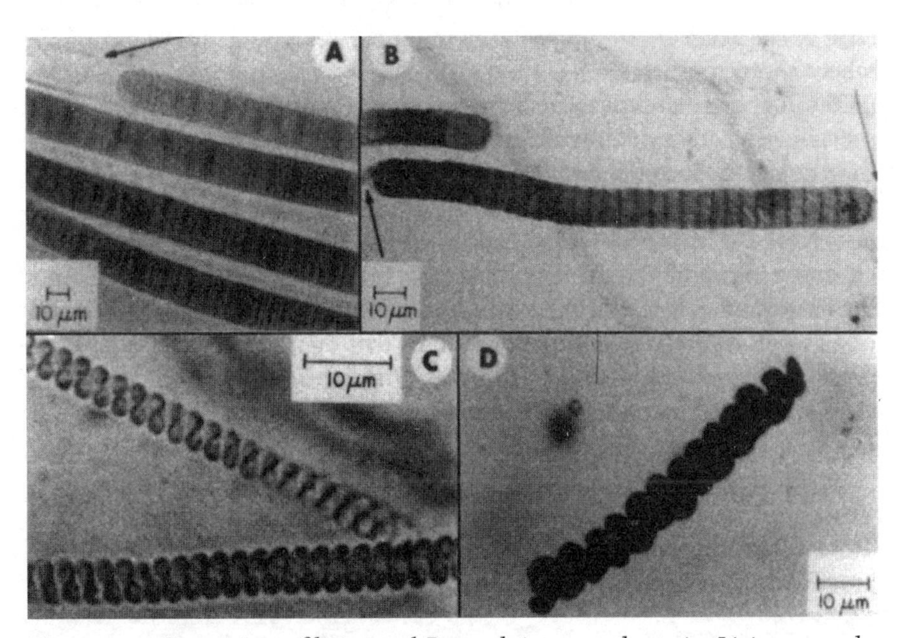

Figure 43.1: Comparison of living and Precambrian cyanobacteria. Living examples (A and C) are from mat-building stromatolitic communities of northern Mexico. (A) Lyngbya *(Oscillatoriaceae), encompassed by a cylindrical mucilaginous sheath (arrow). (B)* Palaeolyngbya, *similarly ensheathed (arrows), from the ≈* 950-Ma-old Lakhanda Formation of eastern Siberia. *(C)* Spirulina *(Oscillatoriaceae). (D)* Heliconema, *a* Spirulina-*like cyanobacterium from the ≈* 850-Ma-old Miroedikha Formation of eastern Siberia.

significant by a large number of workers in many countries—provide a powerful argument for the plausibility of cyanobacterial hypobradytely.

Plausibility of the concept is similarly indicated by quantitative studies recently carried out on large assemblages of modern and Precambrian microbes. Morphometric data (for such attributes as cell size, shape, and range of variability; colony form; sheath thickness and structure) were compiled for 615 species and varieties of living cyanobacteria as well as for an extensive worldwide sample of Proterozoic cyanobacterium-like microfossils, both filamentous (650 taxonomic occurrences in 160 geologic formations) and spheroidal (1400 occurrences in 259 formations). To avoid confusion stemming from variations in taxonomic practice, fossils having the same or similar morphology (regardless of their binomial designations) were grouped together as informal species-level morphotypes designed to have ranges of morphologic variability comparable to those exhibited by living cyanobacterial species. Of the 143 informal species of filamentous microfossils thus recognized, 37% are essentially indistinguishable in morphology from established species of living (oscillatoriacean) cyanobacteria. Similarly, 25% of the 120 informal taxa of spheroidal fossil species have modern species-level (largely chroococcacean) morphological counterparts. Virtually all of the fossil morphotypes are referable to living genera of cyanobacteria, and the patterns and ranges of size distribution exhibited by taxa of cylindrical sheath-like Proterozoic fossils are essentially identical to those of the tubular sheaths that encompass trichomes of modern oscillatoriacean species. [. . .]

IN-DEPTH STUDIES

In addition to studies of cellular morphology and likely (but not firmly established) broad-scope environmental comparisons, what is needed to move the hypobradytelic hypothesis from the plausible to the compelling are supporting data on the paleoenvironment and taphonomy of the fossils in question. A number of such in-depth studies have been carried out, focusing on fossil representatives of two cyanobacterial families, the Entophysalidaceae and the Pleurocapsaceae, members of which are decidedly more distinctive morphologically than are the oscillatoriaceans and chroococcaceans discussed above. Golubic and Hofmann compared ≈ 2-Ga-old *Eoentophysalis belcherensis* with two modern entophysalidaceans (*Entophysalis major* and *Entophysalis granulosa*). They showed that not only are the fossil and modern species morphologically comparable (in cell shape and in form and arrangement of originally mucilaginous cellular envelopes) and that they exhibit similar frequency distributions of dividing cells and essentially identical patterns of cellular development (resulting from cell division in three perpendicular planes), but also that both taxa form microtexturally similar stromatolitic structures in comparable intertidal to shallow marine environmental settings, that they undergo similar postmortem degradation sequences, and that they occur in microbial communities that are comparable in both species composition and biological diversity. [. . .] Several species of fossil and living pleurocapsaceans have also been compared in detail. [. . .]

These in-depth studies of entophysalidaceans and pleurocapsaceans—involving analyses of environment, taphonomy, development, and behavior, in addition to cellular morphology—provide particularly convincing evidence of species-specific fossil-modern similarities.

CYANOBACTERIA ARE HYPOBRADYTELIC

Thus, numerous workers worldwide have noted and regarded as significant the detailed similarities in cellular morphology between Precambrian and extant cyanobacteria. A substantial fraction of known Proterozoic oscillatoriacean and chroococcacean cyanobacteria have living species-level morphological counterparts, and almost all such fossils are referable to living cyanobacterial genera. And in-depth studies of several fossil-modern species pairs of morphologically distinctive entophysalidaceans and pleurocapsaceans permit detailed comparison of morphology, development, population structure, environment, and taphonomy, all of which show that for at least these taxa, ancient and modern cyanobacteria are essentially indistinguishable in salient characteristics.

Taken together, these observations support an obvious conclusion—the morphology (and evidently the physiology as well) of diverse taxa belonging to major cyanobacterial families evolved little or not at all over hundreds of millions, indeed thousands of millions of years. In comparison with the later history of life, this widespread hypobradytely is surprising. In Phanerozoic evolution, bradytelic stasis is notable principally because of its rarity, but in the Precambrian it seems to have been a general phenomenon characteristic of a group of prokaryotic microorganisms that dominated the Earth's biota, possibly even as early as 3.5 Ga ago. Why have cyanobacteria evidently changed so little over their exceedingly long evolutionary history?

Survival of the ecologically unspecialized

To understand the underlying causes of cyanobacterial hypobradytely, it is instructive to review Simpson's thoughtful analysis in *Tempo and Mode*, for although he was unaware of the Precambrian prokaryotic fossil record, Simpson was much interested in slowly evolving (bradytelic) Phanerozoic lineages. In addition to noting (but dismissing) the possibility that 'asexual reproduction (as inhibiting genetic variability)' might be conducive to slow evolution,[7] he singled out two principal factors: large population size, and ecologic versatility, an exceptional degree of adaptation 'to some ecological position or zone with broad . . . selective limits . . . a particular, continuously available environment'. Because unusually slow evolution involves 'not only exceptionally low rates of [evolutionary change] but also survival for extraordinarily long periods of time', and because 'more specialized phyla tend to become extinct before less specialized', Simpson proposed 'the rule of the survival of the relatively unspecialized'.

Although intended by Simpson to apply to Phanerozoic organisms, chiefly animals, these same considerations (with the addition of asexual reproduction) apply to Precambrian cyanobacteria. First, with regard to reproduction, cyanobacteria are strictly asexual, lacking even the parasexual processes known to occur in some other prokaryotes. Given the remarkable longevity of the cyanobacterial lineage and moderate or even low rates of mutation, however, the absence of sexually generated genetic variability cannot be the sole explanation for their hypobradytely. Second, like virtually all free-living microorganisms, cyanobacteria typically occur in local populations of large size. Coupled with their ease of dispersal (via water currents, wind, and hurricanes, for example) and for many species a resulting very wide (essentially cosmopolitan) geographic distribution, their large populations can also be presumed to have played a role in their evolutionary stasis. Third, and probably most important, however, is the ecologic versatility of the group.

Summarized in Table 2[8] are known ranges of survivability (and of growth under natural conditions) for modern oscillatoriaceans and chroococcaceans, the most primitive and commonly occurring Precambrian cyanobacterial families. Similar tolerance is also exhibited by members of other cyanobacterial families. For example, a nostocacean was revived after more than a century of storage in a dried state and a scytonematacean is reported to have maintained growth at pH 13. Thus, cyanobacteria exhibit notable ecologic flexibility, and even though no single oscillatoriacean or chroococcacean species is known to be capable of tolerating the total range of observed growth conditions (for example, thermophiles dominant in 70°C waters rarely grow below 50°C, and species adapted to highly alkaline lakes do not occur in acid hot springs), both groups include impressive ecologic generalists, able to thrive in virtually all present-day widespread environments. Moreover, many of the oscillatoriacean and chroococcacean genera for which wide ecologic tolerance has been demonstrated are the same as those having species-level Precambrian-extant counterparts. Finally,

[7] Simpson (n. 1 above), 137.

[8] The Table is not included here.

numerous cyanobacteria, including both oscillatoriaceans and chroococcaceans, are capable of fixing atmospheric nitrogen; provided with light, CO_2, a source of electrons (H_2, H_2S, H_2O), and a few trace elements, such cyanobacteria are highly effective colonizers, able to invade and flourish in a wide range of habitats.

The wide ecologic tolerance of cyanobacteria is almost certainly a product of their early evolutionary history. Fossil evidence suggests that oscillatoriaceans and chroococcaceans were extant as early as $\approx$ 3.5 Ga ago. If so, they must have originated and initially diversified in an oxygen-deficient environment, one lacking an effective UV-absorbing ozone layer. In such an environment, the ability to photosynthesize at low light intensities coupled with the presence of gas vesicles to control buoyancy would have permitted planktonic cyanobacteria to avoid deleterious UV by inhabiting the deep oceanic photic zone, just as *Synechococcus* does today. Similarly, numerous characteristics of living benthic mat-building cyanobacteria—effective DNA repair mechanisms, synthesis of UV-absorbing scytonemin, secretion of copious extracellular mucilage, phototactic motility, adherence to substrates, stromatolitic mat formation—initially may have been adaptations to cope with a high UV flux in near-shore shallow water settings. Adaptive radiation in an early oxygen-deficient environment is also suggested by the ability of cyanobacteria to live in either the presence or absence of oxygen, their capability to switch between oxygenic and anoxygenic photosynthesis, the occurrence of oxygen-sensitive nitrogenase in many taxa, and the restriction of nitrogenase-protecting heterocysts to late-evolving members of the group. In addition, both the low affinity of cyanobacterial ribulose-bisphosphate carboxylase for CO_2 and the presence of intracellular CO_2-concentrating mechanisms may reflect initial adaptation of the lineage to a CO_2-rich primordial environment.

Finally, the remarkable hardiness of cyanobacteria—their ability to survive wide ranges of light intensity, salinity, temperature, and pH as well as prolonged desiccation and intense radiation—may be a product of their marked success in competing for photosynthetic space with other early-evolving microbes. Unlike the oxygen-producing photosynthesis based on chlorophyll *a* in cyanobacteria, that in all other photoautotrophic prokaryotes is anoxygenic and bacteriochlorophyll based. Because biosynthesis of bacterio-chlorophyll is inhibited by molecular oxygen, oxygen-producing cyanobacteria would have rapidly supplanted oxygen-sensitive anoxygenic photoautotrophs throughout much of the global photic zone. As a result of the ease of their global dispersal and their success in competing for photosynthetic space, cyanobacteria presumably expanded into a broad range of habitable niches during an early, evidently rapid phase of adaptive radiation, evolving to become exceptional ecologic generalists. Thus, the ecologic versatility of cyanobacteria appears to hark back to an early stage of planetary history when they established themselves as the dominant primary producers of the Precambrian ecosystem.

In view of their evolutionary history, it is perhaps not surprising that Simpson's rule of survival of the (ecologically) relatively unspecialized is applicable to cyanobacteria, numerous taxa of which qualify as so-called living fossils. According to Stanley, such extraordinarily long-lived organisms 'are simply champions at warding off extinction'.[9] If so, as has been previously suggested, the 'grand champions,' over all of geologic time, must be the hypobradytelic cyanobacteria!

A bipartite view of the history of life

In broadbrush outline, biotic history thus seems divisible into two separate phases, each characterized by its own tempo and mode, each by its own set of evolutionary rules.

During the shorter more recent Phanerozoic eon, the history of life was typified by the horotelic evolution of dominantly megascopic, sexual, aerobic, multicellular eukaryotes based on alternating life cycle phases specialized either for reproduction or for nutrient assimilation. Changes in the dominant (commonly diploid) phase resulted chiefly from structural modification of organ systems used to partition and exploit particular environments. In large part as a result of this ecologic specialization, the Phanerozoic was punctuated by recurrent episodes of extinction, each followed by the adaptive radiation of surviving lineages.

In contrast with Phanerozoic evolution, much of the earlier and decidedly longer Precambrian history of life was typified by the hypobradytelic evolution of dominantly microscopic, asexual, metabolically diverse, and commonly ecologically versatile prokaryotes, especially cyanobacteria. Evolutionary innovations were biochemical and intracellular. Once established, lineages exhibited long-term stasis. Extinction occurred rarely among prokaryotic ecologic generalists, evidently becoming a significant evolutionary force only late in the Precambrian and primarily affecting ecologically relatively specialized, large-celled eukaryotic phytoplankters.

Although as yet incompletely documented, this bipartite interpretation of evolutionary history seems consistent with the fossil record as now known.

[*Proc. of the National Academy of Sciences USA*, 91 (1994), 6735–42.]

[9] S. M. Stanley, in S. M. Stanley and N. Eldredge (eds.), *Living Fossils* (New York: Springer, 1984), 280.

ALAN COOPER AND RICHARD FORTEY

44 Evolutionary explosions and the phylogenetic fuse

A literal reading of the fossil record indicates that the early Cambrian (c. 545 million years ago) and early Tertiary (c. 65 million years ago) were characterized by enormously accelerated periods of morphological evolution marking the appearance of the animal phyla, and modern bird and placental mammal orders, respectively. Recently, the evidence for these evolutionary 'explosions' has been questioned by cladistic and biogeographic studies which reveal that periods of diversification before these events are missing from the fossil record. Furthermore, molecular evidence indicates that prolonged periods of evolutionary innovation and cladogenesis lit the fuse long before the 'explosions' apparent in the fossil record. [Authors' summary.]

George Gaylord Simpson, in *Tempo and Mode of Evolution*, made a pioneering attempt to describe patterns of speciation as deduced from the fossil record. Numbers of species within major groups of organisms clearly fluctuated throughout their history. It was evident that there were times when many new major taxa appeared over a geologically short time period; other times when a background rate of origination and extinction prevailed. One of the patterns Simpson identified was a rapid evolutionary 'burst' which was preceded in some cases by an attenuated 'tail'—a greatly extended, but often obscure earlier history before the group of organisms in question expanded into prolific evolutionary creativity. Probably the best example of this pattern was provided by the mammals and birds. Both groups have a prolonged but poorly known period of evolution during the Mesozoic when many 'archaic' groups are observed—the birds exemplified by arguably the most famous fossil of them all, the Jurassic *Archaeopteryx*. After the disappearance of the dinosaurs at the end of the Cretaceous, the Palaeocene and Eocene Epochs saw the appearance of many of the ancient sister-groups of the kinds of mammals and birds that still prosper today—and many more extinct kinds besides. Stanley subsequently showed that the rate of mammalian evolution during the early Tertiary was, indeed, exceptionally fast, and it has become customary to describe such phases as 'radiations', or even as 'explosions'—an analogy of unbridled creation inevitably invoking chain reactions. The first appearance of many animal phyla in the Cambrian is often described in similar terms as the Cambrian 'evolutionary explosion'.

Although such evolutionary radiations have been extensively studied, the earlier histories that preceded them have not received quite the same attention. There has been an assumption that the first appearance of familiar groups in the fossil record does closely approximate their time of origin—that is, the rocks provide a correct narrative of the time of first appearance of taxa. On this account, 'true' birds are largely a post-Cretaceous radiation, and the Cambrian was a time when phyla originated. Thus, a literal reading of the fossil record endorses Simpson's notions of special periods of phylogenetic creativity, though no genetic mechanism has been proposed that adequately explains the rapid rate of morphological evolution. Recently, there have been reasons to examine these assumptions. Well-supported phylogenetic trees derived from a combination of classical morphology and molecular sequences have revealed cases of obvious deficiencies in the fossil record. For example, among the mammals the egg-laying monotremes occupy a relatively basal position—yet their pre-Tertiary fossil record is comprised of only a few fragments. Similarly, the diplopods are often accorded a basal position in arthropod evolution, yet their fossil record commences 100 million years after other arthropod groups. If the record of these groups is so deficient, why should we assume that the record is adequate for other kinds of animals, particularly those less likely to be fossilized, such as the small and rare ancestral bird and mammals? What is evidently needed is a test of the divergence times of major groups which does not solely rely upon the fossil record.

New evidence from molecular data

Thanks to the increased availability of DNA sequence information, molecular tests of distant divergence events are becoming common. Divergence dates from molecular data are generally expected to be earlier than those from palaeontological data because they simply estimate the point at which genetic intermixing ceases while fossils record when a species has developed diagnostic characters, as well as attained fossilization. However, the extent of this discrepancy, as revealed by molecular studies, has been particularly surprising. It is important to note that molecular analyses often lack rigorous tests of lineage-specific rates of change or accurate confidence limits, and make assumptions about clock-like behaviour in the genetic changes that have occurred. However, estimates of key divergence dates, using different genes, calibration points, and methods, are encouragingly similar (Table 44.1). They provide evidence that the explosive phases of evolution so amply demonstrated in the fossil record may, in many cases, have been preceded by an extended period of inconspicuous innovation. Maybe these preliminary phases permitted fundamental structural alterations—key innovations—which could only 'explode' subsequently once conditions became suitable. Suggestions that such a scenario applies to two very different parts of the geological column—the Cambrian 'explosion' 545 million years ago, and the diversification of birds and mammals 65 million years ago—lend credence to the idea that the pattern might prove to be more general.

Table 44.1: Molecular date estimates of key taxonomic divergences associated with the Cambrian and early Tertiary explosive radiations[a]

DIVERGENCE	CALIBRATION (MILLION YEARS)	TYPES OF SEQUENCES	ESTIMATED DATE (MILLION YEARS)[b]
K-T boundary			
Therian mammals			
Earliest ordinal divergences	Range of Palaeozoic/Mesozoic fossils	α and β haemoglobin aa	160–190
	Range of Palaeozoic/Mesozoic fossils	α and β haemoglobin aa	143 ± 20[c]
	Cetacea–artiodactyl split = 60 My	11 mitochondrial	130 ± ?
	Range of primate splits	14 nuclear	220–250
Rodents versus primates/ferungulates	Marsupial–eutheria split = 130 My	8 mitochondrial	114 ± 15
	Diapsid–synapsid split = 310 My	47 nuclear	113 ± 9
	Cetacea–artiodactyl split = 60 My	11 mitochondrial	118 ± ?
	Range of primate splits	57 nuclear	115–129
Primates versus ferungulates	Marsupial–eutheria split = 130 My	8 mitochondrial	93 ± 12
	Diapsid–synapsid split = 310 My	18 nuclear	90 ± 8
	Cetacea–artiodactyl split = 60 My	11 mitochondrial	90 ± ?
Modern birds			
Earliest ordinal divergences	Rhea–ostrich split = 80 My	DNA–DNA hybridization	120–132[c]
	7 early Tertiary fossils	1 mt	70–150
	269 Tertiary fossil calibrated quartets	1 mt, 1 nuclear	120–140
4 basal avian orders	Diapsid–synapsid split = 310 My	16 nuclear	97 ± 12
Cambrian boundary			
Metazoan phyla			
Protostoma versus deuterostoma	Range of Palaeozoic/Mesozoic fossils	α and β haemoglobin aa	1050–1100
	Jawless fish–higher vertebrates = 460 My	α and β haemoglobin aa	900–1000+
	Range of Palaeozoic/Mesozoic fossils	4 mt, 3 nuclear	1100–1250
Echinodermata versus vertebrata	Range of Palaeozoic/Mesozoic fossils	4 mt, 3 nuclear	900–1100

[a] My = million years, mt = mitochondrial DNA, nuclear = nuclear DNA, aa = amino acids.
[b] Errors are routinely underestimated.
[c] Calculated from calibration points and data given in references.

The Cambrian evolutionary explosion

The base of the Cambrian marks not only the appearance in the historical record of animal phyla, but also of many of their component major classes. There is no question that this time marked the nearly simultaneous appearance of hard shells (and hence commonly-preserved fossils) in many animal groups, but this may be decoupled from the appearance of the phyla and classes themselves. Among the arthropods, for example, there are Cambrian records of crustaceans, and stem group arachnoids, as well as the extinct trilobites; recently there has been discovered an unexpected variety of lobopods (cousins of the living velvet worm, *Peripatus*), which most authorities regard as basal relatives of the rest of the arthropods. These were accompanied by a whole series of curious arthropods—notably those preserved in the celebrated Middle Cambrian Burgess Shale, but augmented by discoveries in even earlier strata in China, Australia and Greenland. Cladistic analyses of the phylogenetic relationships of these distinctive fossil arthropods in conjunction with living species have revealed that, for all their peculiarities, most of them lay within stem lineages leading to living animals. Even the strangest animal of all—the giant predator *Anomalocaris*—emerged as close to a primitive sister group of the rest of the arthropods. However, the fact that by the basal Cambrian all these highly varied animals had diverged from a common ancestor, necessitating a number of sequential major evolutionary steps, implies an early, Precambrian history of diversification of which the fossil record has yielded no trace (Fig. 44.1). For example, even at their first appearance the trilobites were already differentiated into separate geographic faunas—surely this required an earlier phase of diversification of the clade. If the times taken for the inevitably even earlier separation of the arthropods from their neighbouring phyla—and those phyla, in turn, from one another—are added into the equation then the circumstantial evidence of an early history of phylogenesis seems persuasive, in spite of an absence of obvious 'ancestors' among the soft-bodied fossils of the late Precambrian Vendian fauna.

But how long was this period of Precambrian diversification? It has been argued that the time immediately prior to the appearance of skeletons, and of other fossils of advanced metazoans in the geological record, must have been one of immensely accelerated evolutionary activity—indeed, if the absence of 'ancestors' in the fossil record of the late Precambrian truly records history then such an accelerated rate of genesis *must* have been the case—however far the morphological distance required to be covered. In the absence of fossil evidence then, the only independent test of timing is provided by the genetic changes recording distant divergences as they are preserved in the genome. This is far from a simple calibration, because after such a long time many subsequent sequence changes will serve to add 'noise'

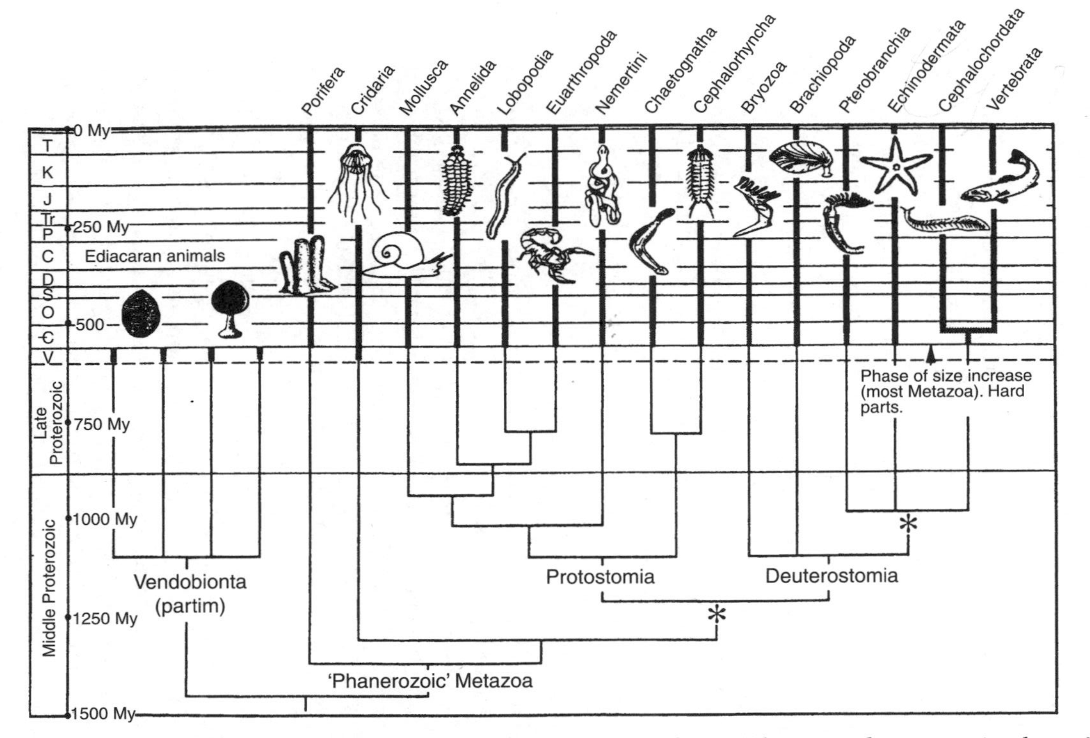

Figure 44.1: Molecular, cladistic, and biogeographic evidence indicates that the Cambrian evolutionary 'explosion' may have been underpinned by a long period of Precambrian diversification. The undoubted diversification at the base of the Cambrian accompanied the acquisition of hard parts—and increase in size and fossilization potential of 'Phanerozoic style' metazoans—but the ground plans of these clades probably had a much longer history than has been claimed. Thick solid lines: known range. [. . .]. Thin solid lines: implied pattern of Precambrian cladogenesis. Animals minute, asterisks denote divergences dated by molecular data. Illustration by M.A. Wills.

which may overwrite the true temporal signal. However, Wray and his colleagues have succeeded in providing molecular evidence of divergence times between the animal phyla, which indeed suggest that the important splits were well before the base of the Cambrian—perhaps more than twice as long ago (Fig. 44.1). They used evidence from the differences in sequences of seven different genes—whose products have very different physiological functions and, they claim, are unlikely to be affected by the same selection pressures. Each gene was independently calibrated, and yet produced encouragingly similar divergence time estimates, while even the comparatively wide margins of error cited all lie within the Precambrian domain. Wray *et al.* also controlled for accelerated rates of molecular evolution in Cambrian metazoa using a relative rates test of the divergences between prokaryotes and either metazoan or non-metazoan taxa. Lastly, factors which were not controlled for, such as variation in the rates of evolutionary change between sequence positions, are actually likely to lead to the underestimation of deep divergence dates, possibly by a considerable margin. Other genetic studies support the result by demonstrating that common developmental pathways were already in place well before the Cambrian. Therefore the molecules consistently support the phylogenetic evidence of 'ghost lineages'—there was a long prehistory of phylogenesis which was decoupled from the Cambrian 'explosion' in fossils.

While recent reports of Precambrian metazoa and trace fossils are a promising sign, why are other obvious 'ancestors' absent from the Precambrian rocks? After all, not only body fossils, but also trace fossils—the scratches and tracks left by the passage of animals through sediment—rapidly increased near the base of the Cambrian. One possibility is that the Precambrian precursors were animals of small size (a few millimetres long) and the Ediacaran animals occupied the large-sized ecological niches, effectively excluding these early metazoans from attaining large size. It is salutary to remember that small animals, be they ever so numerous (aphids, copepods and nematodes would be the most obvious living examples) are virtually unknown as fossils; had the late Precambrian been teeming with such organisms they may have left no trace in the rocks. The Cambrian 'explosion' might, therefore, record rather an increase in size, which was evidently accompanied by the secretion of shells independently in several lineages. The trigger for this transformation—which may have been connected with changes in the atmosphere, or oceanic geochemistry—is currently the subject of intense scrutiny.

The origin of modern bird and placental mammal orders

The history of birds and placental mammals presents many interesting similarities. Modern forms are abundantly represented in the favourable depositional environments of the late Palaeocene/early Eocene, but

Mesozoic bird and mammal fossils are relatively rare and mostly represent 'archaic' forms (e.g. multituberculates and enantiornithines) not closely related to their modern counterparts. However, recent fossil discoveries have started to reveal tantalizing evidence of a Cretaceous diversification. Unexpectedly advanced birds have been found in Chinese and Spanish deposits only slightly younger than *Archaeopteryx*, emphasizing that features associated with sophisticated flight were already present in the early Cretaceous, and that a large swathe of early avian history must therefore be absent from the fossil record. Similarly, recent fossil mammal discoveries have extended the range of orders such as whales and elephants back towards the K–T boundary, and in the case of ungulates, apparently well beyond it. Most modern placental and bird orders are now known from either the late Cretaceous (primates, insectivores, ungulates, loons, ducks, cormorants, and seabirds) or the Palaeogene. Furthermore, phylogenetic analyses of molecular and morphological data (Fig. 44.2) have identified several derived groups (e.g. seabirds, penguins, whales and elephants) which are already present and biogeographically distributed in the earliest Tertiary, supporting the fragmentary Cretaceous fossils. The existence of separate orders, including derived forms, at such an early stage strongly suggests a prolonged, but unobserved, prior stage of diversification.

So how long before the Tertiary did placental mammals and modern birds start to diversify? Molecular sequences have been used to augment the patchy Cretaceous terrestrial fossil record and provide a temporal scale for avian and placental evolution. Studies using long sequences have shown that considerable phylogenetic signal exists between the various orders, which challenges the idea of rapid, star-like radiations in the early Tertiary. In fact, molecular clock estimates consistently place the initial divergences within placental mammals and birds in the middle Cretaceous, some 50–60 million years before they appear, fully diversified, in the early Tertiary. The estimates produce encouragingly similar results despite the use of both Mesozoic and Tertiary fossil calibration points, differing calibration methods, and nuclear and mitochondrial sequences (Table 44.1). They suggest that, as in the Cambrian, a surprisingly long period of early diversification is absent from the known fossil record. Nevertheless, it is important to remember that in both cases the implied period of earlier phylogenesis would be overestimated if rates of molecular evolution were considerably accelerated during large radiations. To minimize this problem, molecular clock studies use conserved sequences from 'housekeeping' genes, because they should be less effected by changes in bodyplans, and test for rate constancy using relative rate tests. These tests reveal that molecular rates in groups such as mammals, birds, and reptiles would need to simultaneously decelerate by a factor of 10–15 after the early Tertiary to explain the molecular data, which seems unlikely. Nevertheless, more studies of

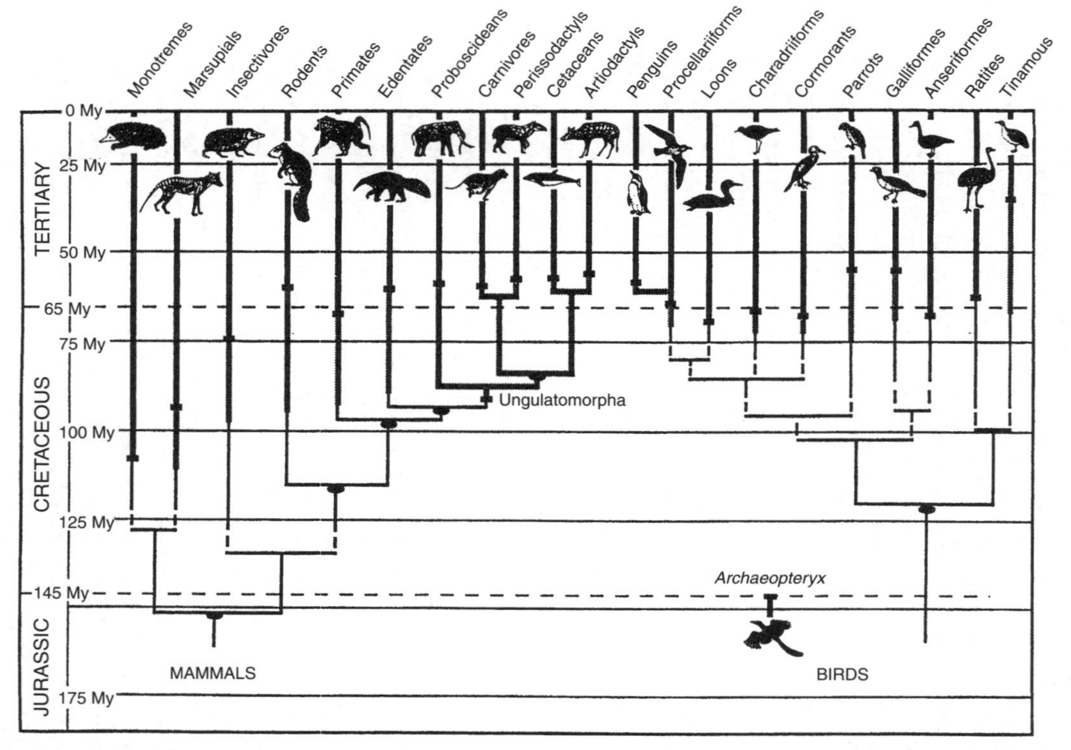

Figure 44.2: Phylogenies of modern mammals and birds plotted against time, based on morphological and molecular data. The mammalian phylogeny is supported by recent research using long molecular sequences. Cladistic and biogeographical data indicate the Early Tertiary 'explosion' of mammalian and avian diversity was preceded by a period of unobserved diversification, which molecular data suggests was surprisingly extensive. Thick solid lines: known range. [. . .] Thin solid lines: ranges calculated from molecular data. Ovals: divergences dated by molecular data. Dotted lines: not accurately dated.

molecular evolution during radiations are required to refine divergence date estimates.

Why then are there not more Cretaceous fossils of ancestral placentals and modern birds? For a start, the earliest differentiation amongst placental (and perhaps avian) taxa may not have been accompanied by recognizable skeletal synapomorphies[1], complicating the identification of ancestral forms. Furthermore, much of the Cretaceous terrestrial vertebrate fossil record is poor and is less likely to contain small-bodied, fragile-boned taxa, especially if they were not common over a geographically widespread area. The Cretaceous niches were dominated by dinosaurs and various reptile groups, as well as ancestral birds and mammals, and it is possible that modern birds and placental mammals were unable to dominate an ecological niche and increase in size and number. Consequently, their habitat and limited body and population sizes may have contributed to a poor fossil record. Lastly, palaeogeographical data suggest that the poorly known Cretaceous environment of the Southern Hemisphere may have been an important centre for terrestrial vertebrate radiation. However, if the molecular data are correct, one would have to predict that it is only a matter of time before a modern bird or placental mammal is discovered in the mid to early Cretaceous.

The phylogenetic fuse

Evolution has proceeded by innovation on the one hand and radiation on the other; and the two are not necessarily coupled. The fastest radiations from the recent past—the insect and bird endemics which evolved on Hawaii over the past five million years or so would be an example—have produced striking opportunistic adaptations but not, apparently, fundamentally new designs. Conversely, there are examples of major *Baupläne* such as represented by phyla today—chaetognaths or echiuroids are two examples—which despite a long history of morphological distinctiveness have never enjoyed a spectacular radiation after the manner of arthropods, molluscs or vertebrates. Key structural innovations do not guarantee a subsequent phase of radiation. Yet where evolutionary 'explosions' are seen in the fossil record it appears that they were frequently preceded by an extended, but comparatively obscure phase of structural innovation, which can still be detected in the genome. Thus, the concept of 'evolutionary explosions' appears to have resulted from the conflation of this earlier phase of phylogenesis with the subsequent phase of adaptive radiation. A far more consistent view of evolution can be gained by conceptually separating the two phases, and accepting that although the scarcity of ancestral forms can mean the earlier

[1] Synapomorphies are a kind of shared character used to identify phylogenetic relations between taxa.

phase is often obscure, the latter is the result of standard microevolutionary processes over extended periods of time.

The Cambrian and Cretaceous–Tertiary examples, separated as they are by several hundred million years and supported by widely different molecular data, suggest that the decoupling of radiation from phylogenesis may prove to be a widespread phenomenon. There are already hints that other crucial events in the history of life may have had a similar pattern. The radiation of terrestrial plants is recorded by well-known fossils of late Silurian to Devonian age (*c.* 400 million years ago), yet cryptospores which suggest air-borne dispersal are known as early as Middle Ordovician (*c.* 460 million years ago). Among the plants, angiosperms are likely to have had a history that was richer than their meagre early Mesozoic fossil record portends, before their mid-Cretaceous explosive radiation. Recent evidence suggests that shark-like or even bony fish may be present as the merest fragments in the Harding Sandstone (Ordovician), yet the generally accepted appearance and subsequent radiation of these vertebrate groups is the Devonian period. It seems that for every evolutionary explosion there may be a phylogenetic fuse. The processes taking place during these prolonged but obscure periods of evolutionary innovation are the next issue to be examined.

[*Trends in Ecology and Evolution*, 13 (1998), 151–6.]

DAVID DILCHER

45 Towards a new synthesis: major evolutionary trends in the angiosperm fossil record

Angiosperm paleobotany has widened its horizons, incorporated new techniques, developed new databases, and accepted new questions that can now focus on the evolution of the group. The fossil record of early flowering plants is now playing an active role in addressing questions of angiosperm phylogeny, angiosperm origins, and angiosperm radiations. Three basic nodes of angiosperm radiations are identified: (i) the closed carpel and showy radially symmetrical flower, (ii) the bilateral flower, and (iii) fleshy fruits and nutritious nuts and seeds. These are all coevolutionary events and spread out through time during angiosperm evolution. The proposal is made that the genetics of the angiosperms pressured the evolution of the group toward reproductive systems that favored outcrossing. This resulted in the strongest selection in the angiosperms being directed during the flower, fruits, and seeds. That is why these organs often provide the best systematic characters for the group. [Author's summary.]

Here I focus on the fossil record of the same plants that Stebbins did in his book, *Variation and Evolution of Plants*, the angiosperms.[1] This contribution

[1] G. L. Stebbins, *Variation and Evolution in Plants* (New York: Columbia University Press, 1950). The extract is from a paper given at a colloquium inspired by the 50th anniversary of Stebbins's book. Angiosperms are flowering plants, the largest division of plants (the other groups of plants are conifers, ferns, mosses, and liverworts).

has the advantage of being written more than 50 years after Stebbins wrote about his view of the fossil record of the angiosperms. His use of the fossil record of angiosperms as a model in his evolutionary synthesis was hampered because in the 1940s the paradigm in angiosperm paleobotany was to match fossils, especially leaves, to extant genera. The successes of the angiosperm paleobotanists (e.g., D. Axelrod, H. Becker, E. W. Berry, R. Brown, R. Chaney, and H. MacGinitie, and many others for 100 years before them) were judged by their ability to match a high percentage of fossils to living genera. Once the identifications were made to living genera, their focus was on questions of phytogeography and paleoclimate. This meant that almost no fossil angiosperms were recognized as extinct; it was quite impossible to focus questions of plant evolution on the fossil record of the angiosperms in 1950 as George Gaylord Simpson had done with the fossil vertebrate record in his classic *Tempo and Mode in Evolution* in 1944. [. . .]

This trend that had dominated angiosperm paleobotany for more than 100 years continued into the early 1970s. The supposed failure of the fossil record to contribute to understanding the evolution of the early angiosperms was still evident in 1974 when Stebbins published *Flowering Plants: Evolution Above the Species Level*. In chapter 10, 'The Nature and Origin of Primitive Angiosperms', there is no substantive use of the fossil record to address this question. The theories and hypothesis presented by Stebbins are based on the comparative morphology and anatomy of living angiosperms considered primitive at that time rather than the fossil record of early angiosperms.

However, at the same time, the early 1970s, special attention was being focused on the fine features of the morphology of angiosperm leaf venation and the cuticular anatomy of living and fossil angiosperms. Most of the early angiosperms from the Cretaceous and early Tertiary were being found to be extinct or only distantly related to living genera. Grades and clades of relationships were being founded on the basis of careful character analysis. During this time, it became scientifically acceptable to be unable to identify a modern genus to match a fossil. Fossil angiosperms were analyzed on the basis of multiple detailed objective characters, and degrees of relationships could be established based on the extent to which these same combinations of characters were found in living families, subfamilies, or genera. Analyses of the fossil angiosperm record were being constructed that included vast amounts of data based on careful anatomical and morphological analysis of the diversity of characters found in living genera and modern families. Large collections of cleared leaves and cuticular preparations were developed, and whole families were surveyed to establish their range of venation and cuticular characters and fruit and seed anatomy and morphology to research the fossil history of a family. This anatomical/morphological style of systematic-based angiosperm paleobotany was a distinct change from the floristic approaches that focused on paleogeographic and paleoclimatic questions and dominated the field before 1970.

This new paradigm shift opened the door for a new synthesis of the fossil record of angiosperms. New questions about the evolutionary biology of the fossil record of angiosperms could now be addressed based on detailed character-based data of living and fossil angiosperms often organized with the help of cladistic analysis.[2] At this same time there was renewed interest in exploring the fossil plant record to determine the origin and early evolutionary history of the angiosperms. The techniques of careful analysis and the concerted effort to open up a new fossil record of early angiosperms by the use of small, often charcoalified plant remains, or fragments of cuticle sieved from sediment from newly collected material of the Jurassic to the Upper Cretaceous were very successful. A whole new area of the study of intermediate-sized fossil plants, often termed mesofossils as opposed to microfossils or megafossils, expanded to occupy the majority of angiosperm research in some laboratories with good success. It is the success of these new techniques applied to the fossil record of angiosperms that now provides a new database from which to analyze some of the major trends in angiosperm evolution and allows us to ask new questions. [. . .]

How has angiosperm reproductive biology changed through time?

EVOLUTION OF THE CLOSED CARPEL[3]

The closed carpel is the one major feature that separates the angiosperms from other vascular seed plants. The closure most often is complete and entirely seals off the unfertilized ovules from the outside environment. Suggestions that this provided protection for the vulnerable ovules from beetles or other herbivores have been proposed as a reason for the closure of the carpel. However, I think that the closure of the carpel may be more directly related to the evolution of the bisexual flower.[4] During the evolution of the flower, as the male and female organs of the flower were brought into proximity, the need for protection against self-fertilization was so important that biochemical and mechanical barriers were developed very early in flowering plant ancestors. The mechanical barrier is the closed carpel and the biochemical barrier is the incompatibility systems that developed to prevent

[2] Cladistic analysis is a method of using the characters of species to infer their phylogenetic relations. 'Character' refers to any observable feature of an organism. For cladistic analysis, characters are distinguished into discrete units called character states.

[3] The carpel is the female reproductive structure; in flowers (that is, in angiosperms) it is enclosed.

[4] Bisexual flowers have both male and female structures in the same flower. Contrast that with unisexual flowers, in which a flower has organs of only one sex.

the successful growth of pollen tubes. Some living angiosperms have loosely closed carpels or lack any firm closure at all. It has been suggested that these have sufficient exudates to fill the carpel opening so that the carpel has a biochemical barrier against self-fertilization.

Although the closed carpel is the fundamental strategy for preventing self-pollination, the addition or loss of sepals, petals, and stamens must have been important events ensuring outcrossing. It is reasonable to assume that the development of attractive colored organs and nectaries, the clustering together of female (ovule-bearing) organs and male (pollen-bearing) organs, and, finally, the association of the female and male organs together on the same axis were all changes designed to increase the effectiveness of insect pollination. The closed carpel and biochemical incompatibility are natural early steps that followed or took place at the same time as the evolution of the floral features just mentioned. The closed carpel in a showy flower ensured outcrossing by animal pollinators while increasing pollen exchange with bisexual flowers. The closed carpel serves as a plant's control mechanism to guarantee that outcrossing happens. Any mechanical protection it offered probably always has been of secondary importance and can be easily overcome by insects.

EVOLUTION OF FLORAL FORM AND PATTERNS

Radial symmetry

The floral organs of all early angiosperms are radially symmetrical, a symmetry exhibited by all of the floral organs and flowers whether they are small or large, unisexual or bisexual. The earliest known angiosperm flowers suggest that individual carpels were borne helically on an elongated axis with pollen organs if present, either subtending and helically arranged on the same axis. Similarly, the early small flowers, unisexual or bisexual, have axes with radially arranged organs. In small flowers the elongation of many early flowering axes is compressed so that the organs appear radially arranged. This organization is clearly seen in larger flowers such as *Archaeanthus*[5] and the Rose Creek flower. This radial arrangement of organs persisted until late into the Late Cretaceous or the Paleocene.

Bilateral symmetry

By Paleocene and Eocene time, there are several evidences in the fossil record of bilateral flowers. This evolution probably began during the Upper Cretaceous. The evolution of bilateral flowers is associated with the evolution of social insects and happened in the angiosperms at different stages in

[5] These are two early fossil angiosperms. *Archaeanthus* is about 100 million years old, and similar to *Magnolia*.

the evolution of several living families. In some angiosperm families, bilateral symmetry may be present in only a part of the family, while in other families the entire family, is characterized by bilateral symmetry. As discussed below, this must relate to the time at which different groups evolved.

EVOLUTION OF SMALL AND LARGE FLOWERS

Flower size in living angiosperms is quite variable. Only during the past 25 years have numerous new fossil flowers been discovered from the Cretaceous. The record that has been developed demonstrates that both medium- and small-sized flowers are present very early. Certainly, flower size must relate to pollinator size. The variability in size of the early flowers suggests that a variety of pollinators were involved in their pollination biology. In addition to insect pollinators, both wind and water were important in the pollination of early angiosperms. Because the wind and the water have changed very little since the Cretaceous, there has been little change in the floral anatomy and morphology of these plants. Therefore, they are examples of some of the most ancient lines of living flowering plants. Those angiosperms that have modified their pollination biology to accommodate insect pollinators have been plants that have undergone the most extensive changes whose fossil ancestors are most different from their modern descendants.

EVOLUTION OF FLORAL PRESENTATION

In flowers that are insect pollinated, the display of the flower is critical. There seem to be clear distinctions between the presentation of the large *Archaeanthus* flower and the small fossil dichasial flowers.[6] The large *Archaeanthus* flower appears to have been terminal on a moderately large axis similar to the flowers of *Liriodendron*[7] or *Magnolia* today. This allows for sturdy support and a colorful display to attract a pollinator. The dichasial flower, in contrast, is small and clustered into an umbel-like arrangement. This allows for a showy display of flowers in different stages of maturity and a broad area of clustered flowers upon which a pollinator can land and move about. However, small unisexual florets such as those of platanoid-like[8] inflorescences and ceratophylloid-like plants have been little affected by animal pollinators. For this reason, they persist today only slightly changed from their form in the Early Cretaceous.

[6] Dichasial flowers have a particular structure. The branches that lead to the flowers give rise to two new branches. The feature is related to the growth habit of the plant, and this has consequences for the flower structure, as discussed later in the extract.

[7] *Lioriodendron* is a genus in the magnolia family. It has two species; the tulip tree is one of them.

[8] Platanaceae (which includes the plane tree, commonly cultivated in cities) and Ceratophyllaceae (aquatic herbs) are two families of angiosperms with deep fossil records.

UNISEXUAL VS. BISEXUAL FLOWERS

The earliest flowers now known appear to be gynodioecious.[9] One axis has only carpels with a clear indication that no other organs subtended them, while an attached axis has both carpels and pollen-bearing organs. So, was the first flower unisexual or bisexual? It appears to have had the potential to be both. Some early flowers, such as the platanoids and ceratophylloids, appear to be unisexual and never to have had a bisexual ancestry. Others such as *Archaefructus*, many of the small flowers from Portugal and the larger flowers from the Dakota Formation, are certainly bisexual. I suggest that the ancestral lineage of the angiosperms was most likely unisexual, and that with the availability of insect pollinators the efficiency of bisexual flowers won the day.

What are the significant nodes of angiosperm evolution?

There are three major nodes or events through time that resulted in major radiations of the angiosperms. These nodes include the evolution of showy flowers with a closed carpel, the evolution of bilateral flowers, and the evolution of nuts and fleshy fruits. At each of these events, there is a burst of adaptive radiation[10] within the angiosperms that can be interpreted as an attempt to maximize the event for all of the diversity possible and to use the event for increased reproductive potential.

The evolution of the closed carpel and the evolution of the showy radial flower must have occurred at nearly the same time. This was the first adaptive node marking a distinct coevolution[11] of early flowering plants and animal (insect) pollinators. The success of this involvement of insects in the reproductive biology of plants was not new. Dating back into the Paleozoic, insects most probably were involved in pollination of some of the seed ferns such as *Medullosa*. During the Mesozoic, several plants were certainly using animals for pollination as part of their reproductive biology. These include plants such as the Cycadoidea, *Williamsonia, Williamsoniella*, and, perhaps, some seed ferns such as *Caytonia*. Insect diversity increased parallel to the increasing diversity of the angiosperms during the Mesozoic. This node of evolution corresponds to the initial coevolution of animals and flowering plants in gamete transport. These early showy flowers came in many sizes,

[9] Gynodioecious: a species in which flowers are either hermaphrodite or exclusively female.

[10] Adaptive radiation refers to a phase in evolution when a small number of ancestral species evolve into a large number of descendant species that show a range of ecological adaptations.

[11] Coevolution means that two species (or taxa) influence each other's evolution. Here Dilcher is concerned with coevolution between angiosperms and insects. Much of his interpretation of angiosperm evolution is in terms of insect coevolution. Janzen, in Chapter 49, also looks at plant–animal coevolution.

were displayed on the plant in many different ways, and were uniform in the types of organs they contained and the radial symmetry of these organs. They must have accommodated many different types of pollinators as evidenced by the variety of their anthers, stigmatic surfaces, nectaries, and the sizes and positions of the floral organs. It was through the success of this coevolution that the angiosperms became the dominant vegetation during the early Late Cretaceous. Ordinal and family clades began to become identifiable during the later Early Cretaceous and the early Late Cretaceous. However, at the same time, some of the angiosperms never developed showy flowers and used other means of gamete transport for cross-pollination such as wind (early platinoids) and water (early ceratophylloids).

The evolution of bilateral flowers happened about 60 million years after the origin of the angiosperms. This node in coevolution never affected the water- or wind-pollinated groups that were already established. The evolution of the bees late in the Late Cretaceous was a coevolutionary event with the evolution of bilateral flowers. This occurred independently in many different clades of flowering plants that were already established by the mid-Late Cretaceous. The potential for flowers to further direct the behavior of insects to benefit their pollination had a profound influence on those clades that evolved during the late Upper Cretaceous and early Tertiary. Flowers not only presented their sex organs surrounded by sterile floral organs with attractive patterns and colors, exuding attractive fragrances and filled with nectar and pollen for food, but the bilateral flowers could show the animals which way to approach them and how to enter and exit them. This allowed flowers to maximize the potential for precise gamete exchange that was impossible with radially symmetrical flowers. Such clades as the Papilionoideae (legume subfamily), Polygalaceae, and Orchidaceae, among others, demonstrate this coevolution. The success of these clades and especially the Orchidaceae, with its vast number of species, demonstrates the potential of this coevolutionary event.

The evolution of large stony and fleshy fruits and seeds is the last major coevolutionary node of the angiosperms. This is not to say that there were not the occasional attractive fruits produced earlier, but a large radiation of fruit and seed types of the angiosperms occurred during the Paleocene and Eocene. The change in angiosperm fruit size was noted by Tiffney who associated this change with the radiation of rodents and birds. This coevolutionary node allowed for both the further radiation of the angiosperms and the radiation of the mammals and birds. Stone noted that there was a tendency to develop animal-dispersed fruit types in the Juglandaceae several times in different clades of this family. Many angiosperm families took advantage of the potential to disperse their fruits and seeds by bird and mammal vectors during the early Tertiary as evidenced by the bursts of the evolution of fruits and seeds during this time. It is interesting to note that at this same time the angiosperms also were experiencing a radiation of wind-dispersed fruits and seeds. This radiation of fruit and seed dispersal strategies in the angiosperms, late in their evolution (early Tertiary), is yet one more example of a means to promote outcrossing for the group.

Why did angiosperms evolve?

Coevolutionary events are largely responsible for the origin and subsequent nodes of evolution and radiation of the angiosperms. As we begin to find reproductive material of very early angiosperms, it becomes clear that some or most angiosperms developed bisexual insect-pollinated flowers very early, while some lines also maintained unisexual flowers with abiotic means of pollination. The coevolution with insects sparked a tremendous potential for plants to outcross by co-opting animals to carry their male gametes (pollen) to other individuals and other populations of the same species.

Each node of angiosperm evolution established genetic systems that favor outcrossing. The showy bisexual flower, the more specialized bilateral flower, and the nutritious nuts and fleshy fruits all are means by which the flowering plants increase their potential for outcrossing. The majority of angiosperm evolution is centered on this increased potential for outcrossing through coevolution with a wide variety of animals. In most cases the animals benefited as well from this coevolutionary association. Wind and water pollination syndromes also allowed for outcrossing and have continued to exist since the Early Cretaceous. However, they have never developed the diversity of those angiosperms pollinated by animals. Also several abiotically pollinated angiosperms, for example the Fagaceae (*Quercus* or oaks) and the Juglandaceae (*Carya* or pecans), later accommodated themselves for animal dispersal of their fruits or seeds. The importance of outcrossing cannot be underestimated as a driving force in the evolution of the angiosperms.

The ability of the angiosperms to accommodate and maximize benefits from animal behavior has been responsible for the evolutionary success of the group. As individual clades made use of particular coevolutionary strategies the diversity of both the angiosperms and animal groups increased. The benefits to the angiosperms were the benefits of the genetics of outcrossing. Because this is a sexual process, it was accomplished by means of evolutionary changes to flowers and fruits and seeds. This is why these particular organs have been centers of angiosperm evolution and why they are so useful in angiosperm systematics today.

[*Proceedings of the National Academy of Sciences, USA*, 97 (2000), 7030–6.]

Section H

Case studies

Some of the following case studies relate mainly to one of the other sections of the book; others relate to more than one section; others do not obviously relate to any section at all.

The extract about sex (Chapter 48) can be thought of as a case study in the analysis of adaptation. Section C looked at adaptation as a general concept. It did not look specifically at examples, in which biologists had tried to understand how particular characters were adaptive. Sex is something of a problem because, when Maynard Smith tried to see how it was adaptive, he created a paradox. Sex should not exist; natural selection will favour asexual reproduction. The solution to the paradox is almost the Holy Grail of a large theoretical sub-branch of evolutionary biology, but it still has not been satisfactorily tracked down.

The first and last extracts in the section, on senescence (Chapter 46) and on mutation rates (Chapter 52), respectively, are on characters that are almost certainly non-adaptive. It would be rather a puzzle if senescence were an adaptation, because it is disadvantageous for the organism that senesces. In Medawar's account, natural selection does not positively favour senescence, but senescence emerges from the way in which natural selection shapes life histories. The real adaptation is to have a high quality body at the time of peak reproductive value, in youth. Natural selection will trade off body quality in age against improvements early on. Medawar's theory provides the starting point for most subsequent evolutionary research on the problem.

Mutation is needed for evolution, and for natural selection, to occur. It has sometimes been suggested that natural selection favours elevated mutation rates, to facilitate evolution. However, Sniegowski *et al.* (Chapter 52) argue that natural selection drives mutation rates down to a minimum. Mutations only occur because it would be too expensive to improve the accuracy of DNA copying any further. However, in some microbes, mutator genes can be favoured in certain circumstances.

Crick's classic paper on the genetic code (Chapter 47), including his 'frozen accident' theory, does not relate to any other main section. It is about the way biochemical constraint and natural selection interacted in the origin of part of the molecular genetic apparatus. The ideas are incidentally important in the evidence that all modern life shares a unique common ancestor (see Chapter 61, Section J).

Chapter 49 by Janzen is a study of coevolution. Coevolution—the process in which evolutionary changes in one species influence the evolution of other species—has probably shaped many ecological relationships between species and acted as a motor in the history of life. Janzen has written a number of virtuoso essays on the fruits, seeds, and the animals that use and are used by them; I picked this one almost at random from the possibilities. It can be related to Dilcher's paper (Chapter 45) on the main themes of angiosperm evolution; Benner *et al.* (Chapter 41) provide some background on the genomics of fruit.

The other two papers look at two ways in which new, complex systems can evolve. Nilsson and Pelger (Chapter 50) studied theoretically whether the eye could evolve by smooth increments of improved optical design (see Section C introduction) and how many evolutionary steps it would take to evolve from a rudimentary beginning to as advanced an eye as those of vertebrates or cephalopods. Nilsson and Pelger arrive at a pessimistic estimate that the full eye could evolve in a few hundred thousand years, a figure that is interesting to juxtapose with the intuition of many naive thinkers, that the eye could not possibly evolve by gradual natural selection. Raff's article (Chapter 40, Section F) is another piece on the evolution of eyes.

Evolution probably often proceeds in the manner of the eye. However, other systems may evolve by putting together previously existing systems that had some other function. Gerhart and Kirschner (Chapter 51) describe a biochemical example. In mammals, the manufacture and digestion of milk are both metabolic innovations. One of the milk-manufacturing enzymes evolved by combining two previously existing enzymes. Neither of them had anything to do with milk before they were put together, by some accidental mutation that had creative consequences.

P. B. MEDAWAR

46 An unsolved problem of biology

Senescence is a puzzle for evolutionary theory: how can natural selection favour a deterioration in the quality of individuals? Medawar argues that natural selection acts less powerfully in the older sector of a population, even in the absence of senescence. Harmful mutations are less likely to be eliminated if they are expressed late, rather than early, in life. Also, natural selection can act to postpone the expression of a harmful mutation, shifting its expression to later in life. Harmful mutations that are expressed only late in life can therefore accumulate in a population, causing senescence. [Editor's summary.]

I now want to discuss the factors that may have played their part in [the] origin and evolution [of senescence]. As a text I shall use a quotation from the works of August Weismann.

> Death takes place because a worn-out tissue cannot for ever renew itself. Worn-out individuals are not only valueless to the species, but they are even harmful, for they take the place of those which are sound . . . by the operation of natural selection, the life of a theoretically immortal individual would be shortened by the amount which was useless to the species.

Weismann's propositions have the great merit of suggesting, for only the second time, that senescence has had a very orthodox evolutionary origin. But Weismann is arguing in what a student of mine once called a viscous circle, or more exactly a vicious figure-of-eight. He assumes that the elders of his race are worn out and decrepit—the very state of affairs whose origin he purports to be inferring—and then proceeds to argue that because these dotard animals are taking the place of the sound ones, so therefore the sound ones must by natural selection dispossess the old! This is all a great muddle, but there is certainly some truth in it, and I shall spend the rest of my lecture in an attempt to find out what that truth may be.

My argument starts with a discussion of certain demographic properties of a population of potentially immortal individuals, and it will be illustrated by an inorganic model which I shall animate step by step. This choice makes it possible to avoid two common traps. The first of these is to argue that senescence in higher animals has come about *because* they have a post-reproductive period; for 'unfavourable' hereditary factors that reveal their action only in the post-reproductive period are exempt from the *direct* effects of natural selection and there is therefore little to stop them establishing themselves and gaining ground. Any such argument is wholly inadmissible. The existence of a post-reproductive period is one of the consequences of senescence; it is not its cause. The second trap, into which Weismann fell headlong, is to suppose that a population of potentially immortal individuals subject to real hazards of mortality consists in high proportion of very aged animals with a relatively small number of no doubt browbeaten youngsters running round between their feet. It will soon be clear that this idea is equally mistaken.

I want you now to consider a population of objects, living or not, which is at risk—in the sense that its members may be killed or broken—but which is potentially immortal in the sense that its members do not in any way deteriorate with ageing. Test-tubes will do, since, they are clearly 'mortal', and I shall peremptorily assume that they do not become more fragile with increasing age.

Imagine now a chemical laboratory equipped on its foundation with a stock of 1000 test-tubes, and that these are accidentally and in random manner broken at the rate of 10 per cent. per month. Under such an exaction of mortality, a monthly decimation, the activities of the laboratory would soon be brought to a standstill. We suppose therefore that the laboratory steward replaces the broken test-tubes monthly, and that the test-tubes newly added are mixed in at random with the pre-existing stock. The steward will obviously be obliged to buy an average of 100 test-tubes monthly, and I am going to assume that he scratches on each test-tube the date at which he bought it, so that its age-in-stock on any future occasion can be ascertained.

Now imagine that this regimen of mortality and fertility, breakage and replacement, has been in progress for a number of years. What will then be the age-distribution of the test-tube population; that is, what will be the proportions of the various groups into which it may be classified by age? The answer is [. . .] the population will have reached the stable or 'life-table' age-distribution in which there are 100 test-tubes aged 0–1 month, 90 aged 1–2 months, 81 aged 2–3 months and so on. This pattern of age-distribution is characteristic of a 'potentially' immortal population, i.e. one in which the chances of dying do not change with age. The [. . .] older the test-tubes are, the fewer there will be of them—not because they become more vulnerable with increasing age, but simply because the older test-tubes have been exposed more often to the hazard of being broken. Do not therefore think of a potentially immortal population as being numerically overwhelmed by dotards. Young animals outnumber old, and old animals those still older.

As a first step in animating this model, I want you to imagine that the test-tubes now do for themselves exactly what the steward has hitherto been doing for them, i.e. they reproduce themselves, no matter how, at an average rate of 10 per cent. per month in order to maintain their numbers. Since the population is potentially immortal, the rate of reproduction of its members will not vary with their age. It follows that each 'living' test-tube of the existing population will make the same average contribution of offspring to the test-tube population of the future. Each test-tube may lay claim to an equal share of the ancestry of future generations, and its reproductive value is invariant with its age.

The next step in the argument is vital. Although each individual test-tube takes an equal share of the ancestry of the future population, each age-group most certainly does not. The older the age-group, the smaller is its overall reproductive value. The group of test-tubes 2–3 months old, for example, makes a very much greater contribution than the group 11–12 months old. This is not because the test-tubes of the senior group are individually less fertile—their fertility is *ex hypothesi* unchanged—but merely because there are fewer of them; and there are fewer of them not because they have become more fragile—their vulnerability is likewise unaltered—but simply because, being older, they have been exposed more often to the hazard of being broken. It is simply the old story of the pitcher and the well.

Some of the consequences of this decline in the reproductive value of older age-groups will be apparent when I take the next step in animating my test-tube model. The test-tubes are no longer to be thought of as immortal; on the contrary, after a certain age, as a result of some intrinsic shortcoming, they suddenly fall to pieces. For the time being we shall assume that they disintegrate without premonitory deterioration. What will be the effect of this genetically provoked disaster upon the well-being of the race of test-tubes? It must be my fault if the answer does not appear to be a truism—that it depends upon the age at which it happens. If disintegration should occur five years after birth, its consequences would be virtually negligible, for under the regimen which we have envisaged less than one in five hundred of the population is lucky enough to live so long. Indeed, if we relied upon evidence derived solely from the natural population of test-tubes, we should probably never be quite certain that it really happened. We could make quite certain, as we do with animals, only by domesticating our test-tubes, shielding them from the hazards of everyday usage by keeping them in a padded box as pets.

If disintegration should occur one year after birth, an age which is reached or exceeded by about one-quarter of the population, the situation would be fairly grave but certainly not disastrous; after all, by the time test-tubes have reached the age of twelve months they have already made the greater part of their contribution of offspring to the future population. But with disintegration at only one month, the consequences would obviously be quite catastrophic.

This model shows, I hope, how it must be that the force of natural selection weakens with increasing age—even in a theoretically immortal population, provided only that it is exposed to real hazards of mortality. If a genetical disaster that amounts to breakage happens late enough in individual life, its consequences may be completely unimportant. Even in such a crude and unqualified form, this dispensation may have a real bearing on the origin of innate deterioration with increasing age. There is a constant feeble pressure to introduce new variants of hereditary factors into a natural population, for 'mutation', as it is called, is a recurrent process. Very often such factors lower the fertility of viability of the organisms in which they make their effects apparent; but it is arguable that if only they make them apparent late enough, the force of selection will be too attenuated to oppose their establishment and spread. Such an argument may have a particular bearing on, for example, the occurrence of spontaneous tumours and the senile degenerative diseases in mice, [. . .] for these affections make themselves apparent at ages which wild mice seldom, perhaps virtually never reach. We only know of their existence through domestication; small wonder if they have no effect on the well-being of mouse populations in the wild. Mice, of course, do already show evidence of deterioration in the course of ageing, but my reasoning does not presuppose it. It applies to 'potentially immortal populations' with only a quantitative loss of cogency.

It is a corollary of the foregoing argument that the postponement of the time of overt action of a harmful hereditary factor is equivalent to its elimination. Indeed, postponement may sometimes be the *only* way in which elimination can be achieved; but I cannot argue this without an appeal to the phenomena of pleiotropy and linkage, which time will not allow.

It is not good enough to say that what happens to very old animals hardly matters and that what happens to youngsters matters a great deal. For the degree to which anything may matter varies in a predictable way with age, and the selective advantage or disadvantage of a hereditary factor is rather exactly weighted by the age in life at which it first becomes eligible for selection. A relatively small advantage conferred early in the life of an individual may outweigh a catastrophic disadvantage withheld until later.[1] Go back to the test-tube model for a moment, and compare two competing test-tube populations. Both suffer the same average monthly mortality of 10 per cent., and one has, as hitherto, the average monthly birth-rate of 10 per cent. The other population has an average monthly birth-rate of 11 per cent., but the price paid for this hardly profligate increase of fecundity is the spontaneous bursting asunder of each member at age two. Which population will increase the more rapidly in numbers—the potentially immortal, or the mortal population with a birth-rate only one-tenth part higher than the other's? The simplest calculations show that it is the latter. [. . .]

The postponement of the time of overt action of 'unfavourable' hereditary factors is not just a good idea which the organism would be well advised to apply in practice; postponement may be enforced by the action of natural selection and senescence may accordingly become a self-enhancing process. Let me give you a real example in which this process appears to be happening at the present time.

Huntington's chorea is a grave and ultimately fatal nervous disability distinguished by apparently compulsive and disordered movements akin to, and perhaps identifiable with, 'St Vitus' Dance'. Its first full clinical description is in George Huntington's own memoir of 1872, though the evidence I shall appeal to comes largely from the fine treatise of Dr Julia Bell. Huntington's chorea is a hereditary affliction of a rather special sort. Its disabling and clinically important effects first become manifest not in youth or old age but

[1] [Medawar's note] By something that is a catastrophic disadvantage to an older animal I mean a change which is personally catastrophic, and which would certainly be catastrophic to the species as well if it made its appearance in younger animals. But in the strict sense, the verdicts 'advantageous' and 'disadvantageous' can be delivered only after trial by selection, and in this sense to speak of 'catastrophic disadvantages' which don't in fact much matter is self-contradictory.

at an intermediate period, its time of onset—later in men than in women—being most commonly in the age-group 35–39. Its age of onset does however vary, and I want you to assume (what is almost certainly true, though it would be hard to collect the evidence for it) that its age of onset, like the disease itself, is also genetically determined.

If differences in its age of onset are indeed genetically determined, then natural selection *must* so act as to postpone it: for those in whom the age of onset is relatively late will, on the average, have had a larger number of children than those afflicted by it relatively early, and so will have propagated more widely whatever hereditary factors are responsible for the delay. But as the age of onset approaches the end of the reproductive period, so the direct action of selection in postponing it will necessarily fade away. [. . .]

With Huntington's chorea as a lucky concrete example, I can now propound the following general theorem. If hereditary factors achieve their overt expression at some intermediate age of life; if the age of overt expression is variable; and if these variations are themselves inheritable; then natural selection will so act as to enforce the postponement of the age of the expression of those factors that are unfavourable, and, correspondingly, to expedite the effects of those that are favourable—a recession and a precession, respectively, of the variable age-effects of genes. This is what I mean by saying that senescence is a self-enhancing process. The theorem in the form in which I have just put it does not depend upon the existence of a post-reproductive period; it only requires that the reproductive value of each age-group should diminish with increasing age. I have argued that this must necessarily diminish even with a population of potentially immortal and indeterminately fertile individuals, provided only that they are subject to real dangers of mortality. In such a population a younger age-group must necessarily outnumber an older, for the older represents the residue of those who have been longer exposed to mortal hazards. If you should have, as I believe, unjustified qualms about an argument based upon combining an innate potential immortality with a contingent real mortality, I would recall to you my earlier distinction between senscence of sorts (*a*) and (*b*). Senescence of sort (*b*) is not innate or 'laid on' developmentally; it represents the outcome of the cumulative effects of recurrent physical damage, physiological stress, or faulty cellular replication. If you will admit that senescence of this sort is a means by which, irrespective of any genetical background, the reproductive value of each individual in a population is caused to diminish with increasing age, then my argument is quantitatively strengthened, because the numerical preponderance of the younger age-groups will become so much the more pronounced. And if, further, a post-reproductive period of life is already established, then indeed it becomes, as it were, a dustbin for the effects of deleterious genes. But these propositions are mere glosses or refinements. The argument must stand or fall on the case which I first proposed.

I have now suggested three agencies which may have played a part in the evolution of 'innate' senescence: (1) the inability of natural selection to counteract the feeble pressure of repetitive mutation when the mutant genes make their effects apparent at ages which the great majority of the members of a population do not actually reach; (2) the fact that the postponement of the time of action of a deleterious gene is equivalent to its elimination, and may sometimes be the only way in which elimination can be achieved; and (3) the fact that natural selection may actually enforce such a postponement, and, conversely, expedite the age of onset of the overt action of favourable genes. All these theorems derive from the hypothesis that the efficacy of natural selection deteriorates with increasing age. [. . .]

[*An Unsolved Problem of Biology* (London: H. K. Lewis, 1952).]

F. H. C. CRICK

47 The origin of the genetic code

The genetic code (the relation between the nucleotide sequence in DNA and the amino acid sequence in proteins) is effectively universal in life. This could be because the relation between particular triplets of nucleotides and particular amino acids is required for some stereochemical reason: they somehow fit together. Alternatively, it could be a frozen accident, in which the code was established early in life and subsequent changes have been selected against. The stereochemical theory is arguably unlikely, at least as a general explanation. The primitive code may have had fewer amino acids, and less precise relations between DNA and amino acids, than in the modern code. Some authors have argued that the code evolved a form that minimizes the damage done by mutational errors, but Crick argues that this effect is inappreciably small. [Editor's summary.]

Why is the code universal?

Two extreme theories may be described to account for this, though, as we shall see, many intermediate theories are also possible.

The stereochemical theory

This theory states that the code is universal because it is necessarily the way it is for stereochemical reasons. Woese has been the main proponent of this point of view. That is, it states that phenylalanine *has* to be represented by UU^{U}_{C}, and by no other triplets, because in some way phenylalanine is stereochemically 'related' to these two codons. There are several versions of this theory. We shall examine these shortly when we come to consider the experimental evidence for them.

The frozen accident theory

This theory states that the code is universal because at the present time *any change would be lethal*, or at least very strongly selected against. This is because in all organisms (with the possible exception of certain viruses) the code determines (by reading the mRNA) the amino acid sequences of so many highly evolved protein molecules that any change to these would be highly disadvantageous unless accompanied by many simultaneous mutations to correct the 'mistakes' produced by altering the code.

This accounts for the fact that the code does not change. To account for it being the same in all organisms one must assume that all life evolved from a single organism (more strictly, from a single closely interbreeding population). In its extreme form, the theory implies that the allocation of codons to amino acids at this point was entirely a matter of 'chance'.

The stereochemical theory—experimental evidence

In its extreme form, the stereochemical theory states that the postulated stereochemical interactions are still taking place today. It should therefore be a simple matter to prove or disprove such theories.

Pelc and Welton have suggested from a study of models that there is in many cases a specific stereochemical fit between the amino acid and the base sequence of its codon on the appropriate tRNA. Unfortunately, their models were all built backwards so their claims are without support. Such a theory implies that the expected codon sequence occurs somewhere on each tRNA. For example, no such sequence occurs in the tRNA for tyrosine either from yeast or from *E. coli*. In our opinion this idea has little chance of being correct.

A more reasonable idea is that the amino acid fits the *anticodon* on the tRNA. At least this has the advantage that it is always present. A model along these lines for proline has been briefly described by Dunnill, but so far no detailed description has been published, nor has he extended his model-building to other amino acids.

The experimental evidence has already established that when the activating enzyme transfers the amino acid to the tRNA, the interaction is not solely with the anticodon and the common . . . CCA terminal sequence. This is shown by the fact that an activating enzyme from one species will not always recognize the appropriate tRNA from a different species although the anticodons must be very similar if not identical in different species. However, this does not preclude the idea that the interaction is partly with the anticodon and partly with some other part of the tRNA.

The best way to disprove the theory (if indeed it is false) would be to change the anticodon of some tRNA molecule and show that nevertheless it accepted the same amino acid from the activating enzyme. This has already been done for the minor tyrosine tRNA of *E. coli* whose anticodon has been changed (in an Su^{+} strain) from GUA to CUA although the experiments need to be done quantitatively. Further examples of such changes are likely to be reported in the near future. Until this is done we must reserve final judgement on the amino acid-anticodon interaction theory; but we consider it unlikely to be correct, except perhaps in a few special cases.

Even if it were established that the activating enzyme recognizes the anticodon, this would not by itself prove that the recognition is done by inserting the amino acid in a cage formed by the anticodon. Notice that the activating enzyme would have to release amino acid from its own recognition cavity and then insert it into the recognition site on the tRNA. Moreover, when the amino acid has been transferred to the tRNA and the activating enzyme has diffused elsewhere, the amino acid could not stay in the anticodon cage without blocking the interaction with the codon on the mRNA. None of this is impossible but it is certainly elaborate.

It is not easy to see at this stage what evidence would be needed to prove that the anticodon does indeed form a cage for the amino acid, though if the tRNA (or perhaps a fragment of it) could be crystallized it might be possible to see the amino acid sitting in such a position.

The present experimental evidence, then, makes it unlikely that every amino acid interacts stereochemically with either its codon or its anticodon. It by no means precludes the possibility that *some* amino acids interact in either of these ways, or that such interactions, even though now not used, may have been important in the past, at least for a few amino acids. We must now leave the system as it is today and turn to the examination of primitive systems.

The primitive system

It is almost impossible to discuss the origin of the code without discussing the origin of the actual biochemical mechanisms of protein synthesis. This is very difficult to do, for two reasons: it is complex and many of its details are not yet understood. Nevertheless, we shall have to present a tentative scheme, otherwise no discussion is possible.

In looking at the present-day components of the mechanism of protein synthesis, one is struck by the considerable involvement of non-informational nucleic acid. The ribosomes are mainly made from RNA and the adaptor molecules (tRNA) are exclusively RNA, although modified to contain many unusual bases. Why is this? One plausible explanation, especially for rRNA, is that RNA is 'cheaper' to make than protein. If a ribosome were made exclusively of protein the cell would need *more* ribosomes (to make the extra proteins, which would not be a negligible fraction of all the proteins in the cell) and thus could only replicate more slowly. Even though this may be true, we cannot help feeling that the more significant reason for rRNA and tRNA is that *they were part of the primitive machinery* for protein synthesis. Granted this, one could explain why their job was not taken over by protein, since

(i) for rRNA, it would be too expensive,
(ii) for tRNA, protein may not be able to do such a neat job in such a small space.

In fact, as has been remarked elsewhere, tRNA looks like Nature's attempt to make RNA do the job of a protein.

If indeed rRNA and tRNA were essential parts of the primitive machinery, one naturally asks how much protein, if any, was then needed. It is tempting to wonder if the primitive ribosome could have been made *entirely* of RNA. Some parts of the structure, for example the presumed polymerase, may now be protein, having been replaced because a protein could do the job with greater precision. Other parts may not have been necessary then, since primitive protein synthesis may have been rather inefficient and inaccurate. Without a more detailed knowledge of the structure of present-day ribosomes it is difficult to make an informed guess. [. . .]

The primitive code

We must now tackle the nature of the primitive code and the manner in which it evolved into the present code.

It might be argued that the primitive code was not a triplet code but that originally the bases were read one at a time (giving 4 codons), then two at a time (giving 16 codons) and only later evolved to the present triplet code. This seems highly unlikely, since it violates the Principle of Continuity. A change in codon size necessarily makes nonsense of *all* previous messages and would almost certainly be lethal. This is quite different from the idea that the primitive code was a triplet code (in the sense that the reading mechanism moved along three bases at each step) but that only, say, the first two bases were read. This is not at all implausible.

The next general point about the primitive code is that it seems likely that only a few amino acids were involved. There are several reasons for this. It certainly seems unlikely that all the present amino acids were easily available at the time the code started. Certainly tryptophan and methionine look like later additions. Exactly which amino acids were then common is not yet clear, though most lists would include glycine, alanine, serine and aspartic acid. However, if stereochemical interaction played a part in the primitive code, this might select amino acids which were available but not particularly common. Again, it seems unlikely that the primitive code could code *specifically* for more than a few amino acids, since this would make the origin of the system terribly complicated. However, as Woese has pointed out, the primitive system might have used *classes* of amino acids. For example, only the middle base of the triplet may have been recognized, a U in that position standing for any of a number of hydrophobic amino acids, an A for an acidic one, etc.

Even though few amino acids (or groups of amino acids) were recognized, it seems likely that not too many nonsense codons existed, otherwise any message would have had too many gaps. There are various ways out of this dilemma. For example, as mentioned above, only one base of the triplet might have been recognized. Another possibility, however, is that the early message consisted not of the present four bases, but perhaps only two of them.

The number of bases in the primitive nucleic acid

The only strong requirements for the primitive nucleic acid is that it should have been easy to replicate, and that it should have consisted of more than one base, otherwise it could not carry any information in its base sequence. One cannot even rule out the possibility that the base sequence of the two chains was complementary (as in the present DNA). Perhaps a structure is possible with only two bases in which the two chains run parallel (rather than anti-parallel) and pairing is like-with-like. It would certainly be of great interest if such a structure could be demonstrated experimentally.

Leaving this possibility on one side and restricting ourselves to complementary structures, we see that the number of bases must be even. If there were only two in the primitive DNA, the question arises as to which two. The obvious choices are either A with U (or T) or G with C. A less obvious possibility (suggested some time ago by Dr Leslie Orgel, personal communication) is A with I (where I stands for inosine, having the base hypoxanthine). It is not certain that a double helix can be formed having a random sequence of A's and I's on one chain and the complementary sequence (dictated by A-I or I-A pairs) on the other chain, but it is not improbable, especially as the RNA polymers poly A and poly I can form a double helix.

Several advantages could be claimed for this scheme. Adenine is likely to be the commonest base available in the primitive soup, and inosine could arise from it by deamination. Thus the supply of precursors might be easier than in the case of the other two alternatives, though how true this is remains to be established. Then again in a random (A, I) sequence I would presumably code in the same way as G does now, at any rate for the first two positions of the triplet. If we can use the present code as a guide (though we shall argue later that this may be misleading), it is noticeable that the triplets containing only A's or G's in their first two bases (the bottom right-hand corner of the Table) do indeed code for some of the more obviously primitive amino acids.

It is important to notice that a scheme of this sort (or even one with like-with-like pairing) does not violate the principle of continuity. To change over from an (A, I) double helix to one like the present one but having A, I, U and C, the only steps required are a change in the replicase to select smaller base-pairs, and a supply of the two new precursors. The message carried (by the 'old' chain) is unaltered by this step. Gradually mutations would produce U's and C's on this chain and the new codons thus produced could be brought into use as the mechanism for protein synthesis evolved. Eventually G would be substituted for I. At no stage would the message become complete nonsense. The idea that the initial nucleic acid contained only two bases is thus a very plausible one. It remains to be seen whether primitive ribosomal RNA and primitive tRNA could be constructed using only two bases.

The stereochemical alternative

As stated earlier, it seems very unlikely that there is any stereochemical relationship between all the present amino acids and specific triplets of bases; but it is by no means ruled out that a few amino acids can interact in this way. If this were possible, it would certainly help in the initial stages of the evolution of the code. However, sooner or later a transition would have had to be made to the present type of system, involving tRNA's, ribosomes, etc. It seems to us that this could only happen easily if the code at that stage was fairly simple and only coded a rather small number of amino acids.

The evolution of the primitive code

Whatever the early steps in the evolution of the code, it seems highly likely that it went through a stage when only a few amino acids were coded. At this stage either the mechanism was rather imprecise and thus could recognize most of the triplets, or only a few triplets were used, perhaps because the message contained only two types of base. We must now consider what would happen next.

A complication should be introduced into this simple picture. It could well be that at this stage the recognition mechanisms were not very precise and that any given codon corresponded to a *group* of amino acids. Thus codons for alanine might also incorporate glycine, those for threonine might also code serine, etc. However, it is by no means certain that this happened. It seems highly likely that a 'cavity' to accept threonine would also accept serine to some extent, but the converse mistake is less likely and could depend on the exact nature of the structure involved. Thus, though the early coding machinery probably produced errors, we can only guess at their extent.

We shall argue that by far the most likely step was that these primitive amino acids spread all over the code until almost all the triplets represented one or other of them. Our reasons for believing this are that too many nonsense triplets would certainly be selected against, so that most codons would quickly be brought into use. In addition, it would be easier to produce a new tRNA, altered only in its anticodon, while still recognizing the amino acid, than to produce both a new anticodon and a new recognition system for attaching a new amino acid. Thus, we can reasonably expect that the intermediate code had two properties:

(i) few amino acids were coded, and
(ii) almost all the triplets could be read.

Moreover, because of the way this primitive code originated, the triplets standing for any one amino acid are likely to be related. At this stage the organism could only produce rather crudely made protein, since the number of amino acids it could use was small and the proteins had probably not evolved very extensively.

The final steps in the evolution of the code would involve an increase in the precision of recognition and the introduction of new amino acids. The cell would have to produce a new tRNA and a new activating enzyme to handle any new amino acid, or any minor amino acid already incorporated because of errors of recognition. This new tRNA would recognize certain triplets which were probably already being used for an existing amino acid. If so, these triplets would be ambiguous. To succeed, two conditions would have to be fulfilled.

(1) The new amino acid should not upset too much the proteins into which it was incorporated. This upset is least likely to happen if the old and the new amino acids are related.
(2) The new amino acid should be a positive advantage to the cell in at least one protein. This advantage should be greater than the disadvantages of introducing it elsewhere.

In short, the introduction of the new amino acid should, on balance, give the cell a reproductive advantage.

For the change to be consolidated we would expect many further mutations, replacing the ambiguous codons by other codons for the earlier amino acid when this was somewhat better for a protein than the later one. Thus, eventually the codons involved would cease to be ambiguous and would code only for the new amino acid.

There are several reasons why one might expect such a substitution of one amino acid for another to take place between structurally similar amino acids. First, as mentioned above, such a resemblance would diminish the bad effects of the initial substitution. Second, the new tRNA would probably start as a gene duplication of the existing tRNA for those codons. Moreover, the new activating enzyme might well be a modification of the existing activating enzyme. This again might be easier if the amino acids were related. Thus, the net effect of a whole series of such changes would be that *similar amino acids would tend to have similar codons*, which is just what we observe in the present code.

It is clear that such a mechanism for the introduction of new amino acids could only succeed if the genetic message coded for only a small number of proteins and especially proteins which were somewhat crudely constructed. As the process proceeded and the organism developed, more and more proteins would be coded and their design would become more sophisticated until eventually one would reach a point where no new amino acid could be introduced without disrupting too many proteins. At this stage the code would be frozen. Notice that it does not necessarily follow that the original codons, of the original primitive code (as opposed to the intermediate code) will necessarily keep their assignments to the primitive amino acids. In other words, the evolution of the code may well have wiped out all trace of the primitive code. For this reason arguments about which base-pair came into use first on the nucleic acid should not depend too heavily on the assignments of the present code.

The idea described above is crucial to the evolution of the code. It seems to me not to be the same as the idea, suggested by several authors, that the code is designed to minimize the effects of mutations. The implication is that the mutations are those occurring in the many proteins of the organism, and in fact are still occurring today. This is not quite the same as the idea that it is the situation produced by the introduction of a new amino acid to the *developing* code that we have to consider. Moreover, the disturbances had to be minimized not to the present day proteins but to the small number of more primitive proteins then existing. The minimizing of the effects of mutations is in any case likely to have only a small selective advantage even at the present time, and I think it unlikely that it could have had any appreciable effect in moulding the genetic code.

An idea rather close to the one presented above has been developed by Woese. He emphasizes in his discussion the fact that the early translation mechanism would probably be prone to errors. This is indeed an important idea and may well be what actually occurred but it is not identical to the idea suggested above, as can be easily seen by making the rather unlikely assumption that the early mechanism was rather accurate. In this case Woese's ideas are irrelevant and one is driven to the scheme outlined above. Nevertheless,

Woese's discussion follows much the same line as that presented here. However, he argues that by this mechanism it is unlikely that the code could reach the truly optimum code. There is no reason to believe, however, that the present code is the best possible, and it could have easily reached its present form by a sequence of happy accidents. In other words, it may not be the result of trying all possible codes and selecting the best. Instead, it may be frozen at a local minimum which it has reached by a rather random path. [. . .]

There is one feature of the process by which new amino acids were added to a primitive code which is far from clear. This is why several versions of the genetic code did not emerge. It is, of course, easy to say that in fact several did emerge and only the best one survived, but the argument is rather glib. A detailed discussion of what was likely to have happened at this period would involve the consideration of genetic recombination. Did it occur at a very early stage, perhaps even before the evolution of the cell, and, if so, what form did it take? Surprisingly enough, no writer on the evolution of the code seems to have raised this point. Naturally only rather simple processes would be expected, but the selective advantages of such a process would be very great. Perhaps a simple fusion process would suffice for the origin of the code (a suggestion made by Dr Sydney Brenner, personal communication). This would provide spare genes for further evolution and in as far as the code for the fusing organisms differed it would produce fruitful ambiguities. One might even argue that the population which defeated all its rivals and survived was the one which first evolved sex, a curious twist to the myth of the Garden of Eden.

[*Journal of Molecular Biology*, 38 (1968), 367–79.]

J. MAYNARD SMITH

48 The maintenance of sex

Sexual reproduction has an inherent twofold disadvantage relative to parthenogenetic (that is, asexual, or clonal) reproduction. The disadvantage applies in species with separate sexes, and in at least some hermaphrodites. Given the disadvantage, it is a puzzle why sex prevails in so much of multicellular life. [Editor's summary.]

In unicellular organisms, the disadvantages of sex are not great. The usual pattern is for asexual multiplication to be interrupted by sexual fusion only when conditions are severe and when continued multiplication would in any case be impossible. In view of this, and also of the fact that microorganisms commonly adapt to changed circumstances by evolutionary change as well as by individual physiological adaptation, it is not difficult to see why sexual processes, once evolved, should have been maintained.

Table 48.1

		ADULTS	EGGS	ADULTS IN NEXT GENERATION
Parthenogenetic ♀♀		$n \longrightarrow$	$kn \longrightarrow$	Skn
Sexual	♀♀	N	$\frac{1}{2}kN$	$\frac{1}{2}SkN$
	♂♂	N	$\frac{1}{2}kN$	$\frac{1}{2}SkN$

But in multicellular organisms with separate male and female individuals the disadvantages of sex are severe. Suppose that in such a species, with equal numbers of males and females, a mutation occurs causing females to produce only parthenogenetic females like themselves. The number of eggs laid by a female, *k*, will not normally depend on whether she is parthenogenetic or not, but only on how much food she can accumulate over and above that needed to maintain herself. Similarly, the probability *S* that an egg will survive to breed will not normally depend on whether it is parthenogenetic. With these assumptions the changes shown in Table 48.1 occur in one generation.

Hence in one generation the proportion of parthenogenetic females increases from $n/(2N + n)$ to $n/(N + n)$; when n is small, this is a doubling in each generation.

It follows that with these assumptions, the abandonment of sexual reproduction for parthenogenesis would have a large selective advantage in the short run.

It is well known that asexual varieties of plants arise quite commonly, and that their distribution, geographical and taxonomic, suggests that they are successful in the short term but in the long term doomed to extinction. Asexual varieties are much rarer among animals, although they do occur. It is not clear why this should be. Some possible reasons for the comparative rarity of asexual reproduction are:

(1) Meiotic parthenogenesis, followed by fusion of egg and polar body, or of the first two cleavage nuclei, is equivalent to close inbreeding. In naturally outbreeding species the decline in vigour caused by inbreeding might counterbalance the advantage of not wasting material on males. This argument does not apply to ameiotic parthenogenesis.
(2) In many mammals and birds, and some other animals, both parents help raise the young. In such cases parthenogenesis would usually be a disadvantage.

At first sight it seems that hermaphroditism, or monoecy in plants, eliminates the selective advantage of parthenogenesis. In a hermaphrodite species, no material is wasted on males, and no more resources need to be expended on sperm than are needed to fertilize the eggs produced. This argument I believe to be erroneous, at least in the case of hermaphrodites with external fertilization, for the following reasons.

In any species the number of eggs laid (or seeds produced) will be limited by some resource *R*. In a hermaphrodite the same individual will also produce sperm. It is reasonable to assume that the production of sperm will make demands on the same resource *R*, which must therefore be shared between eggs and sperm. The argument in the preceding paragraph amounts to saying that the major part of *R* will be devoted to eggs, only enough being devoted to sperm to ensure that the eggs are fertilized. This conclusion is an example of the use of what J. B. S. Haldane once referred to as 'Pangloss' theorem'—that all is for the best in the best of all possible worlds. Unhappily, Pangloss' theorem is false. In this case it assumes that natural selection necessarily produces a result favourable to the species, regardless of selection at the individual level. This is not so, because individual selection is usually more effective than selection favouring one group or species at the expense of another.

In hermaphrodites with external fertilization, or monoecious plants with compulsory cross-fertilization, the resource *R* will normally be divided equally between eggs and sperm, or ovules and pollen. In such cases hermaphrodites would on the average have only half as many surviving offspring as parthenogenetic females. However, the argument in the appendix[1] does not apply to hermaphrodites with internal fertilization, or to self-fertilizing hermaphrodites, because in these cases individual selection will favour a limitation of the amount of sperm or pollen produced to that needed to ensure the fertilization of the available eggs. The conclusions to be drawn therefore vary according to whether a group has internal or external fertilization, as follows:

(1) In groups with external fertilization, hermaphroditism would not increase the reproductive potential of a species. It is the common mechanism of reproduction in land plants, presumably because it has the advantage that an individual can fall back on self-fertilization in the absence of near neighbours. It does not protect a species against the evolution of parthenogenesis.
(2) In groups with internal fertilization, hermaphroditism does increase the reproductive potential of a species. It may be for this reason that it has become the typical method of reproduction in plathyhelminthes and in gastropods. In the former of these groups it has proved to be a pre-adaptation to parasitism. It does protect a species against the evolution of parthenogenesis.

[1] The appendix is not included here.

The argument in this section seems to lead to the conclusion that, except in the special cases of animals in which both parents care for the young, and of hermaphrodites with internal fertilization, metazoan animals would be expected to give rise frequently to parthenogenetic varieties. Since in fact this conclusion is false, the argument must leave something out of account.

['The Origin and Maintenance of Sex'. In G. C. Williams (ed.), *Group Selection* (Chicago: Aldine, 1971).]

D. H. JANZEN

49 A caricature of seed dispersal by animal guts

Various attributes of fruit, and of vertebrate guts, illustrate the concept of coevolution. Fruits ripen in such a way as to be digestible to vertebrates but not to other animals and microbes. Vertebrates vary in the way they feed on seeds, and how long seeds remain in their guts—which in turn influences how far the seeds are dispersed. The treatment of seeds in animal guts, and the place where the seeds are deposited, influence germination success. [Editor's summary.]

The interaction between fruits and frugivores is deceptively easy to visualize. Fruits ripen, a vertebrate goes to them and eats some (swallowing some seeds as contaminants); it moves away; the seeds pass through it; and sometime later, they germinate. This applies to plants from tiny epiphytes and herbaceous riverbank vines to enormous trees and lianes hundreds of meters in length. It occurs from Arctic cranberry bogs to melon patches in Saharan oases, from tropical swamps to the highest Andean paramo. It dates back at least as far as dinosaurs, if the sweet fleshy fruits on Chinese gingko trees and African podocarp trees are as old as we think they are. But although the basic pattern is simple, it bears an extraordinarily diverse overlay of details which, if modified even slightly, will have very visible and large effects on the population and evolutionary biology of the animals and plants involved. Above all, the system has the property that as partners are eliminated by ecological processes or evolution, seed dispersal is passed to the next animal and seed dispersers are passed to another plant. This continuity is promoted by the fact that the interaction benefits both members; vertebrates have many dietary needs in common, with the result that once a structure has been selected to be edible to one species of animal it is likely to be edible to at least some others, and once an animal has been selected to find and eat one species of fruit (or seed) it is likely to find and eat another. It generally does not proceed to a tightly coevolved one-on-one relationship[1] because there are powerful forces acting against the evolution of the year-round high quality fruit production by one species that would be needed to support a completely monophagous vertebrate; in addition, a single vertebrate species is unlikely to reliably generate a seed shadow of as high quality as can certain kinds of disperser coteries made up of several species of vertebrates and with a composition that changes over the geographical range of the plant. Finally, even with a tightly coevolved mutualism, the habitat occupied by a plant is normally of sufficient heterogeneity that in at least some part of the range the plant will be able to survive with no seed dispersal other than that that occurs by abiotic processes or sloppiness of seed predators.

Fruit ripening

Fruits do not merely hang on a plant until temperature-dependent physiological processes have run their course. Fruits and the seeds they contain mature synchronously yet are genetically and biochemically quite different objects; this alone is sufficient evidence that the time to maturation is not dictated by the chemistry of physiological processes. Rather, fruit and seed development time has been evolutionarily adjusted to some optimal duration in the context of seed disperser availability and hunger, photosynthate accumulation and dispensation, predispersal seed predation, postdispersal seed death, and optimal germination times. Fruits and seeds clearly do not ripen at random with respect to the calendar. There is no optimal date for a fruit to ripen within a plant's fruit crop, but rather there is an optimal distribution of times of fruit ripening. A crop of 5000 large indehiscent juicy fruits and a herd of horses may generate a seed shadow with very high fitness for the parent tree if 10 ripe fruits fall each day for 50 days in a tropical dry season; but if all 5000 fruits fall in a week of a tropical rainy season, disperser satiation, microbes, disperser attraction to succulent vegetation, and the five days needed for the horses to first locate the fruits will take a heavy toll of the tree's offspring.

Selective forces on a ripe fruit differ from those on the rest of the plant. Except for flower nectar and some pollen, all other parts of a plant have been under eons of selection for traits to reduce their digestibility and to poison consumers. For example, cellulose is the most widely distributed and abundant coevolved natural product, and almost no animal can digest it. It

[1] The purest kind of coevolution has evolutionary change in one species-lineage influencing and being influenced by evolutionary change in one other species-lineage. Many coevolutionary relationships are not this simple: here Janzen notes that any one fruit producing species will coevolve with a number of vertebrate fruit-eating species.

is very likely not to be an accident that among all the possible structural carbohydrates, an inedible one has become the dominant one; can you imagine the defenses a plant would have to have against herbivores if cellulose were easily digested?

It is not surprising that cellulose is not the structural carbohydrate in most edible fruit pulp (most exceptions are fiber-rich and woody fruits that are eaten by large browsing vertebrates, that is, animals whose guts contain microbial and protozoan specialists at degrading cellulose-rich plant parts). Ripe fruits have been under only slightly fewer eons of selection to be edible to a select subset of the vertebrates in the habitat and to be toxic or indigestible to the remaining animals and microbes.

Ripe fruit also differs from vegetative parts in a strategic sense. When a ripe fruit is removed by an animal or is shed, the parent has no feedback system to evaluate the cause or consequence of that removal; on the other hand, in many situations herbivory or other tissue loss can be evaluated by the plant as to cause, intensity, and kind, and repair and other facultative responses can be conducted.

Fruit ripening requires that a tissue that has been photosynthetic and intensely antibiotic to animals or pathogens abruptly become non-photosynthetic and highly edible to only a very small subset of the enormous array of herbivores and frugivores in the habitat. The intense nature of chemical and morphological defenses of green fruits is the consequence of the nutrient-rich and fitness-rich developing embryos and other tissues within. The need to convert all or part of a well-protected tissue to one that is highly edible puts numerous restraints on the kinds of defenses the immature fruit can have, the kinds of dispersal agents that can be in the disperser coterie, and the speed with which a fruit can ripen. [. . .]

There are enough dietary needs in common among species that process fruit that when a tissue evolves to where it is edible to one species, the probability is high that its edibility will rise for some other species, ranging from the seed disperser's congenerics to microbes. A ripe fruit cannot therefore be expected to have the life span displayed by other parts of the parent. Fruit ripening is a game of positioning highly perishable items in space and time rather than simply a matter of putting out a bowl of cherries and letting the seed dispersers take their fill as their fancy and numbers dictate. Seed dispersers must locate and harvest a highly desirable and perishable object of capricious and short-term fixed location, and they do not clean it of its contaminants in the harvest. Humans have a very biased view of fruits; even though recognized as perishable, domestic fruits have been artifically selected to be less perishable and have been provided when ripe with the same chemical protection (pesticides) that was evolutionarily discarded in nature in order to get the seeds into the right animals.

Disperser arrival and feeding

Animals arrive at a plant's fruit crop in response to memory, location cues (odor, color, locality), instructions and cues from other animals, hunger, proximity of other fruit crops (allospecific as well as conspecific), fruit quality, random movement, and so forth. The wind—the other prominent disperser of seeds—arrives irrespective of a fruit crop's traits, except for fruiting phenology and plant (and fruit) location. The wind cannot be satiated. Animals may prey on seeds, eat fruit and discard or ignore seeds, eat fruit- and seed-eaters, socialize, eat fruit and cache seeds, and eat fruit and ingest seeds as contaminants. Although only the latter two categories are the focus of this chapter, the others have a strong influence on the composition of the disperser coterie (and hence seed shadow) of any particular plant or species bearing ripe fruit. (A plant's disperser coterie is that array of animal individuals and species that generate its seed shadow; when most narrowly defined, each member of a plant population may have a slightly different disperser coterie.) An individual animal often does several of the preceding acts on a given day, to a given fruit crop, during the year, or during its life cycle, and may do a different combination of these things to each of an array of species of plants. Most seed dispersers are simultaneously dispersing several species of seeds in a particular habitat. Within a time span of a few minutes, a Central American agouti (*Dasyprocta punctata*) may prey on seeds, cache seeds, defecate seeds, eat fruit pulp, fight over fruits, and eat smaller seed predators found in the fruit. The result of this complexity is that dispersal cannot be characterized by merely listing visitors to fruiting plants, by recording how many fruits are 'eaten' by which animals, or by listing plants with fruits whose seeds appear to be dispersed by this or that vertebrate.

However, all is not chaos. Certain species and individuals are much more likely to treat a certain fruit crop in a certain manner than are others. It is safe to speak of parrots (Psittacidae) and peccaries (Tayassuidae) as usually being seed predators, and toucans (Ramphastidae) and horses as usually being seed dispersers. However, understanding any particular plant and its seed disperser coterie requires a detailed understanding of how different animals treat the fruit crop and where the seeds go. There is a large procedural difference between generating generalizations on dispersal and in manipulating or understanding the idiosyncracies of a particular plant and its disperser coterie. The statements about 'average' individuals are likely to be very far out of focus unless the data are collected from individuals known to be representative of a particular distribution.

Animals at a fruit crop display many behaviors that strongly influence their subsequent dispersal of seeds. They select among fruits, become satiated, are sloppy, spit out seeds, fight and flee, defecate seeds swallowed long ago, and damage uneaten fruits. However, although seeds dropped, spit, regurgitated, or defecated below the parent plant may be abortions of the fruit's dispersal mission, in some habitats this seed placement is only the first step in a complex and somewhat serendipitous further dispersal that may generate a portion of a seed shadow quite different from that anticipated by mere consideration of the fruit and vertebrate for which the fruit appears to have been evolutionarily designed. In addition, seeds below the parent are often picked up and carried off by a variety of seed predators who later lose them or never get around to eating them.

But the variation in the seed shadow is not generated solely by the variation in animal responses to fruits and seeds. Fruits and seeds are highly variable within and between species, and many of these variations can be easily interpreted in the context of influencing the ensuing seed shadow. Why do large seeds and nuts often have variously strong fibrous connections between them and the fruit pulp eaten by a vertebrate seed disperser? Why do fruits contain highly variable numbers of seeds within and between crops? Why does fruit seediness vary within and between seed crops? Why do seeds vary 2- to 3-fold in weight and volume within a fruit crop? Such questions have never received botanical attention equivalent to the attention bestowed on frugivorous and seed dispersing vertebrates.

Disperser departure from the fruit crop

Vertebrates move away from fruit crops to search for other kinds of food, to escape predators, to return to a nest or to distant feeding perches, to visit superior fruit crops elsewhere, and for many other reasons. Many of the details of a seed shadow will depend on when and how an animal leaves a fruit crop. The easiest detail to understand is that the longer an animal stays at the fruit crop, the more likely it is to spit, drop, defecate, or abandon the seeds directly below the parent plant. If there were ever an evolutionary prediction, it is that plants that bear fruits containing animal-dispersed seeds are likely to have chemical and morphological traits that make it unpleasant or difficult for seed dispersers to perch in them. Fruit-free seeds that end up below the parents plant not only have to compete with one of the worst competitors in the habitat but are also usually in an area of high concentration of conspecific seeds that is the focal area of search by seed predators and a hotbed of highly competitive siblings.

The patterns of movement of seed dispersers to and from a fruit crop are influenced by many fruit and seed traits such as size and other nutrient rewards of fruits, pattern of fruit ripening within the plant's crop, seediness of fruits, difficulty of getting at the fruit reward, conspicuousness of the crop, size of the crop, location of the crop within the plant crown, and habitat in general.

Disperser defecation and voiding of seed

Once a seed has been carried away from the fruit crop, its misery is by no means ended. If inside the animal, it may be ground up by molars or gizzard, digested anywhere from stomach to caecum to colon, scarified and thereby seduced into (probable) lethal germination (owing to the anaerobic conditions in animal guts), or just detained for the wrong amount of time. If outside the animal, it may be discarded in an intact fruit (often lethal owing to fruit decomposition or a fruit hull that the seedling cannot penetrate), chipped up, buried and later recovered, or buried in a microhabitat where it has little or no chance of survival as a seed or seedling even if forgotten or lost.

What happens to a seed in an animal (or in its paws) depends not only on the animal's traits, but very strongly on detailed seed traits such as seed coat texture, hardness, contour and toughness, volume, specific gravity, seed numbers per fruit and seed/fruit pulp ratios, weight, tenacity to fruit parts such as fibers and hard endocarp, elasticity, and permeability to germination cues. In addition, the fruit does more than just attract the disperser. Laxative chemicals may speed the seed's transit through the animal; hard items in the fruit may prevent molar occlusion on a soft seed; lubricating chemicals in the fruit pulp may render the seed more slippery and therefore difficult to sort out of the chewy fruit mass by the tongue (for spitting or more intense grinding); mildly astringent or otherwise annoying chemicals in the fruit may cause the animal to consume only small amounts at a particular feeding.

Even if the seed passes unharmed through the animal, many animals will defecate it in inappropriate sites (e.g., dry caves, cavities in trees, ponds or rivers, sun-baked soil). Sites that are good for the seeds of species A may be horrible for the seeds of species B; thus, desirable members of the seed disperser coterie will be different for different plants. There is no selection for seed dispersal to oceanic islands; most seeds that start the trip are defecated into the ocean, and those that make it do not report their success back home.

The high percentage of mortality often recorded for the seeds passing through an animal's gut may suggest that the animal is a low-quality disperser. However, virtually all dispersal systems kill most seeds by putting them in inappropriate habitats; highly directed seed flow (in animal dung) may in fact result in just as many surviving saplings as the much more gentle but blind generation of seed shadows by wind or water.

The most is not necessarily, or even likely, the best with respect to how far a disperser carries seeds and how many seeds a disperser carries away from the fruit crop. Beyond the immediate vicinity of the parent plant, the quality of sites isn't correlated with distance. A species of seed disperser that carries off the largest number of seeds from a fruit crop will not necessarily distribute them over scattered suitable sites. Ten seeds defecated by a tapir

(*Tapirus*) in a flowing creek may be worth 100,000 of the same species defecated by a horse in adjacent drought- and fire-prone grassland.

Seed shadows, just like subsequent germination regimes, have a temporal component. Animals with transit times of weeks or months may not only lay down a seed shadow in many different habitats (simultaneously with the seed shadows of seeds eaten at other times), but may also lay down a seed shadow in a different season from the one in which the fruits were eaten. This is especially apparent with large fruits eaten toward the end of a tropical dry season by large mammals that have long retention times for large seeds.

And sometime later, germination

Seeds are subjected not only to competitive and physical environments, but to severe postdispersal seed predation. Large genera of insects and vertebrates make their living digesting seeds, and no small number of fungi kill dispersed seeds. For many species of plants in mainland habitats, the majority of seeds in a crop are killed by animals and disease rather than by competition with other plants. The traits of the seed shadow, and especially the location and intensity of peaks in it, have a strong influence on the degree and kind of postdispersal seed predation. Seed predators search no more at random than do any other foragers. Even the timing of behavioral exit from the seed shadow (germination) may be determined by animals through the degree of scarification of the seed coat during passage through the disperser.

Germination needs a few words in passing. Perhaps one of the most poorly conceived and used concepts in plant biology is *germination rate*. First, it is usually not a rate but rather a percentage. Second, what is usually reported may matter to farmers, but it is quite irrelevant to the biology of wild plants. Third, the ungerminated seeds are often assumed to be dead. I am often asked, 'But doesn't passage through the animal enhance germination?'; 'Don't plants need dispersal to improve their germination success?' Or, 'If germination is not enhanced by animals, why do plants need dispersers?' The implication is that somehow an increased percentage of immediate germination is somehow better for the plant, that ungerminated seeds are dead, and that the behavioral trait of germination is not subjected to the same evolutionary influence as are all other plant traits. Quite to the contrary, seed coats (and nut walls in drupes) are evolutionarily designed to withstand the rigors of a variety of animals' molars, gizzards, intestinal acids and enzymes, anaerobic atmospheres, and so forth and to pass out the other end with enough seed coat remaining to protect the embryo from weather, false germination cues, and seed predators until the appropriate germination conditions come along. The same applies to other fruit traits such as laxatives that speed seed passage through the animal and hard objects in the fruit that prevent molar occlusion. A seed coat that is too tough survives all conditions, but the embryo cannot know when to germinate. One that is too weak fails one of the many challenges inside or outside the animal. The extreme case is that where the parent plant pays offspring in the form of edible seeds (e.g., acorns, pine nuts, herbaceous legume seeds) for the dispersal of its surviving offspring. Here there is selection for a seed coat or nut wall that is penetrable enough that the animal remains interested but yet impenetrable enough not to make the seed available to all potential seed predators. An alternative solution is mass seeding, as in bamboos and many conifers. Here there is little protective seed coat or dispersal by animals, but escape occurs through seed predator satiation and dispersal occurs by wind or nothing.

Seeds that are sufficiently scarified by transit through the disperser to start germinating usually either die in the animal through digestion or in the dung outside of the animal through pathogens, dung processors, or seed predators searching in dung. Seedlings growing out of cattle dung are picturesque, but they are photographs of dead plants. Those who think it amazing that holly berry seeds or nutlets (*Ilex*) require either passage through a bird or treatment with strong acids to respond to germination cues seem unaware that the normal cycle of a holly seed (*Ilex*) is to survive a short trip through a hostile environment or a much longer stay in a more benign one (the litter). We certainly do not know enough about what yields maximum holly tree recruitment to know what pattern of germination in time is best for a holly tree seed crop.

['Dispersal of Seeds by Vertebrate Guts'. In D. Futuyma and M. Slatkin (eds.), *Coevolution* (Sunderland, Mass.: Sinauer, 1983), ch. 11.]

DAN-E. NILSSON AND SUSANNE PELGER

50 A pessimistic estimate of the time required for an eye to evolve

Eyes appear to be such complex structures that it might naively be thought that the evolution by natural selection of an eye would take a long time. The paper estimates that eyes could evolve in less than half a million generations. The estimate comes from a model, in which natural selection favours improvements in visual acuity. Visual acuity can be improved as an initially light-sensitive region of the external surface of the animal evolves into a pinhole camera, and then by the evolution of a lens. [Editor's summary.]

When Charles Darwin (1859) presented his theory of evolution he anticipated that the eye would become a favourite target for criticism. He openly admitted that the eye was by far the most serious threat to his theory, and he wrote: 'that the eye . . . could have been formed by natural selection seems, I freely confess, absurd in the highest possible degree'. Although the problem is principally important, it gradually lost its scientific potency, and has now almost become a historical curiosity. But eye evolution continues to fascinate, although the question is now one of process rate rather than one of principle.

Estimates of the number of generations required to make a certain change to a simple quantitative character[1] are easily made if the phenotypic variation, selection intensity and heritability of the character are known. The evolution of complex structures, however, involves modifications of a large number of separate quantitative characters, and in addition there may be discrete innovations and an unknown number of hidden but necessary phenotypic changes. These complications seem effectively to prevent evolution rate estimates for entire organs and other complex structures. An eye is unique in this respect because the structures necessary for image formation, although there may be several, are all typically quantitative in their nature, and can be treated as local modifications of pre-existing tissues. Taking a patch of pigmented light-sensitive epithelium as the starting point, we avoid the more inaccessible problem of photoreceptor cell evolution. Thus, if the objective is limited to finding the number of generations required for the evolution of an eye's optical geometry, then the problem becomes solvable.

We have made such calculations by outlining a plausible sequence of alterations leading from a light-sensitive spot all the way to a fully developed lens eye. The model sequence is made such that every part of it, no matter how small, results in an increase of the spatial information the eye can detect. The amount of morphological change required for the whole sequence is then used to calculate the number of generations required. Whenever plausible values had to be assumed, such as for selection intensity and phenotypic variation, we deliberately picked values that overestimate the number of generations. Despite this consistently pessimistic approach, we arrive at only a few hundred thousand generations!

A model of eye evolution

The first and most crucial task is to work out an evolutionary sequence which would be continuously driven by selection. The sequence should be consistent with evidence from comparative anatomy, but preferably without being specific to any particular group of animals. Ideally we would like selection to work on a single function throughout the entire sequence. Fortunately, spatial resolution, i.e. visual acuity, is just such a fundamental

[1] A quantitative character is one that varies continuously: body height, for instance. It can be measured simply, and there are usually small differences between individuals, as some people are taller than others. Contrast this with discrete characters, such as gender, that have a number of discrete states. In the next line, selection intensity and heritability are both technical terms; but they can be informally understood as how strongly selection is acting on a character, and to what extent differences between individuals are due to genes.

aspect and it provides the sole reason for an eye's optical design. Spatial resolution requires that different photoreceptor cells have different fields of view. A comparison of their signals then gives information about the direction of the incident light. The smaller the field of view of each individual intensity channel, the better is the potential for accurate spatial information. It does not matter if the spatial resolution is used for measuring self-motion, detection of small targets, or complicated pattern recognition, the fundamental aspect of information is the same, and so are the demands on eye design.

We let the evolutionary sequence start with a patch of light-sensitive cells, which is backed and surrounded by dark pigment, and we expose this structure to selection favouring spatial resolution. We assume that the patch is circular, and that selection does not alter the total width of the structure. The latter assumption is necessary to isolate the design changes from general alterations of the size of the organ. There are two ways by which spatial resolution can be gradually introduced: (i) by forming a central depression in the light-sensitive patch; and (ii) by a constriction of the surrounding pigment epithelium. Both these morphological changes reduce the angle through which the individual light-sensitive cells receive light. [. . .] Initially, deepening of the pit is by far the most efficient strategy, but when the pit depth equals the width, aperture constriction becomes more efficient than continued deepening of the pit. We would thus expect selection first to favour depression and invagination of the light-sensitive patch, and then gradually change to favour constriction of the aperture. During this process a pigmented-pit eye is first formed, which continues gradually to turn into a pinhole eye.

As the aperture constricts, the optical image becomes increasingly well resolved, but constriction of the aperture also causes the image to become gradually dimmer, and hence noisier. It is the random nature of photon capture that causes a statistical noise in the image. When the image intensity decreases, the photon noise increases in relative magnitude, and the low contrast of fine image details gradually drowns in the noise. If we assume that the retinal receptive field, $\Delta\rho_r$, and the optical blur spot, $\Delta\rho_{lens}$, are identical Gaussians with half-widths being the angle subtended by the aperture at a central point in the retina (this effectively means that the retinal sampling density is assumed always to match the resolution of the optical image), then we can [. . .] obtain the maximum detectable spatial frequency, ν_{max}, as:

$$\nu_{max} = (0.375P/A)\,[\ln(0.746A^2\sqrt{I})]^{\frac{1}{2}}, \tag{1}$$

where A is the diameter of the aperture, P is the posterior nodal distance, or pit depth and I is the light intensity in normalized units of 'photons per nodal distance squared per second per steradian'. We can now use this relation to plot resolution against aperture diameter. For a given ambient intensity and eye size there is an optimum aperture size where noise and optical blur are balanced in the image. A large eye or high light intensity makes for an optimum aperture which is small compared with the nodal distance. When the aperture has reached the diameter which is optimal for the intensity at which the eye is used, there can be no further improvement of resolution unless a lens is introduced.

In a lensless eye, a distant point source is imaged as a blurred spot which has the size of the imaging aperture. A positive lens in the aperture will converge light such that the blur spot shrinks, without decreasing the brightness of the image. Most biological lenses are not optically homogeneous, as man-made lenses normally are. In fact, a smooth gradient of refractive index, like that in fish or cephalopod lenses, offers a superior design principle for making lenses: the optical system can be made more compact, and aberrations can be reduced considerably. A graded-index lens can be introduced gradually as a local increase of refractive index. As the focal length becomes shorter, the blur spot on the retina will become smaller. The effect this has on resolution was calculated [. . .] for an ideal graded-index lens. Even the weakest lens is better than no lens at all, so we can be confident that selection for increased resolution will favour such a development all the way from no lens at all to a lens powerful enough to focus a sharp image on the retina.

Camera-type eyes of aquatic animals typically have a spherical graded-index lens which is placed in the centre of curvature of the retina. With this arrangement they achieve virtually aberration-free imaging over a full 180° visual field. Another typical feature is that the focal length of the lens is 2.55 times the lens radius. This relation, called Mattiessen's ratio, represents the ideal solution for a graded-index lens with a central refractive index of 1.52, a value close to the upper limit for biological material. However, the best position for a lens to be introduced in a pinhole eye is in the aperture, clearly distal to the centre of curvature of the retina. Because the central and peripheral parts of the retina will then be at different distances from the lens, there is no need for the lens to be spherical. In fact, an ellipsoid lens is better because it can compensate optically for the difference in retinal distance. Furthermore, the size of the first-appearing lens is determined by the aperture, and need not have the size which will finally be required. As the lens approaches focused conditions, selection pressure gradually appears to move it to the centre of curvature of the retina, to make it spherical, and to adjust its size to agree with Mattiessen's ratio.

Based on the principles outlined above, we made a model sequence of which representative stages are presented in Figure 50.1. The starting point is a flat light-sensitive epithelium, which by invagination forms the retina of a pigmented pit eye. After constriction of the aperture and the gradual

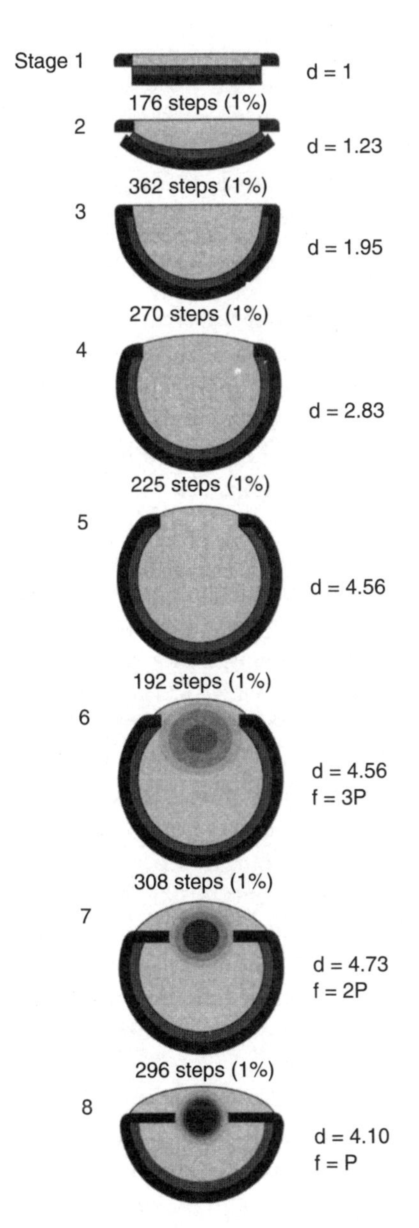

Figure 50.1: Representative stages of a model sequence of eye evolution. In the initial stage (1) the structure is a flat patch of light-sensitive cells sandwiched between a transparent protective layer and a layer of dark pigment. In stages 2 and 3 the photoreceptor layer and pigment layer (hereafter collectively termed the retina) invaginates to form a hemisphere. The protective layer deepens to form a vitreous body which fills the cavity. The refractive index of the vitreous body is assumed to be 1.35, which is only slightly higher than that of water, and not enough to give the vitreous body any significant optical effect. In stages 4 and 5 the retina continues to grow, but without changing its radius of curvature. This causes a gradual shift from deepening of the retinal pit to constriction of the distal aperture. The aperture size in stage 6 was chosen to reflect the typical proportions in real eyes of this type. In stages 6–8 a graded-index lens appears by a local increase in refractive index. The central refractive index of the lens grows from the initial value of 1.35 to 1.52 in the final stage. Simultaneously the lens changes shape from ellipsoid to spherical and moves to the centre of curvature of the retina. As the lens shrinks, a flat iris gradually forms by stretching of the original aperture. The focal length (f) of the lens gradually shortens, and in stage 8 it equals the distance to the retina (P), producing a sharply focused system. The relative change in receptor diameter, required to keep sensitivity constant throughout the sequence, is indicated by the normalized receptor diameter d. *The anatomical change between model stages is given as the number of 1% modification steps.*

formation of a lens, the final product becomes a focused camera-type eye with the geometry typical for aquatic animals (e.g. fish and cephalopods).

The changes in size and position of the aperture cause variations in image brightness in the model sequence. To account for this we have assumed that the receptor diameter is continuously modified such that the photon catch per receptor, and thus the signal to noise ratio, is kept constant throughout the sequence. As the model is of arbitrary size, we have used a normalized receptor diameter (d) which is 1 at stage 1 in Figure 50.1.

The model sequence of Figure 50.1 contains a number of structural elements whose shape and size are gradually modified. To quantify these changes we calculate the number of sequential 1% steps of modification it takes between each stage in Figure 50.1. For example, a doubling of the length of a structure takes 70 steps of 1% ($1.01^{70} \approx 2$). Note that the last step is twice as long as the first in this example. The principle of 1% steps can be applied to changes of any quantitative character. Each structure of the model eye was analysed individually, as if no change follows passively from any other.

There are unavoidable ambiguities in measuring morphological change, because a product will have to be compared with a subjectively chosen origin. It would thus be possible to claim that a doubling in length of a structure is really a three times stretching of the outer half. Both views are correct, but they give different quantifications of the change. Measurements of phenotypic variation in a population suffer from the same type of subjectiveness. As we are going to relate our measures of morphological change only to general estimates of phenotypic variation, we will be safe as long as we avoid unorthodox and strange ways of comparing origin and product. Our principles have been to use whole length measurements of straight structures, arc length of curved structures, and height and width of voluminous structures. Changes in the radius of curvature were accounted for by calculating the arc length of both the distal and proximal surfaces of the structure. Refractive index was related to protein concentration, by assuming that values above 1.34 are due to proteins alone.

The calculated number of 1% changes required between each of the stages (see Fig. 50.1) was plotted against the optical performance of each stage, which was calculated as the number of resolvable image points within the eye's visual field. The [. . .] spatial resolution improves almost linearly with morphological change. There are thus no particularly inefficient parts of the sequence, where much change has to be made for little improvement of function.

Altogether 1829 steps of 1% are needed for the entire model sequence. Natural selection would act simultaneously on all characters that positively affect the performance. In our model there are several transformations that would speed up the improvement of function if they occurred in parallel. True to our pessimistic approach, we deliberately ignored this and assumed that all 1829 steps of 1% change occur in series. This is equivalent to a single structure becoming 1.01^{1829} or 80129540 times longer. In terms of morphological modification, the evolution of an eye can thus be compared to the lengthening of a structure, say a finger, from a modest 10 cm to 8000 km, of a fifth of the Earth's circumference.

The number of generations required

Having quantified the changes needed for a lens eye to evolve, we continue by estimating how many generations such a process would require. When natural selection acts on a quantitative character, a gradual increase or decrease of the mean value, m, will be obtained over the generations. The response, R, which is the observable change in each generation is given by the equation

$$R = h^2 i \sigma_p \quad \text{or} \quad R = h^2 i V m, \tag{2}$$

where h^2 is the heritability, i.e. the genetically determined proportion of the phenotypic variance, i is the intensity of selection, V is the coefficient of variation, which measures the ratio between the standard deviation, σ_p, and the mean, m, in a population. For our estimate we have chosen $h^2 = 0.50$, which is a common value for heritability, while deliberately low values were chosen for both i (0.01) and V (0.01). The response obtained in each generation would then be $R = 0.00005m$, which means that the small variation and weak selection cause a change of only 0.005% per generation. The number of generations, n, for the whole sequence is then given by $1.00005^n = 80129540$, which implies that $n = 363992$ generations would be sufficient for a lens eye to evolve by natural selection.

Discussion

Eyes closely resembling every part of the model sequence can be found among animals existing today. From comparative anatomy it is known that molluscs and annelids display a complete series of eye designs, from simple epidermal aggregations of photoreceptors to large and well-developed camera eyes. The structural components of our model have counterparts with different embryological origin in different groups. For modelling purposes these differences are irrelevant because selection for the various functions will operate on whatever tissue is present at the place where the function is needed.

The development of a lens with a mathematically ideal distribution of refractive index may at first glance seem miraculous. Yet the elevation of refractive index in the lenses of both vertebrates and cephalopods is caused by proteins that are identical or similar to proteins with other cellular functions.[2] Selection has thus recruited gene products that were already there. Assuming that selection operates on small but random phenotypic variations, no distribution of refractive index is inaccessible to selection. It is an inevitable consequence of selection for improved resolution that the population average is continuously adjusted towards the ideal distribution of refractive index. The lens should thus be no more difficult to evolve than any other structure of the model.

It is important that the model sequence does not underestimate the amount of morphological change required. The only real threat to the usefulness of our model is that we may have failed to introduce structures that are necessary for a functional eye. Features of many advanced eyes, such as an adjustable iris and structures for distance accommodation, may in this context seem to be serious omissions from the model sequence. The function of these structures is to make the eye more versatile so that it can perform maximally over a greater range of distances and ambient intensities. However, the improved function brought about by the sequence of modifications in our model does not in any way depend on the existence of these auxiliary structures. It is in fact the other way around: evolution of these refinements requires the existence of the structures that develop in our model.

Vertebrates and cephalopods have a vascularized layer, the choroid, and a supporting capsule, the sclera, in their eyes. The demand for blood supply and structural support comes from the general lifestyle and size of these animals, and the demands involve the entire body, not just the eyes. The complete absence of choroid and sclera in the well-developed camera eyes of polychaetes and gastropod molluscs shows that such structures are not mandatory for eye evolution, and thus are not needed in our model.

The organization and specialization of the cells in the retina require some attention. The photoreceptor cells of advanced eyes are certainly not identical to those of simple light-sensitive spots, but even primitive photoreceptors should be good enough to make improvements of optical resolution worth-while. Improvements of efficiency, and specializations for polarization sensitivity and colour vision, are in no way required for selection to favour improvements of spatial resolution. The nervous tissue in the vertebrate retina can also be ignored as this is clearly a part of the nervous system which just happens to reside in the eye. No invertebrate eyes have an arrangement of this kind.

As far as we can tell, no structure of the eye has been omitted whose presence or development would in any way impede the evolutionary process. Further, we can be sure that real selection would outperform our model

[2] See Chapter 40 on lens evolution.

sequence in finding the modifications that give the best improvement of function for the least morphological change.

Can we be sure that our calculations do not underestimate the number of generations required for the optical structures to evolve? Throughout the calculations we have used pessimistic assumptions and conservative estimates for the underlying parameters. Should one or perhaps even two of these assumptions or estimates in fact be optimistic, we can trust that the remaining ones will at least compensate for the errors made. It is more likely, though, that the complete calculation substantially overestimates the number of generations required.

If we assume a generation time of one year, which is common for small and medium-sized aquatic animals, it would take less than 364000 years for a camera eye to evolve from a light-sensitive patch. The first fossil evidence of animals with eyes dates back to the early Cambrian, roughly 550 Ma [million years] ago. The time passed since then is enough for eyes to evolve more than 1500 times!

If advanced lens eyes can evolve so fast, why are there still so many examples of intermediate designs among recent animals? The answer is clearly related to a fact that we have deliberately ignored, namely that an eye makes little sense on its own. Although reasonably well-developed lens eyes are found even in jellyfish, one would expect most lens eyes to be useless to their bearers without advanced neural processing. For a sluggish worm to take full advantage of a pair of fish eyes, it would need a brain with large optic lobes. But that would not be enough, because the information from the optic lobes would need to be integrated in associative centres, fed to motor centres, and then relayed to the muscles of an advanced locomotory system. In other words, the worm would need to become a fish. Additionally, the eyes and all other advanced features of an animal like a fish become useful only after the whole ecological environment has evolved to a level where fast visually guided locomotion is beneficial.

Because eyes cannot evolve on their own, our calculations do not say how long it actually took for eyes to evolve in the various animal groups. However, the estimate demonstrates that eye evolution would be extremely fast if selection for eye geometry and optical structures imposed the only limit. This implies that eyes can be expected to respond very rapidly to evolutionary changes in the lifestyle of a species. Such potentially rapid evolution suggests that the eye design of a species says little about its phylogenetic relationship, but much about its need for vision. It follows that the many primitive eye designs of recent animals may be perfectly adequate, and simply reflect the animal's present requirements. In this context it is obvious that the eye was never a real threat to Darwin's theory of evolution.

[*Proc. of the Royal Society of London*, ser. B, 256 (1994), 53–8.]

JOHN GERHART AND MARK KIRSCHNER

51 Evolutionary novelty: the example of lactose synthetase

Milk is a key evolutionary innovation in mammals, and is rich in a sugar called lactose. Lactose is manufactured by the enzyme lactose synthetase. Lactose synthetase originated by combining two unrelated enzymes, galactosyl transferase and α-lactalbumin (which evolved from lysozyme). [Editor's summary.]

Milk production is a defining characteristic of mammals. It probably arose in the earliest mammal-like reptiles, and anatomically complex mammary glands are not only present in eutherians and marsupials, but also in non-placental egg-laying monotremes. Lactation allowed important changes to take place in reproductive strategy and ultimately in social development. The marsupial fetus, which develops a late shallow placenta in the uterus, is born when still very small and undeveloped, and suckles in the pouch at a teat that grows in size with the young and delivers milk of a changing composition as the baby gets bigger. The duration of embryonic development is short, and much development that would occur *in ovo* in reptiles and birds or *in utero* in placental animals now occurs after birth.

The secretions of the first mammal-like animals may originally have been a precaution against bacterial infection. Milk contains a variety of microbial inhibitors that are also found in bird's eggs, such as the enzyme lysozyme and the metal chelator lactoferrin. In the early mammals, these secretions may have become more and more a source of nutrition for the young. Along with the modification of a secretory mammary gland and changes in developmental strategy, distinct biochemical changes occurred to generate milk. The major milk proteins—the caseins—are secreted by the mammary gland as aggregates or micelles. They have diverged rapidly, and κ-casein, which is involved in milk clotting, may have been generated from γ-fibrinogen about 250 million years ago during the divergence of mammals from reptiles.

Eutherian milk (as opposed to that of marsupials and monotremes) is also rich in lactose, a disaccharide produced only in the female mammary gland during lactation and not found in large amounts in any other animals. Lactose is secreted from mammary gland cells along with casein and other proteins by the classic pathway of exocytosis. Human milk has an unusually high lactose content (0.2 M) and relatively low protein and fat. The synthesis of lactose depends on the enzyme lactose synthetase, which makes lactose from UDP-galactose and glucose. Lactation is unique to mammals, and it is interesting to ask how the enzyme lactose synthetase originated.

Origin of lactose synthetase

Lactose synthetase is composed of two subunits that form a tight non-covalent complex. One subunit is a galactosyl transferase, which on its own will transfer galactose from UDP-galactose to *N*-acetylglucosamine residues on proteins in the Golgi apparatus. The addition of galactose is an important step in directing secreted proteins to the proper subcellular compartment in all eukaryotes. Galactosyl transferase can also use glucose as an acceptor and synthesize lactose, but this reaction is very inefficient because the affinity of the enzyme for glucose is very low, requiring syrupy concentrations of glucose of 0.45 g/ml for half saturation. The other subunit of lactose synthetase is α-lactalbumin, a protein found only in milk. The complex of α-lactalbumin and galactosyl transferase increases the affinity of the transferase for glucose by 1000-fold. At the same time, α-lactalbumin inhibits the transfer of galactose to *N*-acetylglucosamine, thus driving the pathway toward lactose synthesis.

The key evolutionary step for lactose production in eutherian mammals must have been the generation of α-lactalbumin, a protein that modifies the substrate affinity of galactosyl transferase. Although α-lactalbumin would seem to be a novel mammalian protein, its sequence and structure are very close to that of hen egg white c-type lysozyme (calcium-binding lysozyme), an enzyme that cleaves oligosaccharides on bacterial cell walls, and thus defends against bacterial attack. This particular class of lysozymes originated at least as early as fish and may have diverged from the other lysozyme types before the onset of vertebrate evolution. The α-lactalbumin subunit of lactose synthetase does not, however, retain lysozyme activity. There is about a 35–40% identity between α-lactalbumin and lysozyme sequences from mammals, birds, and lower vertebrates, and extraordinary similarity in three-dimensional structure. The α-lactalbumins from baboon and human milk superimpose over much of the structure of the pigeon egg white lysozyme as well as the lysozymes from the echidna (the spiny anteater, a monotreme) and from horse milk. This is yet another example of a stronger conservation of structure than of amino acid sequence.

There must have been a slow evolution of lysozymes in the direction of α-lactalbumin in early mammals without reference to lactose production; lactose production only arose after milk production had been established. The echidna may represent a revealing example of this early state of affairs. It makes little if no lactose, and presumably has no α-lactalbumin. The other extant monotreme species, the platypus, does have α-lactalbumin and makes lactose. The selection to produce a better and better α-lactalbumin as a subunit for lactose synthetase may have been associated with the value of more and more nutritious lacteal secretions. Whether or not the echidna is a living fossil that contains the missing bifunctional lysozyme/α-lactalbumin is unclear at present, but it seems very likely that α-lactalbumin evolution began with the fortuitous combination of lysozyme with galactosyl transferase, and that the activity of α-lactalbumin as a subunit of lactose synthetase was well established by the time of the ancestral monotreme line. Marsupials have diversified in slightly different ways. The tammar wallaby contains both lactose synthetase and a species-specific galactosyl transferase, and an α-lactalbumin that together convert lactose and UDP-galactose to trisaccharides. Thus the marsupials seem to have gone beyond the eutherian mammals in using α-lactalbumin as a subunit that modifies two different galactosyl transferases.

Much of the change that occurred during the evolution of lysozyme into α-lactalbumin had no apparent relation to its ultimate function in the mammary gland. From the perspective of protein structure very little modification was required to produce lactose synthetase. The two major protein constituents and their substrate binding sites were present generally in vertebrates. Once combined into a complex modest change would have produced the modern enzyme.

[*Proceedings of the Royal Society of London*, series B, 256 (1994), 53–8.]

PAUL D. SNIEGOWSKI, PHILIP J. GERRISH, TOBY JOHNSON, AND AARON SHAVER

52 The evolution of mutation rates: separating causes from consequences

Natural selection can adjust the rate of mutation in a population by acting on allelic variation affecting processes of DNA replication and repair. Because mutation is the ultimate source of the genetic variation required for adaptation, it can be appealing to suppose that the genomic mutation rate is adjusted to a level that best promotes adaptation. Most mutations with phenotypic effects are harmful, however, and thus there is relentless selection within populations for lower genomic mutation rates. Selection on beneficial mutations can counter this effect by favoring alleles that raise the mutation rate, but the effect of beneficial mutations on the genomic mutation rate is extremely sensitive to recombination and is unlikely to be important in sexual populations. In contrast, high genomic mutation rates can evolve in asexual populations under the influence of beneficial mutations, but this phenomenon is probably of limited adaptive significance and represents, at best, a temporary reprieve from the continual selection pressure to reduce mutation. The physiological cost of reducing mutation below the low level observed in most populations may be the most important factor in setting the genomic mutation rate in sexual and asexual systems, regardless of the benefits of mutation in producing new adaptive variation. Maintenance of mutation rates higher than the minimum set by this 'cost of fidelity' is likely only under special circumstances. [Authors' summary.]

Introduction

[. . .] There is ample evidence for genetic variation in the general DNA repair and replication processes that affect mutation, and thus there can be little doubt that the genomic rate of mutation in a population can be influenced by natural selection. Because mutation is required for long-term adaptive evolution, it might seem obvious that the need for new beneficial mutations is of primary importance in setting the genomic mutation rate. The problem with this viewpoint is that the vast majority of mutations that have any phenotypic effect are likely to be deleterious to individual fitness. The influx of deleterious mutations maintains continual selection pressure within populations in favor of lower genomic mutation rates. Indeed, it is likely that this selection to decrease mutation rates was the cause of some major evolutionary transitions such as the use of DNA instead of RNA as the hereditary molecule and the evolution of complex enzymatic proofreading and repair systems. As we discuss in more detail below, selection to decrease the deleterious mutation rate is likely to be much stronger than selection to increase the beneficial mutation rate under most circumstances.

Nonetheless, mutation persists in populations. A.H. Sturtevant thus highlighted the essential problem of mutation rate evolution in 1937 when he asked 'Why does the mutation rate not evolve to zero?' In the simplest possible analysis, a modern answer to Sturtevant's question relies on either (1) physicochemical or physiological constraints on evolutionary reductions in mutation rate to below the observed levels, or (2) selection for an increased rate of production of beneficial mutations. The challenge in understanding mutation rate evolution lies in evaluating the relative importance of these factors.

Theoretical progress in understanding the genetical evolution of mutation rates has been achieved by explicitly considering the effect of natural selection on the frequencies of alleles that modify the mutation rate (mutation rate modifiers) in populations. Selection on a mutation rate modifier can be classified as either direct or indirect. Direct selection is theoretically straightforward and depends on the effect (if any) of the modifier allele on fitness through factors other than its effect on mutation. Indirect selection, in contrast, depends on nonrandom association (termed 'linkage disequilibrium') between the modifier allele and alleles at other loci affecting fitness. Because linkage disequilibrium in a population is rapidly eroded by recombination, the efficacy of such indirect selection on the mutation rate is highly dependent on the recombination rate. In particular, because beneficial mutations are expected to be rare compared with deleterious mutations, indirect selection to increase the mutation rate is greatly weakened by recombination (see Figure 52.1), whereas indirect selection to decrease the mutation rate is less affected.

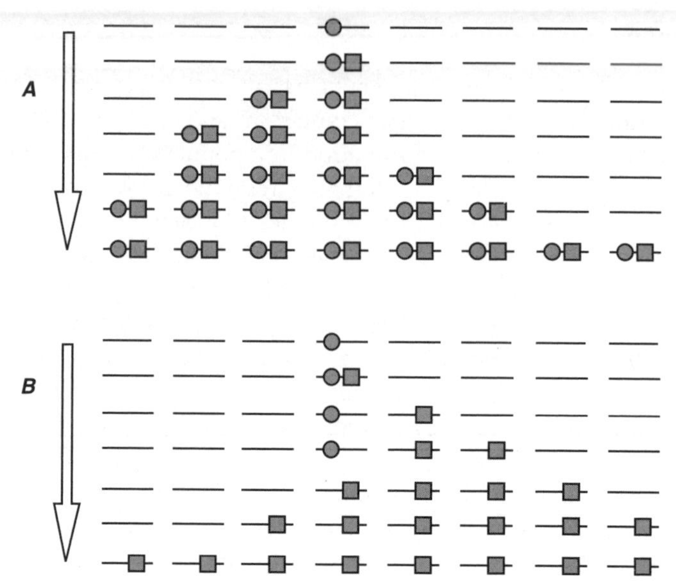

Figure 52.1: Indirect selection on the mutation rate. A modifier that increases the mutation rate (circle) tends to be preferentially associated with (in positive linkage disequilibrium with) a beneficial allele arising by mutation (square). **A:** *With complete linkage between the two loci, the modifier can hitchhike along with the beneficial allele as it sweeps to fixation in the population.* **B:** *Recombination disrupts the association between the modifier and the beneficial allele and decreases the probability of hitchhiking. Note that deleterious mutations are not shown; their prevalence creates a continual indirect selection in favor of reduced mutation rates.*

The mutation rate that evolves in a population thus depends on direct and indirect selection on modifier alleles, and the strength of indirect selection depends on the rate of recombination in the population. In the remainder of this review, we consider in more detail how these factors affect the mutation rate. We begin with a discussion of the evolution of equilibrium genomic mutation rates in both sexual and asexual populations; this is the area in which most theoretical work has been concentrated. Next we consider the sporadic evolution of high mutation rates in asexual populations; most empirical studies have been concentrated in this area. Then we briefly discuss some proposed ways in which the adaptive advantage of a high beneficial mutation rate may be achieved without a substantial increase in the rate of deleterious mutation. Finally, we suggest avenues for further research.

Equilibrium genomic mutation rates

As mentioned above, most theoretical work has concentrated on calculating the genomic mutation rate (hereafter μ_g) expected at equilibrium in an evolving population. This focus on genomic mutation rates was motivated by early analyses that suggested that selection would be too weak for most individual loci to evolve specific mutation rates. Empirical motivation for the theoretical interest in genomic mutation rates has come from the remarkable observation of Drake that in a range of DNA-based microbes with genome sizes spanning almost four orders of magnitude, estimated μ_g varied by considerably less than one order of magnitude (Fig. 52.2). This pattern suggests an evolutionary equilibrium value of μ_g that is independent of genome size in these taxa. In eukaryotes, there is no evidence for a constant value of μ_g (see Figure 52.2). [. . .]

In a seminal early paper, Kimura proposed two hypotheses for how, in the absence of beneficial mutations, the ultimate evolution of μ_g to zero under

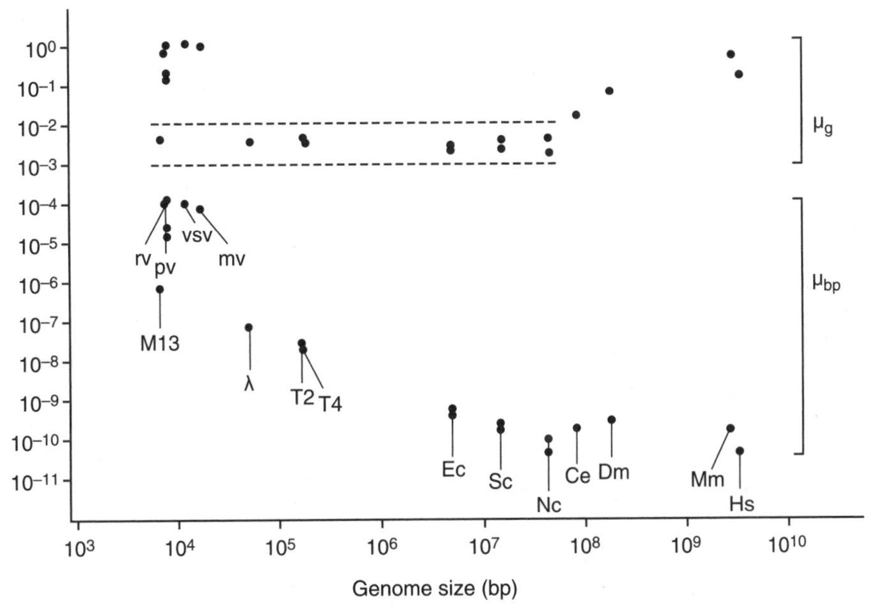

Figure 52.2: Estimates of mutation rates per-genome (μ_g, upper points) and per-base-pair (μ_{bp}, lower points), plotted against genome size on a log-log scale. The data are taken from Drake and Holland and Drake et al. *Multiple symbols are drawn where independent estimates from different loci in the same organism were available. No error bars are shown but the errors are probably large. RNA viruses: rv, rhinovirus; pv, poliovirus; vsv, vesicular stomatitis virus; mv, measles virus. Bacteriophages (DNA-based) are shown according to their usual epithets M13, T2, T4, and λ.* E. coli *(Ec),* Saccharomyces cerevisiae *(Sc), and* Neurospora crassa *(Nc) are shown. Higher eukaryotes: Ce,* C. elegans*; Dm,* D. melanogaster*, Mm,* Mus musculus*; Hs,* Homo sapiens*. Outliers thought to reflect estimates from non-representative loci. Drake's observation of a conserved μ_g amongst DNA-based microbes is highlighted with dashed lines.*

indirect selection due to deleterious mutations could be prevented. These hypotheses are: (1) that further reductions of μ_g (below prevailing values) are physicochemically impossible, and (2) that further reductions are physiologically costly and hence would impose on individuals a prohibitively high direct selective cost. Neither hypothesis can be ruled out for all taxa at present. While the relatively narrow range of per-base-pair mutation rates observed in eukaryotes across a large range of genome sizes (Fig. 52.2) might indicate that μ_g has evolved to some universal physicochemical minimum in these taxa, there is no clear theoretical prediction as to what that minimum should be, and some empirical evidence (see below) suggests that further reductions in per-base-pair mutation rate remain possible. The observation that per-base-pair mutation rates vary widely among microbial species (Fig. 52.2) clearly suggests that lower values of μ_g are physically possible in at least some prokaryotic taxa. The isolation of antimutator mutants (bearing alleles that lower mutation rates) in bacteriophage T4[1] and in *Escherichia coli* also hints that further reductions in mutation rate may be possible. The T4 antimutators have been shown to decrease mutation rates along certain mutational pathways while actually increasing rates along other pathways, yielding no net reduction in μ_g. At least one *E. coli* antimutator, however, appears to decrease μ_g approximately twofold below the prevailing wild-type level.

Limited available theory suggests that the physiological cost of fidelity in DNA replication (in terms of time and energy) should increase monotonically as the mutation rate approaches zero, thus potentially imposing a direct fitness cost on individuals bearing lower mutation rates. In vitro studies of polymerase enzyme purified from antimutator and wild-type strains of phage T4 have shown that increased fidelity incurs both time and energy costs, but there has been no measurement or estimation of the effects of such costs on fitness in any system. There is only indirect evidence suggesting in general that the fitness cost of increasing fidelity could be substantial. In *E. coli*, there is a weak negative relationship between growth rate on minimal medium and genome size. Indeed, the rarity of noncoding DNA in the genomes of many microbes may reflect pervasive selection to increase the rate of genome replication. These observations, however, do not demonstrate that the further increased cost of an antimutator phenotype would have a substantial effect on fitness. In *Drosophila*, an experiment in which populations were exposed to different levels of X-irradiation for up to 600 generations documented stable evolutionary decreases in the rate of

[1] T4 is a phage (i.e. a kind of virus) that can take over *E. coli* cells. It degrades the host *E. coli* cell's DNA and uses the cellular machinery to copy itself. T4 has been much used in research on molecular genetics.

X-ray-induced mutation, followed by an evolutionary return to wild-type mutation rates in some populations after irradiation was stopped. This result suggests that the level of investment in repair of radiation damage is set by a tradeoff between deleterious mutational effects and the cost of repair. In mice, the X chromosome appears to have a lower mutation rate than the autosomes as estimated from data on neutral substitution rates. This is consistent with a tradeoff between the expected greater deleterious effect of recessive mutations expressed on the (hemizygous) X chromosome in males and the cost of fidelity on the X chromosome.

The cost of fidelity is thus a credible factor for preventing the evolution of μ_g to zero, although it remains possible that some taxa are already at or near the physicochemical limit to fidelity of replication. What about the need for new beneficial mutations to facilitate evolutionary adaptation: can it too prevent the evolution of μ_g to zero? A fundamental point to bear in mind when considering this question is that mutation cannot be maintained in a population for the sake of its *future* adaptive utility; natural selection can only maintain mutation as a consequence of its *past* adaptive utility.

The fundamental population genetic process underlying evolutionary adaptation is the rise in frequency of a beneficial allele. As selection increases the frequency of a beneficial allele, it indirectly increases the frequencies of linked alleles; this process is referred to as 'genetic hitchhiking'. A modifier that increases the mutation rate has an enhanced probability of association with a beneficial allele (that is, it is more likely to exist in *linkage disequilibrium* with a beneficial allele), and hence it has an enhanced probability of hitch-hiking to fixation. However, this hitchhiking process is extremely sensitive to recombination, which acts to randomize the associations between alleles in a population (see Fig. 52.1). A major conclusion that has emerged from studies of modifier models is that hitchhiking is unlikely to be an important factor in adjusting μ_g in sexual populations because of the strong effect of recombination in eroding linkage disequilibrium between mutation rate modifiers and fitness alleles. [. . .]

In sexual populations, therefore, it is likely that μ_g is not affected by the occurrence of beneficial mutations, but is instead determined by a tradeoff between indirect selection due to deleterious mutations and the direct selective cost of increasing the fidelity of replication. [. . .]

In asexual populations, beneficial mutations can have a strong effect on the fate of mutation rate modifiers because of linkage (Fig. 52.1); indeed, the evolution of μ_g in asexual populations is a matter of competition among clonal lineages with different mutation rates. R.A. Fisher appears to have been the first to articulate the possibility that an equilibrium mutation rate can evolve in an asexual population under the influence of deleterious *and* beneficial mutations, although his published treatment of this subject was purely verbal. In essence, Fisher's approach considers an ensemble of asexual clones distinguished by their respective values of μ_g. It is assumed that each clone attains a distribution of fitness under the influence of deleterious and beneficial mutations as if it were an independent population at equilibrium and the clone with the mutation rate conferring highest mean fitness prevails. Fisher's conjecture was that this process would yield an optimal compromise between deleterious and beneficial mutation in an asexual system.

Numerous mathematical models have subsequently been developed for the evolution of an optimal equilibrium μ_g in asexual populations. [. . .]

The complete linkage between mutation rate modifiers and fitness loci in asexual populations thus may produce a situation in which μ_g is unlikely to settle into a long-term equilibrium value. When beneficial mutations are not occurring, selection drives μ_g toward a minimal value constrained by the cost of fidelity, regardless of what the long-term optimal value might be; when beneficial mutations are occurring, however, hitchhiking of mutation rate modifiers sporadically elevates the mutation rate above the optimal level. The intuitive picture that emerges is one in which the mutation rate is continually buffeted about by the effects of indirect selection on modifiers. In the next section, we consider the evidence supporting this nonequilibrial view of mutation rates in asexuals.

Evolution of high mutation rates in asexual populations

Most experimental work on the evolution of mutation rates has been conducted in laboratory populations of bacteria, whose large population sizes and short generation times facilitate direct observation of long-term evolutionary phenomena. Numerous studies have documented the evolution of high mutation rates in such experimental populations. Mutator phenotypes have also been observed at substantial frequencies among natural bacterial isolates and in some cancerous somatic cell lineages, indicating that the evolution of high mutation rates in asexual populations is not merely an artifact of laboratory conditions. None of the experimental studies has uncovered evidence that alleles responsible for high mutation rates (mutator alleles) confer direct fitness benefits, and hence their increase in populations is most likely a result of hitchhiking. [. . .] Thus it appears that adapting asexual populations have a propensity to evolve sharply elevated mutation rates under hitchhiking rather than to evolve optimal mutation rates.

Because asexual populations cannot generate variation by recombination, it might be hypothesized that high mutation rates in asexual populations are an adaptation for generating variation. A mutator phenotype that has hitchhiked to fixation is properly regarded as a *consequence* of adaptation, however, not an adaptation itself. The adaptive value of a mutator phenotype depends on whether it increases the rate of *subsequent* adaptation in an asexual population, and the circumstances under which this is likely are limited. On one hand, computer simulation studies have shown that, if a small number of beneficial mutations is to be substituted in a finite asexual population during adaptation, then under some circumstances a population that fortuitously substitutes a mutator allele by hitchhiking early in the bout of adaptation will reach this goal sooner. On the other hand, an analytical model of adaptive evolution in asexual populations predicts a diminishing increase in rate of adaptation with increased mutation rates. At higher rates of beneficial mutation, the speed of adaptation becomes increasingly limited by the rate of selective sorting among clones bearing different beneficial mutations rather than by the rate at which new beneficial mutations arise in the population. Experimental work has shown that this 'clonal interference' effect constrains the adaptive usefulness of a high mutation rate to situations in which beneficial mutations are extremely infrequent, such as when population sizes are very small (perhaps due to bottlenecks) or when a population is initially well-adapted. High mutation rates can substantially accelerate adaptation in asexual populations, but do not necessarily do so.

The available empirical evidence is consistent with this cautious view of the adaptive significance of mutators. In a study of twelve replicate experimental populations of *E. coli* propagated in the laboratory for 10,000 generations, three populations evolved mutator phenotypes but there was no evidence that these mutator populations adapted faster than the remaining nine populations. Some, but not all, cancer cell lineages exhibit a mutator phenotype, and simulation studies show that clonal selection without elevated mutation rates can be sufficient to promote carcinogenesis. A survey of *Salmonella* and *E. coli* isolates classified as pathogenic or nonpathogenic found a suggestive (but nonsignificant) association between mutator phenotype and pathogenicity, hinting at a selective advantage to mutator phenotypes under the rigors of host immune surveillance. A later survey, however, failed to document such an association. Finally, a recent study documented a high frequency (20%) of mutators among *Pseudomonas aeruginosa* strains infecting the lungs of cystic fibrosis (CF) patients. This finding provides circumstantial evidence that mutator phenotypes have a selective advantage in colonizing the CF lung. The alternative view that selection for colonization ability merely increases the probability of a mutator hitchhiking event, however, remains a possibility; the study does not compare the rates of adaptation of mutator and wild-type strains *after* colonization.

Even if a high mutation rate increases the rate of adaptation in an asexual population in the short term, over the evolutionary long term it is clear that indirect selection due to deleterious mutations eventually favors a decrease in mutation rates. Otherwise, mutator phenotypes would be the rule rather than the exception in asexual populations, which is clearly not the case. Given the propensity for asexual populations to substitute mutator alleles by hitchhiking, there is a need for more theory and experimentation on how low mutation rates are restored. Three processes could potentially reduce mutation rates within an asexual population that is fixed for a mutator allele. (1) The most obvious is outright reversion of the mutator allele and substitution of the new wild-type allele back into the population. (2) Compensatory evolution at additional modifier loci could also reduce the mutation rate, as observed in experimental populations of *E. coli* that were fixed for a very strong mutator allele. (3) Rare horizontal genetic exchange events could replace the mutator allele with its wild-type counterpart.

A major alternative to the reduction of mutation rates by selection acting within clones is the possibility that mutator clones are evolutionary dead ends which, despite a possible short-term adaptive advantage, are destined to be out-competed by their wild-type counterparts. Theoretical studies have explored the evolutionary 'meltdown' of finite asexual populations under deleterious mutation pressure, and there is direct genetic evidence for accelerated accumulation of deleterious mutations within mutator clones of *E. coli*. The interaction of mutator hitchhiking and deleterious mutation accumulation in asexual populations is an interesting area for further research.

[. . .]

Conclusion

[. . .] Where recombination is substantial, indirect selection to raise the mutation rate is likely to be ineffective. Where recombination is minimal or absent, indirect selection can raise the mutation rate, but this does not necessarily result in a mutation rate that maximizes or even increases the long-term rate of adaptation. Mutation itself, however, is likely to be a key factor in the maintenance of sexual recombination; major alternative theories for the evolution of sex invoke the clearance of deleterious mutations and the combining of beneficial mutations as ultimate causal factors. Seen in this light, the adjustment of mutation rates by natural selection is but one in a number of intricately related evolutionary processes.

[*Bioessays*, 22 (2000), 1057–66.]

Section I

Human evolution

The section includes readings on human evolution, as a historical problem; on the genetics of modern human populations; and two other topical pieces, about ideas that have long been of interest but are enjoying something of a modern revival—the application of evolution to human medicine and the evolution of language.

The first paper, by Sarich and Wilson (Chapter 53), re-visits a theme of Section G: the use of molecules and fossils in inferring the history of life. Indeed, human evolution provides the classic conflict between the two kinds of evidence. Before the molecular evidence of the 1960s, fossil evidence had suggested that the branch between the lineage leading to modern humans and the lineage leading to our closest modern relatives (chimpanzee) was at least 15 million years old. The molecular clock forced a dramatic revision of that date, forwards to more like 5 million years ago; Sarich and Wilson's 1967 paper (Chapter 53) is a classic, and easily readable notwithstanding its (pioneering at the time) molecular technology.

The molecular evidence pointed to a recent (approx. 5 million years ago) origin for humans, because humans and their near relatives such as chimpanzees have very similar molecules. King and Wilson (Chapter 54) followed this question up. The earlier paper by Sarich and Wilson (Chapter 53) was based on one gene (albumin). The later paper looked at more general measures of molecular-genetic similarity between chimpanzees and humans. King and Wilson's work is the basis of the claim, soon to become widely known, that the DNA of humans and chimpanzees is 99% identical. King and Wilson contrasted this molecular near-identity with the clear morphological difference between the two species. A small number of genetic changes has produced a large morphological change. King and Wilson suggested the genetic changes were mainly in regulatory genes. Their idea has remained a hypothesis for over a quarter of a century, but Carroll (Chapter 39, Section F) discusses how the hypothesis may soon be tested with genomic evidence. (As I said in the introduction to Section F, the allocation of readings 38, 39, 54, and 55 between Sections F and I is fairly arbitrary.)

The basic claim that human and chimp DNA is 99% identical still stands, but Britten (Chapter 55) recently added a proviso. Any one letter in human DNA is the same as the equivalent letter in chimp DNA with 99% (more like 98.5% to be exact) probability. But the DNA of the two species also differs in insertions and deletions (stretches of DNA that have been added or subtracted). They cause a further 3.5% difference, reducing the total similarity to 95%.

The two readings on human genetics are about race and about relaxed selection. Some species are made up of genetically distinct races, but modern humans (perhaps because of their recent origin) are not one of them. Human races are more or less culturally defined and biologically non-existent. Livingstone famously pointed this out (Chapter 57). Mayr (Chapter 22, Section D) makes a related point. Modern humans in rich countries may be approaching a new evolutionary phase in which natural selection is relaxed, due to medical advances and family planning. (In the extreme case in which almost all pre-senile diseases are cured and almost all families have two children, both of whom survive, natural selection would have ceased to operate. It is unlikely things will ever go that far, but we are moving that way and it is reasonable to imagine that the trend will continue.) This fact, long appreciated, has taken on a new interest as our mutation rates have become apparent. When Muller (Chapter 56) wrote, he thought that each human had less than one new mutation. We now know the number is more like 200 (see Kimura, Chapter 13, Section B). We do not know how many of the 200 are selectively neutral and how many deleterious, but even if only a per cent or two are deleterious, then in the absence of selection future generations will increasingly move toward randomized DNA. I share the standard view of evolutionary theorists, that the future of evolution cannot be predicted. However, subject to obvious strong provisos about the uncertainties of facts now and the contingencies of future human cultural practices, I believe that the randomization of our DNA is one reasonable prediction about the human evolutionary future. No doubt our descendants will be able to cope with it, but it bears thinking about how. Muller's piece annihilates one of the sloppier arguments that are sometimes made about this subject. (Let me remark that I have not included Muller out of any enthusiasm for eugenics. There are opportunities for reading eugenics into Muller's piece. I do not think this is the interesting feature of his argument: I like the way he sets out the genetic consequences of relaxed selection, a topic he had thought about more clearly than most other biologists; but what we should, and shall, do about it I prefer to leave as an open question.)

Chapter 58 by Krogman is a classic early paper about the medical consequences of our evolutionary history. The paper can be thought of as an early contribution to the 'new science of Darwinian medicine' recently sketched out by Nesse and Williams. (I give the reference in the Select bibliography section.) Some other topics in Darwinian medicine are discussed by Palumbi (Section J, Chapter 25). Finally, Pinker (Chapter 59) looks at the evolution of human language. Language has often been argued to create a break-point between humans and the rest of life, and not to have evolved by gradual natural selection. Pinker (Chapter 59) elegantly and wittily makes the case that language probably evolved just like any other character. Pinker's book *The language instinct* is one of the most accessible works in a new field, which often calls itself evolutionary psychology and which uses (with a sophistication lacking in most prior work) the theory of natural selection to understand the human mind.

One terminological point may be worth drawing attention to: the difference between hominoids and hominids. Hominoids are the group of all apes (lesser apes—gibbons; great apes—orang-utan, gorilla, and chimpanzees; and humans). Hominids are a subgroup of hominoids, containing humans and all fossils that are more closely related to humans than to other apes: in other words, humans plus australopithecines. The word hominid is being replaced in some but not all quarters by hominin (for instance, see Carroll (Chapter 39)). The reason for the two terms is that *Homo* and *Australopithecus* used to be classified as the family Hominidae. Now, not least because of the kind of molecular evidence seen in Chapters 53 and 54, they are often classified in a subfamily Homininae.

A brief background on the fossil evidence of human evolution may be useful for some of the readings. Before the arrival of molecular methods, human evolution was mainly the story of fossil discoveries. Darwin said nothing about the topic in *The Origin of Species* (1859), but by the time he wrote *The Descent of Man* (1871) there was some controversial evidence from the recently exhumed Neanderthal fossils. *Homo erectus*, the immediate ancestor of *Homo sapiens* was first discovered in Java in 1891 and further important fossils of *H. erectus* were excavated in China between the wars. The Piltdown hoax of 1911–12 mattered historically but was not of lasting influence. The first australopithecine was discovered in South Africa in 1924 but was for long not accepted as any more closely related to humans than to other apes. The picture then grows more complex, but important discoveries were made in East Africa by the Leakeys in the 1950s, 60s, and 70s, and by another team including Johanson and White in the 1970s.

After the branch perhaps 5 or more million years ago there is a gap in the record until 3 1/2 million years ago when we have the fossils of *Australopithecus afarensis*, including the extraordinarily complete fossil known as Lucy. *A. afarensis* shows how the distinctive features of humans arose; australopithecines were mainly like nonhuman apes above the neck (small brain and prognathic mouth) and largely like modern humans below the neck (bipedalists). The evolution of bipedal locomotion therefore preceded the evolution of a big brain and delicate jaw. Some time later there was a split into two lineages, one of robust australopithecines and the other of gracile austrolopithecines. We are more closely related to the gracile forms. Around 1–2 million years ago, brain size was clearly expanding in the human lineage, and some fossils are classified in the genus *Homo* (*Homo* is, simplifying somewhat, defined by large brain size). Early members of *Homo* are found with tools and classified as *Homo habilis*. The formal names and ancestor-descendant relations become confusing at this point, but the next important species is *Homo erectus*. It originated in Africa and was the first of our ancestors to migrate out of Africa—to Asia and Europe. *Homo sapiens* is descended from one or more populations of *Homo erectus*.

VINCENT M. SARICH AND ALLAN C. WILSON

53 Immunological time scale for hominid evolution

The degree of immunological cross-reactivity between the proteins of different species depends on the molecular difference of the proteins. If the proteins evolve in a clock-like manner, the time of origin of a lineage can be inferred. This study suggests that the hominid lineage originated about 5 million years ago, a much more recent figure than fossil evidence then suggested. [Editor's summary.]

It is generally agreed that the African apes are our closest living relatives. However, the time of origin of a distinct hominid lineage has been a subject of controversy for over 100 years. The absence of an adequate fossil record has forced students of hominid evolution to evaluate the phylogenetic significance of anatomical and behavioral characteristics in the living primate species in order to attempt a solution to that controversy. The nature of the problem is such, however, that no definitive answer has yet been given. Current estimates range from a date in the late Pliocene to one in the late Oligocene or early Miocene for the origin of the hominids. This great range (4 million to 30 million years) effectively negates any meaningful discussion of the nature of our pre-Australopithecine ancestors, for the early dates bring us near to a primitive prosimian stock, while the late ones would suggest that a common ancestor for man and the African apes might well resemble a small chimpanzee.

One solution to this question lies in the measurement of the degree of genetic relationship which exists between man and his closest living relatives. As it has recently become clear that the structure of proteins closely reflects that of genes, it is to be expected that quantitative comparative studies of protein structure should aid in providing this measure of genetic relationship.

Proteins appear to evolve over time, as do the organisms of which they are a part. Thus, we may speak of the common ancestor of, for example, the human and chimpanzee serum albumin molecules,[1] this ancestral molecule being present in the common ancestor of man and the chimpanzee. From

[1] Albumins are proteins in blood serum, and therefore relatively easy to take samples of.

the time that the human and chimpanzee lineages separated, their albumins have had the opportunity of evolving independently until today they are recognizably different, but homologously related, molecules. Such homologies may be studied by immunological techniques, the magnitude of the immunological cross-reaction serving as a measure of the degree of structural similarity between the two kinds of albumin.[2] [. . .] Serum samples were obtained from all the living genera of apes and from six representative genera of Old World monkeys and stored at −10°C. Albumin was purified from individual chimpanzee, gibbon, and human serums. Groups of three or four rabbits were immunized by three courses of injections with each of the purified albumins. The antiserums were tested for purity by immunodiffusion, immunoelectrophoresis, and microcomplement fixation (MC′F) with whole serum and purified albumin. Antibodies to components of serum other than albumin were always detectable with the first two methods, but they were too low in concentration to interfere with the MC′F analysis of the cross-reactions discussed below. Pooled antiserums were made by mixing the individual antiserums in reciprocal proportion to their MC′F titers. The degrees of cross-reaction were expressed quantitatively as the index of dissimilarity or immunological distance (ID), that is, the relative concentration of antiserum required to produce a complement fixation curve whose peak was as high as that given by the homologous albumin.

These antiserums were used to obtain the data summarized in Table 53.1. With the antiserum pool prepared against human serum albumin, the albumins of the African apes (gorilla and chimpanzee) reacted more strongly than those of the Asiatic apes (orang, siamang, and gibbon). The antiserum pool directed against chimpanzee (*Pan troglodytes*) albumin showed the albumin of the pygmy chimpanzee to be immunologically identical to that of the homologous species. Human and gorilla albumins reacted somewhat less well but more strongly than did the albumins of the Asiatic apes. The antiserum pool directed against gibbon (*Hylobates lar*) albumin reacted most strongly with that of the siamang. The albumins of the other apes and man were appreciably less reactive.

Conclusions about genetic relationships among the albumins of apes and man may be drawn from these data. The albumins of the orang, gorilla, chimpanzee, and man stand as a unit relative to those of the gibbon and siamang. This is evident from the data obtained with the antiserum to

[2] The principle of the method is that antibodies cross-react most strongly with the exact molecule (antigen) they were formed against. If you cause a rabbit to make antibodies against human albumin, for example, those antibodies will cross-react most strongly with human albumin. Those same antibodies will also react with chimpanzee albumin, and to a lesser extent with albumins from more distant species. The chimpanzee albumin is very like human albumin, but a little different in structure. Gibbon albumin is a little more different again, and so on.

Table 53.1: Reactivity of various primate albumins with antiserums prepared against hominoid albumins [. . .]

SPECIES OF ALBUMIN	INDEX OF DISSIMILARITY		
	ANTISERUM TO *HOMO*	ANTISERUM TO *PAN*	ANTISERUM TO *HYLOBATES*
	Hominoidea (apes and man)		
Homo sapiens (man)	1.0	1.09	1.29
Pan troglodytes (chimpanzee)	1.14	1.00	1.40
Pan paniscus (pygmy chimpanzee)	1.14	1.00	1.40
Gorilla gorilla (gorilla)	1.09	1.17	1.31
Pongo pygmaeus (orang-utan)	1.22	1.24	1.29
Symphalangus syndactylus (siamang)	1.30	1.25	1.07
Hylobates lar (gibbon)	1.28	1.25	1.00
	Cercopithecoidea (Old World monkeys)		
Six species (mean ± S.D.)	2.46 ± .16	2.22 ± .27	2.29 ± .10

Hylobates albumin (Table 53.1). The close relationship between the albumins of the gibbon and siamang is consistent with the fact that these two genera of apes are usually placed together in a separate family or subfamily (Hylobatinae). It is also evident that the albumins of the gorilla, chimpanzee, and man stand as a unit relative to the albumins of the Asiatic apes. Moreover, the albumins of the gorilla, chimpanzee, and man appear to be equidistantly related; the gorilla and chimpanzee albumins are no closer to each other than either is to human albumin. These relationships are consistent with those suggested by Goodman on the basis of qualitative immunodiffusion data.

Table 53.1 also shows that with each antiserum pool the Old World monkey albumins gave markedly weaker reactions (mean ID = 23) than those given by any hominoid albumin. The size of the immunological gap between the albumins of hominoids and Old World monkeys is illustrated by the fact that, at an antiserum concentration where all hominoid albumins give strong reactions, Old World monkey albumins give no reaction with antiserums to hominoid albumins. Thus, the albumins of all the living apes are much more similar to each other than any of them is to nonhominoid albumins.

The phylogenetic significance of the above findings, however, is not unequivocal. For example, at least two explanations are possible for the extremely close structural similarity of ape and human albumins. On the one hand, albumin evolution may have been retarded in the ape, human, or both lineages since their separation. [. . .] Alternatively, the close molecular relationship may reflect a more recent common ancestry between ourselves and the living apes than is generally supposed, albumin evolution having proceeded at the usual rate for primates.

Rates of albumin evolution in primates have recently been investigated. The evidence appears to rule out the first explanation. No conservatism was detected in the evolution of any particular hominoid albumins relative to any other, nor, indeed, in the evolution of hominoid albumin relative to those of any other primate group. Albumin evolution appears to have proceeded to the same extent in the various ape, human, and monkey lineages. Therefore, it seems likely that apes and man share a more recent common ancestry than is usually supposed.

The data presented above enable us to calculate how recently this common ancestor lived. We have recently shown that albumin evolution in primates is a remarkably regular process. Lineages of equal time depth show very similar degrees of change in their albumins. The degrees of change shown would therefore seem to be a function of time, and a mathematical relationship between ID and the time of divergence of any two species must exist. Thus, albumin molecules can serve as an evolutionary clock or dating device. The calibration of that clock, that is, the elucidation of the relationship between ID and time, would allow us to calculate the time of divergence between apes and man.

This relationship is likely to be rather simple. If the amino acid sequences of proteins also evolve at steady rates, and there is evidence that they often do, then the relationship between ID and time of divergence should be of the same form as the relationship between ID and structural difference (number of amino acid replacements). Direct evidence for a simple correlation between immunological cross-reactivity and structural relatedness is available from studies of hemoglobins and cytochromes *c* of known amino acid sequence. Indirect evidence for such a correlation is provided by the correspondence between cross-reactivity and phylogenetic relatedness which has been demonstrated for a variety of proteins.

It appears likely that log ID is approximately proportional to the time of divergence (T) of any two species, that is, $\log \mathrm{ID} = kT$, where k is a constant. This relationship is evident from several sets of MC′F data obtained with various purified dehydrogenases of fishes, amphibians, reptiles, and birds whose times of divergence can be estimated from the fossil record. Let us suppose that a similar relationship is appropriate for albumin evolution in primates.

Although the primate fossil record is fragmentary, it does, in combination with the available immunological evidence, provide sufficient evidence to suggest that the lineages leading to the living hominoids and Old World monkeys split about 30 million years ago. That is, the ID of 2.3 which is the mean ID observed between the albumins of hominoids and Old World monkeys corresponds to a T value of about 30 in the above equation. If log $2.3 = k \times 30$, then $k = 0.012$. Since the mean ID between the albumins of man and the African apes is 1.13, the time of divergence of man from the African apes is log 1.13 divided by 0.012, that is, 5 million years. Proceeding similarly, we calculate that the lineage leading to the orang separated from that leading to the African apes 8 million years ago, and that the time of divergence of the gibbon and siamang lineage from that leading to the other apes and man is 10 million years (Fig. 53.1).

There are, of course, at this stage in our investigation, uncertainties in these calculations. We may possibly be making erroneous assumptions about (i) the time of divergence of apes and Old World monkeys, and (ii) the nature of the relationship between ID and time of divergence. We feel, however, that these possible errors are unlikely to be of sufficient magnitude to invalidate the conclusion that apes and man diverged much more recently than did the apes and Old World monkeys. In our opinion, the albumin data definitely favor those who have postulated that man and the African apes shared a common ancestor in the Pliocene.

If the view that man and the African apes share a Pliocene ancestor and that all the living Hominoidea derive from a late Miocene form is correct,

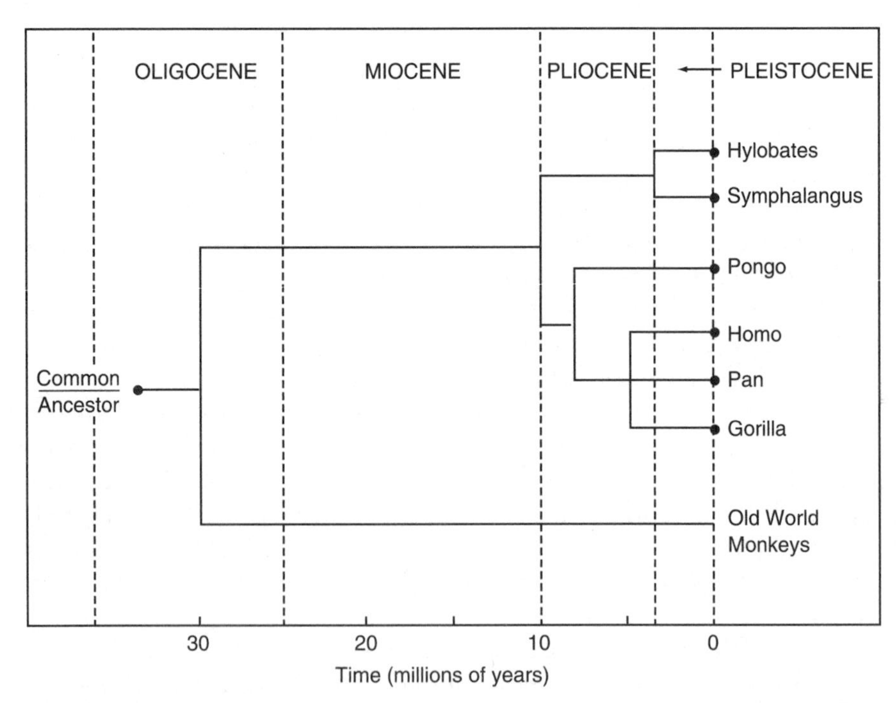

Figure 53.1: Times of divergence between the various hominoids, as estimated from immunological data. The time of divergence of hominoids and Old World monkeys is assumed to be 30 million years.

a number of the problems that have troubled students of this group are resolved. The many features of morphology, particularly in the thorax and upper limbs, which man and the living apes share in varying degrees, but which were not present in the Miocene apes, such as *Dryopithecus* (Proconsul), *Limnopithecus*, and *Pliopithecus*, are then seen as due to recent common ancestry and not, as generally accepted, to parallel or convergent evolution.

We suggest that the living apes and man descended from a small member of the widespread Miocene dryopithecines, which became uniquely successful due to the development of the locomotor-feeding adaptation known as brachiation. The adaptive success of this development and the subsequent radiation of the group possessing it may have made this group the only surviving lineage of the many apes present throughout the tropical and subtropical Miocene forests of the Old World. Possibly the African members of this radiation, in the Middle Pliocene (due perhaps to pressure from the developing Cercopithecinae), began varying degrees of adaptation to a terrestrial existence. The gorilla, chimpanzee, and man appear to be the three survivors of this later radiation. According to this hypothesis, some 3 million years are allowed for the development of bipedalism to the extent seen in the earliest fossil hominid, *Australopithecus*. [. . .]

[*Science*, 158 (1967), 1200–3.]

MARY-CLAIRE KING AND A. C. WILSON

54 Evolution at two levels in humans and chimpanzees

Molecular methods of measuring the similarity between humans and chimpanzees agree in showing that the two species are almost identical—they are no more different, at the molecular level, than are sibling species (that is, pairs of species that are morphologically almost identical). And yet humans and chimpanzees are morphologically very different. Perhaps major morphological changes are driven by evolution in regulatory genes, rather than by evolution in protein-coding genes. [Editor's summary.]

Soon after the expansion of molecular biology in the 1950s, it became evident that by comparing the proteins and nucleic acids of one species with those of another, one could hope to obtain a quantitative and objective estimate of the 'genetic distance' between species. Until then, there was no common yardstick for measuring the degree of genetic difference among species. The characters used to distinguish among bacterial species, for example, were entirely different from those used for distinguishing among mammals. The hope was to use molecular biology to measure the differences in the DNA base sequences of various species. This would be the common yardstick for studies of organismal diversity.

During the past decade, many workers have participated in the development and application of biochemical methods for estimating genetic distance. These methods include the comparison of proteins by electrophoretic, immunological, and sequencing techniques, as well as the comparison of nucleic acids by annealing techniques. The only two species which have been compared by all of these methods are chimpanzees (*Pan troglodytes*) and humans (*Homo sapiens*). This pair of species is also unique because of the thoroughness with which they have been compared at the organismal level—that is, at the level of anatomy, physiology, behavior, and ecology. A good opportunity is therefore presented for finding out whether the molecular and organismal estimates of distance agree.

The intriguing result, documented in this article, is that all the biochemical methods agree in showing that the genetic distance between humans and the chimpanzee is probably too small to account for their substantial organismal differences. [. . .]

Genetic distance and the evolution of organisms

The resemblance between human and chimpanzee macromolecules has been measured by protein sequencing, immunology, electrophoresis, and nucleic acid hybridization. From each of these results we can obtain an estimate of the genetic distance between humans and chimpanzees. Some of the same approaches have been used to estimate the genetic distance between other taxa, so that these estimates may be compared to the human–chimpanzee genetic distance.

First, we consider genetic distance estimated from electrophoretic data [. . .]

With respect to genetic distances between species, the human–chimpanzee *D* value[1] is extraordinarily small, corresponding to the genetic distance between sibling species[2] of *Drosophila* or mammals (Fig. 54.1). Nonsibling species within a genus (referred to in the figure as congeneric species) generally differ more from each other, by electrophoretic criteria, than humans and chimpanzees. The genetic distances among species from dif-

[1] *D* is a measure of genetic distance—of how different two samples are (where in this case the two samples are two species).

[2] Sibling species are pairs of closely related species that are morphologically almost indistinguishable: they do not naturally interbreed but they are identical in appearance. The comparison between human–chimpanzees and sibling species is crucial for King and Wilson's argument. They are concerned with the rates of morphological and of molecular change; in sibling species there has been practically no morphological change.

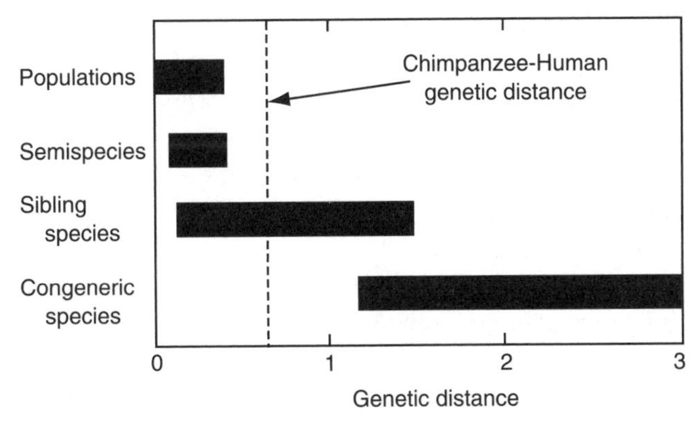

Figure 54.1: The genetic distance, D, *between humans and chimpanzees (dashed line) compared to the genetic distances between other taxa. Taxa compared include several species of* Drosophila, *the horsehoe crab* Limulus polyphemus, *salamanders from the genus* Taricha, *lizards from the genus* Anolis, *the teleost fish* Astyanax mexicanus, *bats from the genus* Lasiurus, *and several genera of rodents.*

ferent genera are considerably larger than the human–champanzee genetic distance.

The genetic distance between two species measured by DNA hybridization also indicates that human beings and chimpanzees are as similar as sibling species of other organisms. [. . .]

Immunological and amino acid sequence comparisons of proteins lead to the same conclusion. Antigenic differences among the serum proteins of congeneric squirrel species are several times greater than those between humans and chimpanzees. Moreover, antigenic differences among the albumins of congeneric frog species (*Rana* and *Hyla*) are 20 to 30 times greater than those between the two hominoids. In addition, the genetic distances among *Hyla* species, estimated electrophoretically are far larger than the chimpanzee–human genetic distance. Finally, the human and chimpanzee β chains of hemoglobin appear to have identical sequences, while the β chains of two *Rana* species differ by at least 29 amino acid substitutions. In summary, the genetic distance between humans and chimpanzees is well within the range found for sibling species of other organisms.

The molecular similarity between chimpanzees and humans is extraordinary because they differ far more than sibling species in anatomy and way of life. Although humans and chimpanzees are rather similar in the structure of the thorax and arms, they differ substantially not only in brain size but also in the anatomy of the pelvis, foot, and jaws, as well as in relative lengths of limbs and digits. Humans and chimpanzees also differ significantly in many other anatomical respects, to the extent that nearly every bone in the body of a chimpanzee is readily distinguishable in shape or size from its human counterpart. Associated with these anatomical differences there are, of course, major differences in posture, mode of locomotion, methods of procuring food, and means of communication. Because of these major differences in anatomy and way of life, biologists place the two species not just in separate genera but in separate families. So it appears that molecular and organismal methods of evaluating the chimpanzee–human difference yield quite different conclusions.

An evolutionary perspective further illustrates the contrast between the results of the molecular and organismal approaches. Since the time that the ancestor of these two species lived, the chimpanzee lineage has evolved slowly relative to the human lineage, in terms of anatomy and adaptive strategy. According to Simpson:

> *Pan* is the terminus of a conservative lineage, retaining in a general way an anatomical and adaptive facies common to all recent hominoids *except Homo*. *Homo* is both anatomically and adaptively the most radically distinctive of all hominoids, divergent to a degree considered familial by all primatologists.

This concept is illustrated in the left hand portion of Figure 54.2. However, at the macromolecular level, chimpanzees and humans seem to have evolved at similar rates (Fig. 54.2, right). For example, human and chimpanzee albumins are equally distinct immunologically from the albumins of other hominoids (gorilla, orangutan, and gibbon),[3] and human and chimpanzee DNAs differ to the same degree from DNAs of other hominoids. Construction of a phylogenetic tree for primate myoglobins shows that the single

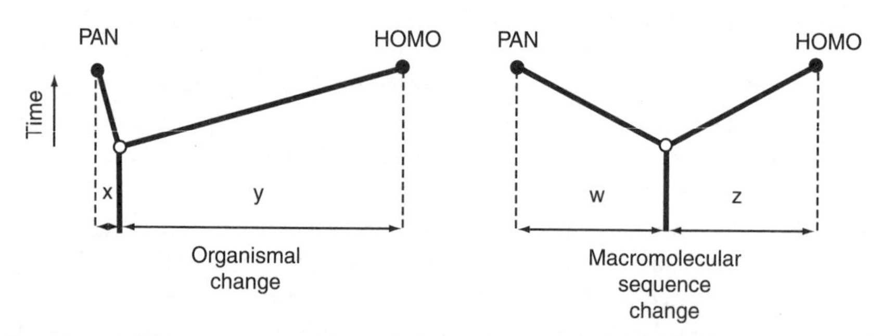

Figure 54.2: The contrast between biological evolution and molecular evolution since the divergence of the human and chimpanzee lineages from a common ancestor. As shown on the left, zoological evidence indicates that far more biological change has taken place in the human lineage (y) than in the chimpanzee lineage (y $\gg$ x); this illustration is adapted from that of Simpson. As shown on the right, both protein and nucleic acid evidence indicate that as much change has occurred in chimpanzee genes (w) as in human genes (z).

[3] See Chapter 53 by Sarich and Wilson.

amino acid difference between the sequences of human and chimpanzee myoglobin occurred in the chimpanzee lineage. Analogous reasoning indicates that the single amino acid difference between the sequences of human and chimpanzee hemoglobin δ chains arose in the human lineage. It appears that molecular change has accumulated in the two lineages at approximately equal rates, despite a striking difference in rates of organismal evolution. Thus, the major adaptive shift which took place in the human lineage was probably not accompanied by accelerated protein or DNA evolution.

Such an observation is by no means peculiar to the case of hominid evolution. It appears to be a general rule that anatomically conservative lineages, such as frogs, have experienced as much sequence evolution as have lineages that have undergone rapid evolutionary changes in anatomy and way of life.

Molecular basis for the evolution of organisms

The contrasts between organismal and molecular evolution indicate that the two processes are to a large extent independent of one another. Is it possible, therefore, that species diversity results from molecular changes other than sequence differences in proteins? It has been suggested by Ohno and others that major anatomical changes usually result from mutations affecting the expression of genes. According to this hypothesis, small differences in the time of activation or in the level of activity of a single gene could in principle influence considerably the systems controlling embryonic development. The organismal differences between chimpanzees and humans would then result chiefly from genetic changes in a few regulatory systems, while amino acid substitutions in general would rarely be a key factor in major adaptive shifts.

Regulatory mutations may be of at least two types. First, point mutations could affect regulatory genes. Nucleotide substitutions in a promoter or operator gene would affect the production, but not the amino acid sequence, of proteins in that operon. Nucleotide substitutions in a structural gene coding for a regulatory protein such as a repressor, hormone, or receptor protein, could bring about amino acid substitutions, altering the regulatory properties of the protein. However, we suspect that only a minor fraction of the substitutions which accumulate in regulatory proteins would be likely to alter their regulatory properties.

Second, the order of genes on a chromosome may change owing to inversion, translocation, addition or deletion of genes, as well as fusion or fission of chromosomes. These gene rearrangements may have important effects on gene expression, though the biochemical mechanisms involved are obscure. Evolutionary changes in gene order occur frequently. Microscopic studies of *Drosophila* salivary chromosomes show, as a general rule, that no two species have the same gene order and that inversions are the commonest type of gene rearrangement. Furthermore, there is a parallel between rate of gene rearrangement and rate of anatomical evolution in the three major groups of vertebrates that have been studied in this respect, namely birds, mammals, and frogs. Hence gene rearrangements may be more important than point mutations as sources for evolutionary changes in gene regulation.

Although humans and chimpanzees have rather similar chromosome numbers, 46 and 48, respectively, the arrangement of genes on chimpanzee chromosomes differs from that on human chromosomes. Only a small proportion of the chromosomes have identical banding patterns in the two species. The banding studies indicate that at least 10 large inversions and translocations and one chromosomal fusion have occurred since the two lineages diverged. Further evidence for the possibility that chimpanzees and humans differ considerably in gene arrangement is provided by annealing studies with a purified DNA fraction. An RNA which is complementary in sequence to this DNA apparently anneals predominantly at a cluster of sites on a single human chromosome, but at widely dispersed sites on several chimpanzee chromosomes. The arrangement of chromosomal sites at which ribosomal RNA anneals may also differ between the two species.

Biologists are still a long way from understanding gene regulation in mammals, and only a few cases of regulatory mutations are now known. New techniques for detecting regulatory differences at the molecular level are required in order to test the hypothesis that organismal differences between individuals, populations, or species result mainly from regulatory differences. When the regulation of gene expression during embryonic development is more fully understood, molecular biology will contribute more significantly to our understanding of the evolution of whole organisms. Most important for the future study of human evolution would be the demonstration of differences between apes and humans in the timing of gene expression during development, particularly during the development of adaptively crucial organ systems such as the brain.

[*Science*, 188 (1975), 107–15.]

ROY J. BRITTEN

55 Divergence between samples of chimpanzee and human DNA sequences is 5%, counting indels

Five chimpanzee bacterial artificial chromosome (BAC) sequences (described in GenBank) have been compared with the best matching regions of the human genome sequence to assay the amount and kind of DNA divergence. The conclusion is that the old saw that we share

98.5% of our DNA sequence with chimpanzee is probably in error. For this sample, a better estimate would be that 95% of the base pairs are exactly shared between chimpanzee and human DNA. In this sample of 779 kb, the divergence due to base substitution is 1.4%, and there is an additional 3.4% difference due to the presence of indels. The gaps in alignment are present in about equal amounts in the chimp and human sequences. They occur equally in repeated and nonrepeated sequences. [Author's summary.]

Many years ago, the hydroxyapatite method for measuring sequence divergence between species was developed by Dave Kohne and me. In this method, hybrid DNA strand pairs were formed from small fragments and the temperature at which they were disassociated determined. This method was used by several groups to compare chimpanzee and human DNA, and the best measurements suggested a divergence of 1.76% of single-copy DNA preparations. This observation led to the widespread quotations that we were 98.5% similar to chimps in our DNA, sometimes, mistakenly, that we had 98.5% gene similarity. I have compared the actual DNA sequences for the chimp BACs[1] that have become available. Previously, two groups have measured the divergence between human and chimp DNA sequences to be 1.25% for base substitutions in about 2 Mb of primarily single-copy DNA. Another estimate of the divergence was 1.23% for sequenced ends of many chimpanzee BACs. These groups did not compute the insertion/deletion contribution to divergence. For the work reported here, there are about 779 kb of BAC sequences in GenBank that can now be aligned with the human genome sequence, although that number will increase, because the full chimpanzee sequence will become available in due course.

Methods

The chimpanzee BAC sequences were downloaded from GenBank, and the 'BLAST the human genome'[2] facility was used to identify the regions of alignment. The available sequence alignment programs are not good at comparing long regions and use heuristic methods to establish gaps in alignment, leading to some uncertainty. Therefore, a FORTRAN program was written to ascertain the actual gaps in alignment. For this purpose, the matching human sequence is trimmed at the beginning to start exactly in alignment with the chimp sequence. The program moves sequentially along the pair of sequences. When it detects a mismatch, if the next 20-nt[3] match is in more than 15 positions, it is counted as a substitution. If the next 20 nt do not match

[1] BAC stands for bacterial artificial chromosome. It is part of the sequencing process.

[2] BLAST is software for comparing DNA sequences (see Chapter 36, footnote 6). GenBank is an electronic archive of DNA sequences.

[3] nt, Nucleotide.

Table 55.1: Divergence of the sample regions

NAME	LENGTH,* NT	CHROM§	SUBSTITUTION,† %	INDELS,‡ %	TOTAL, %
AC006582	72,176	12q14	1.20	0.63	1.87
AC006582	93,769	12p12	1.40	4.12	5.52
AC007214	132,974	12q13	1.31	3.90	5.21
AC097335	148,994	2q14	1.69	3.58	5.27
AC096630	150,877	20p11.2	1.58	2.51	4.09
AC093572	180,352	22q11	1.22	4.50	5.72
Average¶			1.41	3.91	4.85

The total length of these aligned regions is 779,142 bp.
* The length of the region compared.
† The percent of nucleotides replaced by a different nucleotide.
‡ The sum of the length of all gaps in both human and chimpanzee sequences as percent of aligned chimp length.
§ Chromosomal location listed by 'Search the Human Genome'.
¶ Weighted (by length) average.

this well, then the program looks for a nearby region that does match and tests for gaps up to 5 kb long in either sequence to see whether a 20-nt-long test region matches more than 15 nt. It is a fairly specific program, working well for very similar sequences, which do not have gaps larger than 5 kb; this is true for the examples in Table 55.1. It can be adjusted for sequence pairs with more divergence and larger gaps and works for comparison with human sequences of examples of baboon BACs that have been sequenced.

Results

Table 55.1 summarizes the human/chimp mismatch seen in the five BACs examined over the lengths that could be easily aligned. The first example is listed in two rows to represent the two different regions of chromosome 12 that accurately aligned with parts of the chimp sequence. Table 55.1, column 4, shows the percent base substitution, which, on average, is only slightly higher that the previous estimates for the whole single-copy fraction of the genome. There is considerable variation, ranging from 1.2 to 1.69%, indicating what are apparently regional differences in the degree of divergence. The compared parts of the BACs average 127 kb each, and the number of substitutions was about 1,500; thus, the expected small number statistical fluctuation does not account for the observed variation.

Table 55.1, column 5, shows the percent of these alignment lengths that were in gaps in either the human or the chimp sequences. On average, it is much more than twice as great as the base substitution percent. It is these values that have made the total divergence large compared with the old estimates. It appears appropriate to me to consider the full length of the gaps in estimating the interspecies divergence. These stretches of DNA are actually absent from one and present in the other genome. In the past, indels[4] have often simply been counted regardless of length and added to the base substitution count, because that is convenient for phylogenetics.

THE SIZE OF THE INDELS

The length distribution of the gaps is given in Tables 55.2 and 55.3. Table 55.2 shows the smaller gaps for which given sizes occur multiple times in this sample. The second column shows the number of occurrences. Single nucleotide gaps occur with the largest frequency, and the frequency falls

Table 55.2: Lengths of indels

LENGTH, NT	NO.	LENGTH, NT	NO.
1	410	17	6
2	148	18	4
3	108	19	1
4	96	20	3
5	33	21	3
6	33	22	1
7	19	23	1
8	28	24	5
9	12	25	2
10	18	26	2
11	9	27	1
12	14	28	1
13	7	30	1
14	8	31	3
15	3		
16	4		

[4] Indels, insertions and deletions. These are mutations in which extra DNA is inserted into a region of the genome, or in which a bit of the DNA is deleted. The compound word is used because, if you have two DNA sequences (such as chimpanzee and human DNA) and one is shorter than the other, you cannot tell whether one has shrunk by deletion or the other expanded by insertion events. You just know that some kind of 'indel' event has happened. You need more than these two sequences alone to distinguish insertions from deletions.

Table 55.3 The longer indels that occur once in the sample

LENGTH, NT
33, 34, 35, 36, 37, 39, 44, 45, 52, 57, 60, 64, 68, 70, 76, 78, 84, 122, 134, 161, 260, 280, 317, 503, 914, 920, 927, 929, 932, 1,459, 1,480, 2,147, 2,933, 3,545, 4,263

monotonically for larger gaps, except for small number fluctuation. Table 55.3 shows the larger gaps, each of which occurs once in the sample. The fall in frequency with length appears to continue among these cases as their spacing steadily increases with length. The frequency falls somewhat faster than in proportion to the length of the indels but not as fast as the square of the length. [. . .]

Discussion

This is an observation of the major way in which the genomes of closely related primates diverge—by insertion/deletion. More nucleotides are included in insertion/deletion events (3.4%) than base substitutions (1.4%) by much more than a factor of two. However, the number of events is small in comparison. About 1,000 indels listed in Tables 55.2 and 55.3 compared with about 10,000 base substitution events in this comparison of 779,142 nt between chimp and human. Little can be said about the effect of these indel events. There were so few gene regions in this small sample that a statistical analysis of their occurrence did not seem worthwhile. That will have to wait until larger regions for comparison become available. [. . .]

The uncertainty in the estimate of 3.4% indels in Table 55.1 cannot be directly evaluated. In the first place, the sample of 779 kb is small, and the variation between the different BACs is large. Further, there may be large gaps that were missed as part of chimpanzee BAC sequences that could not be aligned with the human genome. Nevertheless, the conclusion is clear that comparison of the DNA sequences of closely related species reflects many events of insertion and deletion. It is the result of a major evolutionary process.

[*Proceedings of the National Academy of Sciences, USA*, 99 (2002), 13633–5.]

H. J. MULLER

56 **Our load of mutations**

Deleterious mutations arise at a certain rate in all living creatures. Normally they are eliminated from a population at approximately the same rate as they arise, by natural selection. If selection is relaxed, mutations will accumulate in the population. Eventually, either the DNA will be randomized by mutation (and the life form will cease to exist), or selection has to be restored. In some modern human populations, medicine and healthy living conditions may be relaxing the force of selection. For humans, we do not know how much (if at all) selection is relaxed, nor do we know the deleterious mutation rate, but Muller calculates the consequences of some illustrative assumptions.

The starting point for this extract is an estimate of how much damage is done by deleterious mutations in humans. Muller estimated, earlier in the paper from which the extract is taken, that enough mutations occur each generation to reduce the survival rate by 20%. [Editor's summary.]

The question now arises: granting this inborn disadvantage which would amount to at least 20% under primitive conditions, may we not regard this as relatively unimportant under our present conditions of living? Furthermore, may we not confidently expect that, with continued advances in general technology, living standards and medicine, the genetic burden will be further lightened and kept very small indeed?

Certainly there is no use in getting morbid over our own natural shortcomings, and it is best not to dwell upon them but to take what steps we can to ameliorate them. Their really scientific diagnosis and treatment is, to be sure, a very recondite and elusive matter indeed, for a surprisingly high proportion of persons, because of the fact that there is such a multitude of different kinds of hereditary ailments, each individually so rare, yet in their collectivity so frequent. Medical men will discover this when they come to take these disorders more seriously and they will then recognize once more that familial complaints, which will then be the main complaints, call for physicians who take the characteristics of the entire family into consideration. In consequence, the pressure of the mutational load[1] will be reduced even more, for the generations immediately treated.

The great trouble with this method is that if (as today) it is unaccompanied by artificial selection it passes down to an indefinite number of future generations the burden that it has spared the treated generation itself. Of course these later generations can be treated in turn. But each successive generation will have not only the mutant genes which have in this way been passed along to it but also its own new crop of mutations. Thus the number of mutant genes will increase unless and until we again let as many die out

[1] Mutational load is a technical term. It is the reduction in the chance that an average individual in the population survives because of the deleterious mutations that are present in the population. A deleterious mutation here refers both to new mutations, that have newly arisen in an individual, and to mutations that occurred a few generations ago and have not yet been eliminated by natural selection. For instance, an individual may be more likely to die because it has inherited a mutation that arose in its grandparents.

as arise.[2] To put the matter in other words, if our ameliorative procedures succeed they must inevitably (barring conscious selection) cause a smaller number of mutant genes to be reproductively eliminated per generation than were eliminated originally. In fact that is today one of the main aims of these procedures. But the number eliminated originally (under primitive conditions) must have been the equilibrium number, i.e. equal to the number of new mutations arising. The number eliminated when ameliorative treatments are given is therefore less than the number of mutant genes arising. Thus the conditions for the application of the basic equilibrium have been violated. [. . .]

The rise in the total frequency of manifestation of mutant genes, must inevitably continue, so long as these circumstances continue, until at last its value becomes so high that a new equilibrium is reached, at which the total frequency of individuals eliminated per generation is again equal to the total frequency of new cases arising. This means that despite all the improved methods and facilities which will be in use at that time the population will nevertheless be undergoing as much genetic extinction as it did under the most primitive conditions. In correspondence with this, the amount of genetically caused impairment suffered by the average individual, even though he has all the techniques of civilization working to mitigate it, must by that time have grown to be as great in the presence of these techniques as it had been in paleolithic times without them. But instead of people's time and energy being mainly spent in the struggle with external enemies of a primitive kind such as famine, climatic difficulties and wild beasts, they would be devoted chiefly to the effort to live carefully, to spare and to prop up their own feeblenesses, to soothe their inner disharmonies and, in general, to doctor themselves as effectively as possible. For everyone would be an invalid, with his own special familial twists.

But, it may be objected, medicine and technology in general will probably continue to make progress, so that after the centuries or millennia needed for getting near to a new equilibrium, adjusted to today's and tomorrow's techniques, the still more advanced methods of that time might be so well able to cope [. . .] as to allow people to suffer from no more net disadvantages, or perhaps even less, than at present. In other words, the

[2] Subsequent work, by Kondrashov in the 1980s, made an important distinction overlooked by Muller. Muller assumed that the removal (or 'death') of one mutant gene would require the death of one individual—the individual who carries the mutation. He may be right. However, it is also possible that multiple mutations may be combined in a single individual by sexual reproduction. Then one individual death may remove multiple mutations. Natural selection would then be more powerful against deleterious mutation. However, Muller's general reasoning still applies, only the numbers are changed.

equilibrium goal set by present practices would by that time be long out of date. And this would, it might be urged, happen repeatedly, in fact continuously, so that an equilibrium for mutant genes would never need to be arrived at and the advancement of technique would always manage to stay ahead of the mutational accumulation process.

The above view, espoused especially by persons with an antipathy to practical applications of genetics in man, is one of blindly optimistic faith in the omnipotence of artificially controlled environmental influences. Its fallacy is of the same kind as found in the view, put forward in kindred circles, that the Malthusian principle is entirely wrong because advances in physical and socio-economic techniques will in the future enable us always to increase our means of subsistence faster than our population can naturally multiply. In both cases the nature and the enormity of the situation have eluded comprehension. It is not realized that the procedure proposed is, in the long run, as effective as trying to push back the flowing waters of a river with one's bare hands.

If the attempt were made to continue indefinitely to substitute a more remote equilibrium by ameliorative practices, it would mean an ever greater heaping up of mutant genes. There would be no limit to this short of the complete loss of all of the genes or their degradation into utterly unrecognizable forms, differing chaotically from one individual of the population to another. Our descendants' natural biological organization would in fact have disintegrated and have been replaced by complete disorder. Their only connections with mankind would then be the historical one that we ourselves had after all been their ancestors and sponsors, and the fact that their once-human material was still used for the purpose of converting it, artificially, into some semblance of man. However, it would in the end be far easier and more sensible to manufacture a complete man *de novo*, out of appropriately chosen raw materials, than to try to refashion into human form those pitiful relics which remained. For all of them would differ inordinately from one another, and each would present a whole series of most intricate research problems, before the treatments suitable for its own unique set of vagaries could be decided upon.

Admitting this to be a *reductio ad absurdum*, our critics might object that no such unlimited continuance of mutational accumulation was intended, but only a reasonable amount of it, whatever that might prove to be. The answer to this is that unless the practice were indefinitely continued, there would be some stopping place (either sudden or gradual) and that, following this, elimination would after all have to be allowed to become equal in frequency to the new mutations, in order to prevent the still further accumulation of mutant genes. Now unless this 'allowed' elimination were brought about by some type of artificial selection, as for instance by voluntary abstention from reproduction, it would mean that those who constituted the proscribed quota—which we have seen to be at least 20% of the population—were, as in early times, dying out as an automatic consequence of their own inadequacy.

If then the eliminated 20% failed involuntarily—that is, despite all their own and their community's efforts, we may be sure that most of the remaining 80%, although they had contrived to reproduce would on the whole differ from the doomed fifth but slightly. That is, they would in the main be 'marginal cases' who had managed to get along, even with the aid of the vastly improved techniques of that time, only barely and with difficulty. For these 'successes' would in fact be encumbered with the great load of those additional mutations that had accumulated during the period in which equilibrium was being postponed. Practically all of them would have been sure failures under primitive conditions and their perpetuation now, after the reattainment of equilibrium, would be contingent on a continuance of the ameliorative practices at that new level of intensity which corresponded with the new equilibrium. This permanent requirement would be the heritage that had been bequeathed by 'debtor generations' like our own. The term 'debtor' is appropriate for such generations because, by instituting for their own immediate benefit ameliorative procedures which delay the attainment of equilibrium and raise the equilibrium level of mutant gene frequency, they transfer to their descendants a price of detriment which the latter must eventually pay in full.

It is very difficult to estimate the rate of the deteriorative genetic process which present practices occasion. No one knows how much less stringent selection is today, in any one particular, than it was in primitive times. But unless we take the naive position that ailments of genetic origin cannot be mitigated by artificial means we must admit that modern methods do result in the saving for reproduction of many mutant genes which otherwise would have been eliminated by the defects they produced. Thus, assuming only our 20% minimum value for the equilibrium frequency of genetic elimination, the fact that the average American now lives beyond 65 is a proof that nothing like the equilibrium quota is eliminated by death before the age of reproduction. This may also be deduced from the fact that an average number of children of not much more than two born per adult is at present about sufficient to maintain the population.

Moreover, we cannot assume that the elimination rate is brought up to the required level by means of a highly selective failure to reproduce on the part of those who live. For a very considerable fraction of those whose lines are dying out today are known to have followed this course more or less purposely, as a result of conditioned behavior that depended primarily upon circumstances of their mode of living, their experiences and their tradition, rather than upon undesirable genetic traits. Making allowance for this major, non-genetic contingent of the relatively 'infertile,' there must be much too little room left for reproductive selection in the multiplication of a population like ours, whose number of children per adult shows a variability so much smaller than in former times.

It is of course possible to calculate readily the rate of deterioration that would result from a given *assumed* amount of relaxation of selection, if values are also assumed for total mutation rate, μ_t, and for the average persistence, $\bar{p}$ (the latter being expressed in terms of the value which it would have had under primitive conditions). Let us, for instance, take the very moderate-seeming assumption that at present one half of those who would have been genetically eliminated in primitive times succeed in perpetuating themselves, and let us at the same time follow our previous assumptions, also chosen on the side of caution, that n_t is only 20% and that $\bar{p}$, for primitive conditions, would be as much as 40 generations. Using these values we find that the increased impairment of the next generation as compared with the present one would be $\frac{1}{2} \times \frac{1}{5} \times \frac{1}{40}$, that is, 0.25%. Examining this reckoning in detail, we see that each person of the next generation would on the average receive $\frac{1}{2} \times \frac{1}{5}$, i.e., $\frac{1}{10}$ of a mutant gene more than this generation had. And if the heterozygous impairment averaged, for primitive conditions, $\frac{1}{40}$ (i.e. 2.5% reduction in the individual's chance of perpetuation per heterozygous gene), the resultant *additional* average impairment per offspring would be $\frac{1}{10} \times \frac{1}{40}$, or 0.25%, from the stand-point of what the effect would have been under primitive conditions.[3] It is evident that if this rate of decline continued (it is much more probable however that, if the mores did not change, the rate of decline would accelerate as techniques improved), it would take some 40 generations, a period of time of the order of a thousand years, to cause the amount of disability—as measured in relation to primitive conditions—to change from the equilibrium level of 20% to a level of 30%. And there would be a corresponding rise from the value of 8 to that of 12 mutant genes per individual.

It is very likely that the combination of values assumed above is a far too cautious one. For example, it is quite conceivable that not merely a half but even three-quarters of the genetically 'proscribed' quota are now perpetuating themselves, that n_t, instead of being 20%, is approximately 100%, and that $\bar{p}$ is, because of an exceptionally high dominance of mutant genes in man, only 8. In that case each individual of the next generation would on the average have ¾ of a mutant gene more than the individuals of this generation, and this would cause the average chance of survival, as measured in relation to primitive conditions, to have a decrement of $\frac{3}{4} \times \frac{1}{8}$, i.e. of

[3] [Muller's note] Our assumption that the selective elimination has been halved implies however that under present conditions, including modern treatments, the impairment would on the average be only one-half as manifest as in primitive times. Thus there would be a lowering of present survival value (i.e. of survival value measured in relation to modern conditions) of only 0.125% per generation.

nearly 10%. In that case, if the average disability were 20% now, as measured in these terms, it would be raised nearly to 30% in the course of a single generation instead of in a thousand years. That this estimate of the change is almost certainly too high is suggested by observations on the similarity in strength, morbidity, etc. of the offspring of savages and of long-civilized peoples, or of upper and lower castes, respectively, when raised under similar conditions. However, it is doubtful whether most ancient and medieval civilizations were much more genetically sparing than conditions of savagery were, for the majority of the people. And certainly the past hundred years have seen more 'progress' in this respect than all the past history of civilization. This makes it the more necessary now to carry out as exact comparative studies as possible, of the kind in question, so that we may be enabled to set an upper limit to our estimate of the possible rate of genetic deterioration. In the meantime, however, we must emphasize our uncertainty concerning the quantitative aspects of this matter, the need of further investigation of them, and the open possibility that the deterioration consequent on the present relaxation of selection may after all be a good deal more rapid than has commonly been imagined even by geneticists.

Whatever the values finally found, it is evident that the natural rate of mutation of man is so high, and his natural rate of reproduction so low, that not a great deal of margin is left for selection. Thus[4] if μ_t has the minimal value of 0.1 (n_t = 0.18) an average reproductive rate of 2.4 children per individual would be necessary to compensate for individuals genetically eliminated, without taking any account whatever of all the deaths and failures to reproduce due to non-genetic causes. But when these are taken into account as well (even though we allow only that reduced number of them that occurs under our modern conditions) it becomes perfectly evident that the present number of children per couple cannot be great enough to allow selection to keep pace with a mutation rate of 0.1. If, to make matters worse, μ_t should be anything like as high as 0.5, a possibility that cannot yet be ignored, our present reproductive practices would be utterly out of line with human requirements.

['Our Load of Mutations', *American Journal of Human Genetics*, 2 (1950), 111–76.]

[4] μ_t is the number of deleterious mutations per reproductive event: that is, the total number of new deleterious mutations in an offspring, contributed by both its parents. Muller uses figures in the 0.1–0.5 range. Life becomes impossible up near the 0.5 figure because of Muller's assumption of one individual death per mutant death μ_t (footnote 2 above). Current estimates of μ_t (now more often symbolized by *U*) range from 2 to 30, far higher than Muller thought possible (see Figure 52.2 (p 331), for instance). How humans exist with this input of deleterious mutation is one of modern biology's unsolved problems.

FRANK B. LIVINGSTONE

57 On the non-existence of human races

It makes sense to refer to 'races' within a species if the species contains a number of relatively distinct forms, in different geographic areas. However, human variation is not like that. Most characters vary continuously in space, and different characters show different patterns of spatial variation. [Editor's summary.]

In this paper I would like to point out that there are excellent arguments for abandoning the concept of race with reference to the living populations of *Homo sapiens*. Although this may seem to be a rather unorthodox position among anthropologists, a growing minority of biologists in general are advocating a similar position with regard to such diverse organisms as grackles, martens, and butterflies. Their arguments seem equally applicable to man. It should be pointed out that this position does not imply that there is no biological variability between the populations of organisms which comprise a species, but just that this variability does not conform to the discrete packages labelled races. The position can be stated in other words as: There are no races, there are only clines.[1]

The term, race, has had a long history of anthropological usage and it can be generally defined as referring to a group of local or breeding populations within a species. Thus, it is a taxonomic term for sub-specific groupings greater than the local population. Most anthropologists today use a genetic definition of races as populations which differ in the frequency of some genes.

The term, race, or its newer synonym, geographical race, is used in a similar way with reference to biological species other than man. Where the term is used, it can be considered as approximately synonymous with the term, subspecies. In 1953 Wilson and Brown first suggested discarding the concept of subspecies since it did not accord with the facts. Their main argument was that the genetic variation among the local populations of a species was discordant.

Variation is concordant if the geographic variation of the genetic characters is correlated, so that a classification based on one character would reflect the variability in any other. Such a pattern of variation is almost never

[1] A cline is a continuous geographical gradient in some biological attribute. For instance, in many species, individuals are bigger nearer the poles and smaller nearer the equator. Provided the gradient is fairly smooth in space, it is called a cline. Clines can also be measured for gene frequencies. In some cases, however, neighbouring populations have different, discrete forms rather than varying smoothly from one extreme to another; then variation is not clinal.

found among the local populations of a wide-ranging species, although it is usually found among related relatively allopatric species.[2]

Thus, although it is possible to divide a group of related species into discrete units, namely the species, it is impossible to divide a single species into groups larger than the panmictic[3] population. The causes of intraspecific biological variation are different from those of interspecific variation and to apply the term subspecies to any part of such variation not only is arbitrary or impossible but tends to obscure the explanation of this variation. If one genetic character is used, it is possible to divide a species into subspecies according to the variation in this character. If two characters are used, it may still be possible, but there will be some 'problem populations,' which, if you are an anthropologist, will be labelled composite or mixed. As the number of characters increases it becomes more nearly impossible to determine what the 'actual races really are.'

In addition to being a concept used to classify human variability, race has also been overworked as an explanation of this variability. When a particular blood group gene or hair form is found to be characteristic of the populations of a particular region, it is frequently 'explained' as being a 'racial' character. This type of explanation means, in other words, that this particular set of human populations possesses this character, while it is absent in the rest of humanity, because of the close common ancestry of the former. At times many characteristics which were thought to be racial have been found in many widely separated populations, so that the explanation in terms of race required the assumption of lengthy migrations. In this way race or common ancestry and migration have been used to explain much of the genetic variability among human populations. Unfortunately such explanations neither accord with our knowledge of the population structure and movements of hunters and gatherers, nor take into consideration the basic cause of biological variation, natural selection.

The incompatibility between race and natural selection has been recognized for a long time; so that if one's major aim was to discover the races of man, one has to disregard natural selection. Thus, non-adaptive characters were sought for and, in some instances, considered found. But the recognition of the role of natural selection has in the past ten years changed the course of research into human variability; or at least it has changed the thinking of the 'aracial ultrapolymorphists.'

If a central problem of physical anthropology is the explanation of the

[2] Allopatric species are different species living in separate geographic regions.

[3] A panmictic population is one in which interbreeding is almost equally probable between any two individuals in the whole population. (It can be contrasted with a subdivided (or 'structured') population, in which interbreeding mainly occurs within local subgroups of the population.) See Wright (Chapter 5) on evolution in structured populations; see Mayr (Chapter 23) for the biological species concept which is not far behind Livingstone's remark.

genetic variability among human population—and I think it is—then there are other methods of describing and explaining this variability which do not utilize the concept of race. This variability can also be described in terms of the concepts of cline and morphism. The variability in the frequency of any gene can be plotted in the same way that temperature is plotted on a weather map. Then one can attempt to explain this variability by the mathematical theory of population genetics. This is a very general theory and is capable of explaining all racial or gene frequency differences, although of course for any particular gene the exact magnitudes of factors, mutation, natural selection, gene drift, and gene flow, which control gene frequency differences are not known. All genes mutate, drift, flow, and for a given environment have fitnesses associated with their various genotypes. Hence differences in the frequency of any gene among a series of populations can be explained by these general factors which control gene frequency change. Gene frequency clines can result from many different types of interaction between the general factors which control gene frequencies. For example, a cline may be due to: (1) the recent advance of an advantageous gene; (2) gene flow between populations which inhabit environments with different equilibrium frequencies for the gene; or (3) a gradual change in the equilibrium value of the gene along the cline. The theoretical analysis of clines has barely begun but there seems to be no need for the concept of race in this analysis.

[*Current Anthropology*, 3 (1962), 279–81.]

WILTON M. KROGMAN

58 The scars of human evolution

Various imperfections, and medical problems, in human bodies can be explained by our evolutionary ancestry. For instance, our recently evolved upright posture has produced stresses in our backs, pelvises, and feet. It has also compromised our circulatory systems. Our flattened snouts have led to problems with our teeth. [Editor's summary.]

It has been said that man is 'fearfully and wonderfully made.' I am inclined to agree with that statement—especially the 'fearfully' part of it. As a piece of machinery we humans are such a hodgepodge and makeshift that the real wonder resides in the fact that we get along as well as we do. Part for part our bodies, particularly our skeletons, show many scars of Nature's operations as she tried to perfect us.

I am not referring to our so-called vestiges—those tag-ends of structures which once were functional, such as the remnant of a tail at the base of the spine, the appendix, the pineal or 'third' eye, the misplaced heart openings of 'blue babies,' or the like. Nor do I mean the freak variations that crop up in uals. I am discussing the imperfect adaptations the human race has n getting up from all fours.

have inherited our 'basic patents,' as W. K. Gregory of the American Museum of Natural History calls them, from a long line of vertebrate (backboned) ancestors: from fish to amphibian to reptile to mammal and finally from monkey to ape to anthropoid to *Homo sapiens*. In all this evolution the most profound skeletal changes occurred when we went from a four-legged to a two-legged mode of locomotion.

Gregory has very aptly called a four-legged animal 'the bridge that walks.' Its skeleton is built like a cantilever bridge: the backbone is the arched cantilever; the vertebrae of the forward part of the backbone are slanted backward and those of the rear forward, so that the 'thrust' is all to the apex of the arch; the four limbs are the piers or supports; the trunk and abdomen are the load suspended from the weight-balanced arch; in front the main bridge has a drawbridge or jointed crane (the neck) and with it a grappling device (the jaws).

When all this was up-ended on the hind limbs in man, the result was a terrific mechanical imbalance. Most of the advantages of the cantilever system were lost, and the backbone had to accommodate itself somehow to the new vertical weight-bearing stresses. It did so by breaking up the single-curved arch into an S-curve. We are born, interestingly enough, with a backbone in the form of the simple ancestral arch, but during infancy it bends into the human shape. When we begin to hold our head erect, at about the age of four months, we get a forward curve in the backbone's neck region; when we stand up, at about a year, we get a forward curve in the lower trunk; in the upper trunk and pelvic regions the backbone keeps its old backward curve.

But we achieve this at a price. To permit all this twisting and bending, Nature changed the shape of the vertebrae to that of a wedge, with the thicker edge in front and the thinner in back. This allows the vertebrae to pivot on their front ends as on hinges, like the segments of a toy snake. On the other hand, it also weakens the backbone, particularly in the lower back region, where the wedge shape is most pronounced. Heavy lifting or any other sudden stress may cause the lowermost lumbar vertebra to slip backward along the slope of the next vertebra. The phrase 'Oh, my aching back' has an evolutionary significance!

There are other ways in which the backbone may literally let us down. The human backbone usually has 32 to 34 vertebrae, each separated from its neighbor by a disk of cartilage which acts as a cushion. Of these vertebrae 7 are cervical (in the neck), 12 thoracic (upper trunk), 5 lumbar (lower trunk), 5 sacral (at the pelvis) and 3 to 5 caudal (the tail). Every once in a while the seventh cervical vertebra may have an unusually long lateral process; if long enough, this protruding piece of bone may so interfere with the big nerves going down to the arm that it has to be sawed off by a surgeon. Most people have 12 pairs of ribs, borne by the 12 thoracic vertebrae, but occasionally the transverse processes of the next lower segment, the first lumbar vertebra, are so exaggerated that they form a 13th pair of ribs. In some people the lowest (fifth) lumbar vertebra is fused with the sacral vertebrae. The latter are usually united into one bone, called the sacrum, but sometimes the first sacral vertebra fails to join with its mates. All these idiosyncrasies can cause trouble.

The 'Achilles' heel' of our backbone is the unstable lower end of the vertebral column. This is where we reap most of the evil consequences of standing up on our hind legs. It is a crucial zone of the body—the pathway for reproduction and the junction point where the backbone, the hind end of the trunk and the legs come together. The skeletal Grand Central Station where all this happens is a rather complicated structure consisting of the sacrum and the pelvis. The pelvis is not only a part of the general skeletal framework of the body but also a channel for the digestive and urogenital systems and the coupling to which the muscles of the hind legs are attached. When we stood up on our hind legs, we burdened the pelvis with still another function, namely, bearing the weight of the upper part of the body. How have we changed our pelvis to adapt it to its new position and burdens?

The pelvic structure is made up of three sets of paired bones, the ilium, the ischium and the pubis. The three bones meet at each side in the hip socket, where the head of the thighbone articulates. In standing erect man tilted the whole structure upward, so that the pelvis is at an angle to the backbone instead of parallel to it. The relative position of the three pelvic bones changed, with the pubis now in front instead of below. The bones also were altered in shape. The iliac bones, formerly elongate and bladelike (in the anthropoids), are now shortened and broadened. They form the crests of our hips, and they help support the sagging viscera, especially the large intestine. The pubic bones help to form the subpubic arch—that 'arch of triumph' beneath which we must all emerge to life and to the world. The ischial bones retreat to the rear; they are the bones that bear the brunt of sitting through a double feature or before a television screen.

The greatest change is in the zone of contact between the iliac bones and the wedgelike sacrum—the so-called sacroiliac articulation. Here are focused the weight-bearing stresses set up by the erect posture. Two things have happened to adapt the pelvic structure for 'thrusting' the weight of the trunk to the legs. The area of contact between the sacrum and the iliac bones has increased, strengthening the articulation. In the process the sacrum has been pushed down, so that its lower end is now well below the hip socket and also below the upper level of the pubic articulation. This has brought trouble, for the sacrum now encroaches upon the pelvic cavity and narrows the birth canal that must pass the fetus along to life. Furthermore, the changes have created an area of instability which far too often results in obscure 'low back pain' and in 'slipped sacroiliacs.'

The shortening of the iliac bones has increased the distance between the 12th (lowest) rib and the top or crest of the ilium. This has given us our waist, but it has also materially weakened the abdominal wall, which now, for about a palm's breadth, has only muscle to support it. The greatest weakness of the upright posture is the lower abdominal wall. In four-legged animals the gut is suspended by a broad ligament from the mechanically efficient convex vertebral arch. The burden of carrying the weight of the viscera is distributed evenly along the backbone. Up-end all this and what happens? First of all, the gut no longer hangs straight down from the backbone but sags parallel to it. Secondly, the supporting ligament has a smaller and less secure hold on the backbone. One result of the shift in the weight-bearing thrust of the abdominal viscera is that we are prone to hernia.

Nature has made a valiant effort to protect our lower belly wall. She invented the first 'plywood,' and made it of muscle. Three sheets of muscle make up the wall, and their fibers crisscross at right and oblique angles. This is all right as far as it goes, but it has not gone far enough: there is a triangular area in the wall which was left virtually without muscular support—a major scar of our imperfect evolution.

The upright posture required a major shift in the body's center of gravity, but here Nature seems to have done a pretty good job. The hip sockets have turned to face slightly forward instead of straight to the sides; the sockets and the heads of the thighbones have increased in size, and the neck of the thighbone is angled a bit upward. As a result of this complex of adjustments the bodily center of gravity is just about on a level with a transverse line through the middle of the hip sockets, and the weight of the trunk upon the pelvis is efficiently distributed on the two legs.

Though it does not directly involve the skeleton, I might mention here that the blood circulation is another factor that is not helped by our upright position. Since the heart is now about four feet above the ground, the blood returned to the heart from the veins of the legs must overcome about four feet of gravitational pull. Often our pumping system and veins find the job too much, and the result is varicose veins. The lower end of the large intestine also is affected, for its veins, when up-ended in a vertical position, become congested more easily, so we get hemorrhoids.

Even more serious is the danger to the circulation along the vertebral column. Two great vessels, an artery and a vein, run down this column. At the level where these vessels divide into two branches, one for each leg, the right-sided artery crosses over the left-sided vein. In a quadruped this presents no problems, but in the erect position the two vessels must cross a sharp promontory of bone at the junction of two vertebrae, and the viscera piled up in the pelvis press down on them. During pregnancy the pressure may increase so much that the vein is nearly pressed shut, making for very poor venous drainage of the left leg. This is the so-called 'milk leg' of pregnancy.

Going back to the skeleton, it is clear that the two-legged posture places a much bigger burden on our feet. They have adapted themselves to this by becoming less of a grasping tool (as in the monkeys) and more of a load-distributing mechanism. We have lost the opposability of the big toe, shortened the other toes and increased the length of the rest of the foot. The main tarsal bones, which form the heel, ankle joint and most of the instep, now account for half the total length of the foot, instead of only a fifth as in the chimpanzee. We have also achieved a more solid footing by developing two crosswise axes, one through the tarsals and the other through the main bones of the toes. The little-toe side of our foot is relatively neglected—the toe is little because it is not so useful. Our fallen-arch troubles, our bunions, our calluses and our foot miseries generally hark back to the fact that our feet are not yet healed by adaptation and evolutionary selection into really efficient units.

Now let us go to the other extreme—to the head. A lot has gone on there, too. We have expanded our brain case tremendously, and there can be no doubt that many of the obstetrical problems of Mrs H. Sapiens are due to the combination of a narrower pelvis and a bigger head in the species. How long it will take to balance that ratio we have no idea. It seems reasonable to assume that the human head will not materially shrink in size, so the adjustment will have to be in the pelvis; *i.e.*, evolution should favor women with a broad, roomy pelvis.

If the head has increased in size, the reverse is true of the facial skeleton. Bone for bone the face has decreased in size as we proceed from anthropoid to man. To put it succinctly, we have a face instead of a snout.

What about the teeth in that face of ours? All mammals have four kinds of teeth: incisors in front, canines at the corner, premolars and molars along the sides. With but few exceptions the mammals have both a milk set and a permanent set of teeth. About 100 million years ago, or maybe a bit more, the first mammals had 66 permanent teeth, of which 44 were molars or premolars. Most mammals today have 44 teeth, including 28 molars and premolars. But man, and the anthropoid, has only 32 teeth—8 incisors (upper and lower), 4 canines, 8 premolars and 12 molars. The loss has been greatest in molars, next in incisors, then in premolars, with the canine a veritable Rock of Gibraltar.

While the face bones have decreased in size, our teeth have remained relatively large. Many orthodontists believe that this uneven evolutionary development may be partly responsible for the malocclusion of teeth in children. Certain it is that some human teeth are apparently on the way out: the third molars ('wisdom teeth') are likely to be impacted or come in at a bad angle, and many people never have them at all. Perhaps in another million years or so we shall be reduced to no more than 20 teeth.

[*Scientific American*, 185 (Dec. 1951), 54–7.]

STEVEN PINKER

59 The big bang

Critics have argued that human language could not evolve by natural selection. However, language is a complex adaptation, and natural selection is the only explanation for complex adaptations. Language probably evolved in small stages, each new stage building on the previous stages. Critics have variously misunderstood language, or natural selection. [Editor's summary.]

The elephant's trunk is six feet long and one foot thick and contains sixty thousand muscles. Elephants can use their trunks to uproot trees, stack timber, or carefully place huge logs in position when recruited to build bridges. An elephant can curl its trunk around a pencil and draw characters on letter-size paper. With the two muscular extensions at the tip, it can remove a thorn, pick up a pin or a dime, uncork a bottle, slide the bolt off a cage door and hide it on a ledge, or grip a cup so firmly, without breaking it, that only another elephant can pull it away. The tip is sensitive enough for a blindfolded elephant to ascertain the shape and texture of objects. In the wild, elephants use their trunks to pull up clumps of grass and tap them against their knees to knock off the dirt, to shake coconuts out of palm trees, and to powder their bodies with dust. They use their trunks to probe the ground as they walk, avoiding pit traps, and to dig wells and siphon water from them. Elephants can walk underwater on the beds of deep rivers or swim like submarines for miles, using their trunks as snorkels. They communicate through their trunks by trumpeting, humming, roaring, piping, purring, rumbling, and making a crumpling-metal sound by rapping the trunk against the ground. The trunk is lined with chemoreceptors that allow the elephant to smell a python hidden in the grass or food a mile away.

Elephants are the only living animals that possess this extraordinary organ. Their closest living terrestrial relative is the hyrax, a mammal that you would probably not be able to tell from a large guinea pig. Until now you have probably not given the uniqueness of the elephant's trunk a moment's thought. Certainly no biologist has made a fuss about it. But now imagine what might happen if some biologists were elephants. Obsessed with the unique place of the trunk in nature, they might ask how it could have evolved, given that no other organism has a trunk or anything like it. One school might try to think up ways to narrow the gap. They would first point out that the elephant and the hyrax share about 90% of their DNA and thus could not be all that different. They might say that the trunk must not be as complex as everyone thought; perhaps the number of muscles had been miscounted. They might further note that the hyrax really does have a trunk, but somehow it has been overlooked; after all, the hyrax does have nostrils. Though their attempts to train hyraxes to pick up objects with their nostrils have failed, some might trumpet their success at training the hyraxes to push toothpicks around with their tongues, noting that stacking tree trunks or drawing on blackboards differ from it only in degree. The opposite school, maintaining the uniqueness of the trunk, might insist that it appeared all at once in the offspring of a particular trunkless elephant ancestor, the product of a single dramatic mutation. Or they might say that the trunk somehow arose as an automatic by-product of the elephant's having evolved a large head. They might add another paradox for trunk evolution: the trunk is absurdly more intricate and well coordinated than any ancestral elephant would have needed.

These arguments might strike us as peculiar, but every one of them has been made by scientists of a different species about a complex organ that that species alone possesses, language. Chomsky and some of his fiercest opponents agree on one thing: that a uniquely human language instinct seems to be incompatible with the modern Darwinian theory of evolution, in which complex biological systems arise by the gradual accumulation over generations of random genetic mutations that enhance reproductive success. Either there is no language instinct, or it must have evolved by other means. Since I have been trying to convince you that there is a language instinct but would certainly forgive you if you would rather believe Darwin than believe me, I would also like to convince you that you need not make that choice. Though we know few details about how the language instinct evolved, there is no reason to doubt that the principal explanation is the same as for any other complex instinct or organ, Darwin's theory of natural selection.

Language is obviously as different from other animals' communication systems as the elephant's trunk is different from other animals' nostrils. Nonhuman communication systems are based on one of three designs: a finite repertory of calls (one for warnings of predators, one for claims to territory, and so on), a continuous analog signal that registers the magnitude of some state (the livelier the dance of the bee, the richer the food source that it is telling its hivemates about), or a series of random variations on a theme (a birdsong repeated with a new twist each time: Charlie Parker with feathers). As we have seen, human language has a very different design. The discrete combinational system called 'grammar' makes human language infinite (there is no limit to the number of complex words or sentences in a language), digital (this infinity is achieved by rearranging discrete elements in particular orders and combinations, not by varying some signal along a continuum like the mercury in a thermometer), and compositional (each of the infinite combinations has a different meaning predictable from the meanings of its parts and the rules and principles arranging them).

Even the seat of human language in the brain is special. The vocal calls of primates are controlled not by their cerebral cortex but by phylogenetically older neural structures in the brain stem and limbic system, structures that are heavily involved in emotion. Human vocalizations other than language, like sobbing, laughing, moaning, and shouting in pain, are also controlled subcortically. Subcortical structures even control the swearing that follows the arrival of a hammer on a thumb, that emerges as an involuntary tic in Tourette's syndrome, and that can survive as Broca's aphasics' only speech. Genuine language is seated in the cerebral cortex, primarily the left perisylvian region.

Some psychologists believe that changes in the vocal organs and in the neural circuitry that produces and perceives speech sounds are the *only* aspects of language that evolved in our species. On this view, there are a few general learning abilities found throughout the animal kingdom, and they work most efficiently in humans. At some point in history language was invented and refined, and we have been learning it ever since. The idea that species-specific behavior is caused by anatomy and general intelligence is captured in the Gary Larson *Far Side* cartoon in which two bears hide behind a tree near a human couple relaxing on a blanket. One says: 'C'mon! Look at these fangs! . . . Look at these claws! . . . You think we're supposed to eat just honey and berries?' [. . .]

How plausible is it that the ancestor to language first appeared after the branch leading to humans split off from the branch leading to chimps? Not very, says Philip Lieberman, one of the scientists who believe that vocal tract anatomy and speech control are the only things that were modified in evolution, not a grammar module: 'Since Darwinian natural selection involves small incremental steps that enhance the present function of the specialized module, the evolution of a "new" module is logically impossible.' Now, something has gone seriously awry in this argument. Humans evolved from single-celled ancestors. Single-celled ancestors had no arms, legs, heart, eyes, liver, and so on. Therefore eyes and livers are logically impossible.

The point that the argument misses is that although natural selection involves incremental steps that enhance functioning, the enhancements do not have to be to an existing module. [. . .] An example of a new module is the eye, which has arisen de novo some forty separate times in animal evolution. It can begin in an eyeless organism with a patch of skin whose cells are sensitive to light. The patch can deepen into a pit, cinch up into a sphere with a hole in front, grow a translucent cover over the hole, and so on, each step allowing the owner to detect events a bit better. An example of a module growing out of bits that were not originally a module is the elephant's trunk. It is a brand-new organ, but homologies suggest that it evolved from a fusion of the nostrils and some of the upper lip muscles of the extinct elephant-hyrax common ancestor, followed by radical complications and refinements.

Language could have arisen, and probably did arise, in a similar way: by a revamping of primate brain circuits that originally had no role in vocal communication, and by the addition of some new ones. The neuroanatomists Al Galaburda and Terrence Deacon have discovered areas in monkey brains that correspond in location, input-output cabling, and cellular composition to the human language areas. For example, there are homologues to Wernicke's and Broca's areas and a band of fibers connecting the two, just as in humans. The regions are not involved in producing the monkeys' calls, nor are they involved in producing their gestures. The monkey seems to use the regions corresponding to Wernicke's area and its neighbors to recognize sound sequences and to discriminate the calls of other monkeys from its own calls. The Broca's homologues are involved in control over the muscles of the face, mouth, tongue, and larynx, and various subregions of these homologues receive inputs from the parts of the brain dedicated to hearing, the sense of touch in the mouth, tongue, and larynx, and areas in which streams of information from all the senses converge. No one knows exactly why this arrangement is found in monkeys and, presumably, their common ancestor with humans, but the arrangement would have given evolution some parts it would tinker with to produce the human language circuitry, perhaps exploiting the confluence of vocal, auditory, and other signals there.

Brand-new circuits in this general territory could have arisen, too. Neuroscientists charting the cortex with electrodes have occasionally found mutant monkeys who have one extra visual map in their brains compared to standard monkeys (visual maps are the postage-stamp-sized brain areas that are a bit like internal graphics buffers, registering the contours and motions of the visible world in a distorted picture). A sequence of genetic changes that duplicate a brain map or circuit, reroute its inputs and outputs, and frob, twiddle, and tweak its internal connections could manufacture a genuinely new brain module. [. . .]

The ancestral brain could have been rewired only if the new circuits had some effect on perception and behavior. The first steps toward human language are a mystery. This did not stop philosophers in the nineteenth century from offering fanciful speculations, such as that speech arose as imitations of animal sounds or as oral gestures that resembled the objects they represented, and linguists subsequently gave these speculations pejorative names like the bow-wow theory and the ding-dong theory. Sign language has frequently been suggested as an intermediate, but that was before scientists discovered that sign language was every bit as complex as speech. Also, signing seems to depend on Broca's and Wernicke's areas, which are in close proximity to vocal and auditory areas in the cortex, respectively. To the extent that brain areas for abstract computation are placed near the centers that process their inputs and outputs, this would suggest that speech is more basic. If I were forced to think about intermediate steps, I might ponder the

vervet monkey alarm calls studied by Cheney and Seyfarth, one of which warns of eagles, one of snakes, and one of leopards. Perhaps a set of quasi-referential calls like these came under the voluntary control of the cerebral cortex, and came to be produced in combination for complicated events; the ability to analyze combinations of calls was then applied to the parts of each call. But I admit that this idea has no more evidence in its favor than the ding-dong theory (or with Lily Tomlin's suggestion that the first human sentence was 'What a hairy back!').

Also unknown is when, in the lineage beginning at the chimp-human common ancestor, proto-language first evolved, or the rate at which it developed into the modern language instinct. In the tradition of the drunk looking for his keys under the lamppost because that is where the light is best, many archaeologists have tried to infer our extinct ancestors' language abilities from their tangible remnants such as stone tools and dwellings. Complex artifacts are thought to reflect a complex mind which could benefit from complex language. Regional variation in tools is thought to suggest cultural transmission, which depends in turn on generation-to-generation communication, perhaps via language. However, I suspect that any investigation that depends on what an ancient group left behind will seriously underestimate the antiquity of language. There are many modern hunter-gatherer peoples with sophisticated language and technology, but their baskets, clothing, baby slings, boomerangs, tents, traps, bows and arrows, and poisoned spears are not made of stone and would rot into nothing quickly after their departure, obscuring their linguistic competence from future archaeologists.

Thus the first traces of language could have appeared as early as *Australopithecus afarensis* (first discovered as the famous 'Lucy' fossil), at 4 million years old our most ancient fossilized ancestor. Or perhaps even earlier; there are few fossils from the time between the human-chimp split 5 to 7 million years ago and *A. afarensis*. Evidence for a lifestyle into which language could plausibly be woven gets better with later species. *Homo habilis*, which lived about 2.5 to 2 million years ago, left behind caches of stone tools that may have been home bases or local butchering stations; in either case they suggest some degree of cooperation and acquired technology. *Habilis* was also considerate enough to have left us some of their skulls, which bear faint imprints of the wrinkle patterns of their brains. Broca's area is large and prominent enough to be visible, as are the supramarginal and angular gyri, and these areas are larger in the left hemisphere. We do not, however, know whether habilines used them for language; remember that even monkeys have a small homologue to Broca's area. *Homo erectus*, which spread from Africa across much of the old world from 1.5 million to 500,000 years ago (all the way to China and Indonesia), controlled fire and almost everywhere used the same symmetrical, well-crafted stone hand-axes. It is easy to imagine some form of language contributing to such successes, though again we cannot be sure.

Modern *Homo sapiens*, which is thought to have appeared about 200,000 years ago and to have spread out of Africa 100,000 years ago, had skulls like ours and much more elegant and complex tools, showing considerable regional variation. It is hard to believe that they lacked language, given that biologically they *were* us, and all biologically modern humans have language. This elementary fact, by the way, demolishes the date most commonly given in magazine articles and textbooks for the origin of language: 30,000 years ago, the age of the gorgeous cave art and decorated artifacts of Cro-Magnon humans in the Upper Paleolithic. The major branches of humanity diverged well before then, and all their descendants have identical language abilities; therefore the language instinct was probably in place well before the cultural fads of the Upper Paleolithic emerged in Europe. Indeed, the logic used by archaeologists (who are largely unaware of psycholinguistics) to pin language to that date is faulty. It depends on there being a single 'symbolic' capacity underlying art, religion, decorated tools, and language, which we now know is false (just think of linguistic idiot savants like Denyse and Crystal,[1] or, for that matter, any normal three-year-old).

One other ingenious bit of evidence has been applied to language origins. Newborn babies, like other mammals, have a larynx that can rise up and engage the rear opening of the nasal cavity, allowing air to pass from nose to lungs avoiding the mouth and throat. Babies become human at three months when their larynx descends to a position low in their throats. This gives the tongue the space to move both up and down and back and forth, changing the shape of two resonant cavities and defining a large number of possible vowels. But it comes at a price. In *The Origin of Species* Darwin noted 'the strange fact that every particle of food and drink which we swallow has to pass over the orifice of the trachea, with some risk of falling into the lungs.' Until the recent invention of the Heimlich maneuver, choking on food was the sixth leading cause of accidental death in the United States, claiming six thousand victims a year. The positioning of the larynx deep in the throat, and the tongue far enough low and back to articulate a range of vowels, also compromised breathing and chewing. Presumably the communicative benefits outweighed the physiological costs.

Lieberman and his colleagues have tried to reconstruct the vocal tracts of extinct hominids by deducing where the larynx and its associated muscles could have fit into the space at the base of their fossilized skulls. They argue that all species prior to modern *Homo sapiens*, including Neanderthals, had a standard mammalian airway with its reduced space of possible vowels.

[1] In an earlier chapter of *The Language Instinct* Pinker described individual people who are seriously mentally retarded but have normal, even overdeveloped, linguistic ability.

Lieberman suggests that until modern *Homo sapiens*, language must have been quite rudimentary. But Neanderthals have their loyal defenders and Lieberman's claim remains controversial. In any case, e lengeege weth e smell nember ef vewels cen remeen quete expresseve, so we cannot conclude that a hominid with a restricted vowel space had little language.

Thoughtful evolutionary theorists since Darwin have been adamant that not every beneficial trait is an adaptation to be explained by natural selection. When a flying fish leaves the water, it is extremely adaptive for it to reenter the water. But we do not need natural selection to explain this happy event; gravity will do just fine. Other traits, too, need an explanation different from selection. Sometimes a trait is not an adaptation in itself but a consequence of something else that is an adaptation. There is an advantage to our bones being white instead of green, but there is an advantage to our bones being rigid; building them out of calcium is one way to make them rigid, and calcium happens to be white. Sometimes a trait is constrained by its history, like the S-bend in our spine that we inherited when four legs became bad and two legs good. Many traits may just be impossible to grow within the constraints of a body plan and the way the genes build the body. The biologist J. B. S. Haldane once said that there are two reasons why humans do not turn into angels: moral imperfection and a body plan that cannot accommodate both arms and wings. And sometimes a trait comes about by dumb luck. If enough time passes in a small population of organisms, all kinds of coincidences will be preserved in it, a process called genetic drift. For example, in a particular generation all the stripeless organisms might be hit by lightning or die without issue; stripedness will reign thereafter, whatever its advantages or disadvantages.

Stephen Jay Gould and Richard Lewontin[2] have accused biologists (unfairly, most believe) of ignoring these alternative forces and putting too much stock in natural selection. They ridicule such explanations as 'just-so stories,' an allusion to Kipling's whimsical tales of how various animals got their body parts. Gould and Lewontin's essays have been influential in the cognitive sciences, and Chomsky's skepticism that natural selection can explain human language is in the spirit of their critique.

But Gould and Lewontin's potshots do not provide a useful model of how to reason about the evolution of a complex trait. One of their goals was to undermine theories of human behavior that they envisioned as having rightwing political implications. The critiques also reflect their day-to-day professional concerns. Gould is a paleontologist, and paleontologists study organisms after they have turned into rocks. They look more at grand patterns in the history of life than at the workings of an individual's long-defunct organs. When they discover, for example, that the dinosaurs were

[2] Gould and Lewontin's paper is Chapter 20.

extinguished by an asteroid slamming into the earth and blacking out the sun, small differences in reproductive advantages understandably seem beside the point. Lewontin is a geneticist, and geneticists tend to look at the raw code of the genes and their statistical variation in a population, rather than the complex organs they build. Adaptation can seem like a minor force to them, just as someone examining the 1's and 0's of a computer program in machine language without knowing what the program does might conclude that the patterns are without design. The mainstream in modern evolutionary biology is better represented by biologists like George Williams, John Maynard Smith, and Ernst Mayr, who are concerned with the design of whole living organisms. Their consensus is that natural selection has a very special place in evolution, and that the existence of alternatives does *not* mean that the explanation of a biological trait is up for grabs, depending only on the taste of the explainer.

The biologist Richard Dawkins has explained this reasoning lucidly in his book *The Blind Watchmaker*. Dawkins notes that the fundamental problem of biology is to explain 'complex design.' The problem was appreciated well before Darwin. The theologian William Paley wrote:

> In crossing a heath, suppose I pitched my foot against a *stone*, and were asked how the stone came to be there; I might possibly answer, that, for anything I knew to the contrary, it had lain there for ever: nor would it perhaps be very easy to show the absurdity of this answer. But suppose I had found a *watch* upon the ground, and it should be inquired how the watch happened to be in that place; I should hardly think of the answer which I had before given, that for anything I knew, the watch might have always been there.

Paley noted that a watch has a delicate arrangement of tiny gears and springs that function together to indicate the time. Bits of rock do not spontaneously exude metal which forms itself into gears and springs which then hop into an arrangement that keeps time. We are forced to conclude that the watch had an artificer who designed the watch with the goal of timekeeping in mind. But an organ like an eye is even more complexly and purposefully designed than a watch. The eye has a transparent protective cornea, a focusing lens, a light-sensitive retina at the focal plane of the lens, an iris whose diameter changes with the illumination, muscles that move one eye in tandem with the other, and neural circuits that detect edges, color, motion, and depth. It is impossible to make sense of the eye without noting that it appears to have been designed for seeing—if for no other reason than that it displays an uncanny resemblance to the man-made camera. If a watch entails a watchmaker and a camera entails a cameramaker, then an eye entails an eyemaker, namely God. Biologists today do not disagree with Paley's laying out of the problem. They disagree only with his solution. Darwin is history's most important biologist because he showed how such 'organs of extreme perfection and complication' could arise from the purely physical process of natural selection.

And here is the key point. Natural selection is not just a scientifically respectable alternative to divine creation. It is the *only* alternative that can explain the evolution of a complex organ like the eye. The reason that the choice is so stark—God or natural selection—is that structures that can do what the eye does are extremely low-probability arrangements of matter. By an unimaginably large margin, most objects thrown together out of generic stuff, even generic animal stuff, cannot bring an image into focus, modulate incoming light, and detect edges and depth boundaries. The animal stuff in an eye seems to have been assembled with the goal of seeing in mind—but in whose mind, if not God's? How else could the mere *goal* of seeing well *cause* something to see well? The very special power of natural selection is to remove the paradox. What causes eyes to see well now is that they descended from a long line of ancestors that saw a bit better than their rivals, which allowed them to out-reproduce those rivals. The small random improvements in seeing were retained and combined and concentrated over the eons, leading to better and better eyes. The ability of *many* ancestors to see a *bit* better in the *past* causes a *single* organism to see *extremely* well *now*.

Another way of putting it is that natural selection is the only process that can steer a lineage of organisms along the path in the astronomically vast space of possible bodies leading from a body with no eye to a body with a functioning eye. The alternatives to natural selection can, in contrast, only grope randomly. The odds that the coincidences of genetic drift would result in just the right genes coming together to build a functioning eye are infinitesimally small. Gravity alone may make a flying fish fall into the ocean,[3] a nice big target, but gravity alone cannot make bits of a flying fish embryo fall into place to make a flying fish eye. When one organ develops, a bulge of tissue or some nook or cranny can come along for free, the way an S-bend accompanies an upright spine. But you can bet that such a cranny will not just happen to have a functioning lens and a diaphragm and a retina all perfectly arranged for seeing. It would be like the proverbial hurricane that blows through a junkyard and assembles a Boeing 747. For these reasons, Dawkins argues that natural selection is not only the correct explanation for life on earth but is bound to be the correct explanation for anything we would be willing to call 'life' anywhere in the universe.

And adaptive complexity, by the way, is also the reason that the evolution of complex organs tends to be slow and gradual. It is not that large mutations and rapid change violate some law of evolution. It is only that complex engineering requires precise arrangements of delicate parts, and if the

[3] The gravitational example originated with Williams (Chapter 15).

engineering is accomplished by accumulating random changes, those changes had better be small. Complex organs evolve by small steps for the same reason that a watchmaker does not use a sledgehammer and a surgeon does not use a meat cleaver.

So we now know which biological traits to credit to natural selection and which ones to other evolutionary processes. What about language? In my mind, the conclusion is inescapable. Every discussion in this book has underscored the adaptive complexity of the language instinct. It is composed of many parts: syntax, with its discrete combinatorial system building phrase structures; morphology, a second combinatorial system building words; a capacious lexicon; a revamped vocal tract; phonological rules and structures; speech perception; parsing algorithms; learning algorithms. Those parts are physically realized as intricately structured neural circuits, laid down by a cascade of precisely timed genetic events. What these circuits make possible is an extraordinary gift: the ability to dispatch an infinite number of precisely structured thoughts from head to head by modulating exhaled breath. The gift is obviously useful for reproduction—think of Williams' parable of little Hans and Fritz being ordered to stay away from the fire and not to play with the saber-tooth. Randomly jigger a neural network or mangle a vocal tract, and you will not end up with a system with these capabilities. The language instinct, like the eye, is an example of what Darwin called 'that perfection of structure and co-adaptation which justly excites our admiration,' and as such it bears the unmistakable stamp of nature's designer, natural selection.

If Chomsky maintains that grammar shows signs of complex design but is skeptical that natural selection manufactured it, what alternative does he have in mind? What he repeatedly mentions is physical law. Just as the flying fish is compelled to return to the water and calcium-filled bones are compelled to be white, human brains might, for all we know, be compelled to contain circuits for Universal Grammar. He writes:

> These skills [for example, learning a grammar] may well have arisen as a concomitant of structural properties of the brain that developed for other reasons. Suppose that there was selection for bigger brains, more cortical surface, hemispheric specialization for analytic processing, or many other structural properties that can be imagined. The brain that evolved might well have all sorts of special properties that are not individually selected; there would be no miracle in this, but only the normal workings of evolution. We have no idea, at present, how physical laws apply when 10^{10} neurons are placed in an object the size of a basketball, under the special conditions that arose during human evolution.

We may not, just as we don't know how physical laws apply under the special conditions of hurricanes sweeping through junkyards, but the possibility that there is an undiscovered corollary of the laws of physics that causes brains of human size and shape to develop the circuitry for Universal Grammar seems unlikely for many reasons.

At the microscopic level, what set of physical laws could cause a surface molecule guiding an axon along a thicket of glial cells to cooperate with millions of other such molecules to solder together just the kinds of circuits that would compute something as useful to an intelligent social species as grammatical language? The vast majority of the astronomical number of ways of wiring together a large neural network would surely lead to something else: bat sonar, or nest-building, or go-go dancing, or, most likely of all, random neural noise.

At the level of the whole brain, the remark that there has been selection for bigger brains is, to be sure, common in writings about human evolution (especially from paleoanthropologists). Given that premise, one might naturally think that all kinds of computational abilities might come as a by-product. But if you think about it for a minute, you should quickly see that the premise has it backwards. Why would evolution ever have selected for sheer bigness of brain, that bulbous, metabolically greedy organ? A large-brained creature is sentenced to a life that combines all the disadvantages of balancing a watermelon on a broomstick, running in place in a down jacket, and, for women, passing a large kidney stone every few years. Any selection on brain size itself would surely have favored the pinhead. Selection for more powerful computational abilities (language, perception, reasoning, and so on) must have given us a big brain as a by-product, not the other way around!

But even given a big brain, language does not fall out the way that flying fish fall out of the air. We see language in dwarfs whose heads are much smaller than a basketball. We also see it in hydrocephalics whose cerebral hemispheres have been squashed into grotesque shapes, sometimes a thin layer lining the skull like the flesh of a coconut, but who are intellectually and linguistically normal. Conversely, there are Specific Language Impairment victims with brains of normal size and shape and with intact analytic processing [. . .]. All the evidence suggests that it is the precise wiring of the brain's microcircuitry that makes language happen, not gross size, shape, or neuron packing. The pitiless laws of physics are unlikely to have done us the favor of hooking up that circuitry so that we could communicate with one another in words. [. . .]

To be fair, there are genuine problems in reconstructing how the language faculty might have evolved by natural selection, though the psychologist Paul Bloom and I have argued that the problems are all resolvable. As P. B. Medawar noted, language could not have begun in the form it supposedly took in the first recorded utterance of the infant Lord Macaulay, who after having been scalded with hot tea allegedly said to his hostess, 'Thank you, madam, the agony is sensibly abated.' If language evolved gradually, there must have been a sequence of intermediate forms, each useful to its possessor, and this raises several questions.

First, if language involves, for its true expression, another individual, who did the first grammar mutant talk to? One answer might be: the fifty percent of the brothers and sisters and sons and daughters who shared the new gene by common inheritance. But a more general answer is that the neighbors could have partly understood what the mutant was saying even if they lacked the new-fangled circuitry, just using overall intelligence. Though we cannot parse strings like *skid crash hospital*, we can figure out what they probably mean, and English speakers can often do a reasonably good job understanding Italian newspaper stories based on similar words and background knowledge. If a grammar mutant is making important distinctions that can be decoded by others only with uncertainty and great mental effort, it could set up a pressure for them to evolve the matching system that allows those distinctions to be recovered reliably by an automatic, unconscious parsing process. [. . .] Natural selection can take skills that are acquired with effort and uncertainty and hardwire them into the brain. Selection could have ratcheted up language abilities by favoring the speakers in each generation that the hearers could best decode, and the hearers who could best decode the speakers.

A second problem is what an intermediate grammar would have looked like. Bates asks:

> What protoform can we possibly envision that could have given birth to constraints on the extraction of noun phrases from an embedded clause? What could it conceivably mean for an organism to possess half a symbol, or three quarters of a rule? . . . monadic symbols, absolute rules and modular systems must be acquired as a whole, on a yes-or-no basis—a process that cries out for a Creationist explanation.

The question is rather odd, because it assumes that Darwin literally meant that organs must evolve in successively larger fractions (half, three quarters, and so on). Bates' rhetorical question is like asking what it could conceivably mean for an organism to possess half a head or three quarters of an elbow. Darwin's real claim, of course, is that organs evolve in successively more complex forms. Grammars of intermediate *complexity* are easy to imagine; they could have symbols with a narrower range, rules that are less reliably applied, modules with fewer rules, and so on. In a recent book Derek Bickerton answers Bates even more concretely. He gives the term 'protolanguage' to chimp signing, pidgins, child language in the two-word stage, and the unsuccessful partial language acquired after the critical period by Genie and other wolf-children. Bickerton suggests that *Homo erectus* spoke in protolanguage. Obviously there is still a huge gulf between these relatively crude systems and the modern adult language instinct, and here Bickerton makes the jaw-dropping additional suggestion that a single mutation in a single woman, African Eve, simultaneously wired in syntax, resized and reshaped the skull, and reworked the vocal tract. But we can extend the first half of Bickerton's argument without accepting the second half, which is reminiscent of hurricanes assembling jetliners. The languages of children, pidgin speakers, immigrants, tourists, aphasics, telegrams, and headlines show that there is a vast continuum of viable language systems varying in efficiency and expressive power, exactly what the theory of natural selection requires.

A third problem is that each step in the evolution of a language instinct, up to and including the most recent ones, must enhance fitness. David Premack writes:

> I challenge the reader to reconstruct the scenario that would confer selective fitness on recursiveness. Language evolved, it is conjectured, at a time when humans or protohumans were hunting mastodons. . . . Would it be a great advantage for one of our ancestors squatting alongside the embers, to be able to remark: 'Beware of the short beast whose front hoof Bob cracked when, having forgotten his own spear back at camp, he got in a glancing blow with the dull spear he borrowed from Jack'?
>
> Human language is an embarrassment for evolutionary theory because it is vastly more powerful than one can account for in terms of selective fitness. A semantic language with simple mapping rules, of a kind one might suppose that the chimpanzee would have, appears to confer all the advantages one normally associates with discussions of mastodon hunting or the like. For discussions of that kind, syntactic classes, structure-dependent rules, recursion and the rest, are overly powerful devices, absurdly so.

I am reminded of a Yiddish expression. 'What's the matter, is the bride too beautiful?' The objection is a bit like saying that the cheetah is much faster than it has to be, or that the eagle does not need such good vision, or that the elephant's trunk is an overly powerful device, absurdly so. But it is worth taking up the challenge.

First, bear in mind that selection does not need great advantages. Given the vastness of time, tiny advantages will do. Imagine a mouse that was subjected to a minuscule selection pressure for increased size—say, a one percent reproductive advantage for offspring that were one percent bigger. Some arithmetic shows that the mouse's descendants would evolve to the size of an elephant in a few thousand generations, an evolutionary eyeblink.

Second, if contemporary hunter-gatherers are any guide, our ancestors were not grunting cave men with little more to talk about than which mastodon to avoid. Hunter-gatherers are accomplished toolmakers and superb amateur biologists with detailed knowledge of the life cycles, ecology, and behavior of the plants and animals they depend on. Language would surely have been useful in anything resembling such a lifestyle. It is possible to imagine a superintelligent species whose isolated members cleverly negotiated their environment without communicating with one another, but what a waste! There is a fantastic payoff in trading hard-won knowledge with kin and friends, and language is obviously a major means of doing so.

And grammatical devices designed for communicating precise information about time, space, objects, and who did what to whom are not like the proverbial thermonuclear fly-swatter. Recursion in particular is extremely useful; it is not, as Premack implies, confined to phrases with tortuous syntax. Without recursion you can't say *the man's hat* or *I think he left*. Recall that all you need for recursion is an ability to embed a noun phrase inside another noun phrase or a clause within a clause, which falls out of rules as simple as 'NP → det N PP' and 'PP → P NP.' With this ability a speaker can pick out an object to an arbitrarily fine level of precision. These abilities can make a big difference. It makes a difference whether a far-off region is reached by taking the trail that is in front of the large tree or the trail that the large tree is in front of. It makes a difference whether that region has animals that you can eat or animals that can eat you. It makes a difference whether it has fruit that is ripe or fruit that was ripe or fruit that will be ripe. It makes a difference whether you can get there if you walk for three days or whether you can get there and walk for three days.

Third, people everywhere depend on cooperative efforts for survival, forming alliances by exchanging information and commitments. This too puts complex grammar to good use. It makes a difference whether you understand me as saying that if you give me some of your fruit I will share meat that I will get, or that you should give me some fruit because I shared meat that I got, or that if you don't give me some fruit I will take back the meat that I got. And once again, recursion is far from being an absurdly powerful device. Recursion allows sentences like *He knows that she thinks that he is flirting with Mary* and other means of conveying gossip, an apparently universal human vice.

But could these exchanges really produce the rococo complexity of human grammar? Perhaps. Evolution often produces spectacular abilities when adversaries get locked into an 'arms race,' like the struggle between cheetahs and gazelles. Some anthropologists believe that human brain evolution was propelled more by a cognitive arms race among social competitors than by mastery of technology and the physical environment. After all, it doesn't take that much brain power to master the ins and outs of a rock or to get the better of a berry. But outwitting and second-guessing an organism of approximately equal mental abilities with non-overlapping interests, at best, and malevolent intentions, at worst, makes formidable and ever-escalating demands on cognition. And a cognitive arms race clearly could propel a linguistic one. In all cultures, social interactions are mediated by persuasion and argument. How a choice is framed plays a large role in determining which alternative people choose. Thus there could easily have been selection for any edge in the ability to frame an offer so that it appears to present maximal benefit and minimal cost to the negotiating partner, and in the ability to see through such attempts and to formulate attractive counter-proposals.

Finally, anthropologists have noted that tribal chiefs are often both gifted orators and highly polygynous—a splendid prod to any imagination that cannot conceive of how linguistic skills could make a Darwinian difference. I suspect that evolving humans lived in a world in which language was woven into the intrigues of politics, economics, technology, family, sex, and friendship that played key roles in individual reproductive success. They could no more live with a Me-Tarzan-you-Jane level of grammar than we could.

[*The Language Instinct* (New York: Morrow, 1994), ch. 11.]

Section J

Evolution and human affairs

Few areas—perhaps no area—of human thought has been unaffected by the theory of evolution and the six extracts here are only a sample of the intellectual relationships of Darwinism. The six fall into four main categories: religion (and education, particularly in the USA); philosophy of science; ethics; and economics.

'Creationism' is important in politics and law, but not in science; scientists therefore should not ignore it, but also should not think about it in the way that they think about a scientific theory. I begin the section with a paper by Antolin and Herbers (Chapter 60). They describe their experiences in dealing with creationism, and reflect on large issues. I have also included the reflections of Theodosius Dobzhansky. Dobzhansky was one of the most eminent evolutionary biologists of the twentieth century, and we met a short extract from his writings in Section D (Chapter 25). Dobzhansky's influence also lies not far behind much of the work in Section B, particularly Lewontin's (Chapter 12), as well as Mayr's writings about species (Section D (Chapters 22, 23)). You can even find Cain arguing with Dobzhansky about adaptation in Section C (Chapter 19).

The title of Dobzhansky's paper (Chapter 61) in this section has become a catchphrase in the subject. The paper itself combines an authoritative summary of the scientific case for evolution with the argument (of some personal authority—Dobzhansky was a practising Christian as well as a first-rate evolutionist) that evolution and religion, properly understood, can peacefully coexist. I should point out that, unlike many extracts in this book, I have included all of Dobzhansky's essay. Also, Dobzhansky's appreciation for Teilhard de Chardin, reflected in his closing paragraph, is highly idiosyncratic: 'the acceptance of [Teilhard's] world view falls short of universal' must be the biggest understatement in this book. (For a more typical reaction to Teilhard, see Medawar's review—the reference to which is in the 'Select bibliography'.)

The extract from Hume's *Dialogues on Natural Religion* (Chapter 62) is grander in scope. For all human history, since the birth of philosophy in ancient Greece, the 'argument from design' has been one of the main rational arguments for the existence of God. In a way, the most distinctive feature of Darwin's theory is its substitution of a natural for a supernatural, cause for biological adaptation. Hume, however, saw philosophical difficulties in the theological argument before Darwin had been born. Hume also gives a clear summary of the argument from design; Paley put it in much the same form a few years later in his watch analogy. The ideas that Hume discusses provide one additional context for the more straightforwardly scientific discussions of Section C.

Evolution has enjoyed somewhat mixed relations with the philosophy of science. Since Darwin's time, there have been philosophers who have proved (to their own satisfaction) that evolution is not a proper scientific theory. The working biologist's reaction on learning that evolutionary theory does not fit some philosophical criterion for what a scientific theory should be like is 'so much the worse for the philosophy'! Monod's (Chapter 63) piece is taken from a lecture in 1974. Monod was a molecular biologist of exceptional intellectual range. His lecture defies categorization, but it (as well as containing many acute remarks) to some extent tackles the philosophers on their own ground, showing how the theory of evolution makes refutable predictions that have held up well as molecular biology has unfolded. His remarks about blending inheritance pick up a theme of Section A (Chapters 3 and 4), looking at it in a different way. Any reader who is familiar with positivist, and Popperian, philosophy will detect it behind some of the questions that Monod has set himself.

The final two pieces deal with ethics, and the cosmic consequences of Darwinism. Thomas Henry Huxley's (Chapter 64) 'Evolution and ethics' was also originally delivered as a lecture, in 1893. His target was almost certainly Herbert Spencer's evolutionary philosophy. For Spencer, evolution was a general law, applying outside biology as well as within, and providing a justification for ethical action. Few people read Spencer now, even though he was one of the biggest middlebrow popular writers of his day. Huxley's criticism, like many great critical works, rises above the stimulus that produced it: Huxley had thought out a whole alternative system and did not confine himself to merely negative points. Once you think about the workings of nature, and Darwinian natural selection, you soon realize it is almost the opposite of a moral guide.

We finish with Palumbi's paper (Chapter 65). It is about the forces of natural selection that are being imposed on the world by human beings. Many creatures 'evolve back' in relation to human activity. Disease organisms evolve resistance to drugs; fish evolve smaller size because of fishing activity. These evolutionary responses have tended to be overlooked in the past and evolutionary biologists are now emphasizing their importance. We need to think through the evolutionary consequences before we interfere with the environment. Palumbi also estimates the financial cost to the US economy of evolution: his lower bound estimate is $33–50 billion per year. 'Applied evolution' also includes the use of evolutionary principles in engineering and software development. However, Palumbi's sample alone makes a great introduction to the topic.

MICHAEL F. ANTOLIN AND JOAN M. HERBERS

60 Evolution's struggle for existence in America's public schools

The ongoing creation–evolution controversy in North America thrives on the widespread special creationist beliefs of a significant portion of the public. Creation science supports a literal interpretation of the Judeo-Christian Bible, an earth that is no more than 10,000 years old and created ex nihilo *in six days by a monotheistic God, with no new kinds arising since the period of creation, and with a single flood of staggering force shaping layers of rocks and trapping the organisms fossilized within them. Despite decisions in numerous court cases that specifically exclude creationism and creation science from primary and secondary biology classes in America's public schools, creationists now work locally to minimize or remove evolution from science teaching standards. The nationally organized movement to resist the teaching of evolution has proven highly effective, influencing state and district school boards in addition to individual teachers and schools. Thus, if teaching about evolution and the nature of science is to survive in America's primary and secondary schools, scientists must likewise work with teachers and reach out to state and local school boards. In this perspective we outline the typical creationist arguments we encounter from students, teachers, school board members, and neighbors. We explain briefly how knowledge of both microevolution and macroevolution is important in medicine, agriculture, and biotechnology. We describe a science education controversy that arose within our own school district, how we responded, and what we learned from it. Finally, we argue that even modest outreach efforts to science teachers will be richly repaid.* [Authors' summary.]

[. . .] The ongoing creation–evolution controversy has roots in three aspects of American society: widespread scientific illiteracy, a core value of fairness in public discourse, and the prevalence of religious values in American politics. Most Americans (83%) agree that evolution in some form should be taught in schools, yet fewer than half can correctly identify creationism or evolution. Of those supporting evolution, some (17%) think both evolution and creation should be taught as scientific theories. Another 16% support excluding evolution and teaching only creationism, which means that one-third of the American public thinks creationism should be part of the science curriculum in our public schools. Alarmingly, the same proportions occur among primary and secondary school teachers, a third of whom either resist or avoid teaching evolution. Furthermore, positions have hardened and half of those surveyed state that their minds are 'completely made up' on the issue. In America, large numbers of people misunderstand the process of science and therefore fail to understand how evolution fits into mainstream science.

The result is that teaching the science of evolution is endangered in American public schools, despite having survived numerous courtroom challenges. Most aspects of the creation–evolution controversy are not new; discussions about theistic versus materialistic views of nature long preceded the publication of Darwin's *Origin of Species* in 1859. Although scientists have accepted evolutionary theory for more than a century, challenges to teaching evolution in science classrooms recur with depressing regularity. Creationists exploit widespread misconceptions about evolution to wedge their religious views into local curricula, school board policies, and science teaching standards. Our experiences with university students and teachers in our own school district suggest that most evolutionary biologists in America confront creationist thinking during their professional lives. Evolution's struggle for existence in American public schools continues without prospect of an end, and we call on our fellow evolutionary biologists to reach out beyond their laboratories and classrooms to ensure its survival. What is at stake is more than just evolution; this debate is fundamentally about how science is taught in primary and secondary schools.

Here we have four goals. First, we outline the most common arguments creationists use to attack teaching evolution in public schools. Second, we illustrate how the poor understanding of evolution broadly impacts other areas of biological science. Third, we describe a creation–evolution controversy that arose within our own school district in Colorado, how we responded to it, and what we learned from it. Fourth, we make a number of recommendations for how scientists can become engaged in outreach. We write from the perspective that science and religion provide different ways of viewing the world, but they do not necessarily conflict. Overall, we assert that understanding the creation–evolution controversy and engaging in outreach to primary and secondary education is a responsibility of every practicing scientist.

We start with definitions. In order to discuss science education, we must define what makes up scientific enquiry. We find William Overton's (1982) language in the court decision in the case of McLean versus Arkansas Board of Education to be clear, concise, and practical:

> Essential characteristics of science: (1) It is guided by natural law; (2) It has to be explanatory by reference to natural law; (3) It is testable against the empirical world; (4) Its conclusions are tentative, that is, are not necessarily the final word; and (5) It is falsifiable.

The same definition was used in the Supreme Court decision in 1987 to put a permanent injunction against so-called equal time laws, which required creation science to be taught if evolution is included in science classrooms. Overton's five criteria have tremendous heuristic value, although some philosophers of science exclude falsifiability as a necessary criterion. Falsifiability cannot be applied to explanations of single historical events like those from paleontology, geology, or astronomy in the same way it is applied to present-day experiments that can be replicated. Even so, individual predictions and hypotheses based on historical explanations can be falsified if they fail to be supported by observations of the natural world. Thus, Overton's definition of scientific enquiry is extremely useful for teaching about the scientific method, the interdependence of experiments and observations, and the limits of the scientific enterprise.

Within that context, we define the theory of evolution as a series of explanations of natural forces that result in descent with modification of living organisms. A nonexhaustive list of study topics subsumed by the theory of evolution includes: adaptation by natural selection; genetic drift and changes that result from chance events in small populations; mutation and neutral variation within and between populations; rates of change within lineages; rates of divergence between lineages; phylogenetic relationships among populations and species; and analysis of the history of life as recorded by geology, the fossil record, and analysis of DNA. A list of topics addressed by creationists would be much longer because the diversity of opinion on how supernatural forces might shape our world far outstrip the differences among scientists. We define special creationism as the idea that supernatural forces play a direct and leading role in shaping the history of life. Within that rubric, creation science refers to the idea of an Earth that is no more than 10,000 years old and was created ex-nihilo in six days by a monotheistic God; on this Earth no new kinds have arisen since the period of creation and a single flood of staggering force shaped layers of rocks and trapped the organisms that are fossilized within them. Clearly, creation science posits evidence consistent with a literal reading of the Judeo-Christian Bible; it thereby deviates not only from scientific evolution theory but also from every other creation scenario.

Creationist arguments

We repeatedly face creationist challenges. The arguments depend on misunderstandings of science used by creationists to convince the American public that creationism has a place in science education. Although creationists cannot agree on the form that creationism should take (e.g., young-earth Biblical literalists vs. day-age creationists), they often agree on how to oppose evolution. Here we describe commonly repeated challenges to evolution that are used to oppose teaching evolution, and appropriate responses to the challenges.

EVOLUTION IS JUST A THEORY

In vernacular usage, a theory is an 'educated guess' or supposition, like a theory of how a crime was committed or why our favorite sports legend failed to score the winning points. In science, we reserve theory for logically consistent statements about Nature that have withstood multiple empirical tests. Creationists exploit the blurred distinction between vernacular and technical usage of theory to cast doubt on the scientific validity of evolution. A typical example is the disclaimer glued into science textbooks in Alabama: 'Evolution is a controversial theory some scientists present as scientific explanation for the origin of living things, such as plants, animals, and humans. No one was present when life first appeared on earth. Therefore, any statement about life's origins should be considered as theory, not fact.' Denigration of the term 'theory' is a key strategy of the creationists, and it is incumbent on us to use it correctly.

EVOLUTION DESCRIBES ORIGINS OF LIVING MATTER

The origin of life remains poorly understood relative to the history of life after its origin. Evidence from geology, astronomy, and physics shows the earth to be about 4.5 billion years old. Complex life forms first appeared in the first billion years of earth's history, and the theory of evolution mainly describes events and processes in the following 3.5 billion years. It is possible to synthesize amino acids from mixtures of hydrogen, ammonia, methane, and water, as in the famous Miller–Urey experiments. Other experiments demonstrate the plausibility that most complex biological molecules could have arisen from 'primordial soup'. Even so, most hypotheses about the origin of life from nonliving matter lie outside the main body of evolution theory. For example, the contents of volume 54 (2000) of *Evolution* comprise 192 primary research articles, but not one that concerns the origins of life. Regardless, creationists commonly point to the relatively modest evidence about the origins of life to bolster their claim that the entire theory of evolution is poorly supported. The disclaimer in Alabama textbooks exemplifies this tactic. Similarly, intelligent design proponent Michael Behe promotes his idea of 'irreducible complexity' (described below) by defining evolution as 'a process by which life arose from nonliving matter and subsequently developed entirely by natural means.'[1]

Creationists follow their origins of life argument with so-called improbability analyses to 'prove' that life must have been specially created. The starting point is a truism: Given all possible alternative Universes, each of infinite size, the probability of life arising by chance on this planet at the time it did is infinitesimally small. However, they proceed to argue that an infinitesimally small probability at every point in time and space is equivalent to impossibility across all points in time and space. Not only is this conclusion false, calculations of the probability of life arising by chance are meaningless. Uncertainty about early earth history (marginal probability) means that no one can assign overall probabilities for the origin of life with any degree of confidence. While these subtle and perhaps arcane arguments are

[1] Quoted from M. J. Behe, *Darwin's Black Box* (New York: Simon and Schuster, 1996).

not discernable to the average citizen, the creationists' conclusion that life is impossible because it is highly improbable is abysmal science.

EVOLUTION MEANS 'NATURE RED IN TOOTH AND CLAW'

Tennyson's phrase, written ten years before the publication of Darwin's *Origin of Species*, decries a cold and cruel world, just as did Thomas Hobbes two centuries before: 'No arts; no letters; no society; and which is worst of all, continual fear and danger of violent death; and the life of man, solitary, poor, nasty, brutish, and short.' Clearly, the issue of evil in the world preceded Darwin, but to some the idea of natural selection as a mechanism of evolution runs completely counter to notions of a 'peaceable kingdom'. The prospect of a morally bankrupt universe inferred by creationists from this view of evolution has long provoked opponents of evolution, including William Jennings Bryan in the Scopes' trial and the present-day writings of Philip Johnson. There are two responses to this view. First and most important, the theory of evolution says nothing about Nature's purpose or the meaning of life. Evolution is restricted to description and prediction of conditions that promote descent with modification. Second, the natural world contains many examples of cooperation, and in many cases we can predict that cooperation will evolve from competitive, predatory, or parasitic relationships between species.

EVOLUTION MEANS 'SURVIVAL OF THE FITTEST'

Herbert Spencer, an economist, coined this unfortunate phrase to encapsulate the idea of natural selection. 'Survival of the fittest' became a rallying cry of laissez faire capitalists during the 19th century, racists in the 20th century, and others, as an extension of natural law to justify exploiting or exterminating weak and undesirable members of society. The excesses of social Darwinists led to early antievolution laws promoted by William Jennings Bryan. As recently as March 2001, the presumed link between Darwinian theory and social policy was the rationale offered for an antiracism resolution before the Louisiana legislature; fortunately, the resolution passed only after antiDarwin and antievolution statements were removed. Deciding who in human society is fit and deserves to survive is a social, ethical, and perhaps criminal enterprise that reflects values of human culture at particular times in history. Natural selection is one mechanism of evolution; using it to justify social policies is a perversion of science.

EVOLUTION IS ATHEISM

Darwin understood and feared that many people would equate evolutionary theory with the rejection of God, and this claim remains a centerpiece of creationist objections. Philip Johnson, a retired law school professor from

Berkeley, in his book *Darwin on Trial* (1993) describes evolution as purely a philosophical ideal, called 'materialistic naturalism'. Johnson invokes the typical creationist claim that to be an evolutionist one *must* be an atheist, and that materialistic naturalism is the core value that leads evolutionists to reject creationism. Yet many scientists are theists, and most religious and scientific groups recognize that science and faith comprise separate domains. Science is silent concerning religious and moral issues—at most, the theory of evolution is agnostic. [. . .]

Recognizing the clearly distinct domains of science and religion characterizes those scientists who are religious and those religious leaders who accept evolutionary theory as an explanation of the natural world. Scientists, clerics, and philosophers who conflate science and religion do a disservice to both.

ONLY TWO SCIENTIFIC ALTERNATIVES EXIST: EVOLUTION AND CREATION SCIENCE

This tactic appeared in the 1960s in *The Genesis Flood*, by John Whitcomb and Henry Morris (1961) which invigorated the creation-science movement. Creation science, which reframes geology and the history of life in light of biblical literalism and a young earth, became the primary creationist alternative to evolution after the Supreme Court in 1967 struck down state laws banning the teaching of evolution (including the law in Tennessee that resulted in the 1925 Scopes trial). The result was enactment of equal-time laws that required public schools to teach both creation science and 'evolution science'. These laws were overturned in the U.S. Courts of Appeals and the U.S. Supreme Court as 'serving no secular purpose' and violating the Establishment Clause of the U.S. Constitution, which prevents governments from either establishing or restraining religion.

Equal time arguments represent an egregious misunderstanding of the scientific enterprise because they posit *only* two alternatives: creationism versus evolution. Creation science claims that evidence for evolution is weak and therefore the alternative, creation, must be correct. This rhetorical trick, used in debates by creationists like Duane Gish, imposes a dualism that is quite simply bad science. Scientific theories do not gain support purely by refutation of other theories. Rather, good scientific theories gain because of internal logic and because of consistency with observations of nature. To achieve primacy, a theory must be superior to numerous alternatives. To be sure, individual experiments contain within them a dual structure (A or B), designed to lead to strong inference from falsification of one alternative. Yet any single experiment is embedded within a network of data collection that includes many plausible and testable alternatives. We could not imagine a college-level course in evolution that considers only two scientific alternatives, nor could we imagine a public school science curriculum that would be constrained in the same way. Scientific honesty requires creation science to produce data that support their theories of biblical literalism and refute *all other* faith-based stories of life's origin—not to mention refuting evolutionary theories.

TEACHING CREATIONISM IS ONLY FAIR

Fairness is a fundamental value in American society. We have equal-access laws and equal-funding laws, and we strive in our political discourse to allow every voice to be heard. Opponents of evolution have exploited this value. In November 2000, an advertisement for an 'intelligent design' (ID) conference at Yale University entitled 'Science and Evidence for Design in the Universe' began with a call for fairness: 'Scientific inquiry has always included a fundamental openness to new theories and a willingness to explore a wide range of possibilities. A great success of the scientific enterprise is a tribute to this spirit.' The brochure describes fresh evidence for ID and recent attention paid to it by scholars, implying that it is only fair that the 'new' ideas be fully considered. Fairness in science, however, is only afforded to competing ideas that are supported by evidence. Discredited scientific ideas are not given equal time! In science classes we do not teach flat-earth theory, the theory of phlogiston, or pure Lamarckian inheritance, except to give students a historical context for understanding modern science.

SCIENTIFIC DEBATE IS A SIGN OF WEAKNESS

One of the most enduring (though hardly endearing) aspects of science is that all ideas are continually held up to skepticism, testing, and debate. Creationists often interpret criticism and disagreements among scientists to mean the foundations of evolutionary theory are crumbling away and the slightest push will topple the tower. In the early 1900s, fundamentalists used feuds between Mendelians and Biometricians as a reason for promoting laws that banned the teaching of evolution. Today, creationists use scientific disagreements over tempo and mode of evolution as evidence that evolution theory itself is under dispute, often quoting evolutionary biologists out of context. In reality, disagreements about how to study adaptation, how to define species, or how to interpret the fossil record signal that evolution is healthy science. Debate is intrinsic to the culture of science; ultimately it is how authority in science is decided. Theories that stand up to constant scrutiny and are supported by data are the ones that become authoritative.

EVOLUTION HAS NEVER BEEN TESTED

Creationists claim that evolution has never been scientifically tested. In this case, creationists define evolution narrowly as 'Darwinism', Darwin's original idea that natural selection acts on relatively minor differences between individuals, leading to gradual changes over long periods of time. But evolution, even Darwinian gradualism, has been compared by scientists to alternatives and has withstood critical testing [. . .].[2]

MICROEVOLUTION AND MACROEVOLUTION DEFINE NONOVERLAPPING AREAS OF STUDY

In our experience, most people quickly grasp that natural selection can produce changes within species, and creationists generally accept adaptation by natural selection. Evolution within populations, or microevolution, has been demonstrated experimentally and thus is hard to dispute. In contrast, ideas of common descent among life forms and splitting of lineages into divergent species, or macroevolution, are seen as separate and are rejected by creationists. Creationists dispute evidence for common descent and/or speciation because no one has done an experiment that directly recreates these aspects of evolution. There are two responses. First, the theory of evolution forms a continuum from small-scale allele frequency changes within populations to large-scale phylogenetic changes between lineages. The creationists assert a false dichotomy between micro- and macroevolution; data supporting any point in the evolutionary continuum reinforce the general theory of descent with modification. Second, the requirement that sound scientific inference is derived *only* from experimental evidence is misleadingly narrow. Entire branches of science (e.g., astronomy, meteorology, and ecosystem science) are conducted primarily outside the context of experiments. These disciplines use historical inference, logical deduction, and observation-based testing of predictions to study natural systems that cannot be manipulated. Creationists equate the scientific method with 'experimentation', ignoring that science involves testing of explanations against the natural world, whether the data are collected from experiments or observations of nature. In many cases, scientists conduct 'natural experiments' depending on unusual events in nature to test hypotheses. Insisting that 'true science' only proceeds by controlled experimentation in laboratories represents another fundamental misunderstanding of how science proceeds.

The creationist tactic of falsely separating micro- from macroevolution has led to science teaching standards that include language about adaptation and natural selection while omitting language on common descent. Rejection of common descent and speciation causes creationists to reject concepts like homology. For instance, Behe never uses the term in *Darwin's Black Box*, but

[2] In Chapter 61, Dobzhansky looks at a number of tests of evolution.

instead refers to the duplicated genes of the hemoglobin gene family as analogous. [. . .]

'INTELLIGENT DESIGN (ID) THEORY' IS SOMETHING NEW

ID deserves special mention because it is the seemingly most recent and seemingly most sophisticated attack on the role of evolution in mainstream science education. Intelligent design has received respectability to the point that the ID textbook *Of Pandas and People* is being considered by school boards for adoption in biology classes. The Discovery Institute, based in Seattle, Washington, promotes ID by attacking evolution by sponsoring academic conferences, public lectures, congressional meetings, publication of critiques of biology textbooks and 'study guides' for the Public Broadcasting System (PBS) series on evolution that originally aired September 24–28, 2001. However, even a cursory examination of historical tracts on creationism shows that ID is not new—William Paley (of blind watchmaker fame) argued in the early 1800s that complexity in nature is proof of God's existence. The newest twist is from Behe, who claims that cell biology and biochemistry provide rock-solid examples of 'irreducible complexity'. Structures like flagella and biochemical pathways such as the vertebrate blood clotting mechanism are portrayed as irreducibly complex; Behe argues they could not have arisen through Darwinian natural selection because none of the partially functioning intermediates would be adaptive. The inference is that irreducible complexity of this kind must be the work of some designer.

Intelligent design arguments have been sharply criticized on several grounds. For instance, the logic of irreducible complexity invokes intermediate reductionism: study a problem until it becomes really hard, then appeal to faith for answers. This last stage of ID arguments renders them untestable and moves them outside the realm of science. Irreducible complexity also presumes that adaptations evolve for a specific purpose, from their earliest inception until the 'final' product. Scientists discarded the concept of directed evolution 150 years ago: the literature is replete with examples of traits in use today that evolved for other purposes, such as feathers on birds that evolved for thermoregulation, not flight. Thus Behe's analogy of a mousetrap as irreducibly complex misses the point that metal bars, springs, and blocks of wood all function for reasons other than killing rodents. The claim that irreducible complexity is a new perspective that arises from recent advances in biochemistry and cell biology is patently false because irreducible complexity can also be found at higher levels of biological organization (e.g., vertebrate hearts, with valves, vessels, and chambers—what good is half a heart?). The old and parallel argument about the evolution of vertebrate eyes was dispatched by Darwin himself. Finally, for ID to gain scientific respectability, Smith stipulates that it must identify causation (ID has steadfastly refused to name a designer), have internal and external accountability, and generate testable hypotheses. If ID provides detailed theories that remain within the realm of scientific inquiry, scientists will evaluate them accordingly. As currently promoted, ID theory is neither new nor good science.

The most withering criticism of ID theory comes not from scientists, but from philosophers and theologians. We have modest abilities in these areas, thus we give here only a scientist's-eye summary of the arguments. First, the theology of ID theory looks for directed order in nature, but natural theologians risk losing faith when nature shows its disappointing knack for randomness, capriciousness, and unpleasantness. Second, ID is 'God in the gaps' theology, using divine explanations for what is not yet understood; this theological position runs the risk of describing a God whose divine power diminishes with every new scientific discovery. Together, these make ID an uncomfortable theology. [. . .]

Why is specific training in evolution important for science?

Having described the most common creationist arguments we hear from students and neighbors, we return to the question of why should we be concerned about evolution's current struggle for existence in public schools. If teaching evolution is challenged because of religious objections, how might that hamper individuals who are not taught about evolution or are taught that evolution has no place in science? It is unlikely that such individuals will understand the interplay and differences between the domains of science and the domains of religion; nor are they likely to appreciate the role of science in social policy. A public increasingly disconnected from general scientific knowledge and skepticism is susceptible to pseudoscientific claims that play on gullibility. Here we describe how ignorance of evolutionary science has consequences for medicine, traditional agriculture, and use of genetically modified organisms.[3] [. . .]

A case study in Fort Collins, Colorado

It is perhaps not surprising that when Kansas caught the antievolutionary flu in the summer of 1999, Colorado soon after caught a cold. In fall 1999 we were contacted by parents whose daughter attended Liberty Common School, a charter school within our local Poudre School District in Fort

[3] The original discussion has been omitted here. In Chapter 65 Palumbi looks at several of the topics that the authors discussed.

Collins, Colorado. The charter granted Liberty Common governance by a separate board with extensive parental input, under the promise that Liberty Common would provide an alternative (the Core Knowledge curriculum) enriched in science and mathematics. The concerned parents showed us Liberty's policy on teaching evolution; we present the full text here because it is similar to what could be seen in other school districts:

> Principles for Teaching Evolution
>
> As with other topics, we will adhere to the Core Knowledge Sequence for determining when the theory of evolution is introduced to students (7th grade) and which subtopics should be covered. This subject will not be taught in earlier years.
>
> Human evolution is not listed in the Core Knowledge Sequence or the Colorado Model Standards for Science. Therefore, this will not be a topic of instruction at Liberty Common School.
>
> Discussions of evolutionary theory can lead to discussions of whether or not supernatural forces play a role in the mechanism of evolution or the origin of life. These topics extend beyond the scope of science and will not be taught at Liberty Common School. (See also: Colorado Model Standards for Science 3.4, which states: 'This content standard does not define any student expectations related to the origin of life').
>
> This policy is not intended to restrict the teaching of evolution as outlined in the Core Knowledge Sequence or limit the scientific discussion of related topics.

This policy, the day-to-day lessons taught in biology, and discussions with the school administrators convinced the parents that Liberty Common School was backing away from a full presentation of evolution. They worried about the overall state of science teaching if evolution were excluded, particularly in a school emphasizing science and math. The parents therefore filed a complaint with the District.

After consulting with the science coordinator of the Poudre School District, we agreed with the parents' concerns and decided to become involved. We and our colleagues wrote letters and editorials to the local newspaper about the Liberty Common evolution policy. Our department hosted a public lecture on creationism and science education by Dr. Eugenie Scott (Executive Director of the National Center for Science Education), to which we invited local science teachers, principals, and members of the School Board. As a result of our visible involvement in the complaint against Liberty Common School, a member of the District School Board asked to meet with University evolutionary biologists, and four of us spent two hours briefing him on evolution, science education, and legal issues surrounding the creation–evolution debate. We went through the Liberty policy line-by-line, pointing out insidious turns of phrase. For example, the policy's second paragraph is correct in pointing out that Colorado science standards do not mention human evolution. The science standards likewise fail to mention the evolution of whales, maize, mushrooms, and bacteria, but specifically excluding them from a science curriculum would be absurd. The third paragraph of the policy uses a twist of words. Although some may prefer to use religious contexts to discuss or understand nature, religious objections cannot be used to exclude teaching scientific evidence for evolution. We pointed out that this twist was reserved specifically for the teaching of evolution. Chemistry, physics, and mathematics also could lead to discussions of the role of the supernatural, yet these areas of science were not excluded.

We believe our time was well spent. Eventually the School Board ruled that the second and third provisions would restrict the teaching of evolution at the school and would violate Liberty Common School's science education charter. The school has now modified its policy, and has a curriculum that specifically includes evolution. How well the new policy is carried out deserves continual attention.

WHAT DID WE LEARN?

What happened in Kansas in 1999 happened in our back yard a few months later. It is now common for public school officials at all levels to attempt to minimize the teaching of evolution in science curricula. Just like in Kansas, we were successful in defending science education against creationists' incursions in our own schools, but only by becoming directly involved. Attacks on evolution and thus all of science education *can* happen in your state or local school district, even if relatively few individuals push the attack. Creationists are dedicated to their cause and are active in local educational politics. After repeated losses in the courts, their primary strategy is to influence school boards to deemphasize and possibly eliminate evolution in science classes. Defending science-teaching standards now requires local involvement. Confronting individual teachers is difficult, but we scientists can provide expert advice to school boards on science teaching standards, on which textbooks are adopted, and on testing policies (Will there be questions about evolution?). We share these specific lessons:

Colleges and universities have resources that are extremely helpful in these efforts

When we first learned of the Liberty Common School problem, we were at a loss at how to proceed. A few phone calls around campus, however, made us aware of expertise in grades K-12 science education on our own campus. At Colorado State University we have a Center for the Life Sciences, which supports the myriad of life science programs scattered around campus. This Center, funded in part by a grant from the Howard Hughes Medical Institute, includes grades K-12 outreach as part of its mission. The Center staff includes teachers-in-residence who are on leave from the public schools. In addition, Colorado State University supports a Center for Science, Mathematics, and Technology Education with a broader mission to support all K-12 science outreach and a strong connection to the community of teaching professionals. The teachers-in-residence were invaluable sources of advice: they knew how teachers think and work, they knew the state and district science standards, and they knew how to work through the political process to effect curricular changes in the District. We also found considerable help from colleagues in our School of Education. We admit to some hubris in not having made these connections before the local crisis loomed, and we encourage all science faculty to learn about others at their institutions who work on science education and outreach.

K-12 teachers feel alone

Teachers hear consistently from parents who do not think evolution should be taught and less commonly from those who support teaching evolution. Given the multitude of responsibilities teachers have, it is often easiest to give in to the squeaky wheels—a stance reinforced by many administrators in our public schools. Any support we can give to teachers is extremely well received. Examples include volunteering to give presentations in the schools, organizing campus visits for science classes, and working on science teaching standards with local and state school boards.

Teachers want training on contemporary evolutionary science

Many teachers have inadequate backgrounds in evolution and thus do not feel confident in their ability to steer through the creation–evolution morass. An exciting outcome of our local fight was that our Center for the Life Sciences instituted workshops on evolutionary science for teachers. These two-day workshops, offered twice a year, feature presentations by biologists, geologists, physicists, and anthropologists. The topics include the scientific method as employed by evolutionary scientists; how to meet classroom challenges by creationists; and classroom exercises to demonstrate common descent, natural selection, radiocarbon dating, and human evolution. Teachers have incorporated these exercises into their classes, and the Center for Life Sciences will present an 'Evolution Solution' workshop at the 2001 meeting of the National Association of Biology Teachers. The Center also funds summer internships for teachers, some of whom work in our laboratories. This program allows teachers to do research, which gives them a clearer picture of how we work. In turn, the teachers instruct us about the challenges they face in the classroom and community, in some cases by continuing a correspondence after they return to their schools. To paraphrase radio commentator Garrison Keillor, 'Nothing you do for teachers is ever wasted.'

School board members and local politicians appreciate overtures from academics

We invited members of the school board to Dr. Scott's lecture, and they were most pleased to be given the opportunity to learn more about this complex issue. Since then, we have invited members of our City Council and School

Board to attend other high-profile presentations, such as an endowed lecture series that recently brought Paul Ehrlich, a prominent evolutionary biologist from Stanford University, to campus. While many do not attend, they do appreciate being invited—and they tell us so.

This controversy concerns all scientists

We received broad support from colleagues in the physical and social sciences because they understand that an attack on teaching evolution is an attack on all of science education. Strengthening evolution curricula in schools will strengthen teaching of the sciences in general. School teachers tell us of their difficulties teaching how scientific theories are logically structured and how observations relate (or are irrelevant) to theories, and yet these critical thinking skills are central to science. The creation–evolution debate clearly illustrates the limits of science's domain, as well as the power of its methodology.

Improving science curricula in K-12 need not take a lot of time

Every faculty member can make a difference in small ways. Across our department we host approximately 10 visits to departmental research labs from high school classes every year. On average, every laboratory (faculty, postdocs, and graduate students) also makes two to three visits to local schools. Faculty members also blend science education outreach into their courses. For example, honors biology students are required to visit primary schools for presentations. They work with our teachers-in-residence to become familiar with district standards and to draw up lesson plans, which they coordinate with the teacher whose classes they visit. The students initially think this assignment is the easiest part of the course, but they quickly learn otherwise. Not only do they enjoy this assignment, but some also start to consider a career in teaching.

Recommendations for the future

Maintaining high standards for science teaching will require constant vigilance, as creationists continue to make inroads onto school boards. The place to begin is to examine the teaching standards for your state; next, compare the state standards to those of your school districts and individual schools. Individual states allow varying levels of local control: even if your state and school district have standards that include evolution, they may not be enforced or may be circumvented locally. Teaching standards can be manipulated to remove requirements for teaching all or part of evolution theory, especially ideas of common descent, or to minimize requirements by stipulating that particular parts of evolution theory not appear on statewide examinations. In the state of Colorado, in 1996, standards for evolution were quietly changed to 'not define any student expectations related to the origin of life'. Local efforts, however, may not be enough. Recently, a nonbinding resolution was added to an education bill in the United States Senate by Senator Rick Santorum (R-PA), which declares that 'where biological evolution is taught, the curriculum should help students to understand why this subject generates so much continuing controversy, and should prepare the students to be informed participants in public discussions regarding the subject'. The ensuing discussion on the Senate floor made it clear that the intent of the resolution was to include creationism in primary and secondary schools, in accord with the same 'fairness' and 'macroevolution is not scientifically supported' arguments described above.

We also must examine our college-level curricula to ask whether we give our students, especially future primary and secondary teachers, an adequate understanding of scientific methods. Research shows that teachers' attitudes toward evolution and creationism echo those of the general public, with approximately one third agreeing that creationist accounts belong in science classrooms. This study showed that positive attitudes toward teaching evolution were correlated not with age of the teachers or region of the country, but with adherence to teaching standards, familiarity with philosophy of science, and participation in professional societies. Future teachers must be well trained in the scientific method, and their training in evolutionary science must consistently invoke those methodological principles.

The woeful level of scientific illiteracy in American society is partly our fault, and we recommend three ways that colleges and universities can contribute to creating a science-savvy public. First, we must ensure that science courses for nonmajors include training in critical thinking and scientific methodology. We must also stress the limitations of science's domain; we have found that our students are relieved and comforted to learn that science does not require an atheistic philosophy. Second, we must teach future primary and secondary school teachers the process of science. Just as we require that teachers participate in student teaching to hone their skills, we also must give them opportunities to gain hands-on experience in original research. Third, we must provide continuing education for practicing teachers, including workshops, internships, and teachers in residence programs. These pay quick dividends: a teacher who participated in a summer internship several years ago told us the research experience taught him how much science depends upon attention to small details, in addition to broad theoretical frameworks. When it comes to science and science education, the ongoing creation–evolution debate shows that the devil really is in the details.

[*Evolution*, 55 (2001), 1279–88.]

THEODOSIUS DOBZHANSKY

61 Nothing in biology makes sense except in the light of evolution

Life shows a wide diversity, with millions of species, possessing a range of peculiar ecological adaptations. Life also shows a fundamental unity, for instance, in the use of a common genetic code. The theory of evolution is able to account for the unity and diversity of life, whereas the theory of the separate, supernatural creation of species cannot. Evolution also explains such facts as the match between the molecular and taxonomic similarity of species, homologies including embryonic similarities, and endemic species on oceanic islands. The theory of evolution is established beyond reasonable doubt. Moreover, it does not clash with religious faith. [Editor's summary.]

As recently as 1966, sheik Abd el Aziz bin Baz asked the king of Saudi Arabia to suppress a heresy that was spreading in his land. Wrote the sheik:

> The Holy Koran, the Prophet's teachings, the majority of Islamic scientists, and the actual facts all prove that the sun is running in its orbit . . . and that the earth is fixed and stable, spread out by God for his mankind . . . Anyone who professed otherwise would utter a charge of falsehood toward God, the Koran, and the Prophet.

The good sheik evidently holds the Copernican theory to be a 'mere theory,' not a 'fact.' In this he is technically correct. A theory can be verified by a mass of facts, but it becomes a proven theory, not a fact. The sheik was perhaps unaware that the Space Age had begun before he asked the king to suppress the Copernican heresy. The sphericity of the earth had been seen by astronauts, and even by many earth-bound people on their television screens. Perhaps the sheik could retort that those who venture beyond the confines of God's earth suffer hallucinations, and that the earth is really flat.

Parts of the Copernican world model, such as the contention that the earth rotates around the sun, and not vice versa, have not been verified by direct observations even to the extent the sphericity of the earth has been. Yet scientists accept the model as an accurate representation of reality. Why? Because it makes sense of a multitude of facts which are otherwise meaningless or extravagant. To nonspecialists most of these facts are unfamiliar. Why then do we accept the 'mere theory' that the earth is a sphere revolving around a spherical sun? Are we simply submitting to authority? Not quite: we know that those who took time to study the evidence found it convincing.

The good sheik is probably ignorant of the evidence. Even more likely, he is so hopelessly biased that no amount of evidence would impress him. Anyway, it would be sheer waste of time to attempt to convince him. The Koran and the Bible do not contradict Copernicus, nor does Copernicus contradict them. It is ludicrous to mistake the Bible and the Koran for primers of natural science. They treat of matters even more important: the meaning of man and his relations to God. They are written in poetic symbols that were understandable to people of the age when they were written, as well as to peoples of all other ages. The king of Arabia did not comply with the sheik's demand. He knew that some people fear enlightenment, because enlightenment threatens their vested interests. Education is not to be used to promote obscurantism.

The earth is not the geometric center of the universe, although it may be its spiritual center. It is a mere speck of dust in cosmic spaces. Contrary to Bishop Ussher's calculations, the world did not appear in approximately its present state in 4004 B.C. The estimates of the age of the universe given by modern cosmologists are still only rough approximations, which are revised (usually upward) as the methods of estimation are refined. Some cosmologists take the universe to be about 10 billion years old; others suppose that it may have existed, and will continue to exist, eternally. The origin of life on earth is dated tentatively between 3 and 5 billion years ago; manlike beings appeared relatively quite recently, between 2 and 4 million years ago. The estimates of the age of the earth, of the duration of the geologic and paleontologic eras, and of the antiquity of man's ancestors are now based mainly on radiometric evidence—the proportions of isotopes of certain chemical elements in rocks suitable for such studies.

Sheik bin Baz and his like refuse to accept the radiometric evidence, because it is a 'mere theory.' What is the alternative? One can suppose that the Creator saw fit to play deceitful tricks on geologists and biologists. He carefully arranged to have various rocks provided with isotope ratios just right to mislead us into thinking that certain rocks are 2 billion years old, others 2 million, while in fact they are only some 6,000 years old. This kind of pseudo-explanation is not very new. One of the early antievolutionists, P. H. Gosse, published a book entitled *Omphalos* ('the Navel'). The gist of this amazing book is that Adam, though he had no mother, was created with a navel, and that fossils were placed by the Creator where we find them now—a deliberate act on His part, to give the appearance of great antiquity and geologic upheavals. It is easy to see the fatal flaw in all such notions. They are blasphemies, accusing God of absurd deceitfulness. This is as revolting as it is uncalled for.

Diversity of living beings

The diversity and the unity of life are equally striking and meaningful aspects of the living world. Between 1.5 and 2 million species of animals and plants have been described and studied; the number yet to be described is probably about as great. The diversity of sizes, structures, and ways of life is staggering but fascinating. Here are just a few examples.

The foot-and-mouth disease virus is a sphere 8–12 μm in diameter. The blue whale reaches 30 m in length and 135 t in weight. The simplest viruses are parasites in cells of other organisms, reduced to barest essentials—minute amounts of DNA or RNA, which subvert the biochemical machinery of the host cells to replicate their genetic information, rather than that of the host.

It is a matter of opinion, or of definition, whether viruses are considered living organisms or peculiar chemical substances. The fact that such differences of opinion can exist is in itself highly significant. It means that the borderline between living and inanimate matter is obliterated. At the opposite end of the simplicity-complexity spectrum you have vertebrate animals, including man. The human brain has some 12 billion neurons; the synapses between the neurons are perhaps a thousand times as numerous.

Some organisms live in a great variety of environments. Man is at the top of the scale in this respect. He is not only a truly cosmopolitan species but, owing to his technologic achievements, can survive for at least a limited time on the surface of the moon and in cosmic spaces. By contrast, some organisms are amazingly specialized. Perhaps the narrowest ecologic niche of all is that of a species of the fungus family Laboulbeniaceae, which grows exclusively on the rear portion of the elytra of the beetle *Aphenops cronei*, which is found only in some limestone caves in southern France. Larvae of the fly *Psilopa petrolei* develop in seepages of crude oil in California oil-fields; as far as is known they occur nowhere else. This is the only insect able to live and feed in oil, and its adult can walk on the surface of the oil only as long as no body part other than the tarsi are in contact with the oil. Larvae of the fly *Drosophila carcinophila* develop only in the nephric grooves beneath the flaps of the third maxilliped of the land crab *Geocarcinus ruricola*, which is restricted to certain islands in the Caribbean.

Is there an explanation, to make intelligible to reason this colossal diversity of living beings? Whence came these extraordinary, seemingly whimsical and superfluous creatures, like the fungus *Laboulbenia*, the beetle *Aphenops cronei*, the flies *Psilopa petrolei* and *Drosophila carcinophila*, and many, many more apparent biologic curiosities? The only explanation that makes sense is that the organic diversity has evolved in response to the diversity of environment on the planet earth. No single species, however perfect and however versatile, could exploit all the opportunities for living. Every one of the millions of species has its own way of living and of getting sustenance from the environment. There are doubtless many other possible ways of living as yet unexploited by any existing species; but one thing is clear: with less organic diversity, some opportunities for living would remain unexploited. The evolutionary process tends to fill up the available ecologic niches. It does not do so consciously or deliberately; the relations between evolution and the environment are more subtle and more interesting than that. The environment does not impose evolutionary changes on its inhabitants, as postulated by the now abandoned neo-Lamarckian theories. The best way to envisage the situation is as follows: the environment presents challenges to living species, to which the latter may respond by adaptive genetic changes.

An unoccupied ecologic niche, an unexploited opportunity for living, is a challenge. So is an environmental change, such as the Ice Age climate giving place to a warmer climate. Natural selection may cause a living species to respond to the challenge by adaptive genetic changes. These changes may enable the species to occupy the formerly empty ecologic niche as a new opportunity for living, or to resist the environmental change if it is unfavorable. But the response may or may not be successful. This depends on many factors, the chief of which is the genetic composition of the responding species at the time the response is called for. Lack of successful response may cause the species to become extinct. The evidence of fossils shows clearly that the eventual end of most evolutionary lines is extinction. Organisms now living are successful descendants of only a minority of the species that lived in the past—and of smaller and smaller minorities the farther back you look. Nevertheless, the number of living species has not dwindled; indeed, it has probably grown with time. All this is understandable in the light of evolution theory; but what a senseless operation it would have been, on God's part, to fabricate a multitude of species ex nihilo and then let most of them die out!

There is, of course, nothing conscious or intentional in the action of natural selection. A biologic species does not say to itself, 'Let me try tomorrow (or a million years from now) to grow in a different soil, or use a different food, or subsist on a different body part of a different crab.' Only a human being could make such conscious decisions. This is why the species *Homo sapiens* is the apex of evolution. Natural selection is at one and the same time a blind and a creative process. Only a creative but blind process could produce, on the one hand, the tremendous biologic success that is the human species and, on the other, forms of adaptedness as narrow and as constraining as those of the overspecialized fungus, beetle, and flies mentioned above.

Antievolutionists fail to understand how natural selection operates. They fancy that all existing species were generated by supernatural fiat a few thousand years ago, pretty much as we find them today. But what is the sense of having as many as 2 or 3 million species living on earth? If natural selection is the main factor that brings evolution about, any number of species is understandable: natural selection does not work according to a foreordained plan, and species are produced not because they are needed for some purpose but simply because there is an environmental opportunity and genetic wherewithal to make them possible. Was the Creator in a jocular mood when he made *Psilopa petrolei* for California oil-fields and species of *Drosophila* to live exclusively on some body-parts of certain land crabs on only certain islands in the Caribbean? The organic diversity becomes, however, reasonable and understandable if the Creator has created the living world not by caprice but by evolution propelled by natural selection. It is wrong to hold creation and evolution as mutually exclusive alternatives. I am a creationist *and* an evolutionist. Evolution is God's, or Nature's, method of Creation. Creation is not an event that happened in 4004 B.C.; it is a process that began some 10 billion years ago and is still under way.

Unity of life

The unity of life is no less remarkable than its diversity. Most forms of life are similar in many respects. The universal biologic similarities are particularly striking in the biochemical dimension. From viruses to man, heredity is coded in just two, chemically related substances: DNA and RNA. The genetic code is as simple as it is universal.[1] There are only four genetic 'letters' in DNA: adenine, guanine, thymine, and cytosine. Uracil replaces thymine in RNA. The entire evolutionary development of the living world has taken place not by invention of new 'letters' in the genetic 'alphabet' but by elaboration of ever-new combinations of these letters.

Not only is the DNA-RNA genetic code universal, but so is the method of translation of the sequences of the 'letters' in DNA-RNA into sequences of amino acids in proteins. The same 20 amino acids compose countless different proteins in all, or at least in most, organisms. Different amino acids are coded by one to six nucleotide triplets in DNA and RNA. And the biochemical universals extend beyond the genetic code and its translation into proteins: striking uniformities prevail in the cellular metabolism of the most diverse living beings. Adenosine triphosphate, biotin, riboflavin, hemes, pyridoxin, vitamins K and B_{12}, and folic acid implement metabolic processes everywhere.

What do these biochemical or biologic universal mean? They suggest that life arose from inanimate matter only once and that all organisms, no matter how diverse in other respects, conserve the basic features of the primordial life. (It is also possible that there were several, or even many, origins of life; if so, the progeny of only one of them has survived and inherited the earth.) But what if there was no evolution, and every one of the millions of species was created by separate fiat? However offensive the notion may be to religious feeling and to reason, the antievolutionists must again accuse the

[1] The topic of Crick's paper (Chapter 47).

Creator of cheating. They must insist that He deliberately arranged things exactly as if his method of creation was evolution, intentionally to mislead sincere seekers of truth.

The remarkable advances of molecular biology in recent years have made it possible to understand how it is that diverse organisms are constructed from such monotonously similar materials: proteins composed of only 20 kinds of amino acids and coded only by DNA and RNA, each with only four kinds of nucleotides. The method is astonishingly simple. All English words, sentences, chapters, and books are made up of sequences of 26 letters of the alphabet. (They can be represented also by only three signs of the Morse code: dot, dash, and gap.) The meaning of a word or a sentence is defined not so much by what letters it contains as by the sequence of these letters. It is the same with heredity: it is coded by the sequences of the genetic 'letters'—the nucleotides—in the DNA. They are translated into the sequences of amino acids in the proteins.

Molecular studies have made possible an approach to exact measurements of degrees of biochemical similarities and differences among organisms. Some kinds of enzymes and other proteins are quasiuniversal, or at any rate widespread, in the living world. They are functionally similar in different living beings, in that they catalyze similar chemical reactions. But when such proteins are isolated and their structures determined chemically, they are often found to contain more or less different sequences of amino acids in different organisms. For example, the so-called alpha chains of hemoglobin have identical sequences of amino acids in man and the chimpanzee, but they differ in a single amino acid (out of 141) in the gorilla. Alpha chains of human hemoglobin differ from cattle hemoglobin in 17 amino acid substitutions, 18 from horse, 20 from donkey, 25 from rabbit, and 71 from fish (carp).

Cytochrome C is an enzyme that plays an important role in the metabolism of aerobic cells. It is found in the most diverse organisms, from man to molds. E. Margoliash, W. M. Fitch, and others have compared the amino acid sequences in cytochrome C in different branches of the living world. Most significant similarities as well as differences have been brought to light. The cytochrome C of different orders of mammals and birds differ in 2 to 17 amino acids, classes of vertebrates in 7 to 38, and vertebrates and insects in 23 to 41; and animals differ from yeasts and molds in 56 to 72 amino acids. Fitch and Margoliash prefer to express their findings in what are called 'minimal mutational distances.' It has been mentioned above that different amino acids are coded by different triplets of nucleotides in DNA of the genes; this code is now known. Most mutations involve substitutions of single nucleotides somewhere in the DNA chain coding for a given protein. Therefore, one can calculate the minimum numbers of single mutations needed to change the cytochrome C of one organism into that of another.

Minimal mutational distances between human cytochrome C and the cytochrome C of other living beings are as follows:

Monkey	1	Chicken	18
Dog	13	Penguin	18
Horse	17	Turtle	19
Donkey	16	Rattlesnake	20
Pig	13	Fish (tuna)	31
Rabbit	12	Fly	33
Kangaroo	12	Moth	36
Duck	17	Mold	63
Pigeon	16	Yeast	56

It is important to note that amino acid sequences in a given kind of protein vary within a species as well as from species to species. It is evident that the differences among proteins at the levels of species, genus, family, order, class, and phylum are compounded of elements that vary also among individuals within a species. Individual and group differences are only quantitatively, not qualitatively, different. Evidence supporting the above propositions is ample and is growing rapidly. Much work has been done in recent years on individual variations in amino acid sequences of hemoglobins of human blood. More than 100 variants have been detected. Most of them involve substitutions of single amino acids—substitutions that have arisen by genetic mutations in the persons in whom they are discovered or in their ancestors. As expected, some of these mutations are deleterious to their carriers, but others apparently are neutral or even favorable in certain environments. Some mutant hemoglobins have been found only in one person or in one family; others are discovered repeatedly among inhabitants of different parts of the world. I submit that all these remarkable findings make sense in the light of evolution; they are nonsense otherwise.

Comparative anatomy and embryology

The biochemical universals are the most impressive and the most recently discovered, but certainly they are not the only vestiges of creation by means of evolution. Comparative anatomy and embryology proclaim the evolutionary origins of the present inhabitants of the world. In 1555 Pierre Belon established the presence of homologous bones[2] in the superficially very dif-

[2] The topic of De Beer (Chapter 31).

ferent skeletons of man and bird. Later anatomists traced the homologies in the skeletons, as well as in other organs, of all vertebrates. Homologies are also traceable in the external skeletons of arthropods as seemingly unlike as a lobster, a fly, and a butterfly. Examples of homologies can be multiplied indefinitely.

Embryos of apparently quite diverse animals often exhibit striking similarities. A century ago these similarities led some biologists (notably the German zoologist Ernst Haeckel)[3] to be carried by their enthusiasm so far as to interpret the embryonic similarities as meaning that the embryo repeats in its development the evolutionary history of its species: it was said to pass through stages in which it resembles its remote ancestors. In other words, early-day biologists supposed that by studying embryonic development one can, as it were, read off the stages through which the evolutionary development had passed. This so-called biogenetic law is no longer credited in its original form. And yet embryonic similarities are undeniably impressive and significant.

Probably everybody knows the sedentary barnacles which seem to have no similarity to free-swimming crustaceans, such as the copepods. How remarkable that barnacles pass through a free-swimming larval stage, the nauplius! At that stage of its development a barnacle and a *Cyclops* look unmistakably similar. They are evidently relatives. The presence of gill slits in human embryos and in embryos of other terrestrial vertebrates is another famous example. Of course, at no stage of its development is a human embryo a fish, nor does it ever have functioning gills. But why should it have unmistakable gill slits unless its remote ancestors did respire with the aid of gills? Is the Creator again playing practical jokes?

Adaptive radiation: Hawaii's flies

There are about 2,000 species of drosophilid flies in the world as a whole. About a quarter of them occur in Hawaii, although the total area of the archipelago is only about that of the state of New Jersey. All but 17 of the species in Hawaii are endemic (found nowhere else). Furthermore, a great majority of the Hawaiian endemics do not occur throughout the archipelago: they are restricted to single islands or even to a part of an island. What is the explanation of this extraordinary proliferation of drosophilid species in so small a territory? Recent work of H. L. Carson, H. T. Spieth, D. E. Hardy, and others makes the situation understandable.

The Hawaiian islands are of volcanic origin; they were never parts of any continent. Their ages are between 5.6 and 0.7 million years. Before

[3] The topic of Haeckel (Chapter 34) and Garstang's criticism (Chapter 35).

man came their inhabitants were descendants of immigrants that had been transported across the ocean by air currents and other accidental means. A single drosophilid species, which arrived in Hawaii first, before there were numerous competitors, faced the challenge of an abundance of many unoccupied ecologic niches. Its descendants responded to this challenge by evolutionary adaptive radiation, the products of which are the remarkable Hawaiian drosophilids of today. To forestall a possible misunderstanding, let it be made clear that the Hawaiian endemics are by no means so similar to each other that they could be mistaken for variants of the same species; if anything, they are more diversified than are drosophilids elsewhere. The largest and the smallest drosophilid species are both Hawaiian. They exhibit an astonishing variety of behavior patterns. Some of them have become adapted to ways of life quite extraordinary for a drosophilid fly, such as being parasites in egg cocoons of spiders.

Oceanic islands other than Hawaii, scattered over the wide Pacific Ocean, are not conspicuously rich in endemic species of drosophilids. The most probable explanation of this fact is that these other islands were colonized by drosophilids after most ecologic niches had already been filled by earlier arrivals. This surely is a hypothesis, but it is a reasonable one. Antievolutionists might perhaps suggest an alternative hypothesis: in a fit of absent-mindedness, the Creator went on manufacturing more and more drosophilid species for Hawaii, until there was an extravagant surfeit of them in this archipelago. I leave it to you to decide which hypothesis makes sense.

Strength and acceptance of the theory

Seen in the light of evolution, biology is, perhaps, intellectually the most satisfying and inspiring science. Without that light it becomes a pile of sundry facts—some of them interesting or curious but making no meaningful picture as a whole.

This is not to imply that we know everything that can and should be known about biology and about evolution. Any competent biologist is aware of a multitude of problems yet unresolved and of questions yet unanswered. After all, biologic research shows no sign of approaching completion; quite the opposite is true. Disagreements and clashes of opinion are rife among biologists, as they should be in a living and growing science. Antievolutionists mistake, or pretend to mistake, these disagreements as indications of dubiousness of the entire doctrine of evolution. Their favorite sport is stringing together quotations, carefully and sometimes expertly taken out of context, to show that nothing is really established or agreed upon among evolutionists. Some of my colleagues and myself have been amused and amazed to read ourselves quoted in a way showing that we are really antievolutionists under the skin.

Let me try to make crystal clear what is established beyond reasonable doubt, and what needs further study, about evolution. Evolution as a process that has always gone on in the history of the earth can be doubted only by those who are ignorant of the evidence or are resistant to evidence, owing to emotional blocks or to plain bigotry. By contrast, the mechanisms that bring evolution about certainly need study and clarification. There are no alternatives to evolution as history that can withstand critical examination. Yet we are constantly learning new and important facts about evolutionary mechanisms.

It is remarkable that more than a century ago Darwin was able to discern so much about evolution without having available to him the key facts discovered since. The development of genetics after 1900—especially of molecular genetics, in the last two decades—has provided information essential to the understanding of evolutionary mechanisms. But much is in doubt and much remains to be learned. This is heartening and inspiring for any scientist worth his salt. Imagine that everything is completely known and that science has nothing more to discover: what a nightmare!

Does the evolutionary doctrine clash with religious faith? It does not. It is a blunder to mistake the Holy Scriptures for elementary textbooks of astronomy, geology, biology, and anthropology. Only if symbols are construed to mean what they are not intended to mean can there arise imaginary, insoluble conflicts. As pointed out above, the blunder leads to blasphemy: the Creator is accused of systematic deceitfulness.

One of the great thinkers of our age, Pierre Teilhard de Chardin, wrote the following: 'Is evolution a theory, a system, or a hypothesis? It is much more—it is a general postulate to which all theories, all hypotheses, all systems must henceforward bow and which they must satisfy in order to be thinkable and true. Evolution is a light which illuminates all facts, a trajectory which all lines of thought must follow—this is what evolution is.' Of course, some scientists, as well as some philosophers and theologians, disagree with some parts of Teilhard's teachings; the acceptance of his world view falls short of universal. But there is no doubt at all that Teilhard was a truly and deeply religious man and that Christianity was the cornerstone of his world view. Moreover, in his world view science and faith were not segregated in watertight compartments, as they are with so many people. They were harmoniously fitting parts of his world view. Teilhard was a creationist, but one who understood that the Creation is realized in this world by means of evolution.

[*American Biology Teacher*, 35 (1973), 125–9.]

DAVID HUME

62 The argument from design

The argument from design seeks to infer the existence of God from the presence of design in the universe, including the presence of adaptations in living things. Human designed products exist because they have a designer (human beings) to create them. By analogy, designful properties of nature must have a designer too (that is, God). However, the argument relies on analogy. Nature may have some other mechanism that can produce design, without a purposeful designer. (Hume wrote in 1779, when natural selection had not been thought of; but Darwin's theory came to illustrate Hume's general point 100 years later.) The extract comes from a book written in the form of dialogues between characters with names such as Cleanthes, Demea, and Philo; modern scholars tend to think that the character of Philo represents Hume himself. [Editor's summary.]

Not to lose any time in circumlocutions, said Cleanthes, addressing himself to Demea [. . .] I shall briefly explain how I conceive this matter. Look round the world: Contemplate the whole and every part of it: You will find it to be nothing but one great machine, subdivided into an infinite number of lesser machines, which again admit of subdivisions, to a degree beyond what human senses and faculties can trace and explain. All these various machines, and even their most minute parts, are adjusted to each other with an accuracy, which ravishes into admiration all men, who have ever contemplated them. The curious adapting of means to ends, throughout all nature, resembles exactly, though it much exceeds, the productions of human contrivance; of human design, thought, wisdom, and intelligence. Since therefore the effects resemble each other, we are led to infer, by all the rules of analogy, that the causes also resemble; and that the Author of nature is somewhat similar to the mind of man; though possessed of much larger faculties, proportioned to the grandeur of the work, which he has executed. By this argument *a posteriori*, and by this argument alone, do we prove at once the existence of a Deity, and his similarity to human mind and intelligence. [. . .]

What I chiefly scruple in this subject, said Philo, is not so much, that all religious arguments are by Cleanthes reduced to experience, as that they appear not to be even the most certain and irrefragable of that inferior kind. That a stone will fall, that fire will burn, that the earth has solidity, we have observed a thousand and a thousand times; and when any new instance of this nature is presented, we draw without hesitation the accustomed inference. The exact similarity of the cases gives us a perfect assurance of a similar event; and a stronger evidence is never desired nor sought after. But wherever you depart, in the least, from the similarity of the cases, you diminish proportionably the evidence; and may at last bring it to a very weak *analogy*, which is confessedly liable to error and uncertainty. After having experienced the circulation of the blood in human creatures, we make no doubt that it takes place in Titius and Mævius: But from its circulation in frogs and fishes, it is only a presumption, though a strong one, from analogy, that it takes place in men and other animals. The analogical reasoning is much weaker, when we infer the circulation of the sap in vegetables from our experience that the blood circulates in animals; and those, who hastily followed that imperfect analogy, are found, by more accurate experiments, to have been mistaken.

If we see a house, Cleanthes, we conclude, with the greatest certainty, that it had an architect or builder; because this is precisely that species of effect, which we have experienced to proceed from that species of cause. But surely you will not affirm, that the universe bears such a resemblance to a house, that we can with the same certainty infer a similar cause, or that the analogy is here entire and perfect. The dissimilitude is so striking, that the utmost you can here pretend to is a guess, a conjecture, a presumption concerning a similar cause; and how that pretension will be received in the world, I leave you to consider. [. . .]

Nature, we find, even from our limited experience, possesses an infinite number of springs and principles, which incessantly discover themselves on every change of her position and situation. And what new and unknown principles would acturate her in so new and unknown a situation as that of the formation of a universe, we cannot, without the utmost temerity, pretend to determine.

A very small part of this great system, during a very short time, is very imperfectly discovered to us: And do we thence pronounce decisively concerning the origin of the whole?

Admirable conclusion! Stone, wood, brick, iron, brass, have not, at this time, in this minute globe of earth, an order or arrangement without human art and contrivance: Therefore the universe could not originally attain its order and arrangement, without something similar to human art. But is a part of nature a rule for another part very wide of the former? Is it a rule for the whole? Is a very small part a rule for the universe? Is nature in one situation, a certain rule for nature in another situation, vastly different from the former?

[*Dialogues Concerning Natural Religion* (Oxford: OUP, World's Classics, 1993), 45–6, 50–1.]

J. L. MONOD

63 On the molecular theory of evolution

The theory of evolution has a peculiar philosophical status—in its range of influences in human affairs, the range of facts it applies to, and the way it is used to explain unobserved events in the past. The theory of evolution might be thought to be unfalsifiable, but many claims have been made that would (if correct) falsify it, such as a recent age of the Earth, and heredity by certain mechanisms other than DNA-type heredity. Evolution has become a superior theory, following the rise of molecular genetics. Some people dislike evolution because it means that humans exist for historically accidental, rather than obligatory, reasons. [Editor's summary.]

What I wish to talk about today is the present state of the theory of evolution. Let me say right away that when I talk of the theory of evolution, I am speaking strictly of the theory of evolution of living beings within the general framework of the Darwinian theory, a theory which is still alive today. Indeed, it is even more alive than many non-biologists may think.

The theory of evolution is a very curious theory. To begin with it is necessary to recall that in many respects the theory of evolution is the most important scientific theory ever formulated, because of its general implications. There is no question that no other scientific theory has had such tremendous philosophical, ideological, and political implications as has the theory of evolution.

It is also a very curious theory in its status, which is quite different from that of physical theories. The basic aim of the physical theories is to discover universal laws, laws which apply to objects in the whole of the universe, with the hope of being able, from these laws—that is to say from first principles—to derive conclusions, explain phenomena throughout the universe. When a physicist looks at a particular phenomenon it is with the hope that he will be able to show that he can deduce this phenomenon from universal laws, from first principles. The theory of evolution, by contrast, has a different aim. It has a range of application, which is not the universe, but only a tiny corner of the universe, namely the universe of living things as we know them today upon the earth. We can define the aim of the theory as that of accounting for the existence today on the earth of about two million animal species and about a million plant species, plus an unknown number of species of bacteria.

This is a very small corner of the universe, and it is very dubious whether the existence of these very peculiar objects—living beings—can, or ever could, be derived from first principles. I might say now that I do not believe it will ever be possible to do such a thing, for very profound reasons that I will try to explain.

Another curious aspect of the theory of evolution is that everybody thinks he understands it. I mean philosophers, social scientists, and so on. While in fact very few people understand it, actually, as it stands, even as it stood when Darwin expressed it, and even less as we now may be able to understand it in biology. In fact, the first great misunderstander was Spencer himself. He of course was one of the first great evolutionary philosophers, but he was also the very first to show the inadequacy, as he said, of selective evolution to explain evolution.

The other great difficulty about the theory of evolution is that it is what one might call a second-order theory. Second-order, because it is a theory aimed at accounting for a phenomenon that has never been observed, and that never will be observed, namely evolution itself. In the laboratory we are able to set up conditions so that we may be able to isolate mutations of a given bacterial strain, for instance; but to observe a mutation is a very far cry from observing actual evolution. That has never been observed even in its simplest form—the one which is required by the modern theorists to account for evolution, namely the simple differentiation of one species from another. This is a phenomenon that has never been seen. I would not say it never will be, but it seems extremely doubtful. Therefore, if you look at the structure of the theory of evolution, if for instance you open one of these great modern books about evolution, such as the books by Dobzhansky or Mayr, or Simpson, you will see that the discussion always goes the following way—that one starts from the actual data, that is to say the present structure, performances, and anatomy of a given group of animals, and then one looks at the fossil record, and from the fossil record and classical considerations of comparative anatomy, one tries to derive filiation of these forms. With the help of modern biology, such filiations can be also reconstructed by the analysis of sequences in certain molecules such as proteins. What has happened is that the molecular filiations have very beautifully shown that the anatomists were right. Filiations reconstructed in this way are perfectly coherent with filiations reconstructed by the anatomists. So the first thing you do is propose a first-order theory, to answer, for instance, 'How do men descend from fishes?' And you build up a certain set of filiations. And then comes the second-order theory. You want to account for these facts and you introduce all sorts of further considerations, consisting of assumptions about rates of mutations and so on that might have occurred. And finally a reconstruction of the ecology of the groups that are assumed to have come before is required, because in order to account for evolution in terms of the selective theory, you have to assume some sort of selective pressure. Selective pressure is something that develops according to the milieu in which an organism develops, but also according to the performances and, I would say, the personal preferences of the individual. Once you know all that, you have the conditions for a particular application of the theory of evolution to, say, the evolution of man from some fish somewhere in the early secondary era. This is what I mean by second-order theory.

Clearly, no reconstruction of this kind can ever be proved, and worse than that, no reconstruction of this kind can ever be disproved. This might appear to make the whole selective theory of evolution an extremely intellectual construction, especially in view of an epistemology such as that of Karl Popper for whom—and I completely agree with him—the distinctive mark of a truly scientific theory is not that it can be proved—because no theory ever can be proved right—but that a scientific theory must be of such a structure that it can be *dis*proved. In any particular instance this is not the case for the selective theory of evolution. However, this difficulty is not really peculiar to the theory of evolution. Many great physical theories which are corroborated and accepted for their general contents may be very difficult to apply in any given instance. This is the case, for instance, for quantum mechanics. Once one tries to apply quantum mechanics to the description and prediction of the properties of individual chemical molecules of any complexity—even relatively simple ones—one gets into extreme difficulties and is forced to adopt rather arbitrary types of reasoning. However, we accept the quantum theory as forming the basis of all our understanding of chemistry, for different reasons—that is to say, for its general content.

Now what has not always been fully seen is that in fact the selective theory of evolution, even as it was first proposed by Darwin in the first edition of *The Origin of Species*, has (in fact had, at the time), predictive contents much richer than Darwin himself knew or ever found out. Let me give you two examples. What I mean by 'richer content' is that once the theory of selective evolution has been formulated, as Darwin did in 1859, then a certain number of consequences must follow, even though the author of that theory—Darwin in that case—did not (in his days hardly *could* have) seen these consequences. The consequences sometimes go far beyond the selective theory itself or even biology itself. I think one of the most remarkable examples is the famous discussion between Lord Kelvin and Darwin—a discussion which Darwin thought he had lost. Darwin was very much aware of the fact that in order for his theory to be acceptable one needed an enormous expanse of time for the evolution of living beings to have taken place, and he spoke in terms of hundreds of millions of years—many hundreds of millions of years, without being able, of course, to give any sort of precise figure. Kelvin, who was both one of the greatest physicists of his time, the greatest thermodynamicist in any case, and also a deeply religious man (which may have had something to do with his attitudes), proved, by his calculations, that the life of the solar system could not possibly have exceeded say about twenty-five million years. This clearly was not enough for Darwin, and almost reduced him to going back to other interpretations of evolution itself. If there are any Marxists here they will be happy to learn that this can be explained on

Marxist grounds, because Kelvin, though very religious, was a great scientist of nineteenth century England, and had as his model for the energy of the sun, a coal pile. He had no other choice, and by calculating the dissipation of energy from a coal pile of the size of the sun, he could conclude that it was impossible to assume that it could live for more than twenty-five million years—which is pretty good for a coal pile. As we know, even our fuel is not going to last that long. Now, of course, he was wrong, and we know now that solar energy derives from nuclear energy, from fusion in fact, so that we might say that the discovery of nuclear energy, or fusion, or more truly of the famous Einstein equation relating matter to energy, was implicitly contained in the selective theory of evolution of Darwin. This is curious, but it is a fact.

Let me give you another example, which is much more directly related to biology and to the history of the theory of evolution. Around 1871, a mathematician from Edinburgh proved mathematically that Darwin's selective theory of evolution could not possibly be right on the basis of the accepted ideas of the time concerning inheritance—heredity.[1] These accepted ideas that nobody discussed—not even Darwin—were that inheritance was essentially a system of blending and that the offspring from a given couple was a sort of dilution of traits coming from each parent, and that this would go on in successive generations. By fairly simple calculations Jenkins showed that even if a new trait were able to appear by some sort of mutation, 'sport', as Darwin used to say, it would be diluted very fast in the populations which shared, or were destined to share, this heredity, and that therefore within the next two or three generations it would be diluted to the point where it would not possibly have any selective value. Jenkins's reasoning is absolutely without answer. Darwin, I think, probably realized that. It is the reason that in the last edition of his great book during his lifetime he went back on the selective theory, and accepted more and more of the kind of Lamarckism which he had tried to tone down in the first edition.

What Jenkins's remarks called for was a theory of heredity by which inheritance would be essentially discreet, discontinuous, and ensured by units that could be transmitted from generation to generation without losing their somatogenic qualities. Such is the gene. The gene was discovered by Mendel during the lifetime of Darwin, but of course it was not scientific knowledge at the time. It remained virtually unknown until the beginning of the twentieth century. In fact, strictly even more than in regard to Kelvin's objections, the selective theory of evolution as Darwin himself had stated it, required the discovery of Mendelian genetics, which of course, was made. This is an example, and a most important one, of what is meant by the content of a theory, the content of an idea, the content of a *logical* idea concerning the

[1] See also the discussion by Fisher in Chapter 4.

universe, concerning the world. If it fits, a good theory or a good idea will always be much wider and much richer than even the inventor of the idea may know at his time. The theory may be judged precisely on this type of development, when more and more falls into its lap, even though it was not predictable that so much would come of it. [. . .]

To begin with, and this is perhaps one of the most important points, the theory of evolution—in all its forms but especially in its more modern forms—assumes, explicitly or not, that there is a profound basic uniformity among living beings, that the basic machinery is the same in all. This is undoubtedly one of the great results of modern biology—I will not give you examples of it—there are too many. The best example is the fact that the genetic code itself, that is to say the chemical machinery of inheritance, works according to the same basic principles and the same code in every known living being from bacteria to man.

A second assumption of the synthetic theory is that all inheritance and therefore all morphogenetics—all that distinguishes one species from another, or one individual from another—must be referred to information in the genome. Since it is the only information contained in an individual that is transmissible to its descendants, it is the only information upon which selection can have any effects. That is a basic assumption which, you may be surprised to learn, easily could have been considered unproved, and was still disbelieved something like twenty years ago by many biologists. It has been completely accepted only as the result of the discovery of this transcription–translation machinery, and the results which showed that not only does the genome contain information concerning the structure of molecules, but is also a regulatory system, which is both a transmitter and receiver of functional information. This also was implicit in the synthetic theory, to the extent that the synthetic theory considered the population and therefore the whole gene pool as the unit of evolution, and because of the concept of an integrated genome—an integrated system whose properties had to be considered as a whole rather than unit by unit.

The two other conclusions which may now be considered proven are also in this scheme. They refer to the nature of mutations. Darwin had at first talked of sports which happened more or less at random. The geneticists of the classical period observed mutations, and included in the synthetic theory the idea that mutations were spontaneous events that were not controlled from outside. We now know what the nature of mutations is. We can even write chemical formulas for most mutations. We know that they are quantum events, that they occur at the level of single molecules, and therefore that they belong in the realm of microscopic physics—in the realm of events that by their very nature cannot be individually predicted and cannot be individually controlled.

This scheme tells us that the old idea of acquired characters, which had been proposed by Lamarck, not only has never been verified, as you probably all know, but in fact is completely incompatible with all of what we know of the whole structure of transfer of information. First, the spontaneous nature of mutation is incompatible with such an idea and second, we know that this sequence of transfer of information is essentially irreversible. Here, I think, I must correct a wrong idea that has been spreading for the past three or four years. It was discovered some years ago that in some cases, the transcription step from DNA to RNA works in the reverse direction. That is nothing surprising. This is a very simple step and even by the basic principle in physical chemistry of the reversibility of microscopic events, it could be predicted that such events could occur. They do occur, indeed, but this must not be taken to mean that *information* from protein could possibly go back to the genome. I think, in spite of some hesitation even by some very distinguished colleagues, I am ready to take any bet you like that this is never going to turn out to be the case.

So you see that the advance of molecular genetics, of molecular biology, has in fact enriched tremendously, and made explicit a great deal that was implicit in the theory of evolution. It has not revolutionized this concept. On the contrary, it has, if anything, both made it much more precise and much harder. It is a harder theory in its description, it is a harder theory in the sense that it tells us a great deal more, and therefore becomes far more sensitive to the criterion of falsification of Popper. And I would say that we have a complete theory of evolution only now that we have a physical interpretation of these basic steps of phenomena that must be assumed, and that are known to account both for the development of living beings, for the inheritance of their traits, and for evolution.

The upshot of all this is that it is a conceptual error to say that evolution is a law, or even that evolution is a law of living beings. This is wrong. The privilege of living beings is not to evolve but on the contrary to conserve. (I'm sorry to say that, especially in front of a great crowd of undergraduates, but this is the case.) The privilege of living beings is the possession of a structure and of a mechanism which ensures two things: (*i*) reproduction true to type of the structure itself, and (*ii*) reproduction equally true to type, of any accident that occurs in the structure. Once you have that, you have evolution, because you have conservation of accidents. Accidents can then be recombined and offered to natural selection, to find out if they are of any meaning or not. Evolution is not a law; evolution is a phenomenon that occurs when you have structures of this kind. [. . .]

What is very interesting, and it is with this that I should like to end this lecture, is why is there this constant resistance, this rejection of the theory of evolution, of the selective theory of evolution as we understand it. I think the interpretation is perfectly simple. It is a very old and profoundly engrained concept in man that anything that exists, in particular himself, has a very good, an obligatory reason to be there. The aspect of evolutionary theory that is unacceptable to many enlightened people, either scientists or philosophers, or ideologists of one kind or another, is the completely contingent aspect which the existence of man, societies, and so on, must take if we accept this theory. If we accept this theory, we must conclude that the emergence of life on the earth was probably unpredictable before it happened. We must conclude that the existence of any particular species is a singular event, an event that occurred only once in the whole of the universe and therefore one that is also basically and completely unpredictable, including that one species which we are, namely man. We must consider our species as any other species—we are a single species, a single event—and therefore we were unpredictable before we appeared. We are completely contingent in respect, not only to the rest of the universe, but even in respect to the rest of living beings. We might just as well not have been there and not have appeared.

[in R. Harré (ed.), *Problems of Scientific Revolution* (Oxford: OUP, 1974), ch. 2.]

THOMAS HENRY HUXLEY

64 Evolution and ethics

Some authors have extended the theory of evolution to human ethics. However, biological evolution proceeds by a competitive struggle for existence and survival of the fittest. Social progress is more a matter of checking and combating evolutionary processes, rather than encouraging or imitating them. Civilization is a different, almost opposite, process from cosmic evolution. [Editor's summary.]

Hence the pressing interest of the question, to what extent modern progress in natural knowledge, and, more especially, the general outcome of that progress in the doctrine of evolution, is competent to help us in the great work of helping one another?

The propounders[1] of what are called the 'ethics of evolution,' when the 'evolution of ethics' would usually better express the object of their speculations, adduce a number of more or less interesting facts and more or less sound arguments, in favour of the origin of the moral sentiments, in the

[1] The opening of Huxley's section extracted here is particularly directed against Herbert Spencer and his many (at that time, around 1890) followers. Spencer understood biological evolution to be a progressive process, leading to 'higher' forms. He included biological evolution in a more general explanatory system that also explained ethics. However, Huxley's argument has an interest for all attempts, and not just Spencer's, to read ethics into evolution, or deduce ethics from evolution.

same way as other natural phenomena, by a process of evolution. I have little doubt, for my own part, that they are on the right track; but as the immoral sentiments have no less been evolved, there is, so far, as much natural sanction for the one as the other. The thief and the murderer follow nature just as much as the philanthropist. Cosmic evolution may teach us how the good and the evil tendencies of man may have come about; but, in itself, it is incompetent to furnish any better reason why what we call good is preferable to what we call evil than we had before. Some day, I doubt not, we shall arrive at an understanding of the evolution of the æsthetic faculty; but all the understanding in the world will neither increase nor diminish the force of the intuition that this is beautiful and that is ugly.

There is another fallacy which appears to me to pervade the so-called 'ethics of evolution.' It is the notion that because, on the whole, animals and plants have advanced in perfection of organization by means of the struggle for existence and the consequent 'survival of the fittest'; therefore men in society, men as ethical beings, must look to the same process to help them towards perfection. I suspect that this fallacy has arisen out of the unfortunate ambiguity of the phrase 'survival of the fittest.' 'Fittest' has a connotation of 'best'; and about 'best' there hangs a moral flavour. In cosmic nature, however, what is 'fittest' depends upon the conditions. Long since, I ventured to point out that if our hemisphere were to cool again, the survival of the fittest might bring about, in the vegetable kingdom, a population of more and more stunted and humbler and humbler organisms, until the 'fittest' that survived might be nothing but lichens, diatoms, and such microscopic organisms as those which give red snow its colour; while, if it became hotter, the pleasant valleys of the Thames and Isis might be uninhabitable by any animated beings save those that flourish in a tropical jungle. They, as the fittest, the best adapted to the changed conditions, would survive.

Men in society are undoubtedly subject to the cosmic process. As among other animals, multiplication goes on without cessation, and involves severe competition for the means of support. The struggle for existence tends to eliminate those less fitted to adapt themselves to the circumstances of their existence. The strongest, the most self-assertive, tend to tread down the weaker. But the influence of the cosmic process on the evolution of society is the greater the more rudimentary its civilization. Social progress means a checking of the cosmic process at every step and the substitution for it of another, which may be called the ethical process; the end of which is not the survival of those who may happen to be the fittest, in respect of the whole of the conditions which obtain, but of those who are ethically the best.

As I have already urged, the practice of that which is ethically best—what we call goodness or virtue—involves a course of conduct which, in all respects, is opposed to that which leads to success in the cosmic struggle for existence. In place of ruthless self-assertion it demands self-restraint; in place of thrusting aside, or treading down, all competitors, it requires that the individual shall not merely respect, but shall help his fellows; its influence is directed, not so much to the survival of the fittest, as to the fitting of as many as possible to survive. It repudiates the gladiatorial theory of existence. It demands that each man who enters into the enjoyment of the advantages of a polity shall be mindful of his debt to those who have laboriously constructed it; and shall take heed that no act of his weakens the fabric in which he has been permitted to live. Laws and moral precepts are directed to the end of curbing the cosmic process and reminding the individual of his duty to the community, to the protection and influence of which he owes, if not existence itself, at least the life of something better than a brutal savage.

It is from neglect of these plain considerations that the fanatical individualism of our time attempts to apply the analogy of cosmic nature to society. Once more we have a misapplication of the stoical injunction to follow nature; the duties of the individual to the state are forgotten, and his tendencies to self-assertion are dignified by the name of rights. It is seriously debated whether the members of a community are justified in using their combined strength to constrain one of their number to contribute his share to the maintenance of it; or even to prevent him from doing his best to destroy it. The struggle for existence, which has done such admirable work in cosmic nature, must, it appears, be equally beneficent in the ethical sphere. Yet if that which I have insisted upon is true; if the cosmic process has no sort of relation to moral ends; if the imitation of it by man is inconsistent with the first principles of ethics; what becomes of this surprising theory?

Let us understand, once for all, that the ethical progress of society depends, not on imitating the cosmic process, still less in running away from it, but in combating it. It may seem an audacious proposal thus to pit the microcosm against the macrocosm and to set man to subdue nature to his higher ends; but I venture to think that the great intellectual difference between the ancient times with which we have been occupied and our day, lies in the solid foundation we have acquired for the hope that such an enterprise may meet with a certain measure of success.

The history of civilization details the steps by which men have succeeded in building up an artificial world within the cosmos. Fragile reed as he may be, man, as Pascal says, is a thinking reed: there lies within him a fund of energy, operating intelligently and so far akin to that which pervades the universe, that it is competent to influence and modify the cosmic process. In virtue of his intelligence, the dwarf bends the Titan to his will. In every family, in every polity that has been established, the cosmic process in man has been restrained and otherwise modified by law and custom; in surrounding nature, it has been similarly influenced by the art of the shepherd, the agriculturist, the artisan. As civilization has advanced, so has the extent of this interference increased; until the organized and highly developed sciences and arts of the present day have endowed man with a command over the course of non-human nature greater than that once attributed to the magicians. The most impressive, I might say startling, of these changes have been brought about in the course of the last two centuries; while a right comprehension of the process of life and of the means of influencing its manifestations is only just dawning upon us. We do not yet see our way beyond generalities; and we are befogged by the obtrusion of false analogies and crude anticipations. But Astronomy, Physics, Chemistry, have all had to pass through similar phases, before they reached the stage at which their influence became an important factor in human affairs. Physiology, Psychology, Ethics, Political Science, must submit to the same ordeal. Yet it seems to me irrational to doubt that, at no distant period, they will work as great a revolution in the sphere of practice.

The theory of evolution encourages no millennial anticipations. If, for millions of years, our globe has taken the upward road, yet, some time, the summit will be reached and the downward route will be commenced. The most daring imagination will hardly venture upon the suggestion that the power and the intelligence of man can ever arrest the procession of the great year.

Moreover, the cosmic nature born with us and, to a large extent, necessary for our maintenance, is the outcome of millions of years of severe training, and it would be folly to imagine that a few centuries will suffice to subdue its masterfulness to purely ethical ends. Ethical nature may count upon having to reckon with a tenacious and powerful enemy as long as the world lasts. But, on the other hand, I see no limit to the extent to which intelligence and will, guided by sound principles of investigation, and organized in common effort, may modify the conditions of existence, for a period longer than that now covered by history. And much may be done to change the nature of man himself. The intelligence which has converted the brother of the wolf into the faithful guardian of the flock ought to be able to do something towards curbing the instincts of savagery in civilized men.

[*Evolution and Ethics* (London: Macmillan, 1893).]

STEPHEN R. PALUMBI

65 Humans as the world's greatest evolutionary force

In addition to altering global ecology, technology and human population growth also affect evolutionary trajectories, dramatically accelerating evolutionary change in other species, especially in commercially important, pest and disease organisms. Such changes are apparent in antibiotic and human immunodeficiency virus (HIV) resistance to drugs, plant and insect resistance to pesticides, rapid changes in invasive species, life-history change in commercial *fisheries, and pest adaptation to biological engineering products. This accelerated evolution costs at least $33 billion to $50 billion a year in the United States. Slowing and controlling arms races in disease and pest management have been successful in diverse ecological and economic systems, illustrating how applied evolutionary principles can help reduce the impact of humankind on evolution.* [Author's summary.]

Human impact on the global biosphere now controls many major facets of ecosystem function. Currently, a large fraction of the world's available fresh water, arable land, fisheries production, nitrogen budget, CO_2 balance, and biotic turnover are dominated by human effects. Human ecological impact has enormous evolutionary consequences as well and can greatly accelerate evolutionary change in the species around us, especially disease organisms, agricultural pests, commensals, and species hunted commercially. For example, some forms of bacterial infection are insensitive to all but the most powerful antibiotics, yet these infections are increasingly common in hospitals. Some insects are tolerant of so many different insecticides that chemical control is useless. Such examples illustrate the pervasive intersection of biological evolution with human life, effects that generate substantial daily impacts and produce increasing economic burden.

Accelerated evolutionary changes are easy to understand—they derive from strong natural selection exerted by human technology. However, technological impact has increased so markedly over the past few decades that humans may be the world's dominant evolutionary force. The importance of human-induced evolutionary change can be measured economically, in some cases, and is frequently seen in the exposure of societies to uncontrollable disease or pest outbreaks. Attempts to slow these evolutionary changes are widespread but uncoordinated. How well do they work to slow evolution? Can successes from one field be generalized to others?

The pace of human-induced evolution

Paul Müller's 1939 discovery that DDT killed insects won him the 1948 Nobel Prize, but before the Nobel ceremony occurred, evolution of resistance had already been reported in house flies. By the 1960s, mosquitoes resistant to DDT effectively prevented the worldwide eradication of malaria, and by 1990, over 500 species had evolved resistance to at least one insecticide. Insects often evolve resistance within about a decade after introduction of a new pesticide, and many species are resistant to so many pesticides that they are difficult or impossible to control. Similar trajectories are known for resistant weeds which typically evolve resistance within 10 to 25 years of deployment of an herbicide (Table 65.1).

Bacterial diseases have evolved strong and devastating resistance to many antibiotics. This occurs at low levels in natural populations but can become

Table 65.1: Dates of deployment of representative antibiotics and herbicides, and the evolution of resistance

EVOLUTION OF RESISTANCE TO ANTIBIOTICS AND HERBICIDES

ANTIBIOTIC OR HERBICIDE	YEAR DEPLOYED	RESISTANCE OBSERVED
	Antibiotics	
Sulfonamides	1930s	1940s
Penicillin	1943	1946
Streptomycin	1943	1959
Chloramphenicol	1947	1959
Tetracycline	1948	1953
Erythromycin	1952	1988
Vancomycin	1956	1988
Methicillin	1960	1961
Ampicillin	1961	1973
Cephalosporins	1960s	late 1960s
	Herbicides	
2,4-D	1945	1954
Dalapon	1953	1962
Atrazine	1958	1968
Picloram	1963	1988
Trifluralin	1963	1988
Triallate	1964	1987
Diclofop	1980	1987

common within a few years of the commercial adoption of a new drug (Table 65.1). For example, virtually all Gram-positive infections were susceptible to penicillin in the 1940s but in hospitals today, the vast majority of infections caused by important bacterial agents like *Staphylococcus aureus* are penicillin-resistant, and up to 50% are resistant to stronger drugs like methicillin. Treatments that used to require small antibiotic doses now require huge concentrations or demand powerful new drugs. But such solutions are short-lived. For example, vancomycin, one of the only treatments for methicillin-resistant infections, has been overcome by some of the most frequent infectious agents in hospitals. Antibiotics also generate evolution outside hospitals. Resistant strains are common on farms that use antibiotics in livestock production and have been found in soils and groundwater affected by farm effluents.

Retroviruses[1] with RNA genomes evolve even more quickly than bacteria. Every year, vaccinations against influenza must be reformulated, making prediction of next year's viral fashion one of preventative medicine's chief challenges. The virus that causes AIDS, human immunodeficiency virus-1, evolves so quickly that the infection within a single person becomes a quasi-species consisting of thousands of evolutionary variants. Over the course of months or years after HIV infection, the virus continually evolves away from immune system suppression. Evolution in the face of antiviral drugs is just as rapid. For example, the drug nevirapine reduces viral RNA levels for only about 2 weeks. Thereafter, mutations in the HIV reverse transcriptase gene quickly arise that confer drug resistance, and the HIV mutants have a doubling time of 2 to 6 days. This rapid evolution is repeated with virtually all other antiretroviral drugs when given singly, including the inexpensive antiviral drugs zidovudine (azidothymine, AZT), lamivudine (3TC), didanosine (ddI) and protease inhibitors like indinavir.

Rapid evolution caused by humans is not restricted to disease or pest species. Under heavy fishing pressure, fish evolve slower growth rates and thinner bodies, allowing them to slip through gill nets. In hatchery populations of salmon, there is strong selection for dwarf males that return from sea early, increasing their survival. Invading species, transported by humans, have been known to rapidly change to match local selection pressures. For instance, house sparrows, introduced to North America in 1850, are now discernibly different in body size and colour throughout the United States. In some cases, species introduced by humans induce evolution in species around them. For example, after the subtidal snail *Littorina littorea* invaded coastal New England in the late 1800s, native hermit crabs[2] [*Pagurus longicarpus* (Say)] quickly evolved behavioral preference for their shells. The crabs also evolved body and claw changes that fit them more securely in these new, larger shells. Even more quickly, introduced predatory fish have caused rapid evolution of life-history traits and color pattern in their prey species. Rates of human-mediated evolutionary change sometimes exceed rates of natural evolution by orders of magnitude.

[1] A retrovirus is a virus that uses RNA as its hereditary material, but is reproduced via a DNA intermediary stage. The production of the DNA version of the retrovirus, by the enzyme reverse transcriptase, is error-prone, giving retroviruses a high mutation rate. HIV and influenza virus are retroviruses. Many other RNA viruses reproduce without a DNA intermediate stage and are not classified as retroviruses.

[2] Hermit crabs are crabs that live in shells that have been manufactured by snails. They obtain the shells either by occupying an empty shell or by evicting another hermit crab from it.

Causes of evolution

These examples demonstrate pervasive and rapid evolution as a result of human activity. In most cases, the causes of this evolutionary pattern are clear: if a species is variable for a trait, and that trait confers a difference in survival or production of offspring, and the trait difference is heritable by offspring, then all three requirements of evolution by natural selection are present. In such cases, the evolutionary engine can turn, although evolutionary directions and speed can be influenced by factors such as drift, conflicting selection pressure, and correlated characters.

The overwhelming impact of humans on evolution stems from the ecological role we now play in the world, and the industrialization of our agriculture, medicine, and landscape. Successful pesticides or antibiotics are often produced in massive quantities. DDT, for example, was first used by the Allied Army in Naples in 1943, but by the end of World War II, DDT production was proceeding on an industrial scale. Currently, we use about 700 million pounds of pesticide a year in the United States. Antibiotic production is also high, with 25 to 50% going into prophylactic[3] use in livestock feed.

Inefficient use of antibiotics has been cited as a major cause of antibiotic resistance. Partial treatment of infections with suboptimal doses leads to partial control of the infecting cell population, and creates a superb environment for the evolution of resistant bacteria. Up to one-third of U.S. pediatricians report overprescribing antibiotics to assuage patient concerns, particularly in cases of viral childhood congestions that cannot respond to the drug. Failing to complete a course of antibiotics is associated with increased emergence of resistant tuberculosis and HIV infections, and differences in antibiotic use may partly explain differences among nations in antibiotic resistance rates.

Spread of antibiotic resistance has been accelerated by transmission of genes between bacterial species.[4] Recently, biotechnology has applied this acceleration to other species as well, and a new human-mediated mechanism for generating evolutionary novelty has emerged—insertion of exogenous genes into domesticated plants and animals. Taken from bacteria, plants, animals, or fungi, these genes convey valuable commercial traits, and they are placed into new host genomes along with genes that control expression and in some cases allow cell lineage selection. Examples include the insertion of genes for insecticidal proteins, herbicide tolerance or novel vitamins into crop plants; growth hormone genes into farmed salmon; and hormone production genes into livestock 'bioreactors'. These efforts effectively

[3] Prophylactic means that the drug is given to prevent disease, rather than to cure it.

[4] On transmission of genes between species, that is, 'horizontally', see Chapter 36.

increase the rate of generation of new traits—akin to increasing the rate of macromutation. When these traits cross from domesticated into wild species, they can add to the fuel of evolution and allow rapid spread of the traits in natural populations. Genetic exchange from crops has already enhanced the weediness of wild relatives of 7 of the world's 13 most important crop plants although no widespread escape of an engineered gene into the wild has been reported yet.

The economics of human-induced evolution

Evolution is responsible for large costs when pests or disease organisms escape from chemical control. Farmers spend an estimated $12 billion on pesticides per year in the United States. Extra costs due to pest resistance, such as respraying fields, may account for about 10% of these direct expenditures. Despite the heavy use of chemical pesticides, 10 to 35% of U.S. farm production is lost to pest damage. If even 10% of this loss is due to activities of resistant insects (and the figure may be far higher), this represents a $2 billion to $7 billion yearly loss for the $200 billion U.S. food industry. The development of resistance in diamondback moths to *Bacillus thuringiensis* (Bt) toxin in 1989 foreshadows the decline in use of the world's largest selling biopesticide and the need for new approaches. The price of developing a single new pesticide, about $80 million in 1999, is an ongoing cost of agricultural business. Even higher development costs (about $150 million per product) are incurred by pharmaceutical companies. In both sectors, evolution sparks an arms race between human chemical control and pest or disease agent, dramatically increasing costs that are eventually paid by consumers. For example, the new drugs linezolid and quinupristin-dalfopristin were recently approved by the U.S. Food and Drug Administration (FDA) for use on vancomycin-resistant infections. Previously, vancomycin had been used to overcome methicillin resistance, and methicillin was itself a response to the failure of penicillin treatment. This development cascade[5] has been ongoing since the birth of the chemical-control era and represents a poorly quantified cost of evolution.

More direct expenses stem from the increase in drug payments and hospitalization necessary to treat resistant diseases. There are approximately 2 million hospital-acquired infections in the United States each year, a quarter of which are caused by antibiotic-resistant *S. aureus*. Half of these are penicillin-resistant strains that require treatment with methicillin at a cost

[5] 'Development cascade' refers to the way in which the evolution of resistance to one drug (such as penicillin) creates a need for humans to develop a new drug, such as methicillin, to which in turn bacteria evolve resistance, leading to the need to develop a further drug . . . and so on.

of $2 billion to $7 billion. The other half are methicillin-resistant infections, and they cost hospitals an estimated $8 billion per year to cure. Community-acquired, antibiotic-resistant staph[6] infections more than double these costs. These figures are for a single type of infection and do not include other well-known drug-resistant bacteria. For example, in the United States up to 22% of hospital-acquired infections of *Enterococcus faecium* are resistant to vancomycin, and combating such infections drives the price of evolution even higher.

Similar conservative tabulations can be made for the cost of HIV treatment. The current standard of care in the United States is to treat HIV with massive doses of at least three drugs. Because treatment with the inexpensive antiretroviral drug AZT would successfully halt HIV if it did not evolve resistance, the need for more powerful drugs is due to HIV evolution. Drug and treatment prices vary but have recently been estimated at $18,300 per year per patient in the United States. If half the 700,000 HIV patients in the United States receive this level of care, these costs amount to $6.3 billion per year. Costs of lost labor, disruption of health services, development of new drugs, and medical research are not included in this figure, and so the actual cost of HIV evolution is far higher.

The annual evolution bill in the United States approaches $50 billion for these examples (Table 65.2), and probably exceeds $100 billion overall. However, the social price of evolution is far higher. Skyrocketing costs of treating resistant diseases create a situation where effective medical treatment may be

Table 65.2 Examples of the costs of human-induced evolution in insect pests and several disease organisms in the United States

COSTS OF HUMAN-INDUCED EVOLUTION IN SOME INSECT PESTS AND DISEASES

FACTOR	$U.S. BILLIONS PER YEAR
Additional pesticides	1.2
Loss of crops	2 to 7
S. aureus	
Penicillin-resistant	2 to 7
Methicillin-resistant	8
Community-acquired resistant	14 to 21
HIV drug resistance	63
Total for these factors	33 to 50

[6] 'Staph' is informal short-hand for *Staphylococcus*.

economically unattainable for many people. Thus, evolution expands the class of diseases that are medically manageable but economically incurable.

Ways of slowing evolution

Responding to the pervasive reach of evolution in medicine and agriculture, health specialists and agricultural engineers have developed an impressive series of innovative methods to slow the pace of evolution. A large body of theory guides deployment of some of these attempts. Other methods, circulated as guidelines for clinical practices or farming strategies, often appear to be developed through a combination of trial and error and common sense. Independent of their theoretical underpinnings, the following examples show that successful methods often slow evolution for clear evolutionary reasons and that these approaches may be generalizable to other systems.

DRUG OVERKILL AND HIV TRIPLE-DRUG THERAPY

Overkill strategies, the combination of treatments to kill all infectious or invading pests, are common. For example, treatment with a drug cocktail that includes a protease inhibitor and two different reverse transcriptase inhibitors is the Cadillac of AIDS treatment strategies. This approach has been successful longer than any other, because it not only reduces viral levels but also slows the evolution of resistance. The evolutionary biology hidden in this strategy is simple: a strong, multiple-drug dose leaves no virus able to reproduce, and so there is no genetically based variation in fitness among the infecting viruses in this overwhelming drug environment.[7] Without fitness variation, there is no evolutionary fuel, and evolution halts. Lack of HIV variation for growth in this regime is responsible for reduced evolutionary rate and probably drives the current success of triple-drug treatment. However, sequential treatment with single drugs or voluntary drug cessation can foster the evolution of drug resistance, which appears to be increasing. This suggests that the triple-drug overkill strategy will not halt HIV evolution forever but may provide only a brief window for the development of more permanent solutions, such as HIV vaccines.

Overkill strategies have been echoed in pesticide management programs, where they are often termed 'pyramiding', and in treatment of bacterial

[7] Thus, if only one drug is applied, it will kill susceptible viruses, but leave the resistant ones alive. There is genetic variation in fitness, in the sense that the fit forms have one genotype (resistance) and the unfit forms have another genotype (susceptibility). Resistance evolves by natural selection. However, resistant forms to any one drug are initially rare in the population, making it highly improbable that any one virus will be resistant to three different drugs. When three drugs are applied, either all the viruses are killed or, if a few do limp through, it is not because they are genetically resistant. Either way natural selection cannot act to cause resistance to evolve.

infections. However, their use is limited by drug toxicity: extreme doses can have physiological or ecosystem side effects.

DIRECT OBSERVATION THERAPY

Tuberculosis infects one-third of the world's population and is difficult to treat because it requires 6 months of medication to cure. Partial treatment has resulted in evolution of multidrug resistance. To combat this, drug doses are brought individually to patients, who are observed while they take the drugs. This direct-observation therapy has been used to improve patient compliance during the whole treatment regimen, reducing evolution of resistance by ensuring a drug dose long enough and severe enough to completely eradicate the infection from each person. Direct-observation therapy has been credited with snuffing out emerging tuberculosis epidemics and dramatically reducing costs of medical treatment.

WITHHOLDING THE MOST POWERFUL DRUGS

The antibiotic vancomycin has been called the 'drug of last resort', because it is used only when other, less powerful antibiotics fail. Withholding the most powerful drugs lengthens their effective life-span because overall selection pressure exerted by the drug is reduced, slowing the pace of evolution. Although successful in reducing the evolution of resistance to vancomycin by some bacteria, the strategy depends on low use rates in all sectors of the antibiotic industry, including livestock and prophylactic use. Failure to include these sectors in the strategy will engineer its failure.

SCREENING FOR RESISTANCE BEFORE TREATMENT

Screening infections for sensitivity to particular antibiotics before treatment allows a narrow-range antibiotic to be used instead of a broad-spectrum antibiotic. Reduced use of broad-spectrum antibiotics slows evolution of resistance as in the mechanism above. Genotyping of viruses in an HIV infection and prediction of the antiviral drugs to which they are already resistant improves drug usefulness. Similarly, farmers are advised to check their fields after pesticide treatment and then to change the chemical used in the next spraying if many resistant individuals are discovered. Screening for pest susceptibility reduces use of chemicals for which resistance has begun to evolve.

CYCLIC SELECTION DUE TO CHANGING CHEMICAL REGIMES

Farmers are encouraged to follow several simple rules to reduce herbicide resistance: (i) do not use the same herbicide 2 years in a row on the same field, and (ii) when switching herbicides, use a new one that has a different mechanism of action. These guidelines slow evolution through a rapid alteration of selection pressure that sequentially changes the selective landscape. Mutants favored in one generation are not favored in the next, because one mutation is not likely to provide resistance to two herbicides with different mechanisms. Similar cyclic selection regimes have been proposed to limit resistance in intensive-care units and agricultural fields. Mosaic selection, in which different chemicals are used in different places at the same time is a spatial version of this tactic.

INTEGRATED PEST MANAGEMENT

Integrated pest management (IPM) may include chemical control of pests, but does not rely on it exclusively, and is credited with better pest control and with slower evolution of resistance. Slow evolution can come from two sources. First, the multiple control measures used in IPM reduce reliance on chemical treatments, thereby reducing selection for chemical resistance. Second, physical control of populations (e.g. through baiting, trapping, washing, or weeding) reduces the size of the population that is exposed to chemical control. Smaller populations have a reduced chance of harboring a mutation, thereby slowing the evolution of resistance. The term IPM is common only in insect management, but the strategy has appeared independently in hospitals where hand-washing, instead of prophylactic antibiotic use, is encouraged and in weed management, where resistant weeds are pulled by hand.

REFUGE PLANTING

Biotechnology has introduced insecticidal toxin genes into numerous crop species, but resistance to toxins produced by these genes has already evolved in pests, threatening the commercial use of this technology. To reduce the potential for evolution, crop engineers have instituted a program of refuge planting to slow the success of resistant insects. If farmers plant a fraction of a field with non-toxin-producing crop varieties, and allow these to be consumed by insects, a large number of nonresistant pests are produced. These can then mate with the smaller number of resistant individuals emerging from fields of plants producing insecticidal proteins, greatly reducing the number of offspring homozygous for the resistance alleles. In cases where resistance is recessive, refuges slow the spread of resistant alleles, although they require high crop losses in the refuge plantings. This mechanism functions by reducing the inheritance of resistance through increases in the proportion of breeding individuals without resistance alleles.

ENGINEERING EVOLUTION

Using evolution to our advantage may also be possible, although this is seldom attempted. One illustrative exception is the use of the drug 3TC to slow the mutation rate of HIV and thereby, perhaps, to limit its ability to rapidly evolve resistance to other drugs. An ongoing use of evolutionary theory is the prediction of which influenza strains to use for future vaccines. Another is the use of chemical control where resistance includes a severe metabolic cost, making resistant individuals less fit when the chemicals are removed.[8] In such cases, the potential of evolution to lower pest fitness in the absence of a pesticide may be a method of using the power of evolution to our advantage. At unintended evolutionary outcome may be the escape of antibiotic, herbicide, or pesticide resistance genes to natural populations, possibly making them less susceptible to pesticides in the environment. In some agricultural settings, artificial selection for pesticide resistance has been used to protect populations of beneficial insects.

This summary shows that successful control of evolution has followed many different strategies, and that the methods currently used impact all three factors driving evolutionary change (Table 65.3). However, seldom, have all three evolutionary prerequisites been manipulated in the same system, and seldom has the engineering of the evolutionary process been attempted in a systematic way. Instead, in every new case, human-mediated evolution tends to catch us by surprise, and strategies to reduce or stop it are invented from scratch. For example, cyclic selection has been invented at least three times (for control of insects, bacteria, and HIV), IPM at least three times (insects, weeds, and bacteria), and drug overkill at least twice (HIV and tuberculosis).

Overall, three ways to adjust selective pressures are widely used in pest and health management: application of multiple simultaneous chemicals or 'pyramiding', cyclic application of different chemicals, and using different chemicals in different places or 'mosaic application'. Although the principles are exactly the same in all fields, seldom has the literature from one field been used to inform the other. Some strategies that are very successful in one arena have not been tried in others (e.g. no direct-observation therapy has been tried on farms). Yet, the commonality of successful methods (Table 65.3) suggests that lessons in evolutionary engineering from one system may be useful in others and that it may be possible to control evolution far more successfully than is currently practiced. Mathematical models of evolutionary engineering provide some guidance about practical field methods, but this exchange between prediction and practice has only been common in pest management and antibiotic resistance. A critical need is the inclusion of evolutionary predictions in the current debate on global HIV policy. Most

[8] In this example, resistance does evolve, but the resistant forms will have two attributes: (i) resistance; (ii) an ability to cope with the associated chemical that imposes a metabolic cost. The coping ability (ii) becomes a liability when the chemical is not there and being needed to coped with.

Table 65.3: The success of evolutionary engineering mechanisms that reduce evolution can and do work on all three parts of the evolutionary engine

MECHANISMS THAT WORK TO SLOW EVOLUTION

METHOD OF SLOWING EVOLUTION	EXAMPLE
Reduce variation in a fitness-related trait	
Drug overkill with multiple drugs	Triple-drug therapy for AIDS Pesticide pyramiding
Ensure full dosage	Direct observation therapy of tuberculosis
Reduce appearance of resistance mutations	Engineer RT gene of HIV-1
Reduce pest population size	Integrated pest management of resistant mutants
	Nondrug sanitary practices
Reduce directional selection	
Vary selection over time	Herbicide rotation
	Vary choice of antibiotics, pesticides or antiretrovirals
Use nonchemical means of control	Integrated pest management
Limit exposure of pests to selection	Withhold powerful drugs, e.g., restricted vancomycin use
Avoid broad-spectrum antibiotics	Test for drug or pesticide susceptibility before treatment of infections or fields
Reduce heritability of a fitness-related trait	
Dilute resistance alleles	Refuge planting

important, it is seldom realized that a pivotal goal is slowing the evolution of resistance and that, without this, all successful pest and disease control strategies are temporary.

Conclusions and prospects

Rapid evolution occurs so commonly that it is, in fact, the expected outcome for many species living in human-dominated systems. Evolution in the wake of human ecological change should be the default prediction and should be part of every analysis of the impact of new drugs, health policies, pesticides, or biotechnology products. By admitting the speed and pervasiveness of evolution, predicting evolutionary trajectories where possible, and planning mechanisms in advance to slow evolutionary change, we can greatly reduce our evolutionary impact on species around us and ameliorate the economic and social costs of evolution. Ignoring the speed of evolution requires us to play an expensive catch-up game when chemical control agents and medications fail. Because our impact on the biosphere is not likely to decline, we must use our knowledge about the process of evolution to mitigate the evolutionary changes we impose on species around us.

[*Science*, 293 (2001), 1786–90.]

Select bibliography

Section A: From Darwin to the modern synthesis

BOWLER, P. J., *Evolution: The History of an Idea*, revised edn (Berkeley: University of California Press, 1989).

BOX, J. F., *R. A. Fisher. The Life of a Scientist* (New York: John Wiley, 1978).

BROWNE, J., *Charles Darwin*, 2 vols (New York: Simon & Schuster, and London: Jonathan Cape, 1995 and 2002).

CLARK, R. W., *JBS: The Life and Work of J. B. S. Haldane* (New York: Coward-McCann, 1969).

GHISELIN, M. T., *The Triumph of the Darwinian Method* (Berkeley: University of California Press, 1969).

MAYR, E. and PROVINE, W. B. (eds.), *The Evolutionary Synthesis* (Cambridge, Mass.: Harvard University Press, 1980).

PROVINE, W. B., *Sewall Wright and Evolutionary Biology* (Chicago: University of Chicago Press, 1986).

Section B: Natural selection and random drift in populations

BELL, G., *Selection. The Mechanism of Evolution* (New York: Chapman & Hall, 1997).

COOK, L. M., 'Changing Views on Melanic Moths', *Biological Journal of the Linnean Society*, 69, (2000), 431–41.

ENDLER, J. A., *Natural Selection in the Wild* (Princeton: Princeton University Press, 1986).

GRAUR, D. and LI, W-S., *Fundamentals of Molecular Evolution*, 2nd edn. (Sunderland, Mass.: Sinauer, 2000).

JONES, S., *Almost Like a Whale* (London: Doubleday, 1999). Also published as *Darwin's Ghost* (New York: Ballantine Books, 2000).

MAJERUS, M. E. N., *Melanism: Evolution in Action* (Oxford: Oxford University Press, 1998).

PAGE, R. D. M. and HOLMES, E. S., *Molecular Evolution* (Oxford: Blackwell Publishing, 1998).

WEINER, J., *The Beak of the Finch* (New York: Knopf, and London: Cape, 1994).

Section C: Adaptation

DAWKINS, R., *The Blind Watchmaker* (New York: W. W. Norton, and London: Longman and Penguin, 1986).

DAWKINS, R., *Climbing Mount Improbable* (New York: W. W. Norton, and London: Viking Penguin, 1996).

DENNETT, D., *Darwin's Dangerous Idea* (New York: Simon & Schuster, and London: Penguin, 1995).

HAMILTON, W. D., *The Narrow Roads of Geneland*, 2 vols (Oxford and New York: Oxford University Press, 1996 and 2001).

ROSE, M. R. and LAUDER, G. V. (eds.), *Adaptation* (London: Academic Press, 1996).

Section D: Speciation and biodiversity

COYNE, J. A. and ORR, H. A., *Speciation* (Sunderland, Mass.: Sinauer, 2004).

GHISELIN, M. T. *Metaphysics and the Origin of Species* (New York: Columbia University Press, 1998).

HULL, D. L., *Science as a Process* (Chicago: University of Chicago Press, 1988).

SCHILTUIZEN, M., *Frogs, Flies, and Dandelions: the Making of a Species* (Oxford: Oxford University Press, 2001).

Trends in Ecology and Evolution, Special issue, July 2001, on speciation.

Section E: Macroevolution

CARROLL, S. B., GRENIER, J. K., and WEATHERBEE, S. D., *From DNA to Diversity: Molecular Genetics and the Evolution of Animal Design* (Malden, Mass.: Blackwell Science, 2001).

ERWIN, D. H. and ANSTEY, R. L. (eds.), *New Approaches to Speciation in the Fossil Record* (New York: Columbia University Press, 1995).

GOULD, S. J., *Ontogeny and Phylogeny* (Cambridge, Mass.: Harvard University Press, 1977).

GOULD, S. J., *The Structure of Evolutionary Theory* (Cambridge, Mass.: Harvard University Press, 2002).

KEMP, T. S., *Fossils and Evolution* (Oxford: Oxford University Press, 1999).

RAFF, R. A., *The Shape of Life* (Chicago: University of Chicago Press, 1996).

Section F: Evolutionary genomics

GRAUR, D. and LI, W-S., *Fundamentals of Molecular Evolution*, 2nd edn. (Sunderland, Mass.: Sinauer, 2000).

PAGE, R. D. M. and HOLMES, E. S., *Molecular Evolution* (Oxford: Blackwell Publishing, 1998).

Section G: The history of life

HILLIS, D. M., MORITZ, C., and MABLE, B. K. (eds.), *Molecular Systematics*, 2nd edn. (Sunderland, Mass.: Sinauer, 1996).

MAYNARD SMITH, J. and SZATHMÁRY, E., *The Major Transitions of Evolution* (Oxford: W. H. Freeman, 1995).

PARKER, A., *In the Blink of an Eye* (London: Simon & Schuster, 2003).

Section H: Case studies

AUSTAD, S. N., *Why We Age* (New York: Wiley, 1997).

RIDLEY, M., *Mendel's Demon* (London: Weidenfeld & Nicolson, 2000). Also published as *The Cooperative Gene* (New York: Simon & Schuster, 2000).

Section I: Human evolution

Cosmides, L. and Tooby, J. (eds.), *The Adapted Mind* (Oxford and New York: Oxford University Press, 1991).

Jones, J. S., *The Language of the Genes* (London: HarperCollins, 1994).

Klein, R. G., *The Human Career*, 2nd edn. (Chicago: University of Chicago Press, 1999).

Lewin, R., *Principles of Human Evolution* (Oxford and Boston: Blackwell, 2003).

Nesse, R. M. and Williams, G. C., *Why We Get Sick: The New Science of Darwinian Medicine* (New York: Times Books, 1994). Also published as *Evolution and Healing* (London: Weidenfeld & Nicolson, 1994).

Section J: Evolution and human affairs

Adams, M. B. (ed.), *The Evolution of Theodosius Dobzhansky* (Princeton: Princeton University Press, 1994).

Barlow, C. (ed.), *Evolution Extended: Biological Debates on the Meaning of Life* (Cambridge, Mass.: MIT Press, 1995).

Dawkins, R., *A Devil's Advocate* (London: Weidenfeld & Nicolson, 2003).

Grafen, A. (ed.), *Evolution and its Implications* (Oxford: Oxford University Press, 1989).

Medawar, P. B., *Pluto's Republic* (Oxford: Oxford University Press, 1983).

Monod, J., *Chance and Necessity* (London: Fontana, 1970); reprinted (London: Penguin).

Numbers, R. L., *The Creationists: the Evolution of Scientific Creationism* (New York: Knopf, 1992); paperback edn. (Berkeley: University of California Press, 1993).

Paradis, J. and Williams, G. C., *T. H. Huxleys' 'Evolution and ethics' with New Essays on its Victorian and Sociobiological Context* (Princeton: Princeton University Press, 1989).

Pennock, R. T. (ed.), *Intelligent Design Creationism and its Critics* (Cambridge, Mass.: MIT Press, 2001).

Ruse, M., *Darwin and Design* (Cambridge, Mass.: Harvard University Press, 2003).

Biographical notes

ROBERT L. ANSTEY is Professor and Chair of Geological Sciences at Michigan State University.

MICHAEL F. ANTOLIN and JOAN M. HERBERS are in the Department of Biology, Colorado State University, Fort Collins, Colorado.

STEVEN A. BENNER, M. DANIEL CARACO, J. MICHAEL THOMSON, and ERIC A. GAUCHER work in the Departments of Chemistry and of Biology, University of Florida, Gainesville.

ROY J. BRITTEN works in the California Institute of Technology, Corona del Mar, California.

A. J. CAIN is Emeritus Professor of Zoology, University of Liverpool. His books include *Animal Species and their Evolution*.

SEAN B. CARROLL works in the Laboratory of Molecular Biology, University of Wisconsin, Madison, Wisconsin.

L. M. COOK, R. L. H. DENNIS, and G. S. MANI work in the University of Manchester, Manchester.

ALAN COOPER is a Professor in the Department of Zoology, Oxford University, Oxford.

JERRY A. COYNE is a Professor in the Department of Ecology and Evolution, University of Chicago.

F. H. C. CRICK formerly worked at the MRC Laboratory of Molecular Biology, Cambridge, and now works in the Salk Institute, La Jolla, California. He shared the Nobel Prize in 1962 for discovering the structure of DNA. His books include his autobiography *What Mad Pursuit*.

CHARLES DARWIN (1809–1882) went to Edinburgh and Cambridge Universities, circumnavigated the globe on board the *Beagle*, and then lived much of his life at Down House, Kent. His books include *On the Origin of Species*, *The Descent of Man*, and (as it is now usually called) *The Voyage of the 'Beagle'*.

RICHARD DAWKINS is Charles Simonyi Professor of the Public Understanding of Science, Oxford University. His books include *The Selfish Gene*, *The Extended Phenotype*, and *The Blind Watchmaker*.

GAVIN DE BEER (1899–1972) worked at Oxford University, University College London, and the British Museum (Natural History). His books included *Embryology and Evolution*, *The Development of the Vertebrate Skull*, and *Charles Darwin*.

W. J. DICKINSON is a Professor in the Department of Biology, University of Utah.

DAVID DILCHER works in the Florida Museum of Natural History, Gainesville, Florida.

THEODOSIUS DOBZHANSKY (1900–1975) was born in Russia, and migrated to the USA in 1927. He held positions at Columbia University, California Institute of Technology, and Rockefeller University. His books include *Genetics and the Origin of Species*.

DOUGLAS H. ERWIN is Curator in the Department of Paleobiology at the National Museum of National History, Smithsonian Institution, Washington, DC. His books include *The Great Paleozoic Crisis* and (with Derek Briggs) *The Fossils of the Burgess Shale*.

R. A. FISHER (1890–1962) worked mainly at Cambridge University and made fundamental contributions to the theory of evolution and of statistics. His books include *The Genetical Theory of Natural Selection* and *Statistical Methods for Research Workers*.

RICHARD FORTEY works at the Natural History Museum, London. His books inclde *The Hidden Landscape* and *Life: a Biography*.

W. GARSTANG (1868–1949) worked in the Marine Laboratory, Plymouth, Oxford University, and was then Professor of Zoology at the University of Leeds.

JOHN GERHART works in the University of California, Berkeley.

PHILIP J. GERRISH works in the Los Alamos National Laboratory, New Mexico.

H. LISLE GIBBS was in the Department of Biology, University of Michigan, Ann Arbor at the time of the work included here.

S. J. GOULD (1941–2002) worked at Harvard University. His books include *Ontogeny and Phylogeny*, *The Structure of Evolutionary Theory*, and many volumes of popular scientific essays.

A. GRAFEN is Professor of Quantitative Biology, Department of Zoology, Oxford University and the co-author (with R. Hails) of *Modern Statistics for the Life Sciences*.

PETER R. GRANT is class of 1877 Professor of Zoology at Princeton University. His books include *Ecology and Evolution of Darwin's Finches*.

V. GRANT is a Professor at the University of Texas, Austin, Texas.

EDUARDO A. GROISMAN works in the Washington University School of Medicine, St Louis, Missouri.

E. HAECKEL (1834–1919) mainly worked at Jena, Germany. His books included *Die Radiolarien*, *Das System der Medusen*, *Generelle Morphologie der Organismen*, and *Anthropogenie*.

J. B. S. HALDANE (1892–1964) worked in Oxford University, Cambridge University, and University College London, and then in India. His books include *The Causes of Evolution* and many volumes of popular science (an anthology of which is *On Being the Right Size and Other Essays* (ed. J. Maynard Smith), Oxford University Press paperback).

ELLEN E. HOSTERT is in the Biology Board of Studies, University of California, Santa Cruz.

DAVID HUME (1711–76) lived mainly in Edinburgh, Scotland. His books include *A Treatise of Human Nature*.

THOMAS HENRY HUXLEY (1825–95) worked at the Government School of Mines, London. His books included *Evidence as to Man's Place in Nature*, *Lessons in Elementary Physiology*, and *Collected Essays*.

D. H. JANZEN is a Professor of Biology at the University of Pennsylvania, and also works at the Instituto Nacional de Biodiversidad, Costa Rica.

TOBY JOHNSON works in the Institute of Cell, Animal, and Population Biology, University of Edinburgh, Scotland.

MARY N. KARN worked in University College London at the time of the work included here.

H. B. D. KETTLEWELL (1907–79) had a medical career until 1949, working in various hospitals and in general practice. In 1952 he began evolutionary research in the

Department of Zoology, Oxford University. His books include *The Evolution of Melanism*.

MOTOO KIMURA (1924–94) worked in the National Institute of Genetics, Mishima, Japan. His books included *The Neutral Theory of Molecular Evolution* and (with J. F. Crow) *An Introduction to Population Genetics Theory*.

MARY-CLAIRE KING works in the Department of Genome Science and Medicine, University of Washington, Seattle.

MARK KIRSCHNER works at Harvard Medical School, Cambridge, Massachusetts.

WILTON M. KROGMAN was Professor of Physical Anthropology in the University of Pittsburgh, Pennsylvania.

JEFFREY G. LAWRENCE works in the Department of Biological Sciences, University of Pittsburgh, Pennsylvania.

R. C. LEWONTIN is Alexander Agassiz Professor of Biology at Harvard University. His books include *The Genetical Basis of Evolutionary Change*.

FRANK B. LIVINGSTONE is Emeritus Professor of Anthropology at the University of Michigan, Ann Arbor. His books include *Abnormal Hemoglobins in Human Populations*.

JOHN MAYNARD SMITH is Professor of Biology at the University of Sussex. His books include *The Theory of Evolution*, *The Evolution of Sex*, *Evolutionary Genetics*, and (with E. Szathmáry) *The Major Transitions of Evolution*.

E. MAYR is Emeritus Professor of Biology at Harvard University. His books include *Systematics and the Origin of Species*, *Animal Species and Evolution*, and *Evolution and the Diversity of Life*.

P. B. MEDAWAR (1915–87) was Professor of Zoology at Birmingham University and then University College London, and then Director of the National Institute for Medical Research, Mill Hill, London. He shared the Nobel Prize, for his work on immunological tolerance, in 1960. His books include *Pluto's Republic* and *Advice to a Young Scientist*.

J. L. MONOD (1910–76) worked mainly at the Institut Pasteur, Paris, France. He shared the Nobel Prize in 1965 for his work on gene regulation. His books include *Le Hasard et la Nécessité* (translated as *Chance and Necessity*).

H. J. MULLER (1890–1967) worked in universities in Texas (Rice University), Berlin, Leningrad (now St Petersburg), Moscow, Edinburgh, Amherst (Massachusetts), Bloomington (Indiana), and elsewhere. He was awarded the Nobel Prize in 1946 for his work on genetics, particularly on mutation. His mainly wrote papers, but he wrote one popular book, *Out of the Night*.

DAN-E. NILSSON works in the Department of Zoology, Lund University, Sweden.

HOWARD OCHMAN works in the Department of Ecology and Evolutionary Biology, University of Arizona, Tucson.

H. ALLEN ORR is a Professor in the Department of Biology, University of Rochester, Rochester, New York.

STEPHEN R. PALUMBI is a Professor in the Department of Biological Sciences, Stanford University. He is the author of *The Evolution Explosion: How Humans Cause Rapid Evolutionary Change*.

SUSANNE PELGER works in the Department of Genetics, Lund University, Sweden.

L. S. PENROSE (1898–1972) was Professor of Genetics at University College London.

STEVEN PINKER is Professor in the Center for Cognitive Neuroscience, Massachusetts Institute of Technology, Cambridge, Mass. His books include *The Language Instinct*, *How the Mind Works*, and *The Blank Slate*.

R. A. RAFF is Professor of Biology at Indiana University. His books include *The Shape of Life*.

H. K. REEVE is in the Section of Ecology and Systematics, Cornell University, New York.

WILLIAM R. RICE is in the Biology Board of Studies, University of California, Santa Cruz.

VINCENT M. SARICH is Professor in the Department of Anthropology, University of California, Berkeley.

DOLPH SCHLUTER works in the Department of Biology, University of British Columbia, Vancouver, Canada.

J. WILLIAM SCHOPF works in the Center for the Study of Evolution and the Origin of Life, University of California, Los Angeles. His books include *Early Life*.

AARON SHAVER works in the Department of Biology, University of Pennsylvania.

P. W. SHERMAN is in the Section of Neurobiology and Behavior, Cornell University, New York.

PAUL D. SNIEGOWSKI works in the Department of Biology, University of Pennsylvania.

EÖRS SZATHMÁRY works in the Institute for Advanced Study, Budapest, Hungary.

L. TERRENATO works in the Dipartimento di Biologia, Università 'Tor Vergata', Roma, Italy.

L. ULIZZI works in the Dipartimento di Genetica e Biologia Molecolare, Università 'La Sapienza', Roma, Italy.

TODD J. VISION, DANIEL G. BROWN, and STEVEN D. TANKSLEY worked at Cornell University, Ithaca, New York at the time they wrote the paper extracted here.

G. C. WILLIAMS is Emeritus Professor in the Department of Ecology and Evolution, State University of New York, Stony Brook, New York. His books include *Adaptation and Natural Selection, Sex and Evolution, and Natural Selection*, and *Plan and Purpose in Nature*.

ALLAN C. WILSON (1934–91) worked mainly at the University of California, Berkeley.

SEWALL WRIGHT (1889–1988) worked at the University of Chicago. His books include the four-volume treatise *Evolution and Genetics of Populations*.

Acknowledgements

The chapters in the anthology were extracted from the following sources, which are listed in the same order as that of the chapters.

A. From Darwin to the modern synthesis

Darwin, C., extract from an unpublished work on species, and 'Abstract of a letter from C. Darwin, Esq., to Prof. Asa Gray, Boston, 1857', in *Journal of the Proceedings of the Linnean Society (Zoology)*, 3 (1859), 45–62, reproduced by kind permission of The Linnean Society of London.

Maynard Smith, J., 'Weismann and modern biology', *Oxford Surveys in Evolutionary Biology*, 6 (1989), 1–12. Oxford University Press, 1989.

Fisher, R. A., *The Genetical Theory of Natural Selection*, Oxford University Press, Oxford, 1930.

Wright, S., 'The roles of mutation, inbreeding, crossbreeding, and selection in evolution', *Proceedings of the VI International Congress of Genetics*, 1 (1932), 356–66.

Haldane, J. B. S., 'Disease and evolution', *La ricercha scientifica*, 19, supplement (1949), 68–76 (Associazione Pedagogica Italiana, Bologna).

B. Natural selection and random drift in populations

Kettlewell, H. B. D. (1956). 'A résumé of investigations of the evolution of melanism in the Lepidoptera', *Proceedings of the Royal Society of London*, series B, 145 (1946), 297–303. © The Royal Society, London.

Cook, L. M., Dennis, R. L. H., and Mani, G. S., 'Melanic morph frequency in the peppered moth in the Manchester area', *Proceedings of the Royal Society of London*, series B, 266 (1999), 293–7. © The Royal Society, London.

Karn, M. N. and Penrose, L. S., 'Birth weight and gestation time in relation to maternal age, parity, and infant survival', *Annals of Eugenics*, 16 (1951), 147–64. Cambridge University Press, 1951.

Ulizzi, L. and Terrenato, L., 'Natural selection associated with birth weight. VI. Towards the end of the stabilizing component', *Annals of Human Genetics*, 56 (1992), 113–18. Cambridge University Press, 1992.

Gibbs, H. L. and Grant, P. R., 'Oscillating selection in Darwin's finches'. *Nature*, 327 (1987), 511–13.

Lewontin, R. C., (1974). *The Genetic Basis of Evolutionary Change*, Copyright © 1974 by Columbia University Press. Reprinted with permission of the publisher.

Kimura, M., 'Recent developments of the neutral theory viewed from the Wrightian tradition of theoretical population genetics', *Proceedings of the National Academy of Sciences, USA*, 88 (1991), 5969–73, reproduced by permission of the National Academy of Sciences, Washington, DC.

C. Adaptation

Fisher, R. A., *The Genetical Theory of Natural Selection*, Oxford University Press, Oxford, 1930.

Williams, G. C., *Adaptation and Natural Selection*, Princeton University Press, Princeton, 1966.

Grafen, A., 'On the uses of data on lifetime reproductive success', in T. H. Clutton-Brock (ed.), *Reproductive Success*, 454–71, University of Chicago Press, 1988.

Reeve, H. K. and Sherman, P. W., 'Adaptation and the goals of evolutionary research', *Quarterly Review of Biology*, 68 (1993), 1–32, University of Chicago Press, 1993.

Orr, H. A. and Coyne, J. A., 'The genetics of adaptation: a reassessment', *American Naturalist*, 140 (1992), 725–42, University of Chicago Press, 1992.

Cain, A. J., 'The perfection of animals', in J. D. Carthy and C. L. Duddington (eds.), *Viewpoints in Biology*, Vol. 3, 36–63. Butterworth & Co., London, 1964.

Gould, S. J. and Lewontin, R. C., 'The spandrels of San Marco and the panglossian paradigm: a critique of the adaptationist program', *Proceedings of the Royal Society of London*, series B, 205 (1979), 581–98, © The Royal Society, London.

Dawkins, R., *The selfish gene*, Oxford University Press, Oxford, 1976.

D. Speciation and biodiversity

Mayr, E., (1976). *Evolution and the Diversity of Life*, Chapter 3, 26–9. Reprinted by permission of the publisher from *Evolution and the Diversity of Life* by Ernest Mayr, Harvard University Press, Cambridge, Mass. Copyright © 1976 by the Presidents and Fellows of Harvard College. (This book is an anthology of previously published papers; the one used in Chapter 22 was first published in 1958.)

Mayr, E. (1963). *Animal Species and Evolution*. Reprinted by permission of the publisher from *Animal Species and Evolution* by Ernest Mayr, Harvard University Press, Cambridge, Mass. Copyright © 1963 by the Presidents and Fellows of Harvard College.

Darwin, C., *The Origin of Species*, John Murray, London, 1859.

Dobzhansky, T., *The Genetic Basis of Evolutionary Change*, Columbia University Press, New York, 1970.

Rice, W. R. and Hostert, E. E., 'Laboratory experiments on speciation: what have we learned in 40 years?' *Evolution*, 47 (1993), 1637–53, Society for the Study of Evolution, Allen Press, St Louis, Mo., 1993.

Coyne, J. A. and Orr, H. A., 'The evolutionary genetics of speciation', in R. S. Singh and C. Krimbas (eds.), *Evolutionary Genetics*, Cambridge University Press, New York, 2000. (The original paper was published in 1998.)

Schluter, D., *The Ecology of Adaptive Radiation*. Oxford University Press, Oxford, 2000.

Grant, V., *Plant Speciation*, 2nd edn., 1981. Copyright © 1974 by Columbia University Press. Reprinted with permission of the publisher.

E. Macroevolution

Erwin, D. H. and Anstey, R. L., 'Speciation in the fossil record', in D. H. Erwin and R. L. Anstey (eds.), *New Approaches to Speciation in the Fossil Record*, 11–38, Columbia University Press, New York, 1995.

De Beer, G. R., *Homology: an Unsolved Problem*, Oxford Biology Readers, Oxford University Press, Oxford, 1971.

Dawkins, R., *Climbing Mount Improbable*, Viking Press, London, 1996.

Dickinson, W. J., 'Molecules and morphology: where's the homology?' *Trends in Genetics*, 11 (1995), 119–20, with permission from Elsevier Science Ltd., The Boulevard, Langford Lane, Kidlington, OX5 1GB.

Haeckel, E., *Evolution of Man*, 2 vols., Watts, London, 1905.

Garstang, W., *Larval Forms*, Basil Blackwell, Oxford, 1951.

F. Evolutionary genomics

Ochman, H., Lawrence, J. G., and Groisman, E. A., 'Lateral gene transfer and the nature of bacterial innovation', *Nature*, 405 (2000), 299–304.

Vision, T. J., Brown, D. G., and Tanksley, S. D., 'The origins of genomic duplications in *Arabidopsis*', *Science*, 290 (2000), 2114–16.

International Human Genome Sequencing Consortium, 'Initial sequencing and analysis of the human genome', *Nature* 409 (2001), 860–921.

Carroll, S. B., 'Genetics and the making of *Homo sapiens*', *Nature*, 422 (2003), 849–57.

Raff, R. A., *The Shape of Life*, University of Chicago Press, Chicago, 1996.

Benner, S. A., Caraco, M. D., Thomson, J. M., and Gaucher, E. A., 'Planetary biology—paleontological, geological, and molecular histories of life', *Science*, 296 (2002), 864–8.

G. The history of life

Maynard Smith, J. and Szathmáry, E., *The Origins of Life*. Oxford University Press, Oxford, 1999.

Schopf, J. W., 'Disparate rates, differing fates: tempo and mode of evolution changed from the Precambrian to the Phanerozoic', *Proceedings of the National Academy of Sciences, USA*, 91 (1994), 6735–42. The National Academy of Sciences, Washington, DC, 1994.

Cooper, A. and Fortey, R., 'Evolutionary explosions and the phylogenetic fuse', *Trends in Ecology and Evolution*, 13 (1998), 151–6.

Dilcher, D., 'Toward a new synthesis: major evolutonary trends in the angiosperm fossil record', *Proceedings of the National Academy of Sciences, USA*, 97 (2000), 7030–6. The National Academy of Sciences, Washington, DC, 2000.

H. Case studies

Medawar, P. B., *An Unsolved Problem of Biology*, H. K. Lewis, London, 1952.

Crick, F. H. C., 'The origin of the genetic code'. *Journal of Molecular Biology*, 38 (1968), 367–79. Reprinted by permission of the publishers Academic Press Limited, London, and the author.

Maynard Smith, J., 'The origin and maintenance of sex', in G. C. Williams (ed.), *Group Selection*, Aldine Publishers Inc., Chicago, 1971.

Janzen, D. H., 'A caricature of seed dispersal by vertebrate guts', in D. Futuyma and M. Slatkin (eds.), *Coevolution*, 232–62, Sinauer Associates Inc., Sunderland, Mass., 1983.

Nilsson, D-E. and Pelger, S., 'A pessimistic estimate of the time required for an eye to evolve', *Proceedings of the Royal Society of London*, series B, 256 (1994), 53–8. © The Royal Society, London.

Gerhart, J. and Kirschner, M., *Cells, Embryos, and Evolution*, Blackwell Science, Boston, Mass., 1997.

Sniegowski, P. D., Gerrish, P. J., Johnson, T., and Shaver, A., 'The evolution of mutation rates: separating causes from consequences', *Bioessays* 22 (2000), 1057–66.

I. Human evolution

Sarich, V. and Wilson, A. C., 'Immunological timescale for human evolution', *Science*, 158 (1967), 1200–3, AAAS, Washington, DC, 1967.

King, M-C. and Wilson, A. C., 'Evolution at two levels in humans and chimpanzees', *Science*, 188 (1975), 107–15.

Britten, R. J., 'Divergence between samples of chimpanzee and human DNA sequences is 5%, counting indels', *Proceedings of the National Academy of Sciences, USA*, 99 (2002), 13633–5. The National Academy of Sciences, Washington, DC, 2000.

Muller, H. J., 'Our load of mutations', *American Journal of Human Genetics*, 2 (1950), 111–76, University of Chicago Press, 1950.

Livingstone, F. B., 'On the non-existence of human races', *Current Anthropology*, 3 (1962), 279–81, University of Chicago Press, 1962.

Krogman, W. M., 'The scars of human evolution', *Scientific American*, 185 (Dec. 1951), 54–7. Reprinted with permission. Copyright © 1951 by Scientific American, Inc. All rights reserved.

Pinker, S., *The Language Instinct*, William Morrow Inc., New York, 1994.

J. Evolution and human affairs

Antolin, M. F. and Herbers, J. M., 'Evolution's struggle for existence in America's public schools', *Evolution*, 55 (2001), 2379–88.

Dobzhansky, T., 'Nothing in biology makes sense except in the light of evolution', *American Biology Teacher*, 35 (1973), 125–9, © The National Association of Biology Teachers Inc.

Hume, D., *Dialogues Concerning Natural Religion*, Oxford University Press, Oxford, 1993.

Monod, J., 'On the molecular theory of evolution', in R. Harré (ed.), *Problems of Scientific Revolution*, Oxford University Press, 1974.

Huxley, T. H., *Evolution and Ethics*, Macmillan, London, 1893.

Palumbi, S. R., 'Humans as the world's greatest evolutionary force', *Science*, 293 (2001), 1786–90.

Index

Abd el Aziz bin Baz 400
Adam, his navel 401
adaptation 83–130 *passim*; and ancestral characters 103–106; and architectural constraint 121–122; and by-products 361; case studies 217–218, 292–336 *passim*; evolution of 11–12; in genomes 248–249; by mutation? 26–27; gradually? 83–84; yes 85–87, 370; maybe not 96–100; examples 82; an important concept? 84; yes 100–113; no 113–123; meaning of 82, 88–94; does not contradict neutralism 73; and selection 91–93; uncommon 72; mentioned 2–3, 7, 8; *see also* design, non-adaptive evolution
adaptationist programme, its general procession 114–115; exemplified 115–117; alternatives to 120–123
adaptive surface/topography 30–31; and asexual species 146; *see also* multiple peaks
ageing, *see* senescence
albumin 341–344
alcohol metabolism 221; in yeast genome 253–254
aldehyde dehydrogenase, also a crystallin 250
allometry 120, 121
allopatric model, characterized 155, 176n; discussed 155–174 *passim*; experiments on 157–160; genetics of 161–174; *see also* punctuated equilibrium, speciation
Amphioxus 213, 219
anaemia, sickle-cell 47, 73
analogy, contrast with homology 197–198; *see also* convergence, homology
angiosperms, evolution of 259, 284–291; evolutionary genomics of 220–221, 231–236; flower homologies in 199–200
Anstey, R. L., extract from 185–197; introduced 182–183
anteaters 110
antibiotic resistance, case-study in human-induced evolution 422–433 *passim*; and lateral gene transfer 220, 228
Antolin, M. F., extract from 385–399; introduced 383
apes, and human ancestry 338–354
applied evolution 384, 421–433; *see also* Darwinian medicine
Arabidopsis, evolutionary genomics of 231–236, 239
archetype, in Owen's zoology 102, 198; Darwin's reinterpretation 103
argument from design, *see* design
Aristotle 198
armadillos 110
art, Cro-Magnon 373
artificial selection 12, 13, 44
asexual reproduction, in cyanobacteria 272; natural selection of 19–20, 307–310; and species 145–146
atheism, relation with evolution 389–390, 409
australopithecines, characteristics of 339; language in 372
Australopithecus, *see* australopithecines
axolotl 218

backache 364
backbone, human 364–366, 374
bacteria, transfer genes laterally 220, 221–231; possibly even to humans 240; species in 231; resisted by mice 40; in ruminants 255; cause disease 37, 322–433 *passim*; *see also* Cyanobacteria, *Escherichia coli*, *Salmonella*, *Staphylococcus*
bacteriophage, *see* phage
Baer, K. E. von 203
balance theory 70–71
banana clone, infected by fungus 40
Barash, P. 117–118
Benner, S., extract from 250–257; introduced 220
Bickerton, D. 379
biodiversity 131
biogenetic law, *see* recapitulation
biological species concept 138–141, 162; difficulties for 141–146; mentioned 362n; *see also* typology
bipedality, in human evolution 339; and medicine 367
birds, explosive evolution of 275, 281–283; fruit association of 290

birth weight, selection on 46, 59–63; and human evolution 367
Biston betularia 33–34, 49–56
BLAST 227, 232, 238, 351
blending inheritance 1–2, 7; Fisher on 20–26; a predicted error 415–416
blood groups, and disease resistance 41; puzzling polymorphism 69; selected 106
blood circulation, *see* circulation
Bloom, P. 378
blue-green algae 266n; *see* cyanobacteria
body plans, adaptive 101, 112–113; non-adaptive 100, 101–104
brain size, in human evolution 339, 373, 378; and medicine 367; mentioned 402
Brenner, S. 307
Britton, R., extract from 350–354; introduced 337–338
Broca's area 371
Brown, G. D., extract from 231–236; introduced 220–221
Burgess Shale 278

Caenorhabditis elegans 208–209, 236; in genomic comparisons (as "worm") 239, 240–241, 242; mutation rate of 331
Cain, A. J., extract from 100–113; introduced 84; mentioned 383
Cambrian, explosion 275, 276, 278–280
carpel 286–287
Carroll, S. B., extract from 244–249; introduced 221, 337
Caraco, M. D., extract from 250–257; introduced 220
Carson, H. L. 407
cattle dung, dead seedlings in 317
cellulose, indigestible 311–312
central dogma 417
Cepaea, adapted snails 107
Cerion, non-adapted snails 123; challenge species concepts 173
Cetacea, polyphyletic 111
character displacement 154
Charlesworth, B. 97, 173
Cheney, D. L. 372
chimpanzee, genetic similarity to humans 337–354 *passim*
chin, pendentive 114
Chinese whispers 263
choking, human 373
Chomsky, N. 369, 374, 377
chromosomal rearrangement, in genome evolution 232; in speciation 178–181
circulation, of blood, in humans 366
cladism 194, 286
Classical theory 69–72
classification, biological 3, 131; its evolutionary explanation 14–15; and subversion 113; folk 138
classificatory groups, nature of 3
clock, molecular, *see* molecular clock
coelocanth (fish) 267
coevolution, explained 289n, 293, engine of angiosperm evolution 284–291; between seeds and herbivores 310–317; and parasites and hosts 40–41
communication, non-human, and language 369
competition, ecological 11, 38–39
conformity to type 101–102, 199
conservatism, in molecular evolution 80; a sorry privilege 417
continuity, a Darwinian criterion 302, 304, 317–326; *see also* co-option
convergence, and developmental genetics 210; mammal examples 109–111; ubiquitous 111
Cook, L. M., extract from 53–56; introduced 45; mentioned 107
Cooper, A., extract from 275–284; introduced 259
co-option 249–250; *see also* continuity
Copernicanism, just a theory 400
Coyne, J. A., extracts from 96–100, 161–175; introduced 83–84, 133
creationism 383–409 *passim*; their arguments 387–394
Creator, absent-minded 408; better conceived 102; in Darwin's marriage 260; a deceitful trickster 401, 403, 409; in jocular mood 404
Crick, F. H. C., extract from 299–307; mentioned 84
crocodile 250, 267
Cro-Magnons 373
Crustacea, definition 112
crystallins, co-option of 221, 249–250; evolutionary rate of 70; *see also* lens protein
curious aspects 412–413
Cuvier, G. 198
cyanobacteria 266; ecological generalists 271–274; their evolutionary mode 274; living fossils 268–271
cytochrome C 405–406

Darwin, C., extracts from 9–15, 147–150; accepts blending 21–23; on choking 373; his contribution 134; studies parasitic crustaceans 112; threatened by eye 317–318; misunderstands homology 103–104; or understands it? 198–199; on human origins 337; his interests 91; a micromutationist 96–97; on origin of life 260; on perfection 377; a pluralist 119–120; his true predictions 414–416; on speciation 132–133, 147–151; to be surprised 70; though remarkably prescient 409; as theorist, classical 69–71; and shame-faced 20; ideas about variation 21–23, 25; Wallace's spark 7; mentioned 186, 211, 369, 384, 385
Darwin's finches 36–47, 63–67, 177, 190
Darwinian medicine 338, 363–367, 373, 384, 421–433 *passim*
dates, inference of 251–253, 276
Dawkins, R., extracts from 123–130, 205–207; introduced 3, 84, 183; on adaptation 375
DDT 422
De Beer, G., extract from 197–205; introduced 183–184
De Candolle, A-P. 9, 14
definition, of classificatory groups 112
Dennis, R. L. H., extract from 53–56-; introduced 45
density-dependence 38
dentition, *see* teeth
design, and adaptation 82, 89–90, 375–377; argument from 2, 83, 375, 383–384, 410–412; intelligent design theory 393–394
destroy the hybrids 152, 160–161
development (individual), evolution as 1; homology and 203–205, 207–210; and recapitulation 184–185, 210–216; *see also* evo-devo
Dickinson, W. J., extract from 207–211; introduced 184
dicotyledenous plants, origin of 234–235
Dilcher, D., extract from 284–291; introduced 259
dinosaurs 275, 283
directed variation, theories of 8, 25; by mutation 26–27
disease, Haldane on 37–43; and evolution 9; coevolution 39–42; ecological competition 38–39; resistance to 39–40; an antisocial force 42
displacements, *see* takeovers
distance methods, immunological 341–344
divergence, principle of 7, 14–15; *see also* speciation
Dobzhansky, T., extracts from 151–155, 400–410; introduced 133, 383; on chromosomal inversions 106; on non-adaptation 107–108; a micromutationist 96–97; on species 140; mentioned 413; *see also* Dobzhansky-Muller theory
Dobzhansky-Muller theory 133, 151, 168–171, 174, 175
domains, protein 241–242
domain shuffling, explained 209n
domestication, variation under 21, 26; and reproductive isolation 150
Down House 7
drift, *see* genetic drift
drongo 143
Drosophila, *see* fruitfly
duplication 220; identification of 232–233; dates of 233–236; in human genome 237–243 *passim* 247

E. coli, *see Escherichia coli*
ear bones 184, 200–202
East, R. 29
Earth, roundness of 400
ecological genetics 44–45, 92, 106–109, 120, evidence of power of selection 133–134, *see also* peppered moth
economic cost, of evolution 426–427
editorial procedures, of this book 4–6
education, in theory of evolution 385–399 *passim*
effect, not adaptation 89–90, 374
effective population size, *see* population size
Eigen, M. 264
Eldredge, N. 182–183, 185–197 *passim*
El Niño 47, 63–66, 177
electrophoresis 47, 71, 346–347
elephant 14; its trunk 368–369; fossils of 281
embryology, *see* development
engineering, concept of adaptation 82, 89; reverse 84; of evolution 430–431
epistasis 162; explained 162n
error rates 264; *see also* mutation rates
error threshold 263–265
Erwin, T., extract from 185–197; introduced 182–183
Escherichia coli 222, 301; genomics of 224–230 *passim*; mutator genes of 335; mutation rate of 331; phages in 332
essentialism, *see* typology

ethics, and evolution 384, 418–421
eugenics 338
Eurytoma curta, parasitizes gallfly 38
evo-devo 184, 185
evolution, concepts of 1; evidence for 392, 400–410; not a law 417; unlike a physical theory 412, 414; *see also* molecular evolution
evolutionary genomics, *see* genomics, evolutionary
experiments, on speciation 133, 152, 155–161, 164–166, 178–181
explosions, evolutonary 275
extinction, caused parasitically? 39
eye, evolution of 293, 317–326, 393; lens protein 221, 249–250; mentioned 28, 91; *see also ey*
ey (gene) 205–207, 210; mentioned 48

falsifiability, *see* Popper
females, struggled for by males 12–13
finches, Darwin's (Galápagos) 46–47, 63–66, 177, 190
fish, flying 90, 374, 376
Fisher, R. A., extracts from 20–29, 85–88; contrasted with Wright 8, 91; founder of Modern Synthesis 7, 106; and theoretical biology 20; a micromutationist 83, 85–87, 98–100; on mutation rates 333–334; on sex 19; on selection 106, 107; cross-referenced 11n
fishing, evolutionary effects of 424
Fitch, W. M. 405
fitness, meaning of 44; in peppered moth 50–52; in humans 58, 62; in Darwin's finches 64; *see also* reproductive success, survival of the fittest
fitness surface 30n
flower, evolution of 287–291; homologous with leaf 199–200
flowering plants, *see* angiosperms, plant speciation
flying fish 90, 374, 376
Ford, E. B. 93–94
Fortey, R., extract from 275–284; introduced 259
forelimb, homologies of 101–105, 199
fossils, human 339–340; microbial 269–272; plants 285–291; South American mammals 109–111; living 267–274; Precambrian 267–274; speciation in 185–197; dates of 252; concordance with molecular dates 252–257; and conflicts 276–281
frequency-dependent fitness, in disease resistance 41
frogs, genetic similarity among 347
frozen accident 292, 300
fruits, and alcohol metabolism 221, 254; origin of 290; ripening of 311–312; eating of 313–314; seed shadow of 314–317
fruit trees, difficult to perch in 314
fruitfly, competitive in lab 38; at least for now 40; and developmental genetics 209–210; and homology 205, 206–207; mutations in 27, 29, 206–207; mutation rate of 331; selection on 106; a specialized larva 402; which eat yeast 254; experimental speciation in 152, 153, 156–161, 164–165, 171–174; speciation in Hawaii 407–408; genetic similarity among sibling species 346–347; variation in 75; in genomic comparisons (as "fly") 239, 240–241, 242; *see also ey, hedgehog*
future, evolutionary 241, 321–322

Galápagos, finches in 46–47, 63–66, 177, 190
Galeopsis, hybrid speciation 179–181
Garstang, W., extract from 216–219; introduced 185
gastropods, torsion in 216–217
Gaucher, E. A., extract from 250–257; introduced 220
Gehring, W. 206–207
gel electrophoresis 47, 71, 346–347
gene, naming convention 206; number of per organism 29, 74; unit of selection 123–129; interact in development 129–130; *see also ey, hedgehog*, Homeobox
gene duplication 220; *see also* duplication
gene flow, models of speciation 156; *see also* speciation
genetic code, Crick on 292, 299–307; universal 404
genetic drift 8, 70–81 *passim* 115, 120, 121; and Hardy-Weinberg ratio 24; and speciation 156–157; Fisher on 24–25; Wright on 31–32, 35–36
genetic load, *see* load
genetic uniqueness of individual 30, 127
genetics 25; of adaptation 96–100; disease resistance 39–40; and studies of selection 44–45; and speciation 157, 162, 164–175; *see also* antibiotic resistance, heredity, lateral gene transfer, Mendelism, particulate inheritance

genome 220
genome size, limit on 265; and mutation rate 331
genomics, evolutionary 220–257; of bacteria 221–231; of flowering plants 231–236; of humans 237–249
Geoffroy Saint-Hilaire, E. 198
Gerhart, J., extract from 326–328; introduced 18, 185; and domain shuffling 241n
germ cell line 16–18; explained 16n
germ layers 203–204
germination, of seeds 316–317
Gerrish, P. J., extract from 328–336; introduced 292
Ghiselin, M. T. 131
Gibbs, H. L., extract from 63–66; introduced 46–47
Gillespie, J. H. 81
giraffe 85, 203
Glossina (tsetse fly) 39
glyptodonts 110
GM crops 425–426
God 263, 376, 400, 401, 403; *see also* Creationiam, Creator
Goldschmidt, W. 83, 97–98
Gorer, P. A., on mutable mice 41
Gosse, P. H. 401
Gould, S. J., extract from 114–123; introduced 9, 84; criticized 374; on punctuated equilibrium 182–183, 185–197 *passim*
gradualism, in adaptation 317–325, 375–377; phyletic 191–196; *see also* continuity
Grafen, A., extract from 91–94; introduced 82–83
grammar, universal 377
Grant, P. R., extract from 63–66; introduced 46–47
Grant, V., extract from 178–181; introduced 133–134; studies speciation 152, 154
grass, proliferation of 255
Gray, A. 13
Great Chain of Being 1
Gregory, W. K. 364
Groisman, E. A., extract from 221–231; introduced 220
group selection, and evolutionary biology 2–3; false 309; Panglossist? 9, 41, 309; and sex 19, 309; in Wright's theory 35–37

Haeckel, E., extract from 211–216; introduced 184–185; mentioned 407
haemoglobin, variants of 72; evolution of 76, 405–406; sickle cell 47, 73
Haldane, J. B. S., extracts from 37–43; a founder of Modern Synthesis 7, 8, 106; and of theoretical biology 20; on angels 374; on Dr Pangloss 9, 41, 309; on origin of life 260; on peppered moth 54; his rule 171–174; on selection intensity 60n 61
Haldane's Rule 171–174
Hamilton, W. D. 3, 94
Hardy-Weinberg ratio, interest of 24
Hardy, D. E. 407
Harris, H. 72
Hawaii, its fruitflies 176, 190, 407–408; and snails 35
heat shock protein, also a cystallin 250
hedgehog (gene) 209
hemorrhoids 366
Herbers, J. M., extract from 385–399; introduced 383
Herbert, W. 13, 178
herbicides, resistance to 422–433 *passim*
heredity, blending and particulate 1–2; Fisher on 20–26; Darwin needs a theory of 1, 70; Mendelian 2, 20–72; *see also* genetics, inheritance of acquired characters
heritability 323
hermaphroditism, in flowers 289; and in theory 19, 308–309
hernia 366
heterozygosity 68
heterozygous advantage, in sickle-cell haemoglobin 47, 73; in theory 68
high schools, evolution in 385–399 *passim*
higher animals 1
higher taxa (above species level), nature of 3, 112–113
history, and adaptation 94; of life 258–291
hitch-hiking, explained 160n; in speciation 157–160 *passim*; and mutation rates 333–335
HIV 424, 427, 431
holly berries 317
homeobox genes 208, 210
hominids, and hominoids 339
hominoids, and hominids 339
Homo erectus 339, 340, 372
Homo habilis 340
Homo sapiens, see human beings
homology, abused by molecular biologists 184, 207, 224; edifying 102; evidence of evolution 406–407; important but practically elusive 4, 183; level of reference

207–211; origin of concept 197–199; in albumins 340; plants 199–200; in vertebrates 200–203; universal 404; relation with adaptation (design) 101–104; with embryology 203–205; with genetics 205; with genomics 230; and developmental genetics 207–211; *see also* cladism, convergence, orthology, paralogy, phylogeny
horizontal gene transfer, *see* lateral gene transfer
horses, convergence in 111; hybrids of 149
Hostert, E. E., extract from 155–161; introduced 133
Hox genes 185, 208
Hull, D. L. 157
human beings, existential contingency of 418; genetic evolution of 244–247; mutation rate of 331; origin of 337–340, 340–345; races of 136, 338, 361–363; recapitulation in 214–215; relaxed selection in 59–62, 338, 354–360; selection in 46, 47–62; mentioned 4; *see also* blood groups, language
human genome, evolution of 221, 237–239; similarity to chimpanzee's 244–249, 337–354 *passim*
Hume, D., extract from 410–412; introduced 383–384; mentioned 2, 83
Huntingdon's disease (chorea) 297–298
Huxley, J. S. 97
Huxley, T. H., extract from 418–421; introduced 384; letter from Darwin 21; review of Darwin 96
hybrid fitness, by inviability and sterility 147–150; ecology of 177–178; genetics of 163–175; *see also* postzygotic isolation
hybrid speciation 133–134, 178–181
hypobradytely 267–274 *passim*

immunological distance, in phylogenetic inference 337, 340–345
immunoglobulin genes 242
inbreeding, and drift 32, 34–35; depression 19
indels 350–354; explained 353n
individualism, fanatical 420
industrial melanism 43–34, 49–56
infectious disease, *see* disease
influenza A virus 79
information analogy 8, 18, 416, 417
inheritance, *see* genetics, heredity
inheritance of acquired characters, rejected by Mendelians 1, 7, 8, 403, 415, 416–417; Fisher on 26; Weismann on 15, 16–18
insertions 236; *see also* indels
integrated pest management 430
intelligent design creationism 393–394
intensity of selection 34, 60–61, 323
interbreeding, and species 3, 140; *see also* biological species concept, reproductive isolation
International human genome sequencing consortium, extract from 237–244; introduced 221
introns 77
inversions 244
isolation, *see* biological species concept, reproductive isolation

Jacob, F. 18
Janzen, D. H., extract from 310–317; introduced 293
jaw shape, evolution of, *see* chin, ear bones
Jenkin, F. 415; *see also* blending inheritance
Johanson, D. C. 339
Johnson, T., extract from 328–336; introduced 292
jumping genes 224n

Karn, M. N., extract from 57–59; followed up 59–62; introduced 46
Kelvin, Lord 414–415
Kettlewell, H. B, extract from 49–53; followed up 47–48, 53–56; introduced 44–45
Kimura, M., extract from 76–84; introduced 47–48; on Fisher's argument 99; on mutation rates 331–332; his irrelevance 92; mentioned 338
King, M-C., extract from 345–350; introduced 337; mentioned 185
Kirschner, M., extract from 326–328; introduced 18, 185; and domain shuffling 241n
kittiwake 108
knapweed 38
knapweed gallfly 38
Kondrashov, A. 19, 356n
Krogman, W. M., extract from 363–368; introduced 338

LRS, *see* reproductive success
lactate dehydrogenase, also a crystallin 250
lactose synthetase 326–328

lactation 3, 198, 326
Lamarck, J. B. 217, 417
Lamarckian inheritance, Lamarckism, *see* inheritance of acquired characters
lancelet, *see Amphioxus*
Lande, R. 121
language, human, evolution of 338–339, 368–382; mentioned 4, 258
La Plata, swarming with Malthusian mice 10
Larson, Gary 370
larval adaptation 215–216, 217
laryngeal nerve, recurrent 202–203
lateral gene transfer 220, 221; amounts of 225–227; detection of 223–225; effects of 230–231; and mutation rates 336; rate of 230; sequences that go in for it 227–228; traits acquired by 228, 230; and antibiotic resistance 228, 425; in human genomics 238–239
Lawrence, J. G., extract from 221–231; introduced 220
leaf, homologies of 199–200
Leakey family 339
lens protein, evolutionary rate of 79; co-opted 221, 249–250; mentioned 49
leptin 256
Lewontin, R. C., extracts from 67–76, 114–123; introduced 47, 49, 84; criticized 374–375
Lieberman, P. 370, 373–374
limbs, Owen on 101–103
Linnaeus, C. 112, 139, 178
Linnaean classification, *see* classification
Litopterna 39
living fossils 267–274
Livingstone, F. B., extract from 361–363; introduced 337
load, genetic, calculated 75; increasing 354–360; intolerable 79
lower animals 1
Lucy, huminoid fossil 339, 372
Lyell, C. 13
lysozyme, in ruminants 255

Macaulay, Lord, his agony 378
MacBride, E. W. 218
macroevolution, relation with microevolution 4, 182, 188, 392–393
macromutation, and genetics 425–426; and language 379–380; in human evolution 245; in theory 85–87, 98–100
Majerus, M. E. N. 45, 56
malaria 47–73
males, struggle for females 12–23
Malthus, T. 7, 9–10, 357
mammals, classified 3, 326; their ear bones 200–202; and milk 326; explosive evolution of 275, 281–283; South American 109–111
Mani, G. S., extract from 53–56; introduced 45
Maniola jurtina 93
Manton, S. M. 109, 112–113
Margoliash, E. 405
mark-release experiment 50–51
marsupials 109–111
Marxists, happy 414
maternal-foetal incompatibility 41–42, 69
Mather, K. 161
Mattiessen's ratio 320
Maynard Smith, J., extracts from 15–20, 259–265, 307–310; introduced 8, 258, 292; mentioned 84, 100, 375
Mayr, E., extracts from 134–137, 137–147; introduced 131; micromutationist 97; population thinker 3; on speciation 183, 188; on species 337, 348, 383; mentioned 375, 413
meaningful mutations *see* non-synonymous mutations
Medawar, P. B., extract from 293–299; introduced 292; mentioned 84, 378, 383
medicine, Darwinian 338, 363–367, 373, 384, 421–433 *passim*
melanism, industrial 33–34, 49–56
Mendelism, history of 2, 8, 36, 391; contribution to Darwinism 70–71, 126–127, 415; *see also* blending inheritance
Metazoa, phylogeny of 279
mice, Malthusian swarm of 10; resist bacteria but not cats 40; mutation rates of 331
microevolution, relation with macroevolution 4, 182, 188, 392–393
micromutation 83–84, 96–100
microscope, Fisher's analogy 87
milk 326
Miller, S. 260
misseltoe 14
Modern Synthesis 2, 8, 183, 416; *see also* neo-Darwinism
mole rat, blind 79
molecular biology, and evolution 205–211, 384, 412–418; *see also* genomics
molecular clock, introduced 47, 48, 75–76, 221; in influenza A virus 79; how constant? 80–81; unlike phenotypic evolution 80; in phylogenetic inference 48, 220–221; including hominids 337, 340–345; in

genomics 234, 244, 252–266; and times of radiations 276–281
molecular evolution 67–81 *passim* 220–257 *passim*; *see also* genomics
molluscs, defined 112; *see also* gastropods
monkeys, Old World 248, 341–342
monocot-dicot divergence 234–235
Monod, J., extract from 412–418; introduced 2, 384
morphospecies 186
Muller, H. J., extract from 354–360; introduced 338; on sex 19; a classical theorist 47, 72; a micromutationist 97; on speciation 151–152 *see also* Dobzhansky-Muller theory; mentioned 46, 79
multiple peaks, ignored 99; not ignored 123; inevitable 30–31; shifts among 33–35
mustard weed, *see Arabidopsis*
mutability, of resistance genes 41
mutation, deleterious 296; accumulate in modern humans 354–360; action postponed 297–298; and error threshold 263–265; loads of 79, 355; not minimized by genetic code 306; mutation-selection balance 30, 71
mutation, and evolution, in proteins 405–406; as cause of evolution 26–27, 29; directed at DNA level? 26–27, 29; role in evolution 33
mutation, molecular nature of 416; kinds of, in DNA 48
mutation, neutral 72; *see also* neoclassical therory, neutralism
mutation rates, evolution of 292, 328–336; indirect selection on 330; under theory of blending 21–23; and particulate inheritance 25–26; observed 27–28; total rate of 78, 331

Natura non facit saltum 13, 953
natural selection, research on 2, 44–47, 49–67; on haemoglobin *S* 47, 73; on human birth weight 46, 59–63; on fruitflies 106; Galápagos finches 46–47, 63–66, 177; peppered moths 33–35, 49–56; on eye 317–326; moulds senescence 292, 293–299; and sex? 19, 307–310; at work in human genome 248–249; acts on genes 123–130; and adaptation 83, 91–93; and antibiotic resistance 422–433 *passim*; compared with artificial selection 12; Darwin's account of 9–15; evolutionarily important? 8; morals of 384, 418–421; and nature of evolution 1; and molecular evolution 1, 67–81 *passim*; interacts with other factors 32–35; mutation-selection balance 30, 69–71; no plan in 403; influences variation 67–68, *see also* polymorphism; relaxed 59–62, 338, 354–360; units of 2, 3, 123–129; *see also* group selection, units of selection
Neanderthals, discovery of 339; language in 373–374
nematode, *see Caenorhabditis elegans*
neoclassical theory 47, 71–74
neo-Darwinism 8, 76, 81, 96–97; *see also* Modern Synthesis
neoteny 218–219
Nesse, R. M. 338
neutral drift, *see* genetic drift
neutral evolution, rate of 74–77; *see also* rates of evolution
neutralism 67–81 *passim*; an unfortunate term? 72; explained 72–74, 76–77; irrelevant 92
neutral theory *see* neutralism
New Guinea, native ornithologists 138
New Systematics 3
Nicotiana (tobacco plants), amphiploid speciation 179; self-sterile genotypes 41
Nilsson, D-E., extract from 317–326; introduced 293
non-adaptive evolution, edifying 102; relegated 115; in speciation 35–36, 42, 156–157
non-synonymous, mutations 7, 77
Noor, M. 167
Notoungulata 39
novelty, evolutionary 326–328

Ochman, H., extract from 221–231; introduced 220
Ohta, T. 78
Old World monkeys 248, 341–342
olfactory genes 243
Oligocene cooling 255
ontogeny, relation with phylogeny 226–227, 247–252, 377; *see also* development, evo-devo
Onychophora 113
open reading frame, *see* ORF
opossum, evolutionary slow-coach 267
optics, of eye 319–320
ORF 225, 232, 243
Orgel, L. 264, 303
Orr, H. A., extracts from 96–100, 161–175; introduced 83–84, 133
orthogenesis, Wright on 37

orthology, explained 240n; identified 240–241, 247
ostracod 19
Owen, R., misleads Darwin 101–103; introduces homology 197–198

Paleolithic, *see* Upper Paleolithic
Paley, W., his watch 375, 383–384
Palumbi, S., extract from 421–433; introduced 384; noted 394
Panaxia dominula 106
Pangloss, Dr, a versatile stick 9; a group selectionist 41, 309; or adaptationist? 115, 117
panmictic population 362
paralogy, explained 240n; form of homology 209; in human genome 240–241; in yeast 253
parasites, limit host populations 38, 39; degenerate 112
parsimony 405–406
parthenogenesis 19, 308; *see also* asexual reproduction
particulate inheritance 24–26, 126–127; *see also* blending inheritance, genetics, heredity
Pascal, B. 420
Pelger, S., extract from 317–326; introduced 293
pelvis, in human evolution 364–365, 367; *see also* birth weight
Penrose, L. S., extract from 57–59; followed up 59–62; introduced 46
peppered moth 44–45, 49–56
perfection 100–114, 377; *see also* adaptation
Peripatus, adapted 113; has Cambrian relatives 278
pesticide, *see* herbicide
phage, and bacterial genomics 224; and mutation rates 331, 332
Phanerozoic, evolutionary mode of 258–259, 266, 274
philosophy, and evolution 4, 384, 414, 417
phylogenetic fuse 259
phyletic gradualism 191–196; *see also* gradualism
phylogenetic inertia 114
phylogeny, named 211; inference of, the ambition 258; the problem 205; and the principles 200–201, 207–208; embryological evidence 185, 211–215; fossil evidence, *see* fossils; genomic evidence 220; and molecular evidence 6, 383; of birds and mammals 282; of humans 344; of Metazoa 279; *see also* human beings, molecular clock
physics, bungles 414, 415; optical 319–320; theories in 412, 414
Pinker, S., extract from 368–382; introduced 338–339; mentioned 258
pizza, *see* primitive pizza
plant disease 39–40
plant evolution, *see* angiosperms, flowering plants
plant speciation 148–150
Plato 135, 138
pleiotropy 121, 157, 168
politics, and evolution 4
pollinators, cause reproductive isolation 167
polymorphism, favoured by disease 42; of sickle-cell haemoglobin 47, 73; of peppered moth 52; theory of 67–75; mentioned 89
polyploidy, and speciation 143, 168n, 179–181
Popper, K. 384, 414, 417
population control 9–11, 37–38
population thinking 3, 131, 134–137
postmating isolation 167–168
postzygotic isolation 148n, 153; explained 156n; ecological theory of 175–178; genetic theory of 133, 151, 168–171; and Haldane's rule 171–174; *see also* hybrid fitness, reproductive isolation
Precambrian, evolutionary mode 265–274; fossil evolution in 278–280
Premack, D. 357–358
prezygotic isolation 148n; explained 156n; experiments on 159; *see also* reinforcement, reproductive isolation
primitive pizza 261–262
primitive soup 260–261
Primula (primrose), hybrid speciation 180; semiglabrous mutant 40
principle of divergence 7, 14–15
progress, evolutionary 1, 8
proteome, human 237–244, 248
Provine, W. B. 76n, 81
pseudogenes 78; olfactory 243
psychology, evolutionary 339
punctuated equilibrium 183, 185–197

quantitative trait loci (QTL) 84

race 36, 136, 337, 361–363
radiation, evolutionary 265, 275, 277–284 *passim* 289

radioisotope dating 251
Raff, R. A., extract from 249–250; introduced 221
random processes in evolution 8; *see also* genetic drift, neutralism
rates of evolution, of eye 317–326; neoclassical 74; neutral 77–78; rodents v. primates 78–79; viral 79; during fossil speciation 185–197
Ray, J. 139
recapitulation 184–185, 211–216, 218, 407
recombination, generates variation 67–68
recurrent laryngeal nerve 202–203
Reeve, H. K., extract from 93–94; introduced 82–83
regulatory genes, and evolution 185, 337, 345, 349–350
reinforcement 132–133; experiments on 152, 154, 160–161; Darwin implicitly critical of 147, 149; Dobzhansky supportive of 153–155; and Coyne & Orr 166–167; *see also* speciation
relative growth, *see* allometry
relaxed selection 59–62, 338, 354–360
religion, and design 383–384, 388–390, *see also* design; and evolution 385–410 *passim*
replication, origin of 262–263
reproductive community 3, 140
reproductive isolation 140, 144–145, 156–161; *see also* postzygotic isolatin, prezygotic isolation, speciation
reproductive success 82–84, 94–96; *see also* fitness
resistance, to disease 40; and antibiotics 422–433 *passim*
reverse engineering 84
reverse transcription 417
reverse translation 17–18
rhesus blood group 41–42
ribonuclease, in ruminants 255–256
Rice, W. M., extract from 155–161; introduced 133
RNA world 262–263
RNA viruses 79; mutation rates of 331; rapid evolution of 424; *see also* HIV
Robertson, A. 93, 153
rock thrush, dancing, of Guiana 12
rodents 79, associated with fruit 290
Roux, W. 16
ruminants 255
rust (fungal) 39
sabre-tooted tigers, convergent 110; not to be played with 377
Saccharomyces cerevisiae, *see* yeast
St Vitus' dance 297
Salmonella enterica 222, 224, 225, 226, 229, 231
sampling drift, *see* genetic drift
Sarich, V., extract from 340–345; introduced 337
Schluter, D., extract from 175–178; introduced 133
schools, and evolution 385–399 *passim*
Schopf, J. W., extract from 265–274; introduced 258–259
Schuster, P. 264
science, school 394–399
secondary theorem of natural selection 93
seeds, coevolve with animal guts 310–317
selection, natural *see* natural selection
selection intensity 60n
senescence 3, 84, 292, 293–299
sex 3, 19, 84, 292, 307–310
sexual selection 12–13; and genomics 247; and speciation 168
Seyfarth, R. M. 372
Shaver, A., extract from 328–336; introduced 292
Shaw, G. B. 15
Sherman, P. W., extract from 93–94; introduced 82–83
sibling species 346–347
sickle cell anaemia 47; a tired old Bucephalus 73
silent mutations, *see* synonymous mutations
Simpson, G. G., in Modern Synthesis 184; on South American mammals 109–111; on species 142; on tempo and mode 266–267, 271–272, 274, 275, 276, 285; confused 99; a great author 413; on human evolution 348
sloths 110
Sniegowski, P., extract from 328–336; introduced 292; noted 265
Spalax ehrenbergi 79
speciation 132–134, 147–181; by-product theory of 132, 151; biogeographic evidence 153–154; Darwin on 147, 149; by reinforcement 132–133, 152–155, 166–167; chromosomes and 152, 168n, 178–181; experiments on 152, 154, 155–161, 164–166, 178–181; adaptive, experimental evidence 157–159; non-adaptive, experimentally uncertain 156–157; Wright's theory 35–36; and Haldane's 42; genetics of 162–175; in plants 153, 178–181; *see also* allopatric speciation, punctuated

equilibrium, reproductive isolation, reinforcement, sympatric speciation
species 3, 114–147; bacterial 231; biological 131, 137–147; important 146–147; non-dimensional 138–139; not arbitrary 138–139; typological 140–141
Spencer, H. T. 137, 384, 389, 413, 418n
Spieth, H. T. 407
spine, *see* backbone
stabilizing selection 46, 57–62, 66
Staphylococcus 427
Stanley, S. M., on very slow evolution 274; and very fast 275
stasis 188n
Stebbins, G. L. 284–285
struggle for existence 10, 14, 39, 41, 419–420
Sturtevant, A. H. 329
subspecies 35, 348
survival of the fittest 137, 389, 419
sympatric speciation 176
synonymous, mutations 77, 252
Synthesis, Modern, *see* Modern Synthesis
Synthetic Theory of Evolution, *see* Modern Synthesis
systematics, *see* classification, species
Szathmáry, E., extract from 259–265; introduced 258

Tahiti, its snails 35
Tansley, S. M., extract from 231–236; introduced 220–221
taxa, *see* classificatory groups
taxonomy, *see* classification
teachers 397
teaching, of evolution 385–399
teeth, Darwinian medicine of 367
Teilhard de Chardin, P. 383, 409
Templeton, A. R. 166
Terrenato, L., extract from 59–62; introduced 46
theoretical biology 20
Thompson, D'Arcy 121
Thomson, J. M., extract from 250–257; introduced 220
Thomson, W. 119–120
thrush species 139
Tinbergen, N. 51
torsion 217–218
Tragopogon, hybrid speciation 179
transitions and transversions 252n
translocatable elements 224
trypanosomes 39
Turner, J. R. G., on Fisher 98; on jerks 188n
typological thinking, *see* typology
typology 3, 131, 134–137, 187, 198; and species 137–138
Tyrannosaurus 116

Ulizzi, L., extract from 59–62; introduced 46
undergraduates, apologised to 417
units of selection 123–130; *see also* group selection, natural selection
universal, biochemical 405–406, 416; genetic code, *see* genetic code; grammar 377
Upper Paleolithic, linguistic date demolished 373
Urophora jaceana 39
Ussher, Bishop 401, 404

valley-crossing 31, 33, 34–36
values, human, and evolution 4, 418–421
variation (among individuals within a population), blended away 1–2, 20–24; discordance among characters 361–362; under domestication 21, 26; factors determining 29–37; paradox of 67–76; and population thinking 135–137; *see also* directed variation, neutralism, polymorphism, race
varicose veins 366
Varley, G. C. 38
veliger 219–220
virulence 229
viruses, mutation rates of 331; resistance to 39; HIV 424, 427, 431; influenza 79; *see also* phage
Vision, T. J., extract from 231–236; introduced 220–221
viviparity 3
Volkswagen syndrome 268

Wäctershäuser, G. 261–262
Wallace, A. R., sends a letter 7; a Panglossian 114, 115; on speciation 133, 152
warm blood 3
wedges, a Darwinian image 11
Weiner, J. 46
Weismann, A., introduced 1, 8; extract about 15–20; a Panglossian 114; on senescence 294
White, T. D. 339
Williams, G. C., extract from 89–91; introduced 82; on adaptation 91, 96, 375; on Darwinian medicine 338; on genes 124

Wilson, A. C., extracts from 340–345, 345–350; introduced 337
Wilson, E. O. 154, 348
wing, homologies of 208
Woese, C. 303, 306–307
woodpecker 14
Wray, G., and colleagues 280
Wright, S., extract from 29–37; contrasted with Fisher 8, 91; founder of Modern Synthesis 7, 106; and theoretical biology 20; a micromutationist 97; who influences neutralism 48; and is admired by Kimura 76, 81; mentioned 362n

yeast, and alcohol and fruit 220; genome of 253; in genomic comparisons 239, 240–241, 242; mutation rate of 331
Yiddish expression 380

Zuk, M. 94